Springer Optimization and Its Applications

VOLUME 91

Aims and Scope

Optimization has been expanding in all directions at an astonishing rate during the last few decades. New algorithmic and theoretical techniques have been developed, the diffusion into other disciplines has proceeded at a rapid pace, and our knowledge of all aspects of the field has grown even more profound. At the same time, one of the most striking trends in optimization is the constantly increasing emphasis on the interdisciplinary nature of the field. Optimization has been a basic tool in all areas of applied mathematics, engineering, medicine, economics, and other sciences.

The series *Springer Optimization and Its Applications* publishes undergraduate and graduate textbooks, monographs and state-of-the-art expository work that focus on algorithms for solving optimization problems and also study applications involving such problems. Some of the topics covered include nonlinear optimization (convex and nonconvex), network flow problems, stochastic optimization, optimal control, discrete optimization, multi-objective programming, description of software packages, approximation techniques and heuristic approaches.

For further volumes:
http://www.springer.com/series/7393

Nicholas J. Daras
Editor

Applications of Mathematics and Informatics in Science and Engineering

Springer

Editor
Nicholas J. Daras
Department of Mathematics and Engineering Sciences
Hellenic Military Academy
Vari Attikis, Greece

ISSN 1931-6828 ISSN 1931-6836 (electronic)
ISBN 978-3-319-34963-3 ISBN 978-3-319-04720-1 (eBook)
DOI 10.1007/978-3-319-04720-1
Springer Cham Heidelberg New York Dordrecht London

Mathematics Subject Classification (2010): 11M06, 11T71, 11YXX, 14H52, 34A30, 34A34, 49M30, 60G25, 60J10, 65C20, 65C50, 68U20, 81P94, 90C90, 91A40, 91A80, 94AO5, 94A15

Printed on acid-free paper

Springer is part of Springer Science+Business Media (www.springer.com)

Preface

Mathematics has been vital to the development of civilization. From ancient to modern times mathematics has been fundamental to advances in science, engineering, and philosophy. Informatics, as an applied scientific area, represents a leading and broadly oriented base of technology, within key products of contemporary and future engineering trends in many scientific fields of knowledge, such as automation, machinery, computers, mechanics, robotics, telecommunications, electronic components, high tech, industrial science, and technical knowledge.

There is a major link between mathematics and informatics. These are disciplines with the same basic properties and interactions that positively affect the development of both disciplines. Thanks to informatics, mathematical approaches and mathematical methods established themselves in many other disciplines. The interdisciplinary applications of mathematics and informatics are the subject of the present book.

The applications presented here are sometimes difficult to classify mathematically, since tools from several areas of mathematics may be applied. We focus on these applications not by discussing the nature of their discipline but rather their interaction with mathematics.

The 26 papers of the book are independent of each other and they cover many scientific subjects. These are an outgrowth of the 2nd International Conference on Applications of Mathematics and Informatics in Military Sciences (2nd AMIMS), April 12–13, 2013. Hellenic Military Academy, and bring together a wide variety of mathematical methods with applications to science, engineering, and technology. Also studied is the theoretical background required for methods, algorithms, and techniques used in various applications as well as the direction of theoretical results in these applications. Open problems and future areas are also highlighted.

The book presents several results with an extensive discussion on applied operations research, scientific computing and applications, simulation of operations, logistics chain, game theory and allocation strategies, cryptology and computational number theory, security, wireless communications, statistical modeling and applications, invisibility regions and regular meta-materials, unmanned vehicles, modern radar techniques/SAR imaging, satellite remote sensing, coding, geospatial problems, and robotic systems. Furthermore, this work will prove useful as a reference in the respective subjects and as a basis for further study and research.

The key features of the book are the following:

- Working groups meeting in composite sessions to address a wider spectrum of topics, which are of interest to their associated composite group.
- Developing courses of action or methodologies to reconcile issues identified.
- Cooperating prospects between various scientific and technology communities/converging a range of interdisciplinary objects with a large width of applications.

We hope that the book will be especially useful to graduate students and specialists in the interdisciplinary applications of mathematics and informatics, as well as to readers who are working in science and engineering.

Vari Attikis, Greece Nicholas J. Daras

Foreword

Applications of Mathematics and Informatics in Science and Engineering includes both research and survey papers on applied operations research, scientific computing and applications, simulation of operations, logistics, game theory and allocation strategies, cryptology and computational number theory, security, wireless communications, statistical modeling and applications, invisibility regions and regular meta-materials, unmanned vehicles, modern radar techniques, satellite remote sensing, coding, geospatial problems, and robotic weapon systems.

The book will be especially useful to graduate students and specialists in the interdisciplinary applications of mathematics and informatics, as well as to all those who are interested in science and engineering.

Contents

Parametric Design and Optimization of Multi-Rotor Aerial Vehicles

C. Ampatis and E. Papadopoulos

Abstract This work addresses the problem of optimal selection of propulsion components for a multi-rotor aerial vehicle (MRAV), for a given payload, payload capacity, number of rotors, and flight duration. Considering that the main components include motors, propellers, electronic speed controllers (ESC), and batteries, a steady state model is developed for each component using simplified analysis. Based on technical specifications of commercially available batteries, motors and ESCs, component functional parameters identified earlier were expressed as a function of component size, in terms of an equivalent length. Propeller models were developed using available experimental data. Airframe dimensions and total weight were expressed as a function of propeller diameter, number of rotors, and maximum thrust. Using Matlab's "fmincon" function, a program was developed which calculates the optimal design vector using the total energy consumption and vehicle diameter as objective function. Using the developed program, the influence of the payload and of the number of rotors on the design vector and the MRAV size was studied. The results obtained by the program were compared to existing commercial MRAVs.

Keywords Multi-rotor aerial vehicle (MRAV) design • Parametric design • Constrained optimization • Energy and size minimization

C. Ampatis • E. Papadopoulos (✉)
Department of Mechanical Engineering, National Technical University of Athens, Greece
e-mail: christos.ampatis@gmail.com; egpapado@central.ntua.gr

N.J. Daras (ed.), *Applications of Mathematics and Informatics in Science and Engineering*, Springer Optimization and Its Applications 91, DOI 10.1007/978-3-319-04720-1_1,

Introduction

Recently, Multi-Rotor Aerial Vehicles (MRAV) are encountered in an increasing number of military and civilian applications. A particular advantage an MRAV has over other aerial vehicles is its unique ability for vertical stationary flight (VTOL). Micro and mini MRAVs with payload capabilities of up to 100 g and 2 kg respectively [1] offer major advantages when used for aerial surveillance and inspection in complex and dangerous indoor and outdoor environments. In addition, improvements and availability in cost-effective batteries and other technologies are rapidly increasing the scope for commercial opportunities.

In most MRAV configurations, rotors are in the same plane and symmetrically fixed on the airframe. The number of rotors is always even in order to balance the torque produced by the rotors. An exception is the trirotor, where one rotor is placed on a tilting mechanism in order to balance the excess toque. Additional configurations include MRAVs with multiple pairs of coaxial-counter rotating rotors. However, researchers push the limits by studying different configurations where the rotors are not in the same plane but placed arbitrarily in 3D space [2], or even having the ability of thrust vectoring [3, 4].

In any configuration, an MRAV design consists of basic components, such as batteries, electric motors, and propellers, which constitute the vehicle propulsion system. One of the most critical stages in MRAV design is the proper motor–propeller matching. The electric motor market offers a large range of motors for almost any application, thus an MRAV designer does not need to design the motor. Propellers used for MRAV applications are taken from the remote controlled (RC) aircraft market, therefore they are designed for RC aircrafts. However, an MRAV hovers for a great percent of the total flight time, therefore needs propellers designed for maximum hover efficiency. Recently, the MRAV industry produced such propellers but in a limited range. Recent studies resulted in optimized designs of micro and mini rotorcraft vehicle propellers that are easy to manufacture, such as curved plate plastic propellers, [5, 6].

Apart from optimizing each MRAV component separately, an MRAV designer would benefit from an automated design method that would take into account all design requirements to yield an optimized combination of commercially available components. Although studies on automated design methods exist [7, 8], no method exists that takes into account both the propulsion system modeling and the functional parameters of existing components.

In this paper, we propose an MRAV design method, which selects the optimum propulsion system components. Given the MRAV design requirements such as payload, payload capacity, number of rotors, and flight duration, a Matlab program calculates the propulsion system components and MRAV size which leads to an energy-efficient design, or to a design with the smallest size. To achieve this we use simplified models for each component, and expressions of component functional parameters as a function of component size, using their commercially available technical specifications.

Component and System Modeling

The components to be modeled include the electric motors, the electronic speed controller, batteries, propellers, and the airframe. Combining the simplified models will lead to a system model for the MRAV steady state operation.

Electric Motor Model

The electric motors used in MRAV applications are outrunner Brushless Direct Current (BLDC) ones. This is due to their high efficiency and high torque constant (K_T), which allows direct propeller coupling (no gearbox). Although a BLDC motor is a synchronous 3-phase permanent magnet motor, it can be modeled as a permanent magnet DC motor. This leads to a classic three-constant model, see Fig. 1.

In Fig. 1, V_k is the supply voltage (V), i_α is the current through the motor coils (A), e_α is the back-electromotive force (EMF) (V), R_α is the armature resistance (Ω), M is the torque produced by the motor (Nm), and ω is its shaft angular velocity (rad/s). The equations describing the motor are:

$$V_k = e_a + i_a R_a \tag{1}$$

$$e_a = K_e\omega = K_T\omega = N/K_V \tag{2}$$

where K_e is the motor back EMF constant (Vs/rad), K_T is the motor torque constant (Nm/A), N is the motor rpm, and K_V is motor speed constant (rpm/V). The K_T is related to K_V by:

$$K_e = K_T = \frac{30}{\pi}\frac{1}{K_V} \tag{3}$$

The total torque produced by the motor is:

$$M = K_T i_a \tag{4}$$

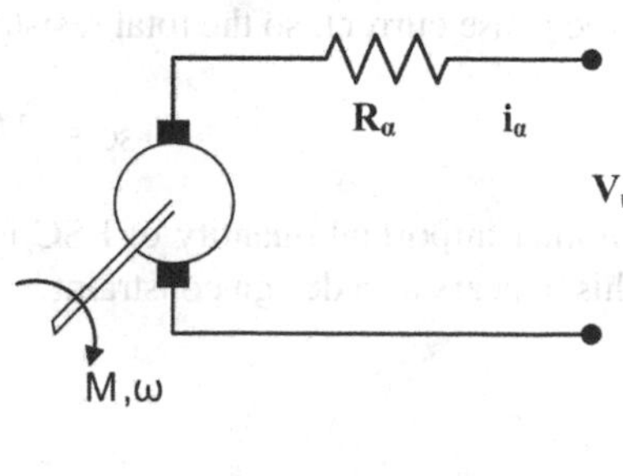

Fig. 1 Electric motor model

The output torque is:

$$M_{\mathrm{mot}} = K_T\,(i_a - i_0) \tag{5}$$

where i_0 is the no-load current. The motor input power is:

$$P_{in} = V_k i_a \tag{6}$$

the motor output power is:

$$\begin{aligned} P_{\mathrm{mot}} &= M_{\mathrm{mot}}\omega = K_T\,(i_a - i_0)\,\omega = e_a\,(i_a - i_0) \\ &= (V_k - i_a R_a)\,(i_a - i_0) \end{aligned} \tag{7}$$

and the motor speed in rpm is:

$$N = (V_k - i_a R_a)\,K_V \tag{8}$$

Given the parameters K_T, R_α, and i_0 we can calculate the performance of the motor.

Electronic Speed Controller Model

Electronic speed controllers regulate motor speed within a range depending on the load and battery voltage. The important quantity here is the ESC power losses, caused by its power MOSFETs. The major parameters that affect ESC power losses are the transistor drain-to-source "ON" state resistance $R_{\mathrm{DS(ON)}}$, transistor characteristics on transient operation, and the frequency switching the transistor "ON" and "OFF." Power losses at full throttle, when transistors are fully "ON," depend only on $R_{\mathrm{DS(ON)}}$, while at partially opened throttle, when the transistors switch between "ON" and "OFF," additional power losses occur.

The range of $R_{\mathrm{DS(ON)}}$ lies between 3 and 15 mΩ and its value is proportional to transistor size. Considering that ESC power losses are a small portion of input power, and the fact that ESC manufacturers do not include in ESC documentation the type of transistors used, we model the ESC as a constant value resistor of $R_{\mathrm{DS(ON)}} = 5\,\mathrm{m}\Omega$. BLDC motor ESCs use three pairs of transistors to manage the three phase current, so the total resistance of the ESC will be:

$$R_{\mathrm{ESC}} = 3R_{\mathrm{DS(ON)}} = 0.015\Omega \tag{9}$$

Another important quantity of ESC is the maximum current i_{ESC} they can handle. This appears as a design constraint.

Battery Model

Due to their high energy density and discharge rate, MRAVs use Lithium Polymer (LiPo) batteries. A LiPo pack consists of identical LiPo cells each with a nominal voltage of 3.7 V. Parallel connection of battery packs raises the battery total capacity, while keeping the nominal total voltage the same. Therefore, the nominal total voltage of a LiPo battery is:

$$V_b = n_c 3.7 \tag{10}$$

where n_c is the number of cells connected in series in a battery pack. The battery has an internal total resistance $R_{\text{bat,tot}}$. When connected to a load its output voltage is:

$$V_{b,\text{out}} = V_b - iR_{\text{bat,tot}} \tag{11}$$

where i is the load current.

Each cell has internal resistance R_{sc}, capacity C_{sc}, and maximum discharge rate DR_c. The total battery capacity is:

$$C_{\text{tot}} = n_p C_{\text{sc}} \tag{12}$$

where n_p is the number of battery packs connected in parallel. Each cell's power is:

$$P_{\text{sc}} = 3.7 DR_c C_{\text{sc}} \tag{13}$$

Each cell's energy is:

$$E_{\text{sc}} = 3.7 C_{\text{sc}} \tag{14}$$

A battery's total power is:

$$P_{\text{bat,tot}} = P_{\text{sc}} n_c n_p \tag{15}$$

while its total energy is:

$$E_{\text{bat,tot}} = E_{\text{sc}} n_c n_p \tag{16}$$

To calculate $R_{\text{bat,tot}}$ we apply Kirchoff's law to a battery consisted of n_p identical packs connected in parallel, each of which consists of n_c identical cells connected in series. Each battery pack has an internal resistance:

$$R_i = n_c R_{\text{sc}},\ i = 1, \ldots, n_p \tag{17}$$

The battery total resistance is:

$$R_{\text{bat,tot}} = \prod_{j=1}^{n_p} R_j \Bigg/ \sum_{i=1}^{n_p} \left(\frac{1}{R_i} \prod_{j=1}^{n_p} R_j \right) = \frac{(n_c R_{\text{sc}})^{n_p}}{n_p (n_c R_{\text{sc}})^{n_p - 1}} = \frac{n_c R_{\text{sc}}}{n_p} \tag{18}$$

Propeller Model

Propellers used on MRAVs are mostly the same propellers used in remote controlled (RC) airplanes. Propeller performance is described by its thrust T(N), power P(W), and torque M(Nm). To model performance in static conditions, we use manufacturer data such as propeller diameter D_p and its pitch p at 75 % of its radius. Performance quantities are then related to propeller speed, diameter, and pitch. This is achieved through a number of coefficients.

The thrust coefficient is given by:

$$C_T = T \Big/ \rho (N/60)^2 D^4 \tag{19}$$

where T is thrust (N), ρ is air density (kg/m^3), N is propeller speed (rpm), and D is the propeller diameter (m).

The power coefficient is given by:

$$C_P = P \Big/ \rho (N/60)^3 D^5 \tag{20}$$

where P is power (W).

The torque coefficient is given by:

$$C_M = M \Big/ \rho (N/60)^2 D^5 \tag{21}$$

where M is torque (Nm). Using the fundamental relation between power, torque, and speed we get:

$$C_M = C_P/2\pi \tag{22}$$

These coefficients are next related to propeller diameter and pitch. Using the Blade Element Momentum Theory (BEMT) and a series of assumptions [9], we get the following equations for thrust and power coefficients:

$$C_T = \frac{\pi^3}{4}\frac{1}{2}\sigma C_{la}\left(\frac{\theta_{0.75}}{3} - \frac{1}{2}\sqrt{\frac{4}{\pi^3}\frac{C_T}{2}}\right) \tag{23}$$

$$C_P = \frac{2}{\pi^2}\frac{C_T^{3/2}}{\sqrt{2}} + \frac{1}{8}\sigma C_{d0} \tag{24}$$

where σ is propeller solidity, $C_{l\alpha}$ is the slope of blade airfoil lift coefficient–incidence angle curve, $\theta_{0.75}$ is propeller pitch angle at 75 % of the propeller radius R, and C_{d0} is a blade's airfoil drag coefficient for zero lift.

To further simplify this model to a restricted propeller size range and geometry, we make the following assumptions. Considering that we refer to geometrically scaled propellers, propeller solidity σ will be constant regardless of propeller size. Additionally, if the propeller size range is no more than one order of magnitude, then the Reynolds number does not change dramatically, so we can assume that the aerodynamic quantities $C_{l\alpha}$ and C_{d0} are constant. Consequently, thrust and power coefficients are only a function of propeller pitch angle $\theta_{0.75}$. From the definition of geometric pitch we get:

$$p = 2\pi R \tan\theta \tag{25}$$

and therefore, the geometric pitch at $0.75R$ will be:

$$p_{0.75} = 2\pi \frac{3}{4} R \tan\theta_{0.75} = \pi \frac{3}{4} D_p \tan\theta_{0.75} \tag{26}$$

Solving Eq. (26) for $\theta_{0.75}$ we get:

$$\theta_{0.75} = \arctan\left(4/3\pi \cdot p_{0.75}/D_p\right) \tag{27}$$

Consequently, using Eqs. (23), (24), and (27) we can relate C_T and C_P to the ratio $p_{0.75}/D_p$ only. Normally, $\theta_{0.75}$ is in the range of 5–30, resulting a $p_{0.75}/D_p$ range of 0.2–1.35. In this region the function $C_T(p_{0.75}/D_p)$ is linear and this can be shown through a numerical solution. Additionally, by observing Eq. (24) we see that C_P is proportional to ${C_T}^{3/2}$, therefore it is proportional to $(p_{0.75}/D_p)^{3/2}$, and this can be also shown through a numerical solution in the $p_{0.75}/D_p$ range.

Consequently, we get the simplified expressions for thrust and power coefficients:

$$C_T = k_1\left(p/D_p\right) + k_2 \tag{28}$$

$$C_P = k_3\left(p/D_p\right)^{3/2} + k_4 \tag{29}$$

where constants k_1 to k_4 can be calculated using experimental data of geometrically scaled propellers.

Note that to obtain energy efficient propellers at hover, the ratio C_T/C_P must be as high as possible. Solving Eqs. (23) and (24) or (28) and (29), we see that this occurs when the ratio p/D_p is as low as possible, i.e., for a given propeller diameter the lowest pitch yields more efficient propellers.

System Model

The system model results from the combination of the propulsion system model and the equilibrium of forces acting on the vehicle. The propulsion system consists of the battery and n_{mot} triples of ESC, and of the motors and propellers connected in parallel, as shown in Fig. 2.

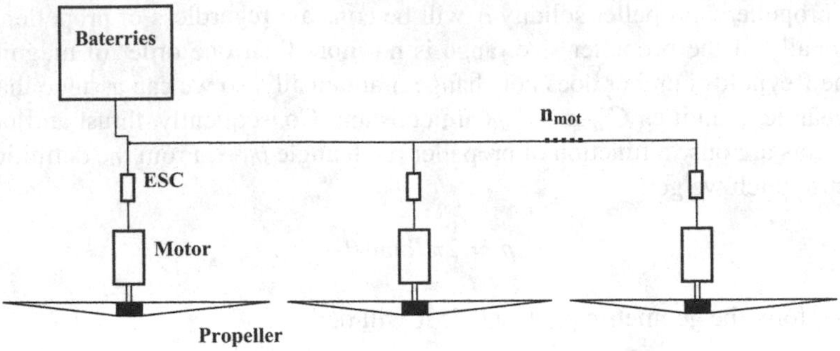

Fig. 2 Propulsion system

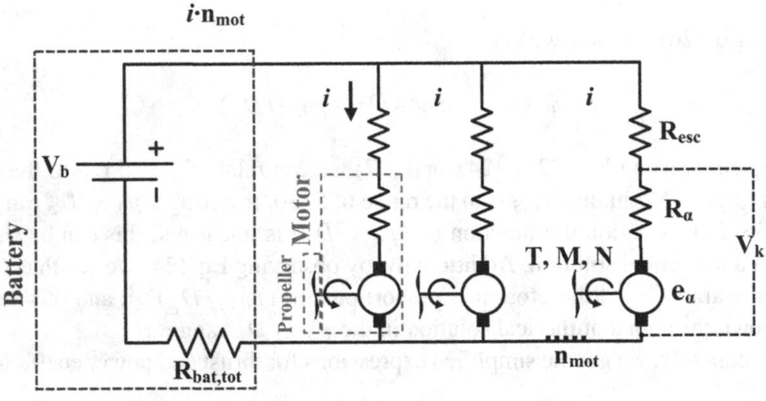

Fig. 3 Propulsion system physical model

The physical model of the propulsion system shown in Fig. 3 combines each component model and outputs the total thrust produced by the n_{mot} rotors. Assuming that all the rotors have the same speed, the current drawn will be the same for each motor.

Applying Kirchoff's law to the circuit of Fig. 3 we get:

$$V_k + iR_{ESC} = V_b - n_{mot} i R_{bat,tot} \tag{30}$$

$$e_a = V_b - i\,(R_a + R_{ESC} + n_{mot} R_{bat,tot}) \tag{31}$$

The rotor speed is given by:

$$N = [V_b - i\,(R_a + R_{ESC} + n_{mot} R_{bat,tot})]\,K_V \tag{32}$$

The above equation is valid only at full throttle, when the ESC transistors are fully on; otherwise, at partially open throttle, the ESC output voltage is less than the maximum, thus the motor voltage will be less than V_k.

Equation (32) shows that the motor equivalent resistance is:

$$R_{\text{tot}} = R_a + R_{\text{ESC}} + n_{\text{mot}} R_{\text{bat,tot}} \tag{33}$$

In this paper we examine the case where the vehicle during a total flight time t_{tot} has two operational modes. (a) A *maximum thrust mode* for a percentage *ATP* of the total flight time t_{tot}, in which motors are at full throttle state producing the maximum static thrust, and (b) *a hover mode*, in which the vehicle hovers for the rest of the flight time. At maximum thrust, the vehicle has the ability to accelerate with an instantaneously maximum acceleration, therefore it has the ability to lift its total weight f_w times.

(a) Maximum thrust mode: The rotor speed is:

$$N_{\text{acc}} = [V_b - i_{\text{acc}} R_{\text{tot}}]\, K_V \tag{34}$$

which is equivalent to the following:

$$N_{\text{acc}} = [V_{k,\text{acc}} - i_{\text{acc}} R_a]\, K_V \tag{35}$$

where $V_{k,\text{acc}}$ is the motor supply voltage equal to the maximum ESC output voltage.

A balance of forces, with a the acceleration, yields:

$$\begin{aligned} \Sigma F = m_{\text{tot}}\text{a} \Rightarrow n_{\text{mot}} T_{\text{acc}} - m_{\text{tot}} g = m_{\text{tot}}\text{a} = (f_w - 1)\, m_{\text{tot}} g \\ \Rightarrow n_{\text{mot}} C_T \rho\, (N_{\text{acc}}/60)^2\, D_p^4 = f_w m_{\text{tot}} g \end{aligned} \tag{36}$$

The maximum instantaneous linear acceleration will be:

$$\text{a} = (f_w - 1)\, g \tag{37}$$

The total mass of the vehicle is:

$$m_{\text{tot}} = m_{\text{bat,tot}} + \left(m_{\text{mot}} + m_p + m_{\text{ESC}}\right) n_{\text{mot}} + m_{\text{frm}} + m_{pl} \tag{38}$$

where $m_{\text{bat,tot}}$ is the battery total mass, m_{mot} is the motor mass, m_p is the propeller mass, m_{ESC} is the ESC mass, m_{frm} is the airframe mass, and m_{pl} is the payload mass.

The equation of motor–propeller power is:

$$P_m = P \Rightarrow \{V_b - i_{\text{acc}} R_{\text{tot}}\}\,(i_{\text{acc}} - i_0) = C_P \rho\, (N_{\text{acc}}/60)^3\, D_p^5 \tag{39}$$

The motor–propeller torque balance yields:

$$M_m = M \Rightarrow K_T\,(i_{\text{acc}} - i_0) = C_P \rho\, (N_{\text{acc}}/60)^2\, D_p^5/2\pi \tag{40}$$

The system input power is:

$$P_{\mathrm{IN,acc}} = V_b i_{\mathrm{acc}} n_{\mathrm{mot}} \tag{41}$$

while the system energy consumption is:

$$E_{\mathrm{IN,acc}} = P_{\mathrm{IN}} t_{\mathrm{tot}} ATP \tag{42}$$

(b) Hover mode: In this mode, the motor speed is:

$$N_{\mathrm{hov}} = [V_{k,\mathrm{hov}} - i_{\mathrm{hov}} R_a]\, K_V \tag{43}$$

where $V_{k,\mathrm{hov}}$ is ESC output voltage that satisfies $V_{k,\mathrm{hov}} < V_{k,\mathrm{acc}}$.

The balance of forces yields:

$$\begin{aligned} \Sigma F = 0 &\Rightarrow n_{\mathrm{mot}} T_{\mathrm{hov}} = m_{\mathrm{tot}} g \Rightarrow \\ n_{\mathrm{mot}} C_T \rho \,(N_{\mathrm{hov}}/60)^2\, D_p^4 &= m_{\mathrm{tot}} g \end{aligned} \tag{44}$$

The equation of motor–propeller power is:

$$P_m = P \Rightarrow$$

$$\{V_{k,\mathrm{hov}} - i_{\mathrm{hov}} R_a\}\,(i_{\mathrm{hov}} - i_0) = C_P \rho \,(N_{\mathrm{hov}}/60)^3\, D_p^5 \tag{45}$$

while the motor–propeller torque balance gives:

$$M_m = M \Rightarrow$$

$$K_T\,(i_{\mathrm{hov}} - i_0) = (1/2\pi)\, C_P \rho \,(N_{\mathrm{hov}}/60)^2\, D_p^5 \tag{46}$$

The system input power is:

$$P_{\mathrm{IN,hov}} = V_b i_{\mathrm{hov}} n_{\mathrm{mot}} \tag{47}$$

and the system energy consumption is:

$$E_{\mathrm{IN,hov}} = P_{\mathrm{IN,hov}} t_{\mathrm{tot}}\,(1 - ATP) \tag{48}$$

Battery total power is constrained by:

$$P_{\mathrm{IN,acc}} \le P_{\mathrm{bat,tot}} \tag{49}$$

while the battery total energy is given by:

$$E_{\mathrm{IN,hov}} + E_{\mathrm{IN,acc}} = E_{\mathrm{tot}} = E_{\mathrm{bat,tot}} \tag{50}$$

Parameterization

The system equations given in the previous section depend on the functional parameters, which define components performance. Here, these parameters are expressed as a function of component length. This length is taken as the cubic root of a component's volume (cubic length) and is referred to as the *equivalent length*. We do the same with propellers using available experimental measurements. Furthermore, we develop equations that correlate airframe size as a function of propeller diameter, number of rotors, and maximum thrust.

Electric Motor

The electric motors we chose for parameterization are the outrunner BLDC motors from AXI manufacturer. The choice is based on the technical specifications available and on the reliability and performance of these motors.

Here, the equivalent length of each motor is related to the outer dimensions of the motor and not to its stator dimensions. The parameters we want to relate to the equivalent length are the motor armature resistance R_{α}, torque constant K_T, no load current i_0, and motor mass m_{mot}. Additionally, motor maximum sustained current (or current capacity) $i_{\max}$ and motor maximum speed $N_{m,\max}$ are parameters that limit motor performance and must be related to equivalent length.

Consequently, we need to develop five equations as functions of equivalent length. After investigation of various correlations of these parameters to the equivalent length, we concluded the following functions due to their optimal fit to manufacturer data. Below, R^2 refers to coefficient of determination, and l_{mot} to motor equivalent length (m).

$$K_T/R_a = 2.6533 \cdot 10^4 l_{\text{mot}}^{3.6032},\ R^2 = 0.902 \tag{51}$$

$$K_T^2/R_a = 1.7548 \cdot 10^5 l_{\text{mot}}^{5.4833},\ R^2 = 0.94 \tag{52}$$

$$M_0 = K_T i_0 = 5.7721 \cdot 10^2 l_{\text{mot}}^{3.1888},\ R^2 = 0.908 \tag{53}$$

$$M_{\max} = K_T\,(i_{\max} - i_0) = 4.5004 \cdot 10^5 l_{\text{mot}}^{4.2222},\ R^2 = 0.96 \tag{54}$$

$$N_{m,\max} = (n_{c,\max} 3.7 - i_0 R_a)\, K_V \Rightarrow$$

$$N_{m,\max} = 25604 e^{-17.687 l_{\text{mot}}},\ R^2 = 0.35 \tag{55}$$

where $n_{c,\max}$ is the maximum number of battery cells in series connection that is proposed by manufacturer.

To relate motor mass to motor equivalent length, we calculated the mean motor density ρ_{mot}:

$$\rho_{\text{mot}} = 2942\,\text{kg/m}^3 \tag{56}$$

Using (56), the motor mass is:

$$m_{\text{mot}} = \rho_{\text{mot}} l_{\text{mot}}^3 \tag{57}$$

Electronic Speed Controller

We chose to parameterize ESCs from JETI due to the availability of technical specifications and their performance. Although the ESC is modeled as a constant resistance, additional parameters are needed that relate its operational limit and mass properties to its equivalent length l_{ESC} (m). These parameters are the ESC maximum sustained current i_{ESC} and ESC mean density ρ_{ESC}.

Using ESC technical specifications, correlations of maximum sustained current i_{ESC} and ESC equivalent length l_{ESC} are obtained as:

$$i_{\text{ESC}} = 8.4545 \cdot 10^6 l_{\text{ESC}}^{3.2451},\ R^2 = 0.88 \tag{58}$$

The mean ESC density calculated as:

$$\rho_{\text{ESC}} = 2580\,\text{kg/m}^3 \tag{59}$$

yielding the ESC mass as:

$$m_{\text{ESC}} = \rho_{\text{ESC}} l_{\text{ESC}}^3 \tag{60}$$

Battery

We chose to parameterize batteries from Kokam for the same reasons as before. The parameters to be related to battery total equivalent length l_{bat} include total power $P_{\text{bat,tot}}$, total energy $E_{\text{bat,tot}}$, total resistance $R_{\text{bat,tot}}$, and mass m_{bat}.

Battery technical specifications concern single battery cells of 3.7 V nominal voltage. However, we need information for any combination of parallel and series connected cells. We assume that n_p cells connected in parallel result in a larger single cell with volume B_{vol}, power P_{bat}, energy E_{bat}, and internal resistance R_{bat}.

Assuming that the battery consists of n_pn_c identical cells of volume $\text{B}_{\text{vol,sc}}$ each, then an equivalent battery will consist of n_c equivalent cells each of which has volume:

$$B_{\text{vol}} = n_p B_{\text{vol,sc}} \tag{61}$$

Therefore, each equivalent cell volume will be:

$$B_{\text{vol}} = l_{\text{bat}}^3 / n_c \tag{62}$$

Applying curve fitting to manufacturer data, the following equation for single cell internal resistance was obtained:

$$R_{\text{sc}} = 2.84668 \cdot 10^{-7} B_{\text{vol,sc}}^{-0.951154} \tag{63}$$

Correspondingly, the equivalent cell internal resistance is:

$$R_{\text{bat}} = 2.84668 \cdot 10^{-7} B_{\text{vol}}^{-0.951154}, R^2 = 0.95 \tag{64}$$

Using (18), (63), and (64), the battery total resistance is:

$$R_{\text{bat,tot}} = n_c R_{\text{sc}} / n_p = n_c 2.84668 \cdot 10^{-7} \left(B_{\text{vol}} / n_p\right)^{-0.951154} \Big/ n_p \Rightarrow$$

$$R_{\text{bat,tot}} = n_c R_{\text{bat}} n_p^{-(1-0.951154)} \approx n_c R_{\text{bat}} n_p^{-0.05} \tag{65}$$

However, n_p will never be large; therefore using the approximation $n_p{}^{0.05}$, battery total resistance will be:

$$R_{\text{bat,tot}} = n_c R_{\text{bat}} \tag{66}$$

Applying curve fitting to manufacturer data, we observe that cell energy and power are proportional to its volume. Therefore, using the mean value of the ratios cell energy to cell volume and cell power to cell volume yield:

$$P_{\text{bat}} = 7.0899 \cdot 10^6 B_{\text{vol}} \tag{67}$$

$$E_{\text{bat}} = 9.0833 \cdot 10^8 B_{\text{vol}} \tag{68}$$

Using (67) and (68), the battery total power and energy are:

$$P_{\text{bat,tot}} = n_c P_{\text{bat}} \tag{69}$$

$$E_{\text{bat,tot}} = n_c E_{\text{bat}} \tag{70}$$

The mean battery cell density is calculated as:

$$\rho_{\text{bat}} = 1907.8\,\text{kg}/\text{m}^3 \tag{71}$$

Yielding the battery total mass:

$$m_{\text{bat}} = \rho_{\text{bat}} B_{\text{vol}} n_c \tag{72}$$

Propeller

The propellers we chose to parameterize are taken from APC. The parameters to be related to propeller diameter D_p and geometric pitch p are the thrust and power coefficient, C_T and C_P respectively.

Previously, it was shown through Eqs. (28) and (29) that for zero flight velocity, C_T and C_P are functions of the ratio p/D_p. The constants k_1 through k_4 in these equations depend on propeller design and the Reynolds number. Here, we are interested in propellers with diameter of 80–500 mm, therefore we use experimental data for these dimensions, so as to satisfy Reynolds number.

Experiments on commercially available propellers used in remote controlled aircrafts were conducted at the University of Illinois, Urbana-Champaign (UIUC) in a wind tunnel [10]. Here, data regarding SPORT type APC propellers are used. From the C_T and C_P measurements for these propellers, those that refer to static conditions are used here. We observed that C_T and C_P are not affected much by propeller speed; therefore we calculated mean values of C_T and C_P for various speeds. These measurements concern propeller diameter of 7 in to 14 in. Finally, the C_T and C_P were correlated to the ratio p/D_p, obtaining the following functions:

$$C_T = 0.0266\left(p/D_p\right) + 0.0793,\ R^2 = 0.31 \tag{73}$$

$$C_P = 0.0723\left(p/D_p\right)^{3/2} + 0.0213,\ R^2 = 0.83 \tag{74}$$

The propeller mass is related to propeller diameter D_p as:

$$m_p = 0.97573 D_p^{2.5741},\ R^2 = 0.98 \tag{75}$$

Number of Rotors

The number of MRAV rotors can be even or odd. MRAVs with odd number of rotors need an additional degree of freedom (tilting) for one rotor, so that it can vector its thrust and regulate excess torque produced by the rotors. This requires extra mechanisms (revolute joint) and an extra actuator to move the rotor. To take this into account, we assume that these extra mechanisms increase vehicle mass with a percentage $f_{M,\mathrm{odd}}$ of the mass of one of the rods holding the motors. Additionally, actuator power increases total power with a percentage $f_{P,\mathrm{odd}}$ of one motor power. Reasonable values for these coefficients are $f_{M,\mathrm{odd}} = 0.5$ and $f_{P,\mathrm{odd}} = 0.01$.

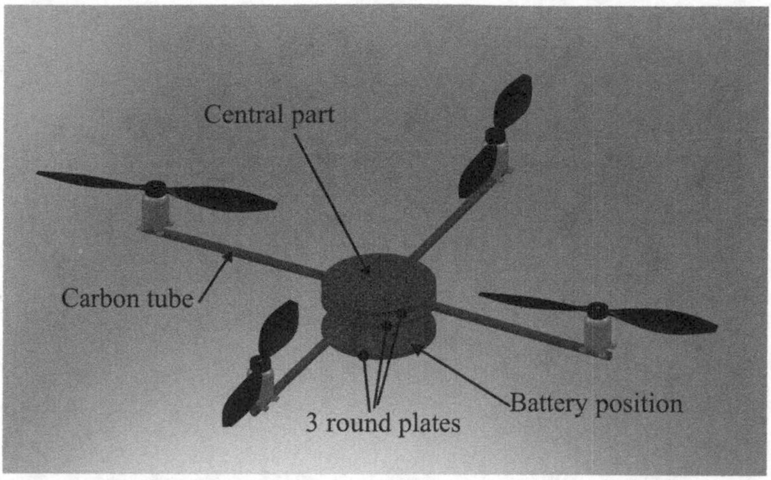

Fig. 4 MRAV airframe components

Airframe

Here, we are interested in the dimensions and mass properties of an MRAV airframe of simple design, with respect to the number of rotors n_{mot}, propeller diameter D_p, and airframe loading during flight.

A common rotor configuration is assumed. All rotors are in the same plane and motors are equidistant lying on a circle with its center coincident to vehicle center. The number of rotors is in the range of 3–8.

To approximate airframe mass its components and material must be assumed. A reasonable design consists of n_{mot} rods to hold the motors, and a central part of the three circular plates holding the rods and enclosing the battery and electronics. Additionally, airframe material is carbon fiber due to its high strength to weight ratio, and the accessories like screws and glue are a percentage $f_{\text{fr,ac}}$ of each rod mass. An illustration of such an airframe is presented in Fig. 4.

Airframe dimensions are defined by propeller diameter and vehicle loading during flight. On Fig. 5, airframe dimensions are shown. These include propeller diameter D_p, rotor spacing r_s, central disk-rotor spacing c_s, center disk radius R_{rc}, motor mounting position radius R_{rm}, and radius R_{rob} of the circle containing the whole vehicle. Note that the radius R_{rm} is the same for each rod. For a given propeller diameter, the dimensions r_s and c_s define the rest airframe dimensions.

The spacing r_s is important for a number of reasons. Primarily, if r_s is too small, there is a danger of adjacent rotor collision during flight due to rod elasticity. As was shown experimentally in [2] and [6], if r_s is too small, then propeller performance deterioration due to adjacent propellers airflow interaction is

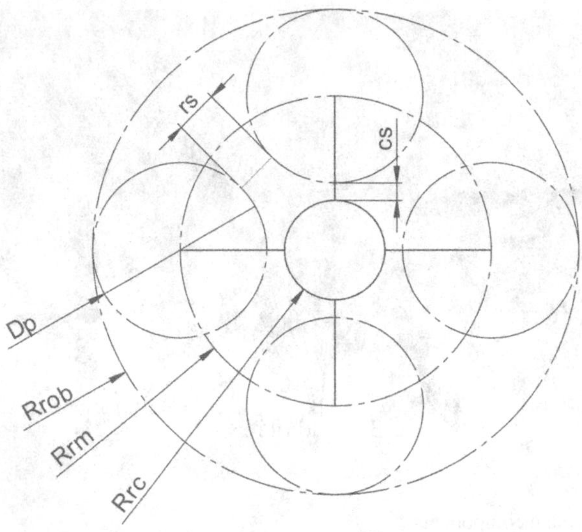

Fig. 5 Airframe dimensions

negligible. Furthermore, r_s cannot be too small because then the central disk will be very small to accommodate the battery and control unit. Additionally, r_s cannot be the same for all multi-rotors, i.e., a quadrotor must have a larger r_s than a hexarotor. For the same reason, c_s must vary for different number of rotors.

Based on the design trials with respect to the above explanation, r_s and c_s were expressed as a function of propeller radius R_p. Central disk thickness was expressed as a reasonable function of R_{rc}. For the calculation of carbon tubes' diameter and thickness, we developed equations that take into account material strength, tube maximum deflection, and tube loading. These equations allow calculation of the airframe mass.

Component Optimal Selection

In the previous sections, component performance was related to component equivalent length. Next, a method is developed for optimal selection of these lengths, which are parameters of the design vector. This vector minimizes an objective function, which is either the vehicle total energy, or the vehicle diameter D_{rob}.

Design Parameters

The design requirements are described by a number of parameters set by the designer. These include the payload m_{pl}, the total flight time t_{tot}, the payload capacity described by f_w indicating how many times the vehicle can lift its own weight, and the factor *ATP* which indicates the percentage of total flight time that the vehicle is at maximum thrust mode.

The design vector consists of the number of battery cells n_c in series, the equivalent battery length l_{bat}, the equivalent motor length l_{mot}, the equivalent ESC length l_{ESC}, the propeller diameter D_p, the ratio p/D_p, and the number of rotors n_{mot}.

Design Vector Domain

The design vector domain results from the size limits of the components that were parameterized earlier. Outside these regions the functions developed earlier may not be valid. Hence, the design vector domain is:

$$0.01 \leq l_{\text{bat}} \leq 0.15 \ \text{(m)} \tag{76a}$$

$$0.01 \leq l_{\text{mot}} \leq 0.08 \ \text{(m)} \tag{76b}$$

$$0.005 \leq l_{\text{ESC}} \leq 0.05 \ \text{(m)} \tag{76c}$$

$$0.05 \leq D_p \leq 0.5 \ \text{(m)} \tag{76d}$$

$$0.2 \leq p/D_p \leq 1.5 \ \text{(m)} \tag{76e}$$

$$1 \leq n_c \leq 10 \tag{76f}$$

Calculation Procedure

In every optimization step, the requirements vector $(m_{\text{pl}}, t_{\text{tot}}, f_w, ATP)$ is constant, while the design vector $(n_c, l_{\text{bat}}, l_{\text{mot}}, l_{\text{ESC}}, D_p, p/D_p, n_{\text{mot}})$ changes until the minimization of objective function is reached.

The calculation procedure follows the following sequence. The battery nominal voltage V_b is calculated using Eq. (10). Using Eq. (36) we get:

$$N_{\text{acc}} = 60 \left(\frac{f_w m_{\text{tot}} g}{n_{\text{mot}} C_T \rho D_p^4} \right)^{1/2} \tag{77}$$

Using Eq. (40) we get:

$$i_{\text{acc}} = K_V C_P \rho \frac{N_{\text{acc}}^2}{60^3} D_p^5 + i_0 \tag{78}$$

Using Eq. (35) we get:

$$V_{k,\text{acc}} = \frac{N_{\text{acc}}}{K_V} + i_{\text{acc}} R_a \tag{79}$$

The motor maximum speed without load is:

$$N_{\max} = [V_b - i_0 R_{\text{tot}}]\, K_V \tag{80}$$

Using Eq. (41), the maximum total input power $P_{\text{IN,acc}}$ is calculated, while using Eq. (42) the total input energy at maximum thrust mode $E_{\text{IN,acc}}$ is calculated. Using Eq. (44) we get:

$$N_{\text{hov}} = 60 \left(\frac{m_{\text{tot}} g}{n_{\text{mot}} C_T \rho D_p^4} \right)^{1/2} \tag{81}$$

Using Eq. (46) we get:

$$i_{\text{hov}} = K_V C_P \rho \frac{N_{\text{hov}}^2}{60^3} D_p^5 + i_0 \tag{82}$$

Using Eq. (43) we get:

$$V_{k,\text{hov}} = \frac{N_{\text{hov}}}{K_V} + i_{\text{hov}} R_a \tag{83}$$

The total input energy at hover $E_{\text{IN,hov}}$ is obtained using Eq. (48), while using Eq. (50) the total input energy E_{tot} is found.

Constraints

The constraints result from the independent variable physical consistency. They are given as follows:

$$\begin{gathered} V_{\text{acc}} - V_b \le 0,\, N_{\text{acc}} - N_{\max} \le 0,\, i_{\max} - i_{\text{ESC}} \le 0 \\ i_{\text{acc}} - i_{\max} \le 0,\, i_{\text{hov}} - i_{\text{acc}} \le 0,\, P_{\text{IN,acc}} - P_{\text{bat,tot}} \le 0 \\ E_{\text{tot}} - E_{\text{bat,tot}} \le 0,\, -i_{\text{acc}} \le 0,\, -i_{\max} \le 0 \end{gathered} \tag{84}$$

Optimization Methodology

For the calculation procedure, a Matlab program was developed that employs the "fmincon" function (minimum of constrained nonlinear multivariable function) which uses one target deterministic constrained optimization method for nonlinear multivariable objective function.

Our target was to determine the most energy-efficient design or the smallest one. Hence, the objectives were the minimization of battery energy $E_{bat.tot}$ or vehicle diameter D_{rob}, respectively.

In order to check that "fmincon" will not be trapped in local minimums, we also developed a program that scans the whole design vector domain, using nested loops. We observed no differences between these methods after some test runs. Consequently, "fmincon" calculates the total minimum for our objective functions.

Design Scenarios

Here we carry out some test runs in order to study the influence of payload and number of rotors on the design vector and the MRAV size. In all design scenarios below, the requirement parameters are set to: $t_{tot} = 15\,\text{min}$, $f_w = 2$, $ATP = 0.1$, $f_{fr,ac} = 0.15$, $f_{M,odd} = 0.5$, and $f_{P,odd} = 0.01$. Finally, we compare our program results to commercially available MRAVs designs.

Study of Parameters Influence

Payload Influence

In this case payload changes from 0 to 1.5 kg, while the number of rotors is constant and equal to 4.

In Fig. 6 the influence of payload on the design vector is shown. In general, we observe that as the payload increases, component equivalent length increases due to power increase. As expected, the ratio p/Dp is always constant and takes the lowest value permitted, indicating that for a given propeller diameter, the propeller pitch should always be the lowest. In addition, total energy minimization yields a more efficient but a larger design than that obtained by minimizing vehicle size. However, these differences are not large.

In Fig. 7, the influence of payload on total mass and on battery mass is illustrated. We observe that the battery mass is always lower for the minimization of total energy. However, vehicle total mass is not sensitive to the two objective functions. This happens because a smaller vehicle has smaller and therefore lighter motors and rotors. Additionally, observing the battery mass chart, we can say that battery mass increases linearly with payload. For the quadrotor, we can say that we need 1.5 kg batteries for 1 kg payload, and because the flight time is 15 min, then we can say that for 1 kg payload we need 100 g batteries for every minute of flight.

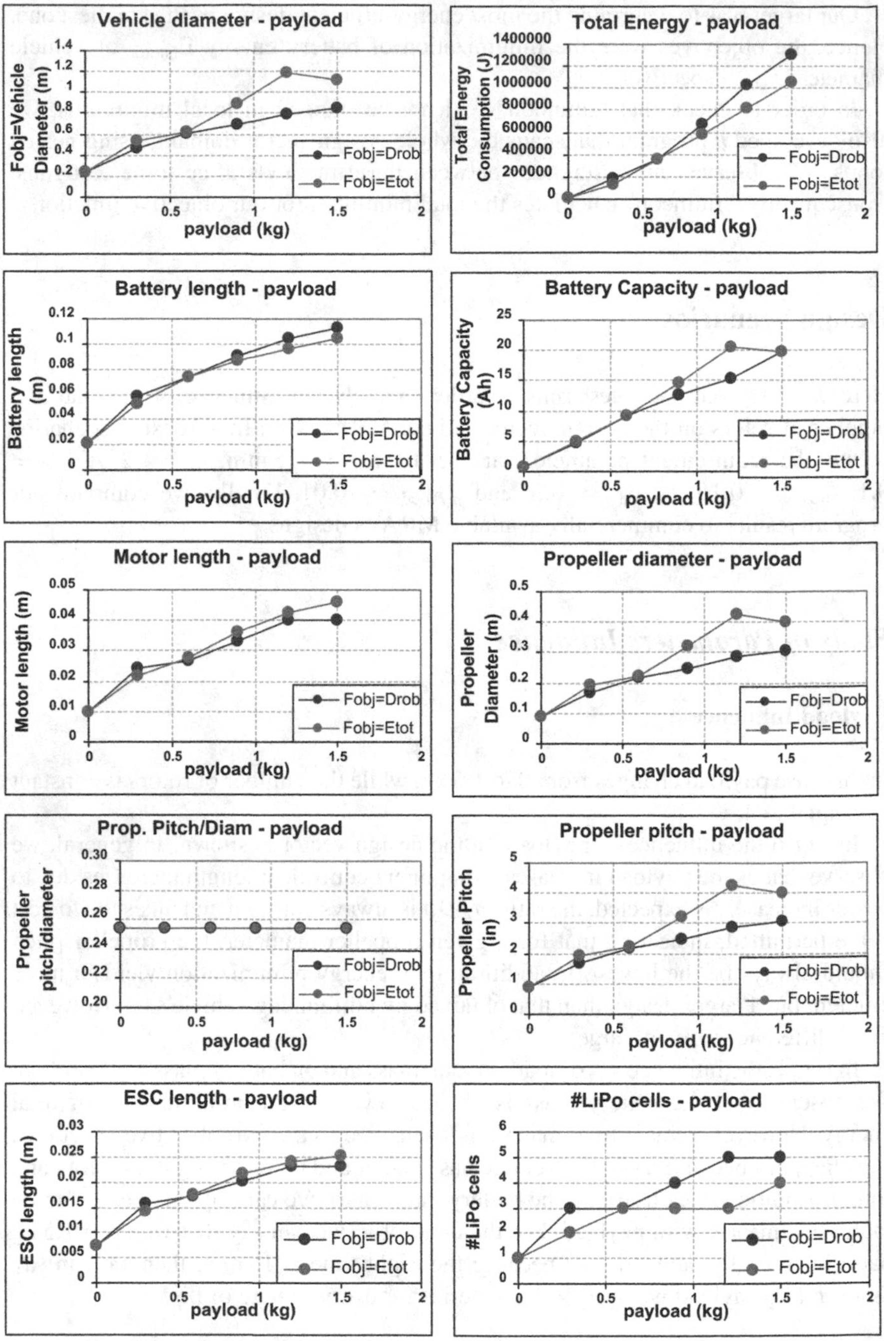

Fig. 6 Influence of payload on the design vector for the number of rotors equal to 4. Objective functions comparison

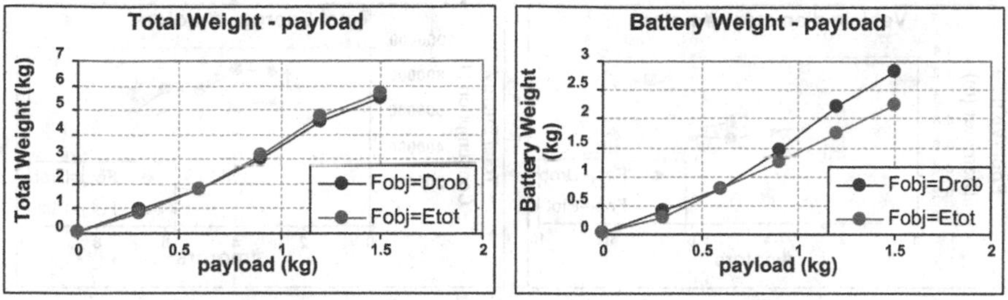

Fig. 7 Influence of payload on the total and battery mass for the number of rotors equal to 4. Objective functions comparison

Number of Rotors Influence

In this case, the number of rotors changes from 3 to 8, while the payload is constant and equal to 1 kg.

In Fig. 8, the influence of rotors number on the design vector is presented. We observe that for energy minimization, the best design has 8 rotors, but this is true for a payload of 1 kg, see Fig. 9. Additionally, we observe the expected decrease in components equivalent length when the number of rotors increases.

Test Cases

To determine whether the developed design methodology is valid and yields designs close to reality, we compare program results to two existing commercial MRAVs. The first is the quadrotor Walkera HM Hoten X Quadcopter, a small MRAV designed for a payload less than 100 g. The other is the Octocopter X88-J2, a large MRAV designed for aerial photography and for payloads up 1.5 kg, see Fig. 10.

Table 1 presents the quadrotor comparison, with data retrieved from [11]. The payload includes the electronics and control unit. We observe that the program yields results very close to reality. The difference lies on battery configuration and mass. The existing vehicle uses two battery cells in series with total energy $2 \times 3.7\text{V} \times 1\text{Ah} = 7.4\text{Wh}$, while the optimized needs $1 \times 3.7\text{V} \times 1.6\text{Ah} = 6\text{Wh}$. Therefore, the optimized vehicle seems to be more energy efficient.

In Table 2 an octorotor comparison is presented, with data taken from [12]. Here we observe that the optimized vehicle is 8 % heavier but 25 % smaller. Also, the optimized vehicle batteries have double capacity because there are two battery cells in series. Thus, the optimized vehicle has total energy $2 \times 3.7\text{V} \times 21.4\text{Ah} = 159\text{Wh}$, while the existing vehicle has total energy $4 \times 3.7\text{V} \times 10.6\text{Ah} = 157\text{Wh}$. We see that the total energy is almost the same for both the designs.

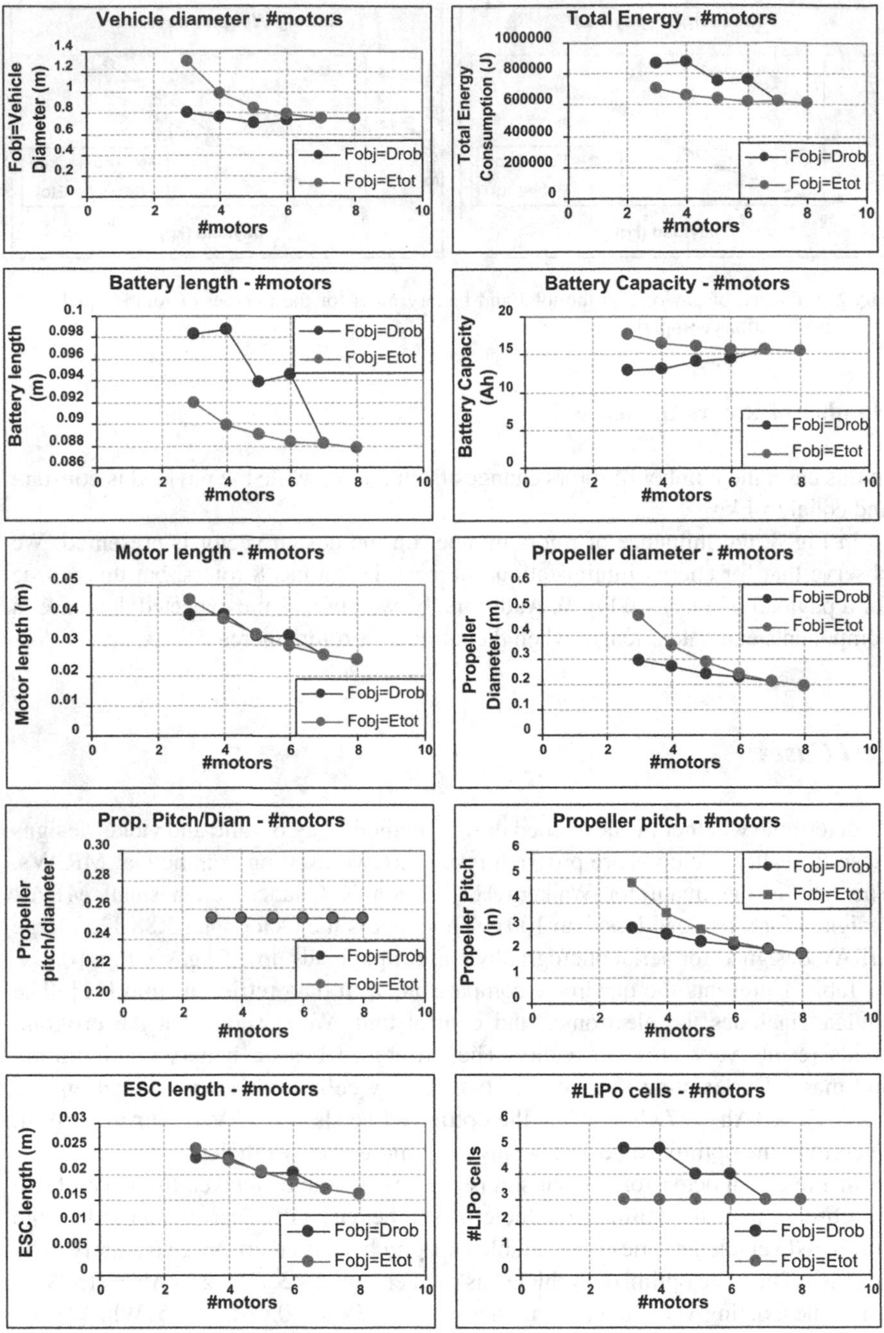

Fig. 8 Influence of the number of rotors on the design vector for payload equal to 1 kg. Objective functions comparison

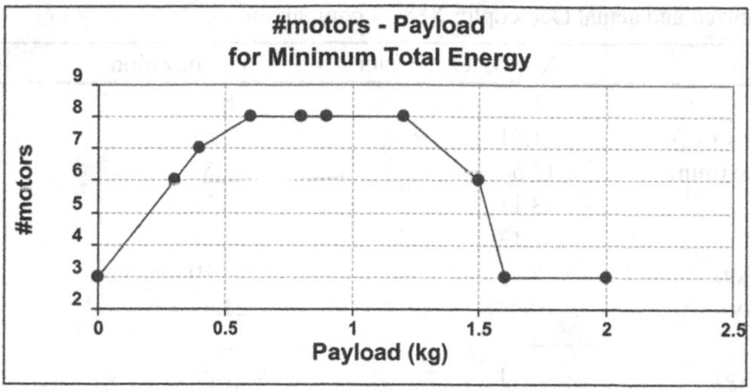

Fig. 9 Influence of payload on the number of rotors for minimum energy

Fig. 10 (*Left*) The quadrotor Walkera HM Hoten X Quadcopter. (*Right*) The Octocopter X88-J2

Table 1 Optimized and actual Walkera HM Hoten X Quadcopter comparison

Model	Walkera Hoten X Quadcopter	Optimization	Difference
#Motors	4	4	0
Payload capacity (f_w)	2	2	0
Total flight time (min)	10	10	0
Total mass (kg)	0.332	0.283	−0.05
Payload (kg)	0.1	0.100	0.00
Vehicle mass (kg)	0.269	0.237	−0.03
Battery capacity (Ah)	1	1.6	0.6
Battery #cells	2	1	−1
Battery mass (kg)	0.064	0.046	−0.02
Propeller diameter (m)	0.186	0.184	0.00
Vehicle diameter (m)	0.500	0.510	0.01

Table 2 Optimized and actual Octocopter X88-J2 comparison

Model	X88-J2 Octocopter	Optimization	Difference
#Motors	8	8	0
Payload capacity (f_w)	1.51	1.51	0
Total flight time (min)	17.5	17.5	0
Total mass (kg)	3.11	3.23	0.12
Payload (kg)	1.13	1.13	0.00
Vehicle mass (kg)	2	2.10	0.10
Battery capacity (Ah)	10.6	21.4	10.8
Battery #cells	4	2	−2
Battery mass (kg)	1.11	1.22	0.11
Propeller diameter (m)	0.305	0.24	−0.07
Vehicle diameter (m)	1.205	0.91	−0.29

Conclusions

This work focused on the parametric design and optimization of a multi-rotor aerial vehicle (MRAV). Using simplified models of propulsion system components such as motors, propellers, electronic speed controllers (ESC), and battery, a total model for an MRAV was created and the whole system performance at hovering and at maximum thrust was described. Additionally, based on the technical specifications of commercially available batteries, motors, and ESCs, component functional parameters were expressed as a function of component size, in terms of an equivalent length. As a result, we were able to calculate system performance as a function of a design vector which consists of each individual component equivalent length. A Matlab program was developed which calculates the optimal design vector using the "fmincon" function. The total energy consumption and the vehicle diameter were considered as objective functions. As a result, for a given payload, payload capacity, number of rotors, and flight duration, the optimal size of each component that minimizes energy or MRAV size was calculated. Finally, using the developed program, we were able to study the influence of the payload, and of the number of rotors, on the design vector and the MRAV size. The results obtained by the program were compared to existing commercial MRAVs, showing that the developed methodology yields designs close to reality. In addition, this methodology provides an MRAV designer with the tools to improve an existing design.

References

1. F. Kendoul, Survey of advances in guidance, navigation, and control of unmanned rotorcraft systems, Journal of Field Robotics, Vol. 29, No 2, pp. 315–378, March/April 2012,

2. Q. Jiang, et al, Analysis and Synthesis of Mult-Rotor Aerial Vehicles, Proceedings of the ASME 2011 International Design Engineering Technical Conferences & Computers and Information in Engineering Conference, IDETC/CIE 2011, August 28–31, Washington, DC, USA, DETC2011-47114, 2011.
3. D. Langkamp, et al, An engineering development of a novel hexrotor vehicle for 3D applications, Proceedings Micro Air Vehicles Conference 2011, Summer edition, 2011.
4. N. Fernandes, Design and construction of a multi-rotor with various degrees of freedom, M.S. Thesis, Technical Univ. of Lisboa, 2011.
5. F. Bohorquez, et al., Design, Analysis and Hover Performance of a Rotary Wing Micro Air Vehicle, Journal of the American Helicopter Society, Vol. 48, No 2, pp. 80–90, 2003.
6. A. Harrington, Optimal Propulsion System Design for a Micro Quad Rotor, M.S. Thesis, University of Maryland, 2011.
7. D. Lundström, K. Amadori, P. Krus, Automation of Design and Prototyping of Micro Aerial Vehicle, AIAA-2009-629, 47th AIAA Aerospace Sciences Meeting, Orlando, FL, USA, Jan. 2009.
8. S. Bouabdallah, Design and Control of Quadrotors with Application to Autonomous Flying, Ph.D. Thesis, École Polytechnique Federale de Lausanne, 2007.
9. J.G. Leishman, Principles of Helicopter Aerodynamics, Cambridge University Press, New York, 2006.
10. J. Brandt, and M. Selig, Propeller Performance Data at Low Reynolds Numbers, 49th AIAA Aerospace Sciences Meeting, AIAA 2011-1255. http://www.ae.illinois.edu/m-selig/props/propDB.html, 2011.
11. Walkera HM Hoten X Quadcopter - 200 size, http://www.helifreak.com/showthread.php?t=452889.
12. X88-J2 Octocopter, http://www.wowhobbies.com/x88octocopter.aspx.

Scattering Relations for a Multi-Layered Chiral Scatterer in an Achiral Environment

Christodoulos Athanasiadis, Evangelia Athanasiadou, Sotiria Dimitroula, and Eleftheria Kikeri

Abstract In this work we study scattering of a plane electromagnetic wave by a multi-layered chiral body in free space. In the interior of the scatterer exists a core which is either a perfect conductor or a dielectric. We obtain integral representations of the scattered fields which consist of a chiral and an achiral counterpart incorporating the boundary and transmission conditions. We introduce a dimensionless version of the scattering problem and we prove the reciprocity principle and a general scattering theorem for the far-field patterns. Finally, we define Herglotz functions and we state the general scattering theorem in terms of the far-field operator which expresses the superposition of the far-field pattern.

Introduction

This paper is concerned with the reciprocity principle and general scattering theorem for the far-field patterns corresponding to the scattering of time-harmonic electromagnetic plane waves upon a multi-layered chiral scatterer with either a perfectly conducting core or a dielectric. This type of scatterer is consisted of a finite number of layers with a homogeneous isotropic chiral medium. On the surfaces of this nested body, transmission conditions are imposed which express the continuity of the medium and the balance of forces acting on it.

A chiral object is a body that cannot be brought into congruence with its mirror by translation and rotation. Chirality is common in a variety of naturally occurring and manmade objects. DNA in a molecular scale, helices, medicine drugs and air defence industry are some examples in which chirality appears. From a technical point of view, chirality is introduced into the classical Maxwell equations via the

---

C. Athanasiadis (✉) • E. Athanasiadou • S. Dimitroula • E. Kikeri
Department of Mathematics, University of Athens, Panepistemiopolis, 15784, Athens, Greece
e-mail: cathan@math.uoa.gr; eathan@math.uoa.gr; sodimitr@math.uoa.gr; ekikeri@math.uoa.gr

N.J. Daras (ed.), *Applications of Mathematics and Informatics in Science and Engineering*, Springer Optimization and Its Applications 91, DOI 10.1007/978-3-319-04720-1_2,

Drude-Born-Fedorov constitutive relations in which the electric and magnetic field are connected through a new material parameter, the well known chirality measure. A large part of the work on scattering problems in chiral media exists in the book of Lakhtakia [20] as well as in [21, 22].

Twersky had made a major contribution to this research area with his papers [23–25] where he proved reciprocity, scattering and optical theorems for acoustic and electromagnetic scattering. Solvability and reciprocity principle for conductive boundary value problems and for the far-field patterns corresponding to an impedance boundary value problem have been proved in [10] and [11] respectively. Reciprocity relation and scattering theorems when the incident wave is a Herglotz function have been proved in [15, 16] for acoustic and electromagnetic scattering. DeFacio developed the impedance theory for electromagnetic scattering by an obstacle with a finite non-intersecting boundaries in [19]. In [5, 6] acoustic scattering amplitudes were constructed for a multi-layered scatterer, and scattering theorems were proved for time-harmonic electromagnetic waves in a piecewise homogeneous medium respectively. Multi-layered ellipsoidal scatterers with sound-soft and -hard core were used in [3] and [4] in which the first three low-frequency coefficients were obtained under ellipsoidal geometry. Dassios et al. proved reciprocity relations and general scattering theorems for far-field patterns in spherical coordinates in elasticity in [18]. Scattering relations for a homogeneous chiral obstacle have been proved in [8] while low-frequency electromagnetic scattering theory for a multi-layered chiral obstacle was developed in [2], where the scattering problem was reduced to an iterative sequence of problems in potential theory and the leading term approximation of the electric far-field pattern was constructed.

In section "Formulation", we proceed by formulating the corresponding scattering problems for the electric field. In section "The Electric Far-Field Pattern", we construct the electric far-field patterns and we determine their chiral and achiral part. In section "Scattering Relations", we restate the problem in terms of a dimensionless version and we prove the reciprocity, general scattering theorem and optical theorem. Finally in section "Herglotz Functions", we introduce Herglotz functions and the far-field operator and we restate a general scattering theorem in terms of Herglotz functions.

Formulation

Let D be a multi-layered chiral scatterer which is a bounded, closed, convex subset of $\mathbb{R}^3$ with a C^2-boundary S_0. The interior of D is divided by means of closed and non-intersecting C^2-surfaces into layers D_j with $S_j = \partial D_j \cap \partial D_{j+1}, j = 1, 2, \ldots, N$. There is one normal unit vector $\hat{\mathbf{n}}(\mathbf{r})$ at each point of any surface S_j pointing at D_j while the surface S_{j-1} surrounds S_j. Each of the layers, D_j, is occupied by a homogeneous,isotropic,chiral medium with electric permittivity ϵ_j, magnetic permeability μ_j and chirality measure β_j for $j = 1, 2, \cdots N$ and vanishing conductivity. The layer D_{N+1} is the core of the scatterer D, within which is the origin and its surface, S_N satisfies the perfect conductor boundary condition or the transmission conditions. The exterior D_0 of the scatterer is an infinite homogeneous isotropic achiral medium with electric permittivity ε_0, magnetic permeability μ_0 and vanishing conductivity. We assume that all physical parameters are real numbers.

We will consider the scattering of time-harmonic electromagnetic plane waves by a multi-layered chiral scatterer D. Let $(\mathbf{E}^i, \mathbf{H}^i)$ be a time-harmonic incident electromagnetic plane wave and $(\mathbf{E}^s, \mathbf{H}^s)$ be the corresponding scattered field. The total electromagnetic field $(\mathbf{E}_0, \mathbf{H}_0)$ in D_0 is given by

$$\mathbf{E}_0 = \mathbf{E}^i + \mathbf{E}^s, \tag{1}$$

$$\mathbf{H}_0 = \mathbf{H}^i + \mathbf{H}^s. \tag{2}$$

The scattered field $(\mathbf{E}^s, \mathbf{H}^s)$ satisfies the Silver-Muller radiation condition [15] while the total electromagnetic exterior field $(\mathbf{E}_0, \mathbf{H}_0)$ satisfies the Maxwell equations in D_0,

$$\nabla \times \mathbf{E}_0 = i\omega\mu_0 \mathbf{H}_0, \tag{3}$$

$$\nabla \times \mathbf{H}_0 = -i\omega\varepsilon_0 \mathbf{E}_0, \tag{4}$$

where ω is the angular frequency. In each layer D_j $j = 1, \cdots N$, the total field satisfies the modified Maxwell equations, in view of Born-Drude-Fedorov [22] constitutive relations

$$\nabla \times \mathbf{E}_j = i\omega\mu_j \frac{\gamma_j^2}{\kappa_j^2} \mathbf{H}_j + \beta_j \gamma_j^2 \mathbf{E}_j, \tag{5}$$

$$\nabla \times \mathbf{H}_j = -i\omega\varepsilon_j \frac{\gamma_j^2}{\kappa_j^2} \mathbf{E}_j + \beta_j \gamma_j^2 \mathbf{H}_j, \tag{6}$$

where $\gamma_j^2 = \dfrac{\kappa_j^2}{1 - \kappa_j^2 \beta_j^2}$ and $\kappa_j^2 = \omega^2 \varepsilon_j \mu_j$ are real physical parameters [22]. Note that the solutions of (3)–(6) are divergence free.

By eliminating the magnetic field, in (3)–(6) we conclude to the following modified Helmholtz type equation

$$\nabla \times \nabla \times \mathbf{E}_j - 2\beta_j {\gamma_j}^2 \nabla \times \mathbf{E}_j - {\gamma_j}^2 \mathbf{E}_j = 0 \text{ in } D_j \text{ for } j = 0, \cdots N, \tag{7}$$

where $\beta_0 = 0$ in free space and $\gamma_0 = \kappa_0 = \omega\sqrt{\varepsilon_0 \mu_0}$ is the free space wave number in the exterior region D_0 of the scatterer. It is easy to see that the following relation holds valid [9],

$$\kappa_j^2 = \frac{\varepsilon_j \mu_j}{\varepsilon_0 \mu_0} \kappa_0^2, \ j = 0, \cdots N. \tag{8}$$

The electric scattered field satisfies the Silver-Muller radiation condition,

$$\lim_{r \to \infty} \left[\mathbf{r} \times \nabla \times \mathbf{E}^s(\mathbf{r}) + i\kappa_0 r \mathbf{E}^s(\mathbf{r}) \right] = \mathbf{0}, \tag{9}$$

uniformly in all directions $\hat{\mathbf{r}} \in S^2$. We introduce the transmission conditions,

$$\hat{\mathbf{n}} \times \mathbf{E}_j = \hat{\mathbf{n}} \times \mathbf{E}_{j+1}, \tag{10}$$

$$\hat{\mathbf{n}} \times \mathbf{H}_j = \hat{\mathbf{n}} \times \mathbf{H}_{j+1}, \tag{11}$$

on $j = 0, 1, \cdots, N-1$. Substituting the magnetic field in (11) from (5) we get

$$\hat{\mathbf{n}} \times \nabla \times \mathbf{E}_j = \frac{\varepsilon_{j+1}}{\varepsilon_j} \frac{\gamma_j^2}{\gamma_{j+1}^2} \hat{\mathbf{n}} \times \nabla \times \mathbf{E}_{j+1} + \gamma_j^2 \left(\beta_j - \frac{\varepsilon_{j+1}}{\varepsilon_j} \beta_{j+1} \right) \hat{\mathbf{n}} \times \mathbf{E}_{j+1}. \tag{12}$$

We assume that the core of the multi-layered scatterer is a perfect conductor with boundary condition,

$$\hat{\mathbf{n}} \times \mathbf{E}_N(r) = \mathbf{0}, \ \text{on } S_N, \tag{13}$$

or the core is a dielectric with transmission conditions

$$\hat{\mathbf{n}} \times \mathbf{E}_N = \hat{\mathbf{n}} \times \mathbf{E}_{N+1}, \tag{14}$$

$$\hat{\mathbf{n}} \times \nabla \times \mathbf{E}_N = \frac{\varepsilon_{N+1}}{\varepsilon_N} \frac{\gamma_N^2}{\gamma_{N+1}^2} \hat{\mathbf{n}} \times \nabla \times \mathbf{E}_{N+1} + \gamma_N^2 \left(\beta_N - \frac{\varepsilon_{N+1}}{\varepsilon_N} \beta_{N+1} \right) \hat{\mathbf{n}} \times \mathbf{E}_{N+1}. \tag{15}$$

From now on, the problem that consists of (7), (9),(10), (12) and (13) will be denoted as (P_1) and the problem that consists of (7), (9), (10), (12), (13) and (14), (15) will be denoted as (P_2). Note that the above transmission problem is well posed and has been studied in [1, 7, 14]. The same problem can be, also, studied by eliminating the electric field in (3)–(6) following an analogous procedure as the one of the electric field.

In this work we will focus on the far-field patterns and the proofs of reciprocity and general scattering theorems.

The Electric Far-Field Pattern

The electric far-field pattern $\mathbf{E}^\infty(\hat{\mathbf{r}})$ is related to the scattered electric field $\mathbf{E}^s$ and it is given by the relation [17]

$$\mathbf{E}^s(\mathbf{r}) = \mathbf{E}^\infty(\hat{\mathbf{r}})h(\kappa_0 r) + O\left(\frac{1}{r^2}\right), \; r \to \infty \tag{16}$$

uniformly in all directions where $h(x) = e^{ix}/(ix)$ is the zeroth-order spherical Hankel function of the first kind. In order to obtain the electric far-field pattern we construct an integral representation of the total exterior electric field where the transmission and boundary conditions are incorporated. We have the following integral representation of the scattered field

$$\begin{aligned}\mathbf{E}^s(\mathbf{r}) = & \int_{S_0} [(\nabla \times \mathbf{E}^s(\mathbf{r}')) \cdot (\hat{\mathbf{n}} \times \tilde{\Gamma}(\mathbf{r},\mathbf{r}')) \\ & - (\hat{\mathbf{n}} \times \mathbf{E}^s(\mathbf{r}')) \cdot (\nabla_{\mathbf{r}'} \times \tilde{\Gamma}(\mathbf{r},\mathbf{r}'))]ds(\mathbf{r}'), \; \mathbf{r} \in D_0\end{aligned} \tag{17}$$

where $\tilde{\Gamma}(\mathbf{r},\mathbf{r}')$ is the free-space dyadic Green's function

$$\tilde{\Gamma}(\mathbf{r},\mathbf{r}') = (\tilde{I} + \kappa_0^{-2}\nabla\nabla)\frac{e^{i\kappa_0|\mathbf{r}-\mathbf{r}'|}}{4\pi|\mathbf{r}-\mathbf{r}'|} \tag{18}$$

with $\tilde{I} = \hat{\mathbf{e}}_1\hat{\mathbf{e}}_1 + \hat{\mathbf{e}}_2\hat{\mathbf{e}}_2 + \hat{\mathbf{e}}_3\hat{\mathbf{e}}_3$ is the identity dyadic and $\hat{\mathbf{e}}_j$, $j = 1, 2, 3$ are the cartesian unit vectors. Inserting (1) in (17) and taking into account that $\mathbf{E}^i$ is a solution of the Eq. (7) for $j = 0$ we obtain

$$\begin{aligned}\mathbf{E}_0(\mathbf{r}) = & \mathbf{E}^i(\mathbf{r}) + \int_{S_0} [(\nabla \times \mathbf{E}_0(\mathbf{r}')) \cdot (\hat{\mathbf{n}} \times \tilde{\Gamma}(\mathbf{r},\mathbf{r}')) \\ & - (\hat{\mathbf{n}} \times \mathbf{E}_0(\mathbf{r}')) \cdot (\nabla_{\mathbf{r}'} \times \tilde{\Gamma}(\mathbf{r},\mathbf{r}'))]ds(\mathbf{r}').\end{aligned} \tag{19}$$

Making use of the transmission conditions (10), (12) on S_0, (19) is equal to

$$\begin{aligned}\mathbf{E}_0(\mathbf{r}) = & \mathbf{E}^i(\mathbf{r}) - \frac{\varepsilon_1\gamma_0^2}{\varepsilon_0\gamma_1^2}\int_{S_0} \hat{\mathbf{n}} \cdot [(\nabla \times \mathbf{E}_1(\mathbf{r}')) \times \tilde{\Gamma}(\mathbf{r},\mathbf{r}')]ds(\mathbf{r}') \\ & + \gamma_0^2\frac{\varepsilon_1}{\varepsilon_0}\beta_1 \int_{S_0} \hat{\mathbf{n}} \cdot [\mathbf{E}_1(\mathbf{r}') \times \tilde{\Gamma}(\mathbf{r},\mathbf{r}')]ds(\mathbf{r}') \\ & - \int_{S_0} \hat{\mathbf{n}} \cdot [\mathbf{E}_1(\mathbf{r}') \times (\nabla_{\mathbf{r}'} \times \tilde{\Gamma}(\mathbf{r},\mathbf{r}'))]ds(\mathbf{r}').\end{aligned} \tag{20}$$

We apply successively the dyadic form of the divergence theorem, taking into account that $\mathbf{E}_j$ and $\tilde{\Gamma}$ are solutions of (7) in D_j, $j = 1, \cdots N$ and D_0, respectively; we introduce the transmission conditions (10), (12) and we obtain (20).

If the core is a perfect conductor, we use the boundary condition (13) and we obtain that

$$\mathbf{E}_0 = \mathbf{E}^i(\mathbf{r}) + \mathbf{E}_c^s(\mathbf{r}) + \mathbf{E}_a^s(\mathbf{r}) \tag{21}$$

where the indexes in $\mathbf{E}_c^s$ and $\mathbf{E}_a^s$ go for the chiral and the achiral parts of the electric scattered field $\mathbf{E}^s$, respectively and are given by

$$\begin{aligned}\mathbf{E}_c^s(\mathbf{r}) = &-\frac{\varepsilon_N}{\varepsilon_0}\beta_N^2\kappa_0^2\int_{S_N}(\nabla\times\mathbf{E}_N(\mathbf{r}'))\cdot(\hat{\mathbf{n}}\times\tilde{\Gamma}(\mathbf{r},\mathbf{r}'))ds(\mathbf{r}')\\ &-\kappa_0^2\sum_{j=1}^{N}\frac{\varepsilon_j}{\varepsilon_0}\beta_j^2\int_{D_j}(\nabla\times\mathbf{E}_j(\mathbf{r}'))\cdot(\nabla_{\mathbf{r}'}\times\tilde{\Gamma}(\mathbf{r},\mathbf{r}'))dv(\mathbf{r}')\\ &-\kappa_0^2\sum_{j=1}^{N}\frac{\varepsilon_j}{\varepsilon_0}\beta_j\int_{D_j}(\nabla\times\mathbf{E}_j(\mathbf{r}'))\cdot\tilde{\Gamma}(\mathbf{r},\mathbf{r}')dv(\mathbf{r}')\\ &-{\kappa_0}^2\sum_{j=1}^{N}\frac{\varepsilon_j}{\varepsilon_0}\beta_j\int_{D_j}\mathbf{E}_j(\mathbf{r}')\cdot(\nabla_{\mathbf{r}'}\times\tilde{\Gamma}(\mathbf{r},\mathbf{r}'))dv(\mathbf{r}'),\ \mathbf{r}\in D_0\end{aligned}\tag{22}$$

and

$$\begin{aligned}\mathbf{E}_a^s(\mathbf{r}) = &\frac{\mu_0}{\mu_N}\int_{S_N}(\nabla\times\mathbf{E}_N(\mathbf{r}'))\cdot(\hat{\mathbf{n}}\times\tilde{\Gamma}(\mathbf{r},\mathbf{r}'))ds(\mathbf{r}')\\ &+\kappa_0^2\sum_{j=1}^{N}\left(1-\frac{\varepsilon_j}{\varepsilon_0}\right)\int_{D_j}\mathbf{E}_j(\mathbf{r}')\cdot\tilde{\Gamma}(\mathbf{r},\mathbf{r}')dv(\mathbf{r}')\\ &+\sum_{j=1}^{N}\left(\frac{\mu_0}{\mu_j}-1\right)\int_{D_j}(\nabla\times\mathbf{E}_j(\mathbf{r}'))\cdot(\nabla_{\mathbf{r}'}\times\tilde{\Gamma}(\mathbf{r},\mathbf{r}'))dv(\mathbf{r}')\ \mathbf{r}\in D_0.\end{aligned}\tag{23}$$

The volume integrals express the contribution of each layer to the exterior field, whereas the surface integrals express the impact of the core.

If the core is dielectric, then in relations (22), (23) the surface integrals on S_N disappear and the volume integrals D_j have an extra term for $j = N + 1$.

Using the asymptotic relations

$$|\mathbf{r}-\mathbf{r}'| = r - \hat{\mathbf{r}}\cdot\mathbf{r}' + O\left(\frac{1}{r}\right),\quad r\to\infty,\tag{24}$$

$$\frac{\mathbf{r}-\mathbf{r}'}{|\mathbf{r}-\mathbf{r}'|} = \hat{\mathbf{r}} + O\left(\frac{1}{r}\right),\quad r\to\infty,\tag{25}$$

in (18) we get the following asymptotic forms

$$\tilde{\Gamma}(\mathbf{r},\mathbf{r}') = \frac{i\kappa_0}{4\pi}(\tilde{I}-\hat{\mathbf{r}}\hat{\mathbf{r}})h(\kappa_0 r)e^{-i\kappa_0\hat{\mathbf{r}}\cdot\mathbf{r}'} + O\left(\frac{1}{r^2}\right),\quad r\to\infty,\tag{26}$$

$$\nabla_{\mathbf{r}'}\times\tilde{\Gamma}(\mathbf{r},\mathbf{r}') = \frac{\kappa_0^2}{4\pi}(\tilde{I}\times\hat{\mathbf{r}})h(\kappa_0 r)e^{-i\kappa_0\hat{\mathbf{r}}\cdot\mathbf{r}'} + O\left(\frac{1}{r^2}\right),\quad r\to\infty.\tag{27}$$

If the core is a perfect conductor then substituting (26)–(27) into (22)–(23) we obtain

$$\mathbf{E}^{\infty}(\hat{\mathbf{r}}) = \mathbf{E}_c^{\infty}(\hat{\mathbf{r}}) + \mathbf{E}_a^{\infty}(\hat{\mathbf{r}}) \tag{28}$$

where

$$\begin{aligned}
4\pi \mathbf{E}_c^{\infty}(\hat{\mathbf{r}}) =& i\kappa_0^3 \frac{\varepsilon_N}{\varepsilon_0} \beta_N^2 \int_{S_N} (\nabla \times \mathbf{E}_N(\mathbf{r}')) \cdot (\hat{\mathbf{n}} \times (\tilde{I} - \hat{\mathbf{r}}\hat{\mathbf{r}})) e^{-i\kappa_0 \hat{\mathbf{r}}\cdot\mathbf{r}'} ds(\mathbf{r}') \\
&+ \kappa_0^4 \sum_{j=1}^{N} \frac{\varepsilon_j}{\varepsilon_0} \beta_j^2 \int_{D_j} (\nabla \times \mathbf{E}_j(\mathbf{r}')) \cdot (\tilde{I} \times \hat{\mathbf{r}}) e^{-i\kappa_0 \hat{\mathbf{r}}\cdot\mathbf{r}'} dv(\mathbf{r}') \\
&+ i\kappa_0^3 \sum_{j=1}^{N} \frac{\varepsilon_j}{\varepsilon_0} \beta_j \int_{D_j} (\nabla \times \mathbf{E}_j(\mathbf{r}')) \cdot (\tilde{I} - \hat{\mathbf{r}}\hat{\mathbf{r}}) e^{-i\kappa_0 \hat{\mathbf{r}}\cdot\mathbf{r}'} dv(\mathbf{r}') \\
&+ \kappa_0^4 \sum_{j=1}^{N} \frac{\varepsilon_j}{\varepsilon_0} \beta_j^2 \int_{D_j} \mathbf{E}_j(\mathbf{r}') \cdot (\tilde{I} \times \hat{\mathbf{r}}) e^{-i\kappa_0 \hat{\mathbf{r}}\cdot\mathbf{r}'} dv(\mathbf{r}'),
\end{aligned} \tag{29}$$

and

$$\begin{aligned}
4\pi \mathbf{E}_a^{\infty}(\hat{\mathbf{r}}) =& - i\kappa_0 \frac{\mu_0}{\mu_N} \int_{S_N} (\nabla \times \mathbf{E}_N(\mathbf{r}')) \cdot (\hat{\mathbf{n}} \times (\tilde{I} - \hat{\mathbf{r}}\hat{\mathbf{r}})) e^{-i\kappa_0 \hat{\mathbf{r}}\cdot\mathbf{r}'} ds(\mathbf{r}') \\
&- i\kappa_0^3 \sum_{j=1}^{N} \left(1 - \frac{\varepsilon_j}{\varepsilon_0}\right) \int_{D_j} \mathbf{E}_j(\mathbf{r}') \cdot (\tilde{I} - \hat{\mathbf{r}}\hat{\mathbf{r}}) e^{-i\kappa_0 \hat{\mathbf{r}}\cdot\mathbf{r}'} dv(\mathbf{r}') \\
&- \kappa_0^2 \sum_{j=1}^{N} \left(\frac{\mu_0}{\mu_j} - 1\right) \int_{D_j} (\nabla \times \mathbf{E}_j(\mathbf{r}')) \cdot (\tilde{I} \times \hat{\mathbf{r}}) e^{-i\kappa_0 \hat{\mathbf{r}}\cdot\mathbf{r}'} dv(\mathbf{r}').
\end{aligned} \tag{30}$$

If the core is dielectric, then the far-field patterns $\mathbf{E}_c^{\infty}$, $\mathbf{E}_a^{\infty}$ are given by (29), (30) where again the surface integrals do not exist and the volume integrals sum up to $j = N+1$. Summarizing the previous results we conclude to the following theorem.

Theorem 1. *The electric far-field patterns of the scattering problems* (P_1) *and* (P_2) *are given by* (28) *where* $\boldsymbol{E}_c^{\infty}$ *and* $\boldsymbol{E}_a^{\infty}$ *are the chiral and achiral counterparts of the corresponding far-field patterns.*

Scattering Relations

In this section we will prove scattering theorems for the scattering problems (P_1) and (P_2). In order to do so, it is more convenient to restate the problem considering a dimensionless version [8] scaling all lengths using α, a typical length scale for the chiral scatterer. Therefore the scattering problems (P_1) and (P_2) take the following form [8]

$$\nabla \times \mathbf{E}_0 - i(\kappa_0\alpha)\mathbf{H}_0 = \mathbf{0} \quad , \nabla \times \mathbf{H}_0 + i(\kappa_0\alpha)\mathbf{E}_0 = \mathbf{0}, \text{ in } D_0 \tag{31}$$

while in the layers $D_j,\ j = 1, 2, \cdots N$

$$\begin{aligned} \nabla \times \mathbf{E}_j - i(\kappa_j\alpha)\left(\frac{\gamma_j}{\kappa_j}\right)^2 \mathbf{H}_j - \beta_j\alpha\gamma_j^2\mathbf{E}_j &= \mathbf{0}, \\ \nabla \times \mathbf{H}_j + i(\kappa_j\alpha)\left(\frac{\gamma_j}{\kappa_j}\right)^2 \mathbf{E}_j - \beta_j\alpha\gamma_j^2\mathbf{H}_j &= \mathbf{0}, \end{aligned} \tag{32}$$

for $j = 1, \cdots N$. The transmission conditions become

$$\begin{aligned} \rho_j \hat{\mathbf{n}} \times \mathbf{E}_j &= \hat{\mathbf{n}} \times \mathbf{E}_{j+1}, \\ \delta_j \hat{\mathbf{n}} \times \mathbf{H}_j &= \hat{\mathbf{n}} \times \mathbf{H}_{j+1} \end{aligned} \tag{33}$$

on $S_j\ j = 0, 1, \cdots N-1$, where $\kappa_0 = \omega\sqrt{\mu_0\varepsilon_0}$, $\rho_j = \sqrt{\dfrac{\mu_j}{\mu_{j+1}}}$ and $\delta_j = \sqrt{\dfrac{\varepsilon_j}{\varepsilon_{j+1}}}$ are real. If the core is dielectric, the Eq. (32) are valid for $j = N+1$ as well while the transmission conditions (33) are also valid for $j = N$ as well.

The scattered field is now a pair $(\mathbf{E}^s, \mathbf{H}^s)$ and satisfy the Silver-Muller radiation condition

$$\hat{\mathbf{r}} \times \mathbf{H}^s + \mathbf{E}^s = o(\frac{1}{r}), \ r \to \infty. \tag{34}$$

We can consider that all lengths have been scaled using α, and thus we can set $\alpha = 1$. It is also more helpful to introduce the fields $\mathbf{U}$ and $\mathbf{U}'$ where $\mathbf{U}'$ is the dual of $\mathbf{U}$. More precisely,

$$\text{if } \mathbf{U} = \mathbf{E} \text{ then } \mathbf{U}' = i\mathbf{H} \tag{35}$$

$$\text{if } \mathbf{U} = \mathbf{H} \text{ then } \mathbf{U}' = -i\mathbf{E}. \tag{36}$$

Therefore, the relation (32) can be rewritten as follows

$$\nabla \times \mathbf{U}_j - \gamma_j^2\beta_j\mathbf{U}_j - \frac{\gamma_j^2}{\kappa_j}\mathbf{U}_j' = \mathbf{0}, \ j = 0, \cdots N. \tag{37}$$

and Eq. (31) for the exterior region D_0 becomes

$$\begin{aligned} \nabla \times \mathbf{U} &= \kappa_0\mathbf{U}', \\ \nabla \times \mathbf{U}' &= \kappa_0\mathbf{U}, \end{aligned} \tag{38}$$

where $(\mathbf{U}')' = \mathbf{U}$. The corresponding far-field pattern $(\mathbf{E}^\infty, \mathbf{H}^\infty)$ is given by

$$\begin{aligned} \mathbf{E}^s(\mathbf{r}) &= \mathbf{E}^\infty(\hat{\mathbf{r}})h(\kappa_0 r) + O\left(\frac{1}{r^2}\right), \ r \to \infty, \\ \mathbf{H}^s(\mathbf{r}) &= \mathbf{H}^\infty(\hat{\mathbf{r}})h(\kappa_0 r) + O\left(\frac{1}{r^2}\right), \ r \to \infty, \end{aligned} \tag{39}$$

uniformly in all directions $\hat{\mathbf{r}} \in S^2$ in the unit sphere. We assume that the incident electromagnetic wave is also dimensionless and it has the form [15]

$$\begin{aligned} \mathbf{E}^i(\mathbf{r}; \hat{\mathbf{d}}, \mathbf{p}) &= i(\kappa_0 \alpha)\mathbf{p}\, e^{i\kappa_0 \hat{\mathbf{d}}\cdot\mathbf{r}}, \\ \mathbf{H}^i(\mathbf{r}; \hat{\mathbf{d}}, \mathbf{p}) &= \hat{\mathbf{d}} \times \mathbf{E}^i(\mathbf{r}; \hat{\mathbf{d}}, \mathbf{p}), \end{aligned} \tag{40}$$

setting $\alpha = 1$. The unit vector $\hat{\mathbf{d}}$ describes the direction of propagation, the vector $\mathbf{p}$ the polarization and they connect with the relation $\hat{\mathbf{d}} \cdot \mathbf{p} = 0$. Henceforth, the dependence of the total, scattered and far-field patterns on the direction of propagation and polarization will be denoted by $(\mathbf{E}_0(\mathbf{r}; \hat{\mathbf{d}}, \mathbf{p}), \mathbf{H}_0(\mathbf{r}; \hat{\mathbf{d}}, \mathbf{p}))$, $(\mathbf{E}^s(\mathbf{r}; \hat{\mathbf{d}}, \mathbf{p}), \mathbf{H}^s(\mathbf{r}; \hat{\mathbf{d}}, \mathbf{p}))$ and $(\mathbf{E}^\infty(\mathbf{r}; \hat{\mathbf{d}}, \mathbf{p}), \mathbf{H}^\infty(\mathbf{r}; \hat{\mathbf{d}}, \mathbf{p}))$, respectively.

Moreover, we have the following notation for the total fields $\mathbf{E}_{m,n}$, $\mathbf{H}_{m,n}$ for $m = 1, 2$ and $n = 0, \cdots N$. The index m goes for the first or the second total field and the index n for the layer S_j of the multi-layered scatterer. Instead of the electromagnetic pair $(\mathbf{E}, \mathbf{H})$, we shall use the fields $\mathbf{U}$ and $\mathbf{U}'$ as defined in (35)–(36) for the total, incident, scattered and far-field pattern. In addition, we introduce the Twersky [25] notation

$$\{\mathbf{U}_1, \mathbf{U}_2\}_{S_0} := \int_S [(\hat{\mathbf{n}} \times \mathbf{U}_1) \cdot \mathbf{U}_2' - (\hat{\mathbf{n}} \times \mathbf{U}_2) \cdot \mathbf{U}_1'] ds. \tag{41}$$

We will proceed by stating and proving the reciprocity principle.

Theorem 1. *The far-field pattern $\mathbf{U}^\infty$ satisfies the reciprocity principle*

$$\mathbf{q} \cdot \mathbf{U}^\infty(\hat{\mathbf{r}}; \hat{\mathbf{d}}, \mathbf{p}) = \mathbf{p} \cdot \mathbf{U}^\infty(-\hat{\mathbf{d}}; -\hat{\mathbf{r}}, \mathbf{q}) \tag{42}$$

for all $\hat{\mathbf{d}}, \hat{\mathbf{r}} \in S^2$ *and* $\mathbf{p}, \mathbf{q} \in \mathbb{C}^3$ *with* $\mathbf{p} \cdot \hat{\mathbf{d}} = \mathbf{q} \cdot \hat{\mathbf{r}} = 0$.

Proof. In view of bilinearity of the form (41) we get

$$\begin{aligned} \{\mathbf{U}_{1,0}, \mathbf{U}_{2,0}\}_{S_0} =& \{\mathbf{U}_1^i, \mathbf{U}_2^i\}_{S_0} + \{\mathbf{U}_1^i, \mathbf{U}_2^s\}_{S_0} \\ &+ \{\mathbf{U}_1^s, \mathbf{U}_2^i\}_{S_0} + \{\mathbf{U}_1^s, \mathbf{U}_2^s\}_{S_0}. \end{aligned} \tag{43}$$

Using the transmission conditions and applying successively the Gauss' theorem we take

$$\{\mathbf{U}_{1,0}, \mathbf{U}_{2,0}\}_{S_0} = \sqrt{\frac{\mu_N}{\mu_0}\frac{\varepsilon_N}{\varepsilon_0}}\{\mathbf{U}_{1,N}, \mathbf{U}_{2,N}\}_{S_N}. \tag{44}$$

For the problem (P_1) taking into account the boundary condition of the core which is now transformed into

$$\hat{\mathbf{n}} \times \mathbf{U}_N = \mathbf{0} \text{ on } S_N, \tag{45}$$

we have

$$\{\mathbf{U}_{1,0}, \mathbf{U}_{2,0}\}_{S_0} = 0. \tag{46}$$

If the core is dielectric, we apply again Gauss' theorem for $\mathbf{U}_{1,N}, \mathbf{U}_{2,N}$ in D_{N+1} and we conclude to (46). Applying again the Gauss' divergence theorem on the exterior region and taking into account the Maxwell equations for the exterior domain (31), setting $\alpha = 1$, we take for the incident part

$$\{\mathbf{U}_1^i, \mathbf{U}_2^i\}_{S_0} = 0. \tag{47}$$

We consider a sphere S_r centered at the origin with radius r large enough to include the scatterer in its interior. Applying the Gauss theorem in the exterior domain we have

$$\{\mathbf{U}_1^s, \mathbf{U}_2^s\}_{S_0} = \{\mathbf{U}_1^s, \mathbf{U}_2^s\}_{S_r}. \tag{48}$$

Letting $R \to \infty$ we pass to the radiation zone and using the radiating condition (34) the surface integral on S_0 becomes zero,

$$\{\mathbf{U}_1^s, \mathbf{U}_2^s\}_{S_0} = 0. \tag{49}$$

Substituting the previous relations in (43) we get

$$\{\mathbf{U}_1^i, \mathbf{U}_2^s\}_{S_0} = -\{\mathbf{U}_1^s, \mathbf{U}_2^i\}_{S_0} \tag{50}$$

which is equal to

$$\{\mathbf{U}_1^i, \mathbf{U}_2^s\}_{S_0} = \{\mathbf{U}_2^i, \mathbf{U}_1^s\}_{S_0}. \tag{51}$$

Taking into account the integral representations of the electric and magnetic fields as in (6.24) in [15] and that $\mathbf{q} \cdot \hat{\mathbf{r}}$ we get

$$\mathbf{q} \cdot \mathbf{U}^{\infty}(\hat{\mathbf{r}}; \hat{\mathbf{d}}, \mathbf{p}) = \frac{-i}{4\pi}\{\mathbf{U}^s(\cdot; \hat{\mathbf{d}}, \mathbf{p}),\, \mathbf{U}^i(\cdot; -\hat{\mathbf{r}}, \mathbf{q})\}_{S_0} \tag{52}$$

Combining it with (51) we get

$$\mathbf{q}\cdot\mathbf{U}^{\infty}(\hat{\mathbf{r}};\hat{\mathbf{d}},\mathbf{p}) = -\frac{i}{4\pi}\{\mathbf{U}^{s}(\cdot;-\hat{\mathbf{r}},\mathbf{q}),\mathbf{U}^{i}(\cdot;-\hat{\mathbf{d}},\mathbf{p})\}_{S_0}$$
$$= \mathbf{p}\cdot\mathbf{U}^{\infty}(-\hat{\mathbf{d}};-\hat{\mathbf{r}},\mathbf{q}). \tag{53}$$

which proves the theorem. □

Theorem 2. *The far-field pattern* $\mathbf{U}^{\infty}$ *satisfies the relation*

$$\overline{\mathbf{q}\cdot\mathbf{U}^{\infty}(\hat{\mathbf{r}};\hat{\mathbf{d}},\mathbf{p})}+\mathbf{\bar{p}}\cdot\mathbf{U}^{\infty}(\hat{\mathbf{d}};\hat{\mathbf{r}},\mathbf{q}) = -\frac{1}{2\pi}\int_{S^2}\overline{\mathbf{U}^{\infty}(\hat{\mathbf{r}}';\hat{\mathbf{d}},\mathbf{p})}\cdot\mathbf{U}^{\infty}(\hat{\mathbf{r}}';\hat{\mathbf{r}},\mathbf{q})ds(\mathbf{r}') \tag{54}$$

for all $\hat{\mathbf{d}},\hat{\mathbf{r}}\in S^2$ *and* $\mathbf{p},\mathbf{q}\in\mathbb{C}^3$ *with* $\hat{\mathbf{d}}\cdot\mathbf{p}=\hat{\mathbf{r}}\cdot\mathbf{q}=0$.

Proof. In view again of the bilinearity of (41) we obtain

$$\{\overline{\mathbf{U}_0(\cdot;\hat{\mathbf{d}},\mathbf{p})},\mathbf{U}_0(\cdot;\hat{\mathbf{r}},\mathbf{q})\}_{S_0} = \{\overline{\mathbf{U}^{i}(\cdot;\hat{\mathbf{d}},\mathbf{p})},\mathbf{U}^{i}(\cdot;\hat{\mathbf{r}},\mathbf{q})\}_{S_0}$$
$$+\{\overline{\mathbf{U}^{i}(\cdot;\hat{\mathbf{d}},\mathbf{p})},\mathbf{U}^{s}(\cdot;\hat{\mathbf{r}},\mathbf{q})\}_{S_0}$$
$$+\{\overline{\mathbf{U}^{s}(\cdot;\hat{\mathbf{d}},\mathbf{p})},\mathbf{U}^{i}(\cdot;\hat{\mathbf{r}},\mathbf{q})\}_{S_0}$$
$$+\{\overline{\mathbf{U}^{s}(\cdot;\hat{\mathbf{d}},\mathbf{p})},\mathbf{U}^{s}(\cdot;\hat{\mathbf{r}},\mathbf{q})\}_{S_0} \tag{55}$$

The term

$$\{\overline{\mathbf{U}^{i}(\cdot;\hat{\mathbf{d}},\mathbf{p})},\mathbf{U}^{i}(\cdot;\hat{\mathbf{r}},\mathbf{q})\}_{S_0} = 0 \tag{56}$$

becomes zero from the divergence theorem and the fact that $\overline{\mathbf{U}^{i}(\mathbf{r}';\hat{\mathbf{d}},\mathbf{p}),\mathbf{U}^{i}(\mathbf{r}';\hat{\mathbf{r}},\mathbf{q})}$ are solutions of (37).

The total fields

$$\{\overline{\mathbf{U}_0(\cdot;\hat{\mathbf{d}},\mathbf{p})},\mathbf{U}_0(\cdot;\hat{\mathbf{r}},\mathbf{q})\}_{S_0} = 0 \tag{57}$$

following the procedure of Theorem (1) and taking into account the fact that all the physical parameters of the scattering problem are real numbers.

Moreover, from (52) we obtain

$$\{\overline{U_1^{i}(\cdot;\hat{\mathbf{d}},\mathbf{p})},\mathbf{U}_2^{s}(\cdot;\hat{\mathbf{r}},\mathbf{q})\}_{S_0} = 4\pi i\mathbf{\bar{p}}\cdot\mathbf{U}^{\infty}(\hat{\mathbf{d}};\hat{\mathbf{r}},\mathbf{q}) \tag{58}$$

and

$$\{\overline{\mathbf{U}_1^{s}(\cdot;\hat{\mathbf{d}},\mathbf{p})},\mathbf{U}_2^{i}(\cdot;\hat{\mathbf{r}},\mathbf{q})\}_{S_0} = 4\pi i\overline{\mathbf{q}\cdot\mathbf{U}^{\infty}(\hat{\mathbf{r}};\hat{\mathbf{d}},\mathbf{p})}. \tag{59}$$

For the scattered fields we consider a sphere S_r centered at the origin with radius r, large enough to include the chiral scatterer in its interior. Applying the Gauss theorem in the region exterior to S_0 and interior to S_r, we obtain

$$\{\overline{\mathbf{U}^s(\cdot;\hat{\mathbf{d}},\mathbf{p})},\mathbf{U}^s(\cdot;\hat{\mathbf{r}},\mathbf{q})\}_{S_0} = \{\overline{\mathbf{U}^s(\cdot;\hat{\mathbf{d}},\mathbf{p})},\mathbf{U}^s(\cdot;\hat{\mathbf{r}},\mathbf{q})\}_{S_r} \tag{60}$$

Letting $r \to \infty$, we can use the asymptotic forms (39) for the scattered fields. Taking into account that $\hat{\mathbf{n}} \times \mathbf{U}'^{\infty} = i\mathbf{U}^{\infty}$ we conclude that

$$\begin{aligned}\{\overline{\mathbf{U}^s(\cdot;\hat{\mathbf{d}},\mathbf{p})},\mathbf{U}^s(\cdot;\hat{\mathbf{r}},\mathbf{q})\}_{S_0} &= \int_{S_\infty} |h(\kappa_o r')|^2 (2i\overline{\mathbf{U}^\infty(\hat{\mathbf{r}}';\hat{\mathbf{d}},\mathbf{p})}\cdot \mathbf{U}_2^\infty(\hat{\mathbf{r}}';\hat{\mathbf{r}},\mathbf{q}))ds(\mathbf{r}') \\ &= 2i\int_{S^2} \overline{\mathbf{U}^\infty(\hat{\mathbf{r}}';\hat{\mathbf{d}},\mathbf{p})}\cdot \mathbf{U}^\infty(\hat{\mathbf{r}}';\hat{\mathbf{r}},\mathbf{q})ds(\hat{\mathbf{r}}')\end{aligned} \tag{61}$$

Substituting (56), (57), (58), (59) and (61) into (55) we conclude to (54). □

Theorem 3. *The following relation holds:*

$$\sigma = -4\pi Re(\bar{\mathbf{p}}\cdot \mathbf{U}^\infty(\hat{\mathbf{d}};\hat{\mathbf{d}},\mathbf{p})) \tag{62}$$

Proof. Since D_0 is achiral, we can follow the same procedure as in [25] where we can see that

$$\sigma = \int_{S^2} |\mathbf{U}^\infty(\hat{\mathbf{r}},\hat{\mathbf{d}},\mathbf{p})|^2 ds(\hat{\mathbf{r}}) \tag{63}$$

Using Theorem 2 and substituting $\hat{\mathbf{r}} = \hat{\mathbf{d}}$ and $\mathbf{p} = \mathbf{q}$ we conclude to relation (62).□

Herglotz Functions

Next, we will prove a general scattering theorem when the incident field is a Herglotz pair with kernel $\mathbf{g}$. Such a pair is of the form

$$\begin{aligned}\mathbf{U}_{\mathbf{g}}(\mathbf{r}) &= i\kappa_0 \int_{S^2} \mathbf{g}(\hat{\mathbf{q}})\, e^{i\kappa_0 \hat{\mathbf{q}}\cdot\mathbf{r}} ds(\hat{\mathbf{q}}), \\ \mathbf{U}_{\mathbf{g}}'(\mathbf{r}) &= -\kappa_0 \int_{S^2} \hat{\mathbf{q}} \times \mathbf{g}(\hat{\mathbf{q}})\, e^{i\kappa_0 \hat{\mathbf{q}}\cdot\mathbf{r}} ds(\hat{\mathbf{q}}),\end{aligned} \tag{64}$$

where $\mathbf{g} \in L^2(S^2)$ and $\mathbf{g}\cdot\hat{\mathbf{q}} = 0$. We note that these functions given by (64) are solutions of (38) [15, 16]. When the incident field is a Herglotz pair of the form,

$$\mathbf{U}_{\mathbf{g}}^i(\mathbf{r}) = \int_{S^2} \mathbf{U}^i(\mathbf{r};\hat{\mathbf{q}},\mathbf{g}(\hat{\mathbf{q}}))ds(\hat{\mathbf{q}}), \tag{65}$$

then the corresponding scattered field and far-field pattern are given by

$$\begin{aligned} \mathbf{U}^s_{\mathbf{g}}(\mathbf{r}) &= \int_{S^2} \mathbf{U}^s(\mathbf{r};\hat{\mathbf{q}},\mathbf{g}(\hat{\mathbf{q}}))ds(\hat{\mathbf{q}}), \\ \mathbf{U}^\infty_{\mathbf{g}}(\hat{\mathbf{r}}) &= \int_{S^2} \mathbf{U}^\infty(\hat{\mathbf{r}};\hat{\mathbf{q}},\mathbf{g}(\hat{\mathbf{q}}))ds(\hat{\mathbf{q}}). \end{aligned} \tag{66}$$

Theorem 1. *For the problems* (P_1) *and* (P_2) *the following scattering relations are valid*

$$\{\mathbf{U}^s_{\mathbf{g}}, \overline{\mathbf{U}^i_{\mathbf{h}}}\}_{S_0} = -4\pi i < \mathbf{U}^\infty_{\mathbf{g}}, \mathbf{h} >, \tag{67}$$

$$\{\mathbf{U}^s_{\mathbf{g}}, \overline{\mathbf{U}^s_{\mathbf{h}}}\}_{S_0} = -2i < \mathbf{U}^\infty_{\mathbf{g}}, \mathbf{U}^\infty_{\mathbf{h}} > \tag{68}$$

where $<,>$ *denotes the inner product in* $L^2(S^2)$.

Proof. Relation (67) comes from (52) while relation (68) comes from (61) when the incident field is $\mathbf{U}^i_{\mathbf{h}}$.

Next, we define the far-field operator that corresponds to the far-field pattern $\mathbf{U}^\infty$ as follows $F : L^2(S^2) \to L^2(S^2)$,

$$(F\mathbf{g})(\hat{\mathbf{r}}) = \int_{S^2} \mathbf{U}^\infty(\hat{\mathbf{r}};\hat{\mathbf{q}},\mathbf{g}(\hat{\mathbf{q}}))ds(\hat{\mathbf{q}}) \tag{69}$$

with far-field equation $F\mathbf{g} = \mathbf{U}^\infty_{\mathbf{g}}$ [15]. The far-field operator is very important in solving inverse scattering problems. Many methods have been developed in this direction as the dual space method [14], linear sampling method [12] and factorization of the far-field operator [13]. The following corollary derives from the general scattering Theorem 2 considering superpositions of the incident and scattered fields on the unit sphere. □

Corollary 1. *The electric far-field operator* F *corresponding to the problems* (P_1), (P_2) *satisfies the following relation*

$$< F\mathbf{g}, \mathbf{h} > + < \mathbf{g}, F\mathbf{h} >= -\frac{1}{2\pi} < F\mathbf{g}, F\mathbf{h} > \tag{70}$$

Proof. In order to arrive at this result we apply theorem (2) for $\mathbf{p} = \mathbf{g}(\hat{\mathbf{d}})$ and $\mathbf{q} = \mathbf{h}(\hat{\mathbf{r}})$; and we integrate over the unit sphere twice following an analogous procedure as in [8, 16]. □

Conclusions

The above study can lead to results for simple scatterers (perfect conductor or a dielectric) when the physical parameters satisfy $\varepsilon_j = \varepsilon_{j+1}, \mu_j = \mu_{j+1}, \beta_j = \beta_{j+1}$ for $j = 0, \cdots, N-1$ [3]. Moreover, if the physical parameters were complex numbers then, in general scattering Theorem 2 and in Corollary 1, there would be an extra term that would derive from the total fields. In optical Theorem 3 the extra term describes the absorbing cross section [8]. We should also note that if chirality measure is zero, $\beta = 0$, then we conclude to already known results for the achiral case.

References

1. Athanasiadis C., Costakis G., Stratis I.G., *Electromagnetic scattering by a perfectly conducting obstacle in a homogeneous chiral environment: Solvability and low frequency theory*, Math. Meth. Appl. Sci. 25(2002), 927–944.
2. Athanasiadis C.*Low frequency electromagnetic scattering theory for a multi-layered scatterer*,Quart. J. Mech. Appl. Math. 44 (1991), 55–67.
3. Athanasiadis C.*The multi-layered ellipsoid with a soft core in the presence of a low frequency acoustic wave.*,Quart. J. Mech. Appl. Math. 47 (1994) 441–459.
4. Athanasiadis C. *The hard-core multi-layered ellipsoid in a low frequency acoustic field* Int., J., Eng. 32, (1994) 1352–1359.
5. Athanasiadis C. *Scattering relations for time-harmonic electromagnetic waves in a piecewise homogeneous medium* Math. Proc. Camb. Phil. Soc. (1998), 123, 179.
6. Athanasiadis C. *On the acoustic scattering amplitude for a multi-layered scatterer* J. Austral. Math. Soc. Ser. B 39(1998), 431–448.
7. Athanasidis C. and Stratis I. *On a transmission problem for the time-harmonic Maxwell equations.*Rend.Math.Appl. 16 (1996), 671–688.
8. Athanasiadis C.,Martin C. and Stratis I. *Electromagnetic scattering by a homogeneous chiral obstacle:scattering relations and the far-field operator.* Math.Meth.Appl.Sci. 22, 1175–1188.
9. Athanasiadis A., Stratis G. *Low frequency electromagnetic scattering theory for a multi-layered chiral obstacle* Methods and Applications of Analysis, Vol. 6, No. 4, pp. 437–450, (1999).
10. Angell T.S., Kirch A. *The conductive boundary condition for Maxwell's equations* SIAM J. Appl. Math. 52(6) (1992), 1597–1610.
11. Angell T.S., Colton D., Kirch A. *Far-field patterns and inverse scattering problems for imperfectly conducting obstacles* Math. Proc. Camb. Phil. Soc. 106 (1989), 553–569.
12. Cakoni F., Colotn D., Monk P. *The linear Sampling Method in Inverse Electromagnetic Scattering*, CBMS Series, SIAM Publications 80, (2011).
13. Cakoni F., Colton D. *Qualitative Methods in Inverse Scattering Theory*, Springer, Series on Interaction of Mathematics and Mechanics (2006).
14. Colton D. and Kress R. *Integral equations methods in scattering theory.* (Wiley, 1983).
15. Colton D. and Kress R. *Inverse acoustic and electromagnetic scattering theory.* (Springer-Verlag,1992).
16. Colton D. and Kress R. *Eigenvalues of the far field operator and inverse scattering theory.* SIAM J. Math.Anal. 26 (1995), 601–615.
17. Dassios G., Kleinman R. *Low Frequency Scattering*, Clarendon Press, 2000.
18. Dassios G.,Kiriaki K.and Polysos D. *On the scattering amplitudes for elastic waves.* Z.Angew.Math.Phys. 38 (1987), 856–873.

19. Defacio B. *Classical,linear electromagnetic impedance theory with infinite integrable discontinuities.*J.Math.Phys. 31 (9) (1990), 2155–2164.
20. Lakhtakia A., V.K. Varadan, V.V. Varadan. *Time-harmonic Electromagnetic Fields in Chiral Media.* Lecture notes in Physics, vol. 335, Springer, 1989.
21. Lakhtakia A. *On the Huygen's principles and the Ewald-Oseen extinction theorems for,and the scattering of,Beltrami fields.* Optik 90 (1992) 35–40.
22. Lakhtakia A. *Beltrami Fields in Chiral Media.* World Scientific, 1994.
23. Twersky V. *On scattering of waves by random distributions.I.Free space scattering formalism.* J. Math. Phys. (3) 4 (1962), 700–715.
24. Twersky V. *On a general class of scattering problems.* J. Math. Phys. (3) 4 (1962) 716–723.
25. Twersky V. *Multiple scattering of electromagnetic waves by arbitary configurations.* J. Math. Phys. (8) 3 (1967), 589–598.

Almost Periodic Solutions of Navier–Stokes–Ohm Type Equations in Magneto-Hydrodynamics

Evagelia S. Athanasiadou, Vasileios F. Dionysatos, Panagiotis N. Koumantos, and Panaiotis K. Pavlakos

Abstract In this paper we construct (Bohl-Bohr- and Stepanoff-) almost periodic solutions of an evolution equation of the form $\left(\frac{d}{dt} + A\right) x(t) = F(t, x(t))$, $t \in \mathbb{R}$, describing the velocity and the magnetic field of a viscous incompressible homogeneous ideal plasma in magneto-hydrodynamics. By $-A$ it is denoted the infinitesimal generator of a C_0-semigroup e^{-tA}, $t \in \mathbb{R}^+$ of operators acting on an ordered Hilbert space E and $F : \mathbb{R} \times E \to E$ is a given function. We also examine the case of the construction of positive almost periodic solutions.

Keywords Evolution equation • Analytic semigroups • Strong and classical solutions • Ordered Banach spaces • Magneto-hydrodynamics

Mathematics Subject Classification (2010): 34K30 · 35R10 · 47H07

Introduction

Let $\Omega := \mathbb{R}^3 \times X$ be a vector bundle over a smooth domain X of $\mathbb{R}^3$.

E.S. Athanasiadou • V.F. Dionysatos • P.K. Pavlakos
Mathematics Department, National and Kapodistrian University of Athens,
Panepistimiopolis, 15784, Athens, Greece
e-mail: eathan@math.uoa.gr; vassilisdionysatos@gmail.com; ppavlakos@math.uoa.gr

P.N. Koumantos (✉)
Mathematics Department, National and Kapodistrian University of Athens,
Panepistimiopolis, 15784, Athens, Greece

Physics Department, National and Kapodistrian University of Athens,
Panepistimiopolis 15783, Athens, Greece
e-mail: pkoumant@phys.uoa.gr

N.J. Daras (ed.), *Applications of Mathematics and Informatics in Science and Engineering*, Springer Optimization and Its Applications 91, DOI 10.1007/978-3-319-04720-1_3,

In the magneto-hydrodynamic approximation, the flow of a viscous incompressible homogeneous ideal plasma with kinematic viscosity $\nu > 0$ can be described in Ω in terms of two 3-vector fields, the plasma velocity $u = u(t, x)$, and the magnetic field $B = B(t, x)$.

We also assume that the plasma has a scalar pressure $p = p(t, x)$, a mass density ρ, a (finite) constant electric conductivity σ, a magnetic constant permeability θ, and a current density $j = j(t, x)$. We will examine two cases $j = \theta^{-1} curl(B)$ and $j = \sigma u \times B$, imposing the boundary condition $u \mid \partial X = 0$.

Consider that the first density is valid in a general quasi-stationary approximation model of the Maxwell equations while the second one requires that the electric field e must be very weak, and this model describes the evolution of the electromagnetic field in magneto-hydrodynamics. Following a common practice in magneto-hydrodynamics, we eliminate e.

Assuming a modification of Ohm's law based on some physical reasoning, u and B fulfill equations of the following form:

$$\frac{\partial u}{\partial t} - \nu \Delta u = -(u \cdot \nabla)u - \frac{1}{\rho}\, grad\ p + \frac{1}{\rho}\, j \times B - f_1 \tag{1}$$

$$\frac{\partial B}{\partial t} - \frac{1}{\theta\sigma} \Delta B = curl\ (u \times B) - (\theta\sigma)^{-1} f_2 \tag{2}$$

$$div(u) = 0,\ div(B) = 0 \tag{3}$$

where $f_i = f_i(t)$, $i = 1, 2$ are suitable continuous functions and Δ is the Laplace operator.

Now we introduce the transformation magnetic field $z = B - b$ with $z \cdot n = 0$ (n denotes the outward unit normal to ∂X), where b is an undisturbed magnetic field b in assuming the case that b is generated by currents outside of $\bar{X}$. Adopting all the currents are slowly variable, so that $\frac{\partial b}{\partial t} = 0$, b is given by the Biot-Savart law, and fulfills the conditions $div(b) = 0$ and $curl(b) = 0$ in X.

Moreover remembering that z is assumed to be tangential we require that for every $t \in \mathbb{R}$ there exists a function $\phi \in W^{1,2}(\partial X, \mathcal{B}(\partial X), \nu)$ such that

$$z = d\phi \tag{4}$$

on ∂X.

Then the equations we actually consider are obtained from the previous ones by replacing B with $z + b$ and since $\frac{\partial b}{\partial t} = 0$ yields

$$\frac{\partial u}{\partial t} - \nu \Delta u = -(u \cdot \nabla)u - \frac{1}{\rho}\, grad\ p + \frac{1}{\rho}\, j \times (z + b) - f_1 \tag{5}$$

$$\frac{\partial z}{\partial t} - \frac{1}{\theta\sigma} \Delta z = curl(u \times (z + b)) - (\theta\sigma)^{-1} f_2 \tag{6}$$

$$div(u) = 0,\ div(z) = 0 \tag{7}$$

Next let H be the Hilbert space

$$H := \left\{ u \in L^2(X, \mathcal{B}(X), \lambda^3) : \int_X u \cdot \nabla\psi(x)\, d\lambda^3(x) = 0 \right\} \tag{8}$$

for every ψ in $W^{1,2}(X, \mathcal{B}(X), \lambda^3)$ and let $Zz = -(\theta\sigma)^{-1}\Delta z$ be the non-negative self-adjoint magneto-hydrodynamic operator from $D(Z) := \{z \in H \cap W^{2,2}(X, \mathcal{B}(X), \lambda^3) : \Delta z \in H, curl\, z \cdot n = 0 \ on\ \partial X, z \mid \partial X = d\phi \mid \partial X\ with\ a\ \phi\ into\ W^{1,2}(\partial X, \mathcal{B}(\partial X), \sigma)\}$ into H.

We also consider the non-negative self-adjoint "magneto-hydrodynamic-Stokes" operator $Y := -\nu P_2 \Delta$ from $D(Y) = W^{2,2}(X, \mathcal{B}(X), \lambda^3) \cap W_0^{1,2}(X, \mathcal{B}(X), \lambda^3)$, into the Hilbert space $L^2(X, \mathcal{B}(X), \lambda^3)$, where P_2 denotes the projection operator from $L^2(X, \mathcal{B}(X), \lambda^3)$ onto H and let e^{-tY}, e^{-tZ}, $t \in \mathbb{R}^+$ be the semigroups generated by Y and Z respectively.

Then for $\varepsilon \geq 0$ we obtain the following system of "approximating" operator-differential equations:

$$\frac{\partial u}{\partial t} + Yu = P_2[-((e^{-\varepsilon Y} u) \cdot \nabla)u + j \times (e^{-\varepsilon Z} z + b) - f_1] \tag{9}$$

$$\frac{\partial z}{\partial t} + Zz = curl(u \times e^{-\varepsilon Z} z) + curl(u \times b) - (\theta\sigma)^{-1} f_2 \tag{10}$$

$$div(u) = 0,\ div(z) = 0 \tag{11}$$

In other words writing our approximating systems in the form

$$\left(\frac{d}{dt} + A\right) x(t) = F(t, x(t)),\ t \in \mathbb{R} \tag{12}$$

we are leading to solve an evolution (***Ohm-Navier-Stokes***) equation of the type (12) in the product ordered Hilbert space $E := H \times H$ with elements $x = (u, z)$, endowed with scalar product $(x_1, x_2) = (u_1, u_2) + a(z_1, z_2)$ and the usual ordering, where $A := (Y, Z)$ is a non-negative self-adjoint operator on $D(Y) \times D(Z)$, a is a suitable positive constant, and F absorbs all nonlinearities.

For more details concerning physical notions, terminology and techniques in thermo-magneto-hydrodynamics, Navier-Stokes and Ohm-Navier-Stokes equations we refer to [1, 4–6, 8–13, 15, 16, 19, 20, 22, 26, 27] and [29–31].

New results concerning the regularity of the (classical or strong) solutions of (12) are obtained by means of a linearization method and of regularity properties of the solutions of the corresponding linear evolution equations of the type:

$$\left(\frac{d}{dt} + A\right) y(t) = f(t) \tag{13}$$

where f is a given E-valued function on $\mathbb{R}$.

Preliminaries and Notation

Throughout this work we use standard notations. $\mathbb{R}$ ($\mathbb{C}$) denotes the real (complex) numbers.

We shall denote by $D(A)$ the domain and by $R(A)$ the range of the operator A acting on E, and by I the identity operator on E. We shall also denote by $\mathcal{B}(X)$ the σ-algebra of Borel subsets of a topological space X.

Further we denote by $(X, \mathcal{B}(X), \mu)$ a Borel measure μ on $\mathcal{B}(X)$ and by $L^p(X, \mathcal{B}(X), \mu)$ the corresponding L^p-spaces, $1 \leq p \leq \infty$. In particular $(\mathbb{R}^n, \mathcal{B}(\mathbb{R}^n), \lambda^n)$ denotes the Borel-Lebesgue measure on $\mathbb{R}^n$.

In what follows we use the exponential notation e^{-tA}, $t \in \mathbb{R}^+$ ($\mathbb{R}^+ := [0, +\infty)$) to denote the analytic semigroup $T(t)$, $t \in \mathbb{R}^+$ acting on E, whose infinitesimal generator is the operator $-A$ acting on the ordered Hilbert space E and satisfying a well-known estimation:

$$\left\| e^{-tA} \right\| \leq M_0 \, e^{-\delta t} \tag{14}$$

for some $M_0 > 0$, $\delta > 0$, whenever $t \in \mathbb{R}^+$. See also [2, 23–25, 28] and [32].

Next we remind the mean of the classical and the strong solution.

A function $u : \mathbb{R} \to D(A)$ is called a ***classical solution*** on $\mathbb{R}$ of (12) or (13) if it is strongly differentiable for every t in $\mathbb{R}$ and satisfies (12) or (13) for every t in $\mathbb{R}$.

By analogy a function $u : \mathbb{R} \to D(A)$ is called a ***strong solution*** on $\mathbb{R}$ of (12) or (13) if it is strongly differentiable for λ^1-almost every t in $\mathbb{R}$ and satisfies (12) or (13) for λ^1-almost every t in $\mathbb{R}$.

Hammerstein operators are closely related to Eq. (12). So in many cases we study (12) by reducing this to a nonlinear Hammerstein-type integral equation: $x = Tx$, where T is a suitable Hammerstein-type integral operator.

Let E be a general Banach space with norm $\|\cdot\|$

We denote by $C_b(\mathbb{R}, E)$ the Banach space of bounded continuous functions $u : \mathbb{R} \to E$ endowed with supremum norm

$$|u| := sup \, \{\|u(t)\| : t \in \mathbb{R}\} \tag{15}$$

and let $C(\mathbb{R}, E)$ be the Fréchet space of continuous functions $u : \mathbb{R} \to E$.

Let also $M^p(\mathbb{R}, E)$ ($1 \leq p < \infty$) be the Banach space of Bochner-measurable functions $u : \mathbb{R} \to E$ for which:

$$\int_t^{t+1} \|u(s)\|^p \, d\lambda^1(s) < +\infty, \; for\ every\ t \in \mathbb{R} \tag{16}$$

under the norm:

$$|u|_{M^p} := sup \left\{ \left(\int_t^{t+1} \|u(s)\|^p \, d\lambda^1(s) \right)^{\frac{1}{p}} : t \in \mathbb{R} \right\} \tag{17}$$

It is well-known that (as linear manifolds) we have:

$$M^p(\mathbb{R}, E) \subseteq M^1(\mathbb{R}, E),\ 1 \le p < \infty \tag{18}$$

An element u in $C_b(\mathbb{R}, E)$ is said to be ***(Bohl-Bohr) almost periodic*** if, given $\varepsilon > 0$, there is a positive real number $l := l(\varepsilon)$ such that any interval of $\mathbb{R}$ of length l contains at least one point τ (called an ε-almost period of u) for which:

$$|u_\tau - u| = sup\ \{\|u(t+\tau) - u(t)\| : t \in \mathbb{R}\} < \varepsilon \tag{19}$$

Bochner's criterion asserts that $u \in C_b(\mathbb{R}, E)$ is almost periodic if the set:

$$H(u) := \{u_\tau,\ \tau \in \mathbb{R}\} \tag{20}$$

of all translates of u is relatively compact in $C_b(\mathbb{R}, E)$.

By $A_B(\mathbb{R}, E)$ will be denoted the closed subspace of $C_b(\mathbb{R}, E)$ of all almost periodic functions in $C_b(\mathbb{R}, E)$.

A function u in $M^p(\mathbb{R}, E)$ is said to be ***Stepanoff-*** (or S^p-) almost periodic if, given $\varepsilon > 0$, there is a positive real number $l := l(\varepsilon)$ such that any interval of $\mathbb{R}$ of length l contains at least a point τ (called an ε-S^p-almost period of u) for which:

$$|u_\tau - u|_{M^p} := sup \left\{ \left(\int_t^{t+1} \|u(s+\tau) + u(s)\|^p \, d\lambda^1(s) \right)^{\frac{1}{p}} < \varepsilon : t \in \mathbb{R} \right\} \tag{21}$$

By $A_{S^p}(\mathbb{R}, E)$ will be denoted the closed subspace of $M^p(\mathbb{R}, E)$ of all S^p-almost periodic functions from $\mathbb{R}$ into E, $1 \le p < \infty$.

Clearly $A_{S^p}(\mathbb{R}, E) \subseteq A_{S^1}(\mathbb{R}, E)$, whenever $1 \le p < \infty$.

More details about almost periodicity (Bohl-Bohr, Stepanoff) can be found in [3, 7, 17] and [18].

Almost Periodic Solutions of Eq. (13)

Theorem 1. *Let* $f \in A_{S^p}(\mathbb{R}, E)$*, for some* p *in* $[1, \infty)$*. Then equation* (13) *has at least one strong solution* u *in* $A_B(\mathbb{R}, E)$*.*

Proof. Clearly it is sufficient to consider the case $p = 1$ since $A_{S^p}(\mathbb{R}, E) \subseteq A_{S^1}(\mathbb{R}, E)$, $1 \le p < \infty$.

Let the function $u : \mathbb{R} \to E$ defined by the formula:

$$u(t) := \int_{-\infty}^{t} e^{-(t-s)A} f(s) d\lambda^1(s) = \int_0^{+\infty} e^{-sA} f(t-s) d\lambda^1(s),\ t \in \mathbb{R} \tag{22}$$

The (Bochner) integral in (22) exists since for every $t \in \mathbb{R}$,

$$\|u(t)\| = \left\| \int_0^{\infty} e^{-sA} f(t-s) d\lambda^1(s) \right\| \leq \int_0^{\infty} \left\| e^{-sA} f(t-s) \right\| d\lambda^1(s)$$

$$\leq M_0 \int_0^{+\infty} e^{-\delta s} \|f(t-s)\| \, d\lambda^1(s) \leq M_0 \sum_{n=0}^{\infty} \int_n^{n+1} e^{-\delta s} \|f(t-s)\| \, d\lambda^1(s)$$

$$\leq M_0 \sum_{n=0}^{\infty} \int_n^{n+1} e^{-\delta n} \|f(t-s)\| \, d\lambda^1(s)$$

$$= M_0 \sum_{n=0}^{\infty} \left(e^{-\delta n} \int_n^{n+1} \|f(t-s)\| \, d\lambda^1(s) \right)$$

$$= M_0 \sum_{n=0}^{\infty} \left(e^{-\delta n} \int_{t-n-1}^{t-n} \|f(p)\| \, d\lambda^1(p) \right) \leq M_0 \left(\sum_{n=0}^{\infty} e^{-\delta n} \right) |f|_{M^1}$$

$$\leq M_0 (1 - e^{-\delta})^{-1} |f|_{M^1} \tag{23}$$

Therefore $u \in C_b(\mathbb{R}, E)$.
From (22) it follows by the chain rule of strong differentiation that there exists:

$$\dot{u}(t) := \frac{d}{dt} \int_{-\infty}^{t} e^{-(t-s)A} f(s) d\lambda^1(s) = \frac{d^+}{dt} \int_{-\infty}^{t} e^{-(t-s)A} f(s) d\lambda^1(s) \tag{24}$$

(strong derivative and right strong derivative), for λ^1-almost everywhere $t \in \mathbb{R}$.
Further using the continuity of $t \mapsto e^{-tA}x$, for all x in E we get:

$$\lim_{h \to 0+} h^{-1} \int_t^{t+h} e^{-(t+h-s)A} \left(f(s) - f(t) \right) d\lambda^1(s) = 0 \tag{25}$$

$$\lim_{h \to 0+} h^{-1} \int_t^{t+h} \left(e^{-(t+h-s)A} - I \right) f(t) d\lambda^1(s) = 0 \tag{26}$$

λ^1-almost everywhere $t \in \mathbb{R}$.

Then making use of (22), adding and subtracting the terms $\int_t^{t+h} e^{-(t+h-s)A} f(s) d\lambda^1(s)$, $\int_t^{t+h} f(t) d\lambda^1(s) = hf(t)$, we have (for $h > 0$):

$$h^{-1}(u(t+h) - u(t))$$

$$= h^{-1} \int_{-\infty}^{t+h} e^{-(t+h-s)A} f(s) d\lambda^1(s) - h^{-1} \int_{-\infty}^{t} e^{-(t-s)A} f(s) d\lambda^1(s)$$

$$= h^{-1} \Big(\int_{-\infty}^{t+h} e^{-(t+h-s)A} f(s) d\lambda^1(s) - \int_{-\infty}^{t} e^{-(t-s)A} f(s) d\lambda^1(s)$$

$$+\int_t^{t+h} e^{-(t+h-s)A} f(t)d\lambda^1(s) - \int_t^{t+h} e^{-(t+h-s)A} f(t)d\lambda^1(s) + hf(t)$$
$$-\int_t^{t+h} f(t)d\lambda^1(s)\Big) \tag{27}$$

Since

$$\int_{-\infty}^{t+h} e^{-(t+h-s)A} f(s)d\lambda^1(s)$$
$$= \int_{-\infty}^{t} e^{-(t+h-s)A} f(s)d\lambda^1(s) + \int_t^{t+h} e^{-(t+h-s)A} f(s)d\lambda^1(s),$$

it follows by (27):

$$h^{-1}(u(t+h) - u(t))$$
$$= h^{-1}\Bigg(\int_{-\infty}^{t} e^{-(t+h-s)A} f(s)d\lambda^1(s) + \int_t^{t+h} e^{-(t+h-s)A} f(s)d\lambda^1(s)$$
$$-\int_{-\infty}^{t} e^{-(t-s)A} f(s)d\lambda^1(s) + \int_t^{t+h} e^{-(t+h-s)A} f(t)\, d\lambda^1(s)$$
$$-\int_t^{t+h} e^{-(t+h-s)A} f(t)d\lambda^1(s) + hf(t) - \int_t^{t+h} f(t)\, d\lambda^1(s)\Bigg)$$
$$= h^{-1}\Bigg(\int_{-\infty}^{t} e^{-(t+h-s)A} f(s)d\lambda^1(s) - \int_{-\infty}^{t} e^{-(t-s)A} f(t)d\lambda^1(s)\Bigg)$$
$$+ h^{-1}\Bigg(\int_t^{t+h} e^{-(t+h-s)A} f(t)d\lambda^1(s) - \int_t^{t+h} e^{-(t+h-s)A} f(t)d\lambda^1(s)\Bigg)$$
$$+ h^{-1}\Bigg(\int_t^{t+h} e^{-(t+h-s)A} f(t)d\lambda^1(s) - \int_t^{t+h} f(t)d\lambda^1(s)\Bigg) + f(t)$$
$$= h^{-1}\int_{-\infty}^{t} \left(e^{-(t+h-s)A} - e^{-(t-s)A}\right) f(s)d\lambda^1(s)$$
$$+ h^{-1}\int_t^{t+h} e^{-(t+h-s)A} (f(s) - f(t))\, d\lambda^1(s)$$
$$+ h^{-1}\int_t^{t+h} \left(e^{-(t+h-s)A} - I\right) f(t)\, d\lambda^1(s) + f(t) \tag{28}$$

On the other hand applying classical arguments we deduce:

$$\begin{aligned}
&\int_{-\infty}^{t} \left(e^{-(t+h-s)A} - e^{-(t-s)A}\right) f(s)\, d\lambda^1(s) \\
&= \int_{-\infty}^{t} \left(e^{-hA} - I\right) e^{-(t-s)A} f(s)\, d\lambda^1(s) \\
&= \left(e^{-hA} - I\right) \int_{-\infty}^{t} e^{-(t-s)A} f(s) d\lambda^1(s) \qquad (29)
\end{aligned}$$

Combining (28) and (29) we find:

$$\begin{aligned}
&h^{-1}\left(u(t+h) - u(t)\right) \\
&= h^{-1}\left(e^{-hA} - I\right) \int_{-\infty}^{t} e^{-(t-s)A} f(s) d\lambda^1(s) \\
&+ h^{-1} \int_{t}^{t+h} e^{-(t+h-s)A} \left(f(s) - f(t)\right) d\lambda^1(s) \\
&+ h^{-1} \int_{t}^{t+h} \left(e^{-(t+h-s)A} - I\right) f(t)\, d\lambda^1(s) + f(t) \qquad (30)
\end{aligned}$$

Now letting $h \to 0+$ in (30) and because of (25), (26) we obtain:

$$\dot{u}(t) = \underset{h\to 0+}{\ell im}\, h^{-1}\left(u(t+h) - u(t)\right) = -Au(t) + f(t), \qquad (31)$$

for almost every $t \in \mathbb{R}$.

Therefore by (31) we conclude that $u : \mathbb{R} \to E$ is a bounded strong solution of (13) in $C_b(\mathbb{R}, E)$.

Now it follows by (22) that:

$$|u_\tau - u| \le M_0 \left(1 - e^{-\delta}\right)^{-1} |f_\tau - f|_{M^1} \qquad (32)$$

which proves that $u \in A_B(\mathbb{R}, E)$. □

Theorem 2. *Let $f \in C(\mathbb{R}, E) \cap A_{S^p}(\mathbb{R}, E)$, for some p in $[1, \infty)$. Then equation* (13) *has exactly one classical solution u in $A_B(\mathbb{R}, E)$.*

Proof. We modify the proof of the preceding Theorem 1. Again it is sufficient to consider the case $p = 1$.

Let the function u in $C_b(\mathbb{R}, E)$ defined by (22) and (23).
Then we have again the inequality (24) and (25), (26) hold for every $t \in \mathbb{R}$.

Hence letting again $h \to 0+$ in (30) we deduce:

$$u'(t) = -Au(t) + f(t), \ \textit{for every } t \in \mathbb{R} \qquad (33)$$

On the other hand by (22) implies that (32) also holds.

Then (32) and (33) yields the fact that $u : \mathbb{R} \to E$ is a bounded classical solution of (13) in $A_B(\mathbb{R}, E)$.

For the uniqueness let u_1 be another classical solution of (13) in $A_B(\mathbb{R}, E)$.

Then $\phi = u - u_1$ is a classical solution in $C_b(\mathbb{R}, E)$ of the homogeneous linear differential equation:

$$\phi'(t) = -A\phi(t) \tag{34}$$

and since $\phi \in C_b(\mathbb{R}, E)$ it follows $\phi(t) = 0$, for every t in $\mathbb{R}$ (cf. [33]), that is $u = u_1$. □

Almost Periodic Solutions of (12)

Let Φ be the corresponding ***Nemytskii operator*** of the nonlinear operator $F : \mathbb{R} \times E \to E$ appearing in (12), i.e., for every $y : \mathbb{R} \to E$, Φy is defined by the formula:

$$\Phi y(t) := F(t, y(t)),\ t \in \mathbb{R}$$

Now we state the following conditions concerning the Nemytskii operator Φ.

Let $(B, \|\cdot\|_B)$, $(B_1, \|\cdot\|_{B_1})$ be Banach spaces of E-valued functions on $\mathbb{R}$.

Condition (Φ_1): $\Phi y \in B$ provided $y \in B_1$ and there exists a constant $\gamma(B, B_1) := \gamma > 0$ such that:

$$\|\Phi y_1 - \Phi y_2\|_B \leq \gamma\ \|y_1 - y_2\|_{B_1},\ \textit{for all}\ y_1, y_2 \in B_1 \tag{35}$$

Condition (Φ_2): $\Phi y \in B$ provided $y \in B_1$ and there exists a real-valued function $\gamma(B, B_1) := \gamma : \mathbb{R} \to \mathbb{R}^+$ such that:

$$\|\Phi y_1(t) - \Phi y_2(t)\| \leq \gamma(t)\ \|y_1(t) - y_2(t)\|,\ \textit{for all}\ y_1, y_2 \in B_1\ \textit{and}\ t \in \mathbb{R} \tag{36}$$

Theorem 3. *Let (Φ_1) hold when $B := B_1 := A_{S^p}(\mathbb{R}, E)$ and $\frac{\gamma M_0}{(p\delta)^{\frac{1}{p}}} < 1$, for some p in $[1, \infty)$. Then there exists at least one strong solution u of (12) in $A_{S^p}(\mathbb{R}, E)$.*

Proof. Let $p \in [1, \infty)$.

Consider a "pick" Hammerstein-type operator

$$K_p : A_{S^p}(\mathbb{R}, E) \to A_{S^p}(\mathbb{R}, E), \tag{37}$$

which to any y in $A_{S^p}(\mathbb{R}, E)$ associates (according to Theorem 1) a strong solution

$$K_p y(t) := \int_0^{+\infty} e^{-sA} \Phi y(t - s) d\lambda^1(s),\ t \in \mathbb{R} \tag{38}$$

in $A_B(\mathbb{R}, E) \subseteq A_{S^p}(\mathbb{R}, E)$ of the linear evolution equation:

$$\frac{dx(t)}{dt} + Ax(t) = \Phi y(t) \tag{39}$$

Next let $y_1, y_2 \in A_{S^p}(\mathbb{R}, E)$ and $\sigma \in \mathbb{R}$.
Thus applying (14), (Φ_1) and Fubini-Tonelli Theorem we have:

$$\begin{aligned}
&\int_\sigma^{\sigma+1} \|K_p y_2(t) - K_p y_1(t)\|^p \, d\lambda^1(t) \\
&= \int_\sigma^{\sigma+1} \left\| \int_0^{+\infty} e^{-sA} \Phi y_2(t-s) d\lambda^1(s) - \int_0^{+\infty} e^{-sA} \Phi y_1(t-s) d\lambda^1(s) \right\|^p d\lambda^1(t) \\
&\leq \int_\sigma^{\sigma+1} \int_0^{+\infty} \left\| e^{-sA} (\Phi y_2(t-s) - \Phi y_1(t-s)) d\lambda^1(s) \right\|^p d\lambda^1(s)\, d\lambda^1(t) \\
&\leq \int_\sigma^{\sigma+1} \int_0^{+\infty} M_0^p e^{-\delta p s} \left\| (\Phi y_2(t-s) - \Phi y_1(t-s))\, d\lambda^1(s) \right\|^p d\lambda^1(s)\, d\lambda^1(t) \\
&= M_0^p \int_0^{+\infty} e^{-\delta p s} \int_\sigma^{\sigma+1} \|(\Phi y_2(t-s) - \Phi y_1(t-s))\|^p \, d\lambda^1(t)\, d\lambda^1(s) \\
&\leq \gamma^p M_0^p \int_0^{+\infty} e^{-\delta p s} \int_\sigma^{\sigma+1} \|y_2(t-s) - y_1(t-s)\|^p \, d\lambda^1(t)\, d\lambda^1(s) \\
&= \gamma^p M_0^p \int_0^{+\infty} e^{-\delta p s} \int_{\sigma-s}^{\sigma-s+1} \|y_2(\xi) - y_1(\xi)\|^p \, d\lambda^1(\xi)\, d\lambda^1(s) \\
&\leq \gamma^p M_0^p \int_0^{+\infty} e^{-\delta p s} \, d\lambda^1(s)\, |y_2 - y_1|^p_{M^p(\mathbb{R}, E)} \\
&= \gamma^p M_0^p \frac{1}{p\delta} |y_2 - y_1|^p_{M^p(\mathbb{R}, E)}
\end{aligned} \tag{40}$$

This shows that K_p is a contraction operator on $A_{S^p}(\mathbb{R}, E)$, and its fixed point u is a strong solution of (12) in $A_{S^p}(\mathbb{R}, E)$. □

Theorem 4. *Let* (Φ_2) *holds when* $B_1 := A_B(\mathbb{R}, E)$, $B := A_{S^p}(\mathbb{R}, E)$, *for some* $p \in [1, \infty)$ *with* $\gamma \in C_b(\mathbb{R}, \mathbb{R}^+)$ *and* $M_0 |\gamma| \delta^{-1} < 1$. *Then there exists at least one strong solution* u *of* (12) *in* $A_B(\mathbb{R}, E)$.

Proof. Consider a "pick" Hammerstein-type operator

$$T : A_B(\mathbb{R}, E) \to A_B(\mathbb{R}, E) \tag{41}$$

which to any y in $A_B(\mathbb{R}, E)$ associates (according to Theorem 1) a strong solution

$$Ty(t) := \int_0^{+\infty} e^{-sA} \Phi y(t-s)\, d\lambda^1(s), t \in \mathbb{R} \tag{42}$$

in $A_B(\mathbb{R}, E)$ of the linear evolution equation:

$$\frac{dx(t)}{dt} + Ax(t) = \Phi y(t) \tag{43}$$

Now let $y_1, y_2 \in A_B(\mathbb{R}, E)$.
Hence applying (14) and assumption on (Φ_1) we see that:

$$\begin{aligned}
&\|Ty_2(t) - Ty_1(t)\| \\
&= \left\| \int_0^{+\infty} e^{-sA}\Phi y_2(t-s)d\lambda^1(s) - \int_0^{+\infty} e^{-sA}\Phi y_1(t-s)d\lambda^1(s) \right\| \\
&\leq \int_0^{+\infty} \left\| e^{-sA}\left(\Phi y_2(t-s) - \Phi y_1(t-s)\right)\right\| d\lambda^1(s) \\
&\leq M_0 \int_0^{+\infty} e^{-\delta s} \left\|(\Phi y_2(t-s) - \Phi y_1(t-s))\right\| d\lambda^1(s) \\
&\leq M_0 \int_0^{+\infty} e^{-\delta s} \left|\gamma(t-s)\right| \left\|(y_2(t-s) - y_1(t-s))\right\| d\lambda^1(s) \\
&\leq M_0 \left|\gamma\right| \int_0^{+\infty} e^{-\delta s} \left\|(y_2(t-s) - y_1(t-s))\right\| d\lambda^1(s) \\
&\leq M_0 \left|\gamma\right| \delta^{-1} \left|y_2 - y_1\right|
\end{aligned} \tag{44}$$

for every $t \in \mathbb{R}$.

Therefore T is a contraction operator on $A_B(\mathbb{R}, E)$, and its fixed point u is a strong solution of (12) in $A_B(\mathbb{R}, E)$. □

Theorem 5. *Let (Φ_2) holds when $B := B_1 = A_B(\mathbb{R}, E)$ and $\gamma \in C_b(\mathbb{R}, \mathbb{R}^+)$ with $M_0 |\gamma| \delta^{-1} < 1$. Then there exists exactly one classical solution u of (12) in $A_B(\mathbb{R}, E)$.*

Proof. Consider the Hammerstein-type operator:

$$K : A_B(\mathbb{R}, E) \rightarrow A_B(\mathbb{R}, E) \tag{45}$$

which to any y in $A_B(\mathbb{R}, E)$ associates (according to Theorem 2) the unique classical solution

$$Ky(t) := \int_0^{+\infty} e^{-sA}\Phi y(t-s)\, d\lambda^1(s), t \in \mathbb{R} \tag{46}$$

in $A_B(\mathbb{R}, E)$ of the linear evolution equation:

$$\frac{dx(t)}{dt} + Ax(t) = \Phi y(t) \tag{47}$$

Now let $y_1, y_2 \in A_B(\mathbb{R}, E)$ and $t \in \mathbb{R}$.
Hence applying (14) and assumption on (Φ_1) we see that:

$$\begin{aligned}
&\|Ky_2(t) - Ky_1(t)\| \\
&= \left\| \int_0^{+\infty} e^{-sA} \Phi y_2(t-s) d\lambda^1(s) - \int_0^{+\infty} e^{-sA} \Phi y_1(t-s) d\lambda^1(s) \right\| \\
&\leq \int_0^{+\infty} \left\| e^{-sA} \left(\Phi y_2(t-s) - \Phi y_1(t-s) \right) \right\| \, d\lambda^1(s) \\
&\leq M_0 \int_0^{+\infty} e^{-\delta s} \, \|(\Phi y_2(t-s) - \Phi y_1(t-s))\| \, d\lambda^1(s) \\
&\leq M_0 \, |\gamma| \int_0^{+\infty} e^{-\delta s} \, \|(y_2(t-s) - y_1(t-s))\| \, d\lambda^1(s) \\
&\leq M_0 \, |\gamma| \, \delta^{-1} \, |y_2 - y_1| \qquad (48)
\end{aligned}$$

for every $t \in \mathbb{R}$.

Hence K is a contraction operator on $A_B(\mathbb{R}, E)$, and its fixed point u is the unique classical solution of (12) in $A_B(\mathbb{R}, E)$. □

Positive Almost Periodic Solutions of Eqs. (12) and (13) in the Ordered Hilbert Space E

In this section we characterize existence and uniqueness of E^+-valued almost periodic solutions of the Eqs. (12) and (13).

Let E be an ordered Banach space endowed with the usual strong topology τ, the ordering $\leq$ and with the corresponding (strong) closed positive cone E^+ (cf. [14]).

Let also (X, S, μ) be a σ-finite measure on X and let $\mathcal{F}(X, E)$ the ordered vector space of functions from X into E with cone $\mathcal{F}(X, E^+)$.

Firstly we define the ordered vector space $I_{o,\mu}(X, E)$ of ***elementary o-integrable*** functions (with respect to (X, S, μ)) in the subspace $E_{o,\mu}(X, E)$ of ***elementary o-measurable*** functions of $\mathcal{F}(X, E)$.

Then we extend the space $I_{o,\mu}(X, E)$ to the ordered vector space $L^1_{o,\mu}(X, E)$ (with positive cone $P_{o,\mu}(X, E^+)$) of ***o-integrable*** functions in the subspace $M_{o,\mu}(X, E)$ of ***o-measurable*** functions of $\mathcal{F}(X, E)$.

Moreover we shall denote by $\int_X u(t)\, d\mu(t)$ the E-valued well defined ***o-integral*** of an element u in $L^1_{o,\mu}(X, E)$.

In particular if the measure (X, S, μ) denotes the Borel-Lebesgue measure $(J, \mathcal{B}(J), \lambda^1)$ on a finite or infinite subinterval of $\mathbb{R}$ with endpoints a, b we shall use the notations $\int_a^b u(t)\, d\lambda^1(t) := \int_X u(t)\, d\mu(t)$ and $P(J, E^+) := P(X, E^+)$.

Finally we emphasize that $0 \le \int_X u_1(t)\, d\mu(t) \le \int_X u_2(t)\, d\mu(t)$, for every $u_1, u_2 \in L^1_{o,\mu}(X, E)$ with $0 \le u_1(t) \le u_2(t)$, for μ-almost everywhere t in X. For more details see [21].

Concerning Eq. (13) we have the next results.

Theorem 6. *Let f in $P(\mathbb{R}, E^+) \cap A_{S^p}(\mathbb{R}, E)$, for some p in $[1, \infty)$. Then equation* (13) *has at least one positive strong solution u in $A_B(\mathbb{R}, E^+)$.*

Proof. As a consequence of the preceding assumptions formula (22) defines a positive function u in $C_b(\mathbb{R}, E^+)$. From here the claim follows applying similar arguments as in Theorem 1. □

Theorem 7. *Let f in $P(\mathbb{R}, E^+) \cap C(\mathbb{R}, E^+) \cap A_{S^p}(\mathbb{R}, E)$, for some p in $[1, \infty)$. Then Equation* (13) *has exactly one positive classical solution u in $A_B(\mathbb{R}, E^+)$.*

Proof. As in the preceding Theorem we conclude that $e^{-tA} f \in P(\mathbb{R}, E^+)$, whenever $t \ge 0$. Therefore again formula (22) defines a positive function u in $C_b(\mathbb{R}, E^+)$, and the claim is a consequence of Theorem 2. □

Now we shall turn our attention to the Eq. (12) assuming that the convolution $e^{-A} * \Phi y$ belongs in $P(\mathbb{R}, E^+)$, for every $y \in P(\mathbb{R}, E^+) \cap C_b(\mathbb{R}, E^+)$, where:

$$(e^{-A} * \Phi y)(t) := \int_0^{+\infty} e^{-sA} \Phi y(t-s)\, d\lambda^1(s),\ t \in \mathbb{R}. \tag{49}$$

We also consider the following condition.

Condition (Φ_3): $\Phi y \in A_B(\mathbb{R}, E^+) \cap P(\mathbb{R}, E^+)$ whenever $y \in A_B(\mathbb{R}, E^+)$ and there exists $\gamma \in C_b(\mathbb{R}, \mathbb{R}^+)$ such that:

$$\|\Phi y_1(t) - \Phi y_2(t)\| \le \gamma(t) \|y_1(t) - y_2(t)\|,\ \textit{for all } t \in \mathbb{R} \textit{ and } y_1, y_2 \in A_B(\mathbb{R}, E^+) \tag{50}$$

Theorem 8. *Let* (Φ_3) *holds and $M_0\ \delta^{-1}\ |\gamma| < 1$. Then* (12) *has exactly one classical positive solution u in $A_B(\mathbb{R}, E^+)$.*

Proof. The assertion follows since the Hammerstein-type positive operator

$$K : A_B(\mathbb{R}, E^+) \to A_B(\mathbb{R}, E^+) \tag{51}$$

which to any y in $A_B(\mathbb{R}, E^+)$, associates (according to Theorem 7) exactly one positive classical solution Ky in $A_B(\mathbb{R}, E^+)$ of the linear evolution equation:

$$\frac{dx(t)}{dt} + Ax(t) = \Phi y(t) \tag{52}$$

is a contraction operator on the closed cone $A_B(\mathbb{R}, E^+)$ of the ordered Banach space $A_B(\mathbb{R}, E)$, and its fixed point is the unique positive classical solution of (12) in $A_B(\mathbb{R}, E^+)$. □

References

1. AMANN, H. & QUITTNER, P., *Semilinear parabolic equations involving measures and low regularity data*, Trans. Amer. Math. Soc., 356 (3) (2003), 1045–1119.
2. ATHANASIADOU E. S., DIONYSATOS V. F., KOUMANTOS P. N. & PAVLAKOS P. K., A semigroup approach to functional evolution equations, *Mathematical Methods in the Applied Sciences* (*MMA*), 37 (2) (2014), 217–222.
3. BOGDANOWICZ, W. M. & WELCH, J. N., On a linear operator connected with almost periodic solutions of linear differential equations in Banach spaces, *Math. Ann.*, 172 (1967), 327–355.
4. CARATHÉODORY, C., Untersuchungen über die Grundlagen der Thermodynamik, *Math. Ann.*, 67 (1909), 355–386.
5. CHANDRASEKHAR, S., *Hydrodynamic and Hydromagnetic Stability*. Dover Publications, Inc., New York, (1981).
6. COWLING, T. G., *Magnetohydrodynamic*. New York: Interscience Publishers, 1957.
7. DALETSKIĬ Y. L. & KREĬN, M. G., *Stability of Solutions of Differential Equations in Banach Spaces*, Translations of Mathematical Monographs, 43, Amer. Math. Soc. Providence, Rhode Island, 1974.
8. DUVAUT, G., N. & LIONS, J. L., Inéquations en Thermoélastcité et Magnétohydrodynamique, *Arch. Rational Mech. Anal.*, 46 (1972), 241–279.
9. EBIN, D. G. & MARSDEN, J., Groups of diffeomorphisms and the motion of an incompressible fluid, *Ann. of Math.*, 92 (1970), 102–163.
10. FARWIG, R., KOZONO, H., & SOHR, H., An L^q-approach to Stokes and Navier-Stokes equations in general domains, *Acta Math.*, 195 (2005), 21–53.
11. FÖRSTE, J., Über die Grundgleichungen der Plasmadynamik auf der Basis der Zweiflüssigkeitstheorie, *Z. Angew. Math. Mech.*, 59 (1979), 553–558.
12. GIGA, Y. & YOSHIDA, Z., On the Ohm-Navier-Stokes system in MHD, *J. Mathematical Phys.*, 24 (12) (1983), 2860–2864.
13. – On the equations of the two-component theory in magnetohydrodynamics, *Comm. Partial Differential Equations*, 9 (1984), 503–522.
14. KANTOROVICH, L. V., VULIKH, B. Z. & PINSKER, A. G., *Functional analysis in partially ordered spaces*. Gostekhizdat, 1950 (Russian).
15. LADYZHENSKAYA, O.A., & V. A. SOLONNIKOV, The solution of certain magnetohydrodynamics problems for an incompressible viscous liquid, *Tr. Math. Inst. Akad. Nauk. SSSR*, 59 (1969), 115–187.
16. LASSNER, G., Über ein Rand-Anfangswertproblem der Magnetohydradynamik, *Arch. Rational Mech. Anal.*., 25 (1967), 388–405.
17. LEVITAN, B. M. & ZHIKOV, *Almost Periodic Functions and Differential Equations*. Cambridge Univ. Press, Cambridge, 1982.
18. MASSERA, J. L., & SCHÄFFER, J. J., Linear differential equations and functional analysis, I. *Ann. of Math.*, 67 (3) (1958), 517–573.
19. MOFFATT, *Magnetic field generation in electrically conducting fluids*. Cambridge University Press, New York, (1978).
20. NOLL, W., Lectures on the foundations of Continuum Mechanics and Thermodynamics, *Arch. Rational Mech. Anal.*, 52 (1973), 62–92.
21. PAVLAKOS, P. K., On integration in partially ordered groups, *Canad. J. Math.*, 35 (2) (1983), 353–372.
22. PIKELNER, S.B., *Grundlagen der kosmischen Elektrodynamik* (Russ). Moskau (1966).
23. RANKIN, S. M., Semilinear evolution equations in Banach spaces with applications to parabolic partial differential equations, *Trans. Amer. Math. Soc.*, 336 (2) (1993), 523–535.
24. REED, M. & SIMON, B., *Methods of Modern Mathematical Physics*. Vol II, Fourier analysis, selfadjointness, Academic Press, New York, 1975.

25. REED, M. & SIMON, B., *Methods of Modern Mathematical Physics*. Vol IV, Analysis of operators, Academic Press, New York, 1978.
26. SCHLÜTER, A., Dynamik des Plasma, I und II., *Z. Naturforsch.* 5a, 72–78, (1950), 6a,, 73–79 (1951).
27. SCHONBECK, M. E., SCHONBECK, T. P., Süli, E., Large-time behavior of solutions to the magneto-hydrodynamic equations, *Math. Ann.*, 304 (1996), 717–756.
28. SEGAL, I., Non-linear semigroups, *Ann. of Math.*, 78 (1963), 339–364.
29. STRÖHMER, G., About an initial-boundary value problem from magneto-hydrodynamics, *Math. Z.*, 209 (1992), 345–362.
30. – An existence result for partially regular weak solutions of certain abstract evolution equations, with an application to magneto-hydrodynamics, *Math. Z.*, 213 (1993), 373–385.
31. TAYLOR, M. E., *Partial Differential Equations I, II, III*. Vols 115, 116, 117. Springer-Verlag, Berlin-Heidelberg-New York-Tokyo, 1996.
32. YOSIDA, K., *Functional Analysis*. (6th edition) Springer-Verlag, Berlin-Heidelberg-New York-Tokyo, 1980.
33. ZAIDMAN, S. D., Some asymptotic theorems for abstract differential equations, *Proc. Amer. Math. Soc.*, 25 (1970), 521–525.

Admission Control Policies in a Finite Capacity Geo/Geo/1 Queue Under Partial State Observations

Apostolos Burnetas and Christos Kokaliaris

Abstract We consider the problem of admission control in a discrete time Markovian queue with a finite capacity, a single server, and a geometric arrival and departure processes. We prove the threshold structure of the optimal admission policy under full information on the number of customers in the system. We also consider the admission control problem under partial state information, where the decision maker is only informed whether the system is empty, full, or in some intermediate state. We formulate this problem as a Markov Decision Process with the state representing the posterior distribution of the number of customers and apply a heuristic algorithm from the literature to approximate the optimal policy. In numerical experiments we demonstrate that the pair of the mean and variance of the posterior distribution may be effectively used instead of the full distribution, to implement the optimal policy. We also explore the behavior of the profit function and the value of information with respect to several system parameters.

Introduction

We consider the problem of admission control in a finite capacity discrete time Markovian queue under full and partial state information. Admission control is a general approach for dynamically adjusting the input stream to a service system. It models the trade-off between the payoff received by serving more customers and the cost due to higher system congestion. Admission control can be applied directly by allowing or denying entrance to an incoming customer, or indirectly by adjusting an admission fee.

A. Burnetas (✉) • C. Kokaliaris
Department of Mathematics, University of Athens, Panepistemiopolis, 15784, Athens, Greece
e-mail: aburnetas@math.uoa.gr; chrisk.msor@gmail.com

N.J. Daras (ed.), *Applications of Mathematics and Informatics in Science and Engineering*, Springer Optimization and Its Applications 91, DOI 10.1007/978-3-319-04720-1_4,

Many systems in practice implement some form of admission control under various forms. The dynamic decision of whether to admit or reject an incoming customer is based on the information on system congestion at the time of arrival, which is assumed to be readily available to the decision maker. Nevertheless, there exist situations where this is not the case. In many service systems, particularly those related to telecommunications, call centers etc., it may impossible or expensive for the admission controller to constantly keep track of the exact queue length. Instead, it is often more practical to maintain a coarser picture of the true system state. In such cases the decision can only be based on the information available. A new trade-off is introduced in this manner, between on the one hand the lower cost of tracking the system state and on the other hand the losses incurred due to the less efficient implementation of the admission control policy.

In this paper we explore the comparison between full and partial information in a stylized setting which allows analytical solution of the full information problem. In particular, we model a single server finite capacity Markovian queue in discrete time. We consider direct admission/rejection control policies that may be based on the number of customers in the system in the full information case, or in a partial information setting where the system manager is informed when the system is full or completely empty but does not know the actual number of customers waiting in intermediate states. This information is easier to track if it can be detected when the server becomes idle and when customers are turned away because of overflows.

The contribution of the paper is in admission control under partial state observations. Both areas are extensively covered in the literature and our outline cannot be comprehensive. On the one hand, admission control has been studied in depth from many different viewpoints, albeit mostly in a continuous time setting. [16] presents an early overview of admission policies in various queueing settings and considers both central control as well as customer equilibrium strategies. The admission control problem in an M/M/1 queue is used in [5] to motivate the application of event-based dynamic programming. Admission control is also a useful management tool in finite capacity multiclass service systems, where there is the additional trade-off between receiving the admission fee and saving system resources for future arrivals of customers in higher paying classes. [10] and [12] consider a multiclass system with no waiting queue, while recently [9] model a series system with finite buffers where admission control interacts with the blocking effect. In a discrete time setting, [1] analyze both admission and service control problems in a $Geo/Geo/c$ queue with a infinite capacity where instead of an admission reward there is a cost rate for idle servers and the controller selects dynamically between two positive arrival rates.

The issue of various information levels in a queue has also been studied extensively, mostly from the point of view of delay-related information provided to strategically acting customers. [4] consider a $M/M/1$ queue with balking and identify conditions so that providing customers with more accurate information on expected delay may improve or hurt system performance. [2] also consider the effect of various information levels on customer equilibrium in a queue with server vacations and setup times for server restart. On the other hand, the following two papers are closer to the way we use partial information in this paper. [6] analyze the admission control problem in a discrete time queue where there is a k-period delay in the observation of queue lengths and prove a generalized threshold structure for the optimal policy. Finally, [7] consider a multiserver system with no waiting room where the controller observes only the instances where a customer is rejected because the system is full. They prove the form of the optimal policy for the single server case and propose heuristic rules for multiple server systems.

In this paper we model the admission control problem under partial information as a partially observed Markov decision process. The idea of modeling a partially observed system as an MDP with state variable, the posterior distribution of the unobserved state, goes back to the early days of stochastic dynamic programming. [14] and [13] show the fundamental property that the finite horizon value function is piecewise linear and convex in the generalized state and propose a computation algorithm. Because of the computational complexity of the problem there exists a rich literature on both exact and approximate computational approaches. [8] and [17] survey various algorithms. In our paper we employ the Perseus approximation algorithm proposed by [15].

The rest of the paper is structured as follows. In section "Model Description" we describe the control problem and model both the full information and the partially observed systems as Markov decision processes. In section "Complete Information Policy" we establish the threshold structure of the optimal admission policy under full information. To do this we show by induction that the finite horizon value function is concave in the number of customers. In section "Partially Observable System" we turn to the partial observations model and analyze the application of the Perseus algorithm for approximating the optimal policy. In section "Computational Results" we discuss some computational experiments which demonstrate that the mean and variance of the posterior distribution may be effectively used instead of the full distribution, to implement the optimal policy. We also perform sensitivity analysis of the profit function and the value of information with respect to various parameters. Section "Conclusions" concludes.

Model Description

Consider a discrete time finite-capacity $Geo/Geo/1/M$ queue. In each period n there is a single arrival of a new customer with probability p, and a departure of a customer being served with probability q. More than one arrivals or departures in a single period may not occur. The service discipline is first-come-first-served.

The system manager employs admission control. Specifically any arriving customer may be admitted into the system or rejected. Customers who arrive when the system is full are rejected. When a customer is admitted, an admission reward R is obtained. On the other hand, waiting customers incur a holding cost equal to h per period and per customer present in the system. The system manager's objective is to maximize the infinite horizon expected discounted net profit, with discount factor β.

We analyze the admission control problem under two distinct cased regarding the information available to the decision maker. In the first case (complete information), the system manager is aware of the number of customers present in the system at the beginning of every period and can use this information for the admission/rejection decision. In the second case (partially observable) the manager only knows whether the server is busy or idle and whether the system is full or there are empty spaces. In other words, the information on the state of the system is restricted to either $n = 0, n = M$, or $0 < n < M$, where n is the number of customers present at the beginning of the period.

We first consider the full information model. In this case the admission control problem can be expressed as a Markovian Decision Process as follows. The state X_t denotes the number of customers present in the system at the beginning of a period, thus the state space is finite: $S = \{0, 1, \ldots, M\}$. The action set is $A = \{1, 2\}$, where $a = 1, 2$ denote the admission and rejection actions, respectively. We adopt the following convention regarding the decision-making process. At the beginning of period t the decision maker observes the state X_t and makes a decision on whether to admit or reject a potential customer arrival during the period. The events of arrival and/or departure occur at the end of the period. Therefore, the fact that the decision is made before observing whether there is indeed an arrival or not, is without loss of generality,because there is no state change between decision making and arrival.

Given the timing of events described above, let $V(i)$ denote the value function, i.e., the optimal expected discounted net profit under infinite horizon, given initial state i:

$$V(i) = \sup_f E_f \left[\sum_{t=0}^{\infty} \beta^t \left(\beta R\, \mathbf{1}(A_t) - hX_t\right) | X_0 = i \right], \quad i \in S, \tag{1}$$

where f denotes a dynamic policy of admitting or rejecting customers, A_t the event that a customer is admitted in period t, and $\mathbf{1}(A)$ the indicator function. The value function V satisfies the following optimality equations

$$\begin{aligned} V(i) = &- hi + \max\{p\beta R + \beta p(1-q)V(i+1) \\ &+ \beta((1-p)(1-q) + pq)V(i) + \beta(1-p)qV(i-1)), \\ &\beta qV(i-1) + \beta(1-q)V(i)\}, \ i = 1, \ldots, M-1, \\ V(0) = &\max\{p\beta R + \beta pV(1) + \beta(1-p)V(0), \beta V(0)\} \\ V(M) = &- hM + \beta qV(M-1) + \beta(1-q)V(M). \end{aligned} \tag{2}$$

Note that the admission reward R is discounted since it is assumed that the arrival occurs at the end of the period.

The finite horizon version of the optimality equations is similar

$$\begin{aligned} V_n(i) = &- hi + \max\{p\beta R + \beta p(1-q)V_{n-1}(i+1) \\ &+ \beta((1-p)(1-q) + pq)V_{n-1}(i) + \beta(1-p)qV_{n-1}(i-1)), \\ &\beta qV_{n-1}(i-1) + \beta(1-q)V_{n-1}(i)\}, \ i = 1, \ldots, M-1, \end{aligned}$$

$$V_n(0) = \max\{p\beta R + \beta p V_{n-1}(1) + \beta(1-p)V_{n-1}(0), \beta V_{n-1}(0)\}$$
$$V_n(M) = -hM + \beta q V_{n-1}(M-1) + \beta(1-q)V_{n-1}(M) \tag{3}$$

where $V_0(i) = 0$ and $V_n(i)$ denotes the optimal expected discounted profit for horizon n. Since the state and action spaces are finite, it follows immediately from standard results of Markov Decision Process analysis (e.g., [11]) that $\lim_{n\to\infty} V_n(i) = V(i) < \infty$ uniformly in $i = 0, \ldots, M$.

The partially observable case is considered next. Now the timing of events is as above; however, the decision maker is not aware of the exact system state X_t when making the admission/rejection decision. Instead, we assume that at the beginning of period t, after all transitions of the previous period have been completed, the decision maker obtains a signal θ_t about the system state, where $\theta_t = 0, 1, 2$ when $X_t = 0, 0 < X_t < M, X_t = M$, respectively. The acceptance/rejection decision is made after this signal is received. The problem is again to identify an admission policy that makes decision based only on the available state information and maximizes the infinite horizon expected discounted net profit.

In the partial information framework we adopt the Bayesian Dynamic Programming approach (c.f. [13]) according to which the state of the system is defined as the posterior probability distribution of the unobservable true state, given the current partial information and the previous history of decisions and information evolution.

Let $\pi_t(i), i = 0, \ldots, M$ denote the posterior probability that there are i customers in the system at the beginning of period t. The probability vector $\pi_t = (\pi_t(0), \ldots, \pi_t(M))$ is defined as the state of the partially observable system and referred to as the information vector. The state space is the $(M+1)$-dimensional simplex Π.

The one-step payoff and the update of the information vector are defined as follows. Let p_{ij}^a denote the transition probability of the unobservable state (i.e., the number of customers in the system) from i to j under decision a, and $r_{j\theta}^a$ the probability of obtaining signal θ at the end of the period if the decision at the beginning of the period is a and the state after transition is j. Similarly, let $w_{ij\theta}^a$ denote the single period payoff. Given the model description and the timing of events described so far, these quantities can be easily determined. Specifically, the transition probabilities given $a = 1$ are equal to

$$p_{00}^1 = 1-p,\ p_{01}^1 = p,$$
$$p_{M,M-1}^1 = q,\ p_{MM}^1 = 1-q$$
$$p_{i,i-1}^1 = (1-p)q,\ p_{ii}^1 = (1-p)(1-q) + pq,\ p_{i,i+1}^1 = p(1-q), \tag{4}$$

for $i = 1, \ldots, M-1$, while given $a = 2$ they are equal to

$$p_{00}^2 = 1,$$
$$p_{i,i-1}^2 = q,\ p_{ii}^2 = 1-q, \tag{5}$$

for $i = 1, \ldots, M$. The signal probabilities are easily seen to be equal to

$$r^a_{00} = 1, r^a_{M2} = 1, r^a_{j1} = 1, j = 1, \ldots, M-1, a = 1, 2 \tag{6}$$

and zero otherwise.

The single period payoffs are equal to

$$w^a_{ij\theta} = -hi + \beta R \mathbf{1}(i < M, a = 1, j = i+1). \tag{7}$$

Given the above, first let

$$q^a_i = \sum_{j=0}^{M} \sum_{\theta=1}^{2} p^a_{ij} r^a_{j\theta} w^a_{ij\theta} \tag{8}$$

be the expected one-period payoff given that the (unobservable) state is i and action a is taken. Furthermore, the probability of observing signal θ given an information vector π at the beginning of the period and decision a is equal to

$$r(\theta|\pi, a) = \sum_{i=0}^{M} \sum_{j=0}^{M} \pi(i) p^a_{ij} r^a_{j\theta}. \tag{9}$$

From Bayes' theorem it follows that the updated information vector $\tilde{\pi}(\pi, a, \theta)$, i.e., the posterior distribution of the unobserved state at the beginning of a period, given information vector π, decision a and signal θ at the previous period is given by

$$\tilde{\pi}(j|\pi, a, \theta) = \frac{\sum_{i=0}^{M} \pi(i) p^a_{ij} r^a_{j\theta}}{r(\theta|\pi, a)}, \quad j = 0, \ldots, M. \tag{10}$$

The Bellman optimality equations are now derived as follows. Let $v(\pi)$ denote the optimal infinite horizon expected discounted net profit, given that a prior distribution π for the initial state. Then v satisfies:

$$v(\pi) = \max_{a=1,2} \left\{ \sum_i \pi(i) q^a_i + \beta \sum_\theta r(\theta|\pi, a) v(\tilde{\pi}(j|\pi, a, \theta)) \right\} \tag{11}$$

The finite-horizon version of the optimality equations, which also provides a computational approximation for the optimal value function and the optimal policy is

$$v_n(\pi) = \max_{a=1,2} \left\{ \sum_i \pi(i) q^a_i + \beta \sum_\theta r(\theta|\pi, a) v_{n-1}(\tilde{\pi}(j|\pi, a, \theta)) \right\}, n = 1, 2, \ldots, \tag{12}$$

with $v_0(\cdot) = 0$.

Complete Information Policy

In this section we derive the structure of the optimal policy under full information. We show that the optimal policy has a threshold structure. The derivations are based on proving by induction that the finite horizon complete information value function $V_n(i)$ is nonincreasing and concave in i for all n. The approach is analogous to that in concavity proofs for continuous time problems under uniformization. However in the discrete time setting it is possible to have both arrival and departure events in a single period. This introduces a change in the dynamics which requires a new proof.

Let

$$\Delta_n(i) = V_n(i+1) - V_n(i), n \geq 0, i = 0, 1, M-1 \tag{13}$$

denote the benefit, i.e., the net difference in profit induced to the system when one more customer is present in the queue. From the optimality equations (3) it follows that it is optimal to admit a customer at stage n in state $i > 0$, i.e., $a_n^*(i) = 1$ if and only if

$$\beta pR + \beta p(1-q)V_{n-1}(i+1) + \beta((1-p)(1-q) + pq)V_{n-1}(i) \\ +(1-p)qV_{n-1}(i-1)) \geq \beta qV_{n-1}(i-1) + \beta(1-q)V_{n-1}(i)\},$$

which after algebra reduces to

$$H_{n-1}(i) \equiv \beta pR + \beta p(1-q)\Delta_{n-1}(i) + \beta pq\Delta_{n-1}(i-1) \geq 0. \tag{14}$$

Similarly, it is optimal to admit a customer in state $i = 0$, i.e., $a_n^*(0) = 1$ if and only if

$$\beta pR + \beta pV_{n-1}(1) + \beta(1-p)V_{n-1}(0) \geq \beta V_{n-1}(0),$$

or equivalently,

$$H_{n-1}(0) \equiv \beta pR + \beta p\Delta_{n-1}(0) \geq 0. \tag{15}$$

We next analyze the structure of the optimal admission control policy. The benefit of admitting a new customer is due to the admission fee and is constant regardless of the state. On the other hand, the holding cost is increasing in the number of customers in the system. It is thus expected that it is beneficial to admit an arriving customer only when the system is not too congested. In the next proposition it is shown that the value function $V_n(i)$ is concave in i, i.e., the benefit function $\Delta_n(i)$ is nonincreasing in i. This property implies the threshold structure of the optimal policy.

Theorem 1. *1. The benefit function $\Delta_n(i)$ is nonincreasing in i for all n.*
2. There exist integers $c_n, n = 1, 2, \ldots$, such that the optimal policy at stage n is to accept an arrival in state i if and only if $i \leq c_n$.

Proof. We will prove both parts of the theorem by simultaneous induction on n. For $n = 0$ part (1) holds since $\Delta_0(i) = 0$ for all i. For some $n \geq 1$ assume that

$$\Delta_{n-1}(i) \geq \Delta_{n-1}(i+1), i = 0, \ldots, M-2.$$

We will first show part 2 for n and then part 1 to complete the induction step.

Consider the optimal action in state n. From the induction hypothesis it follows that the function $H_{n-1}(i)$, defined in (14) and (15), is nonincreasing in i. Let $c_n = \max(i = 0, \ldots, M : H_{n-1}(i) \geq 0)$, with the convention $\max \emptyset = -1$. From the monotonicity of H_{n-1} it follows that $a_n^*(i) = 1$ if and only if $i \leq c_n$, thus part 2 of the theorem is true for n.

We next show part 1 for n, i.e., that

$$\Delta_n(i) \geq \Delta_n(i+1), i = 0, \ldots, M-2. \tag{16}$$

We first consider $i = 1, \ldots, M-3$. From the form of the optimal policy at stage n, there are only four possible cases for the optimal actions $a_n^*(i), a_n^*(i+1), a_n^*(i+2)$ as listed below. For each case we compute $\Delta_n(i)$ and $\Delta_n(i+1)$ and show that (16) holds.

Case (i): $a_n^*(i) = a_n^*(i+1) = a_n^*(i+2) = 1$. In this case it follows from (3) that

$$\begin{aligned}
V_n(i) = &-hi + p\beta R + \beta p(1-q)V_{n-1}(i+1) \\
&+ \beta((1-p)(1-q) + pq)V_{n-1}(i) + \beta(1-p)qV_{n-1}(i-1), \\
V_n(i+1) = &-h(i+1) + p\beta R + \beta p(1-q)V_{n-1}(i+2) \\
&+ \beta((1-p)(1-q) + pq)V_{n-1}(i+1) + \beta(1-p)qV_{n-1}(i), \\
V_n(i+2) = &-h(i+2) + p\beta R + \beta p(1-q)V_{n-1}(i+3) \\
&+ \beta((1-p)(1-q) + pq)V_{n-1}(i+2) + \beta(1-p)qV_{n-1}(i+1).
\end{aligned}$$

Therefore,

$$\begin{aligned}
\Delta_n(i) = &-h + \beta p(1-q)\Delta_{n-1}(i+1) \\
&+ \beta((1-p)(1-q) + pq)\Delta_{n-1}(i) + \beta(1-p)q\Delta_{n-1}(i-1), \\
\Delta_n(i+1) = &-h + \beta p(1-q)\Delta_{n-1}(i+2) \\
&+ \beta((1-p)(1-q) + pq)\Delta_{n-1}(i+1) + \beta(1-p)q\Delta_{n-1}(i)
\end{aligned}$$

and (16) holds since the coefficients of all $\Delta_{n-1}(\cdot)$ terms are nonnegative.

Case (ii): $a_n^*(i) = a_n^*(i+1) = 1, a_n^*(i+2) = 2$. In this case we have

$$\begin{aligned} V_n(i) = & -hi + p\beta R + \beta p(1-q)V_{n-1}(i+1) \\ & + \beta((1-p)(1-q)+pq)V_{n-1}(i) + \beta(1-p)qV_{n-1}(i-1), \\ V_n(i+1) = & -h(i+1) + p\beta R + \beta p(1-q)V_{n-1}(i+2) \\ & + \beta((1-p)(1-q)+pq)V_{n-1}(i+1) + \beta(1-p)qV_{n-1}(i), \\ V_n(i+2) = & -h(i+2) + \beta(1-q)V_{n-1}(i+2) + \beta qV_{n-1}(i+1). \end{aligned}$$

Therefore,

$$\begin{aligned} \Delta_n(i) = & -h+\beta p(1-q)\Delta_{n-1}(i+1) \\ & +\beta((1-p)(1-q)+pq)\Delta_{n-1}(i)+\beta(1-p)q\Delta_{n-1}(i-1), \\ \Delta_n(i+1) = & -h-\beta pR+\beta(1-p)(1-q)\Delta_{n-1}(i+1)+\beta(1-p)q\Delta_{n-1}(i), \end{aligned}$$

and

$$\begin{aligned} \Delta_n(i+1)-\Delta_n(i) = & -\beta pR-\beta p(1-q)\Delta_{n-1}(i+1)-\beta pq\Delta_{n-1}(i) \\ & + b(1-p)(1-q)(\Delta_{n-1}(i+1)-\Delta_{n-1}(i)) \\ & + \beta p(1-q)(\Delta_{n-1}(i)-\Delta_{n-1}(i-1)) \\ = & -H_{n-1}(i+1) \\ & + b(1-p)(1-q)(\Delta_{n-1}(i+1)-\Delta_{n-1}(i)) \\ & + \beta p(1-q)(\Delta_{n-1}(i)-\Delta_{n-1}(i-1)). \end{aligned}$$

In the last equation $H_{n-1}(i+1) \geq 0$ because $a_n^*(i+1) = 1$, while the remaining two terms are nonpositive from the induction hypothesis. Therefore $\Delta_n(i+1) - \Delta_n(i) \leq 0$ and (16) holds.

Case (iii): $a_n^*(i) = 1, a_n^*(i+1) = a_n^*(i+2) = 2$. Similarly, substituting the corresponding forms of $V_n(i)$, $V_n(i+1)$, $V_n(i+2)$ from (3) and after some algebra we obtain

$$\begin{aligned} \Delta_n(i) = & -h-\beta pR+\beta(1-p)(1-q)\Delta_{n-1}(i)+\beta(1-p)q\Delta_{n-1}(i-1), \\ \Delta_n(i+1) = & -h+\beta(1-q)\Delta_{n-1}(i+1)+\beta q\Delta_{n-1}(i), \end{aligned}$$

thus,

$$\begin{aligned}\Delta_n(i+1)-\Delta_n(i) &= \beta pR+\beta p(1-q)\Delta_{n-1}(i+1)+\beta pq\Delta_{n-1}(i)\\ &\quad + b(1-p)(1-q)(\Delta_{n-1}(i+1)-\Delta_{n-1}(i))\\ &\quad + \beta p(1-q)(\Delta_{n-1}(i)-\Delta_{n-1}(i-1))\\ &= H_{n-1}(i+1)\\ &\quad + b(1-p)(1-q)(\Delta_{n-1}(i+1)-\Delta_{n-1}(i))\\ &\quad + \beta p(1-q)(\Delta_{n-1}(i)-\Delta_{n-1}(i-1)).\end{aligned}$$

In the last equation $H_{n-1}(i+1) < 0$ because $a_n^*(i+1) = 2$, while the remaining two terms are also nonpositive from the induction hypothesis. Therefore $\Delta_n(i+1) - \Delta_n(i) \leq 0$ and (16) holds.

Case (iv): $a_n^*(i) = a_n^*(i+1) = a_n^*(i+2) = 2$. Proceeding similarly as above we obtain

$$\begin{aligned}\Delta_n(i) &= -h+\beta(1-q)\Delta_{n-1}(i)+\beta q\Delta_{n-1}(i-1),\\ \Delta_n(i+1) &= -h+\beta(1-q)\Delta_{n-1}(i+1)+\beta q\Delta_{n-1}(i),\end{aligned}$$

thus, (16) holds.

To complete the proof of (16), we must show that $\Delta_n(0) \geq \Delta_n(1)$ and $\Delta_n(M-2) \geq \Delta_n(M-1)$. These relations follow by considering cases for the corresponding optimal actions in states $0, 1, 2$, and $M-2, M-1$, respectively. The derivations are completely analogous and are omitted.

This completes the induction and the proof of the theorem. ∎

Partially Observable System

In this section we consider the system under partial state observations. The admission policy now cannot depend on the number of customers in the system, which is not observable, but rather on the information vector π at the beginning of the period which is a sufficient statistic summarizing the previous history of actions and signals. Since the state space Π of the partially observable Markov decision process consists of all $(M+1)$-dimensional probability vectors π, thus it is uncountable, the task of computing optimal policies is significantly harder than in the full information case. The main theoretical result on which most computational algorithms are based was derived in [13] and states that the finite horizon value function $v_n(\pi)$ is convex and piecewise linear in π. Specifically, for each $n \geq 1$ there exist vectors $\alpha_k^{(n)} \in \mathbb{R}^{M+1}, k = 1, \ldots, m_n$ such that

$$v_n(\pi) = \max_{k=1,\ldots,m_n} \alpha_k^{(n)} \cdot \pi, \quad \pi \in \Pi, \tag{17}$$

where $\cdot$ denotes inner product. Vectors $\alpha_k^{(n)}$ are computed recursively in n, based on the value iteration of optimality equations (12) as follows. Assume that all vectors $\alpha_k^{(n-1)}, n = 1, \ldots, m_{n-1}$ have been computed for iteration $n-1$. Then, for iteration n, substituting

$$v_{n-1}(\tilde{\pi}(j\,|\pi, a, \theta)) = \max_{k=1,\ldots,m_{n-1}} \alpha_k^{(n-1)} \cdot \tilde{\pi}(j\,|\pi, a, \theta)$$

and the value of $\tilde{\pi}(j\,|\pi, a, \theta)$ from (10) into (12), we obtain

$$v_n(\pi) = \max_{a=1,2} \left\{ \sum_i \pi(i) q_i^a + \beta \sum_\theta \max_{k=1,\ldots,m_{n-1}} \sum_i \sum_j \alpha_{kj}^{(n-1)} p_{ij}^a r_{j\theta}^a \pi(i) \right\}, \quad (18)$$

from which it follows that $v_n(\pi)$ is also piecewise linear in π, and the list of vectors $a_k^{(n)}$ can be derived.

Equation (17) implies that for each n the state space Π can be partitioned in m_n subsets such that in subset k the value function is linear with gradient $\alpha_k^{(n)}$, and the same action is optimal for all vectors in the subset. Of course, since m_n is typically much larger than the number of actions, an action may be optimal in more than one subsets. On the other hand some of the subsets in the partition may be empty, because the iterative computation in (18) may yield vectors that never attain the maximum in (17).

Property (17) introduces a significant simplification in the computation of the optimal policy. However the number of α-vectors generally increases exponentially in n, and identifying them is essentially equivalent to computing all vertices of a convex polyhedron. This is the reason that, although several approaches have been proposed for computing the optimal policy using (17), (e.g., [14], [3]), in all of them the task of exact computation is intractable even for modest size problems.

Because of the complexity of exact computation of the optimal policy, a large number of approximation algorithms have been proposed in the literature. Most of them consider a subset B of states in Π, and approximate the iterative computation in (18) by computing a smaller number of terms in the maximum. This results in a smaller number of vectors and thus in underestimating the value function v_n at each iteration. The various algorithms differ in the way the subset B is constructed and in the method employed to approximate (18).

In this paper we employ the algorithm Perseus proposed and analyzed in [15]. In this algorithm the subset B is constructed by selecting a number of vectors $\pi \in \Pi$ via simulation. In our implementation the vectors are generated from a uniform distribution in Π. The main idea of the Perseus algorithm is that at iteration n the iterative computation of the a vectors is not performed for all points in B separately. Instead, if a new vector α' is computed according to (18) for some $\pi \in B$, then it is applied to all other points in $b \in B$ to check whether $\alpha' \cdot b > v_{n-1}(b)$. For all b that satisfy this inequality the computation in (18) is omitted, since an improvement in the value of the previous iteration has already been achieved. The algorithm capitalizes on the fact that one new vector produced in some iteration generally increases the value function for other points as well. Therefore, by only limiting the complicated iterative computation for a subset of points in B significant computational reduction is achieved, at the expense of missing some vectors α that could be derived if these computations had not been omitted. The algorithm terminates when a convergence criterion on the difference in values between two successive iterations is met.

The output of the Perseus algorithm is a set of vectors $\alpha_k, k = 1, \ldots, m$ and a corresponding set of optimal decisions $a_k^*, k = 1, \ldots, m$ determined for each vector α_k from (18). These are used as follows. At the beginning of each period the decision maker computes the information vector π, i.e., the posterior distribution of the number of customers in the system given the previous history of actions and observations. Then a maximizing vector α_{k_0} is identified such that

$$a_{k_0}\pi = \max_{k=1\ldots m} \alpha_k \cdot \pi.$$

The action taken is $a_{k_0}^*$, corresponding to vector α_{k_0}. After the admission/rejection decision is made, the arrival and/or departure events occur, the decision maker receives the corresponding observations and proceeds to compute the updated information vector for the next period so that the cycle repeats.

Computational Results

In this section we perform two sets of computational experiments to obtain further insights on the behavior of the partial information problem as well as the implication of the lack of full observations on the system profits.

A general remark for the discussion in this section is that the optimal policy and profit under complete information are computed by implementing the value iteration of the dynamic programming algorithm in (3). Since the state and action spaces are finite, the computations are exact at each iteration n. The infinite horizon policy and profit are approximated by using a convergence stopping criterion. On the other hand, the profit and optimal policy under partial information are computed approximately using the Perseus algorithm instead of the exact computation of all α vectors, as discussed in the previous section. Therefore, when we refer to the optimal profit and optimal policy under partial information in this section, we will always mean the approximation obtained by Perseus.

In the first set of experiments we employ the Perseus algorithm for a queue with capacity $M = 10$, arrival and departure probabilities $p = 0.5$ and $q = 0.7$, respectively, holding cost rate $h = 2$, discount factor $\beta = 0.7$, and two values of the admission reward (a) $R = 4$ and (b) $r = 5$. We apply Perseus on a set B of 10^5 information vectors. The algorithm stops when for all points in B the absolute difference in the approximate value function between two successive iterations does not exceed a threshold $\epsilon = 10^{-7}$.

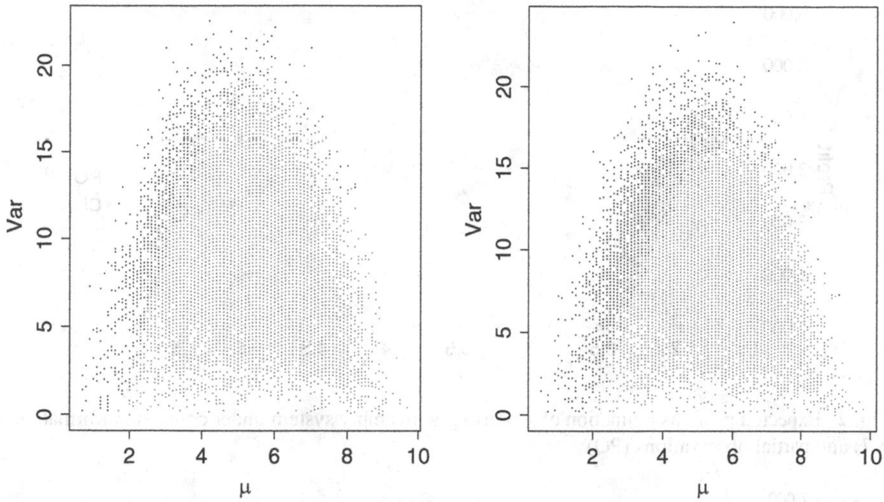

Fig. 1 Optimal policy as a function of the mean and variance of the information vector for $R = 4$ and $R = 5$

In order to present the results graphically, we proceed as follows. For each information vector $\pi \in B$ we compute the expected value $\mu(\pi) = \sum_{j=0}^{M} j\pi(j)$ and the variance $\text{var}(\pi) = \sum_{j=0}^{M}(j - \mu(\pi))^2\pi(j)$. We then plot all points $(\mu(\pi), \text{var}(\pi)), \pi \in B$, in colors that correspond to the optimal decision at each point, black for admission and red for rejection. The two plots corresponding to the two values of the admission reward R are presented in Fig. 1.

In these plots we first observe that, as expected, when the admission reward is increased the admission decision is optimal for a larger set of information vectors. It is also interesting that the admission and rejection regions are almost perfectly separated in both plots. This indicates that the mean–variance pair of the information vector is a good surrogate for determining the optimal action. Specifically, we observe that for each value of the variance, there is a threshold on the mean such that admission is optimal when the estimated mean number of customers in the system is below the threshold. This property can be viewed as a generalization of the complete information policy, where there is a fixed admission threshold on the observed number of customers. Here the threshold is on the expected number of customers and increases as the variance increases. This shows that the increased uncertainty about congestion induces the owner to accept more customers and is somewhat counterintuitive.

The second set of computational experiments is aimed at sensitivity analysis as well as exploring the value of information. We vary certain system parameters and perform a comparison of the expected profit between the complete information and partial observation models, in order to assess the impact of the lack of full information on system performance. The numerical analysis is performed as follows. We consider a baseline case of parameter values $M = 10, p = 0.5, q = 0.7, h = 2, R = 4, \beta = 0.7$. We then vary in turn the admission reward R in

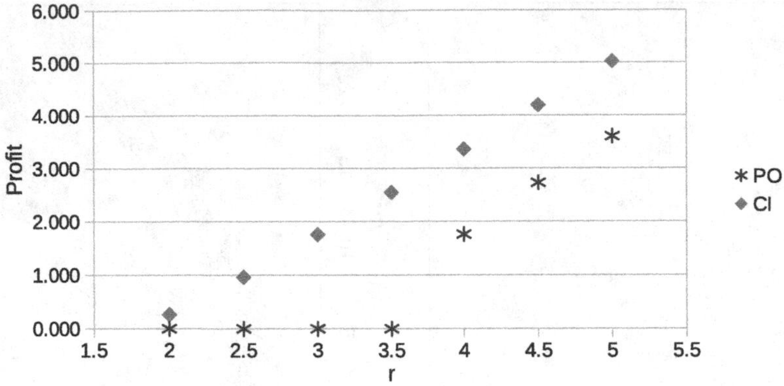

Fig. 2 Expected profit as a function of R starting with empty system under complete information (CI) and partial observations (PO)

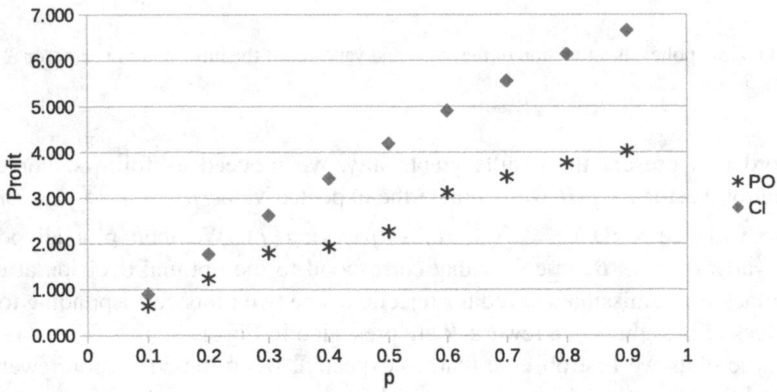

Fig. 3 Expected profit as a function of p starting with empty system under complete information (CI) and partial observations (PO)

the range $[2, 5]$, the arrival probability in the range $[0.1, 0.9]$, and the departure probability in the same range. For each parameter we present $V(0)$, i.e., the optimal expected discounted profit under complete information starting with an empty system and $v((1, 0, \ldots, 0))$, i.e., the same profit value under partial information for the information vector corresponding to zero customers in the system with probability 1. The results are presented in Figs. 2, 3 and 4, respectively.

From these experiments we observe that in general the profit is increasing in R, p, and q. All these properties are intuitive. In particular, as R increases, there is a greater incentive to admit more customers, thus profits increase overall. When the arrival probability p increases, the system has a larger customer base which increases the owner's flexibility in terms of the admission policy. We have also observed in the numerical analysis that the acceptance threshold is decreasing, so that a smaller proportion of arriving customers is accepted at the optimal policy.

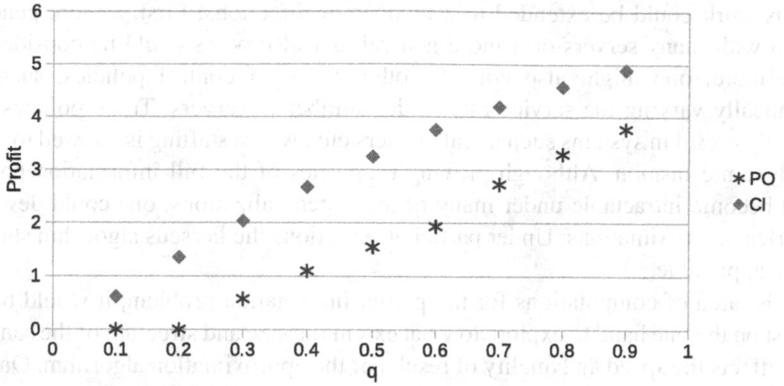

Fig. 4 Expected profit as a function of q starting with empty system under complete information (CI) and partial observations (PO)

Nevertheless, the profit increases overall. Similarly, as the departure probability increases, customers are served more efficiently, and thus the system owner can afford to admit more thus increasing profits. This is also confirmed by the fact that the admission thresholds increase with increasing q.

Regarding the profit reduction due to lack of information, we observe that it is in general nonnegligible. This implies that there may be significant value for the system manager to invest in improving the information available in the system. The fact that the loss is increasing in p and q shows that the value of information is higher when there is a larger customer base as well as when the service is more efficient. With respect to R, the value of information seems to be the highest for intermediate values of the parameter.

Conclusions

In this paper we considered the problem of admission control in a discrete time Markovian queue with partial state observations. We proved the threshold structure of the optimal admission policy under full information and formulated an approximation algorithm for computing the optimal policy under partial information using the posterior distribution as a modified state. Using numerical analysis we demonstrated that the pair of the mean and variance of the posterior distribution may be effectively used instead of the full distribution, to implement the optimal policy. We also explored the behavior of the profit function and the value of information with respect to several system parameters.

This work could be extended to a number of directions. First, a more general system with many servers or a more general arrival process could be considered. Furthermore, one might also consider other types of control policies, such as dynamically varying the service rate or the number of servers. These policies are generally useful in systems such as call centers etc., where staffing is allowed to vary in a dynamic fashion. Although proving properties of the full information policy might become intractable under many of these generalizations, one could develop numerical approximations. Under partial observations the Perseus algorithm should still be applicable.

In the area of computations for the partial information problem, it would be of interest on the one hand to explore to what extent the size and structure of the sample set B affects the speed and quality of results of the approximation algorithm. On the other hand, it would also be interesting to take a closer look at the issue of using the mean–variance, or alternative statistics, to approximate the optimal policy under partial information and quantify the accuracy of the approximation by simulating the system under the original policy and that based on the surrogate information.

Acknowledgments This research has been co-financed by the European Union (European Social Fund - ESF) and Greek national funds through the Operational Program "Education and Lifelong Learning" of the National Strategic Reference Framework (NSRF)-Research Funding Program: Thales-Athens University of Economics and Business-New Methods in the Analysis of Market Competition: Oligopoly, Networks and Regulation.

References

1. J. Artalejo and O. Hernández-Lerma. Performance analysis and optimal control of the geo/geo/c queue. *Performance Evaluation*, 52:15–39, 2003.
2. A. Burnetas and A. Economou. Equilibrium customer strategies in a single server Markovian queue with setup times. *Queueing Systems*, 56(3–4):213–228, 2007.
3. H.-T. Cheng. *Algorithms for Partially Observed Markov Decision Processes*. PhD thesis, University of British Columbia, British Columbia, Canada, 1988.
4. P. Guo and P. Zipkin. Analysis and comparison of queues with different levels of delay information. *Mgt. Sci.*, 53:962–970, 2007.
5. G. Koole. *Monotonicity in Marko Reward and Decision Chains: Theory and Applications*. Now Publishers Inc., 2007.
6. J. Kuri and A. Kumar. Optimal control of arrivals to queues with delayed queue length information. *IEEE Transactions on Automatic Control*, 40:1444–1450, 1995.
7. K. Y. Lin and S. M. Ross. Admission control with incomplete information of a queueing system. *Operations Research*, 51:645–654, 2003.
8. W. Lovejoy. A survey of algorithmic methods for partially observed markov decision processes. *Annals of Operations Research*, 28(1):47–65, 1991.
9. W. P. Millhiser and A. N. Burnetas. Optimal admission control in series production systems with blocking. *IIE Transactions*, 45:1035–1047, 2013.
10. E. L. Örmeci, A. Burnetas, and J. van der Wal. Admission policies for a two class loss system. *Stochastic Models*, 17:513–539, 2001.Wal]
11. M. Puterman. *Markov Decision Processes*. John Wiley and Sons Inc., New York, 1994.

12. S. V. Savin, M. A. Cohen, N. Gans, and Z. Katalan. Capacity management in rental businesses with two customer bases. *Operations Research*, 53:617–631, 2005.
13. R. D. Smallwood and E. J. Sondik. Optimal control of partially observable markov processes over a finite horizon. *Operations Research*, 21(5):1071–1088, 1973.
14. E. J. Sondik. *The Optimal Control of Partially Observable Markov Processes*. PhD thesis, Stanford University, Stanford, California, 1971.
15. M. Spaan and N. Vlassis. Perseus: Randomized point-based value iteration for pomdps. *Journal of Artificial Intelligence Research*, 24:195–220, 2005.
16. S. Stidham. Optimal-control of admission to a queuing system. *IEEE Transactions On Automatic Control*, 30(8):705–713, 1985. ISSN 0018-9286.
17. C. White III. A survey of solution techniques for the partially observed markov decision process. *Annals of Operations Research*, 32(1):215–230, 1991.

Error Bounds for Trapezoid Type Quadrature Rules with Applications for the Mean and Variance

Pietro Cerone, Sever S. Dragomir, and Eder Kikianty

Abstract In this paper, we establish some inequalities of trapezoid type to give tight bounds for the expectation and variance of a probability density function. The approach is also demonstrated for higher order moments.

Introduction

The trapezoid rule is a method to approximate the integral $\int_a^b f(x)\,dx$, by approximating the area under the curve of $f(x)$ as a trapezoid:

$$\int_a^b f(x)\,dx \approx (b-a)\frac{f(a)+f(b)}{2}.$$

Some inequalities have been established to give bounds for the error of this approximation, and we summarised the result in the following proposition (cf. Cerone and Dragomir [5]).

P. Cerone
Department of Mathematics and Statistics, La Trobe University, Bundoora 3086, Australia
e-mail: P.Cerone@latrobe.edu.au

S.S. Dragomir
School of Engineering and Science, Victoria University, Melbourne 8001, Australia
School of Computational and Applied Mathematics, University of the Witwatersrand, South Africa
e-mail: Sever.Dragomir@vu.edu.au

E. Kikianty (✉)
Department of Mathematics, University of Johannesburg, Auckland Park 2006, South Africa
e-mail: ekikianty@uj.ac.za; eder.kikianty@gmail.com

N.J. Daras (ed.), *Applications of Mathematics and Informatics in Science and Engineering*, Springer Optimization and Its Applications 91, DOI 10.1007/978-3-319-04720-1_5,

Proposition 1. *Let $g : I \subset \mathbb{R} \to \mathbb{R}$ be a function and $u, v \in I$ with $u < v$. Consider the approximation of the integral of g on $[u, v]$ by the trapezoid rule, that is, find abound for the quantity:*

$$\left| \frac{g(u) + g(v)}{2}(v - u) - \int_u^v g(t)\, dt \right| . \tag{1}$$

The following bounds for (1) *holds for any $u, v \in I$ with $u < v$:*

a. *If g is of bounded variation on $[u, v]$, then*

$$\left| \frac{g(u) + g(v)}{2}(v - u) - \int_u^v g(t)\, dt \right| \leq \frac{1}{2}|v - u| \left| \bigvee_u^v (g) \right| . \tag{2}$$

b. *If g is Lipschitz continuous with Lipschitz constant L, then*

$$\left| \frac{g(u) + g(v)}{2}(v - u) - \int_u^v g(t)\, dt \right| \leq \frac{1}{4} L(v - u)^2 \tag{3}$$

c. *If g'' exists and bounded, then*

$$\left| \frac{g(u) + g(v)}{2}(v - u) - \int_u^v g(t)\, dt \right| \leq \frac{1}{12}(v - u)^3 \|g''\|_\infty . \tag{4}$$

d. *If g' exists and is absolutely continuous on $[u, v]$, then*

$$\left| \frac{g(u) + g(v)}{2}(v - u) - \int_u^v g(t)\, dt \right| \leq \begin{cases} \frac{1}{4}\|g'\|_\infty (v - u)^2, & \text{if } g' \in L_\infty[u, v]; \\ \frac{\|g'\|_p (v - u)^{1+\frac{1}{q}}}{2(q + 1)^{\frac{1}{q}}}, & \text{if } g' \in L_p[u, v],\ \frac{1}{p} + \frac{1}{q} = 1,\ p > 1; \\ \frac{1}{2}\|g'\|_1 (v - u), & \text{if } g' \in L_1[u, v]. \end{cases} \tag{5}$$

We refer the reader to Cerone and Dragomir [5] for more details on the trapezoid type inequalities.

One of the applications of integral inequalities is to obtain bounds for the expectation, variance and moments of continuous random variables defined over a finite interval [1]. In Barnett et al. [1], it is noted that some Ostrowski type inequalities may be used to obtain these bounds (see, e.g., Brnetić and Pečarić [2]). We refer the readers to the monograph by Barnett et al. [1], for an overview of these inequalities.

There are other inequalities which provide bounds for means and variances. Chernoff [7], for instance, proved that for any Gaussian random variable X and absolutely continuous function G, we have $\operatorname{Var}(G(X)) \leq E(G'(X))^2$. This inequality is then generalised for higher-order derivatives in Houdré and Kagan [9]. A characterisation of distributions (normal, gamma, negative binomial or Poisson) is given in [10] by means of a Chernoff type inequality. We refer to the papers by Cacaoullos [3], Cacoullos and Papadatos [4], Chang and Richards [6] and Dharmadhikari and Joag-Dev [8], for further inequalities involving variances.

In this paper, we aim to provide some inequalities of trapezoid type to give tight bounds for the expectation and variance of a probability density function f. In section "Main Results", we give approximations for the first and second moments of a function $f : [a,b] \to \mathbb{R}$ around the midpoint of the domain, i.e.,

$$\int_a^b \left(x - \frac{a+b}{2}\right) f(x)\,dx \quad \text{and} \quad \int_a^b \left(x - \frac{a+b}{2}\right)^2 f(x)\,dx.$$

We make use of the trapezoid type inequalities to obtain error bounds for the approximation. In section "Applications to Mean and Variance", we apply the results to obtain bounds for the expectation and variance of a probability density function f. Remark 1 demonstrates the applicability of the approach for higher order moments.

Main Results

Firstly, we note that inequality (4) also holds when we weaken the assumption, as presented in the next proposition.

Proposition 2. *Let $g : I \to \mathbb{R}$ be a function and $u, v \in I$ with $u < v$. If g' is absolutely continuous and $g'' \in L_\infty[u,v]$, then,*

$$\left| \frac{g(u)+g(v)}{2}(v-u) - \int_u^v g(t)\,dt \right| \leq \frac{1}{12}(v-u)^3 \|g''\|_\infty,$$

for all $u, v \in I$.

Proof. Since g'' exists almost everywhere, we have

$$\begin{aligned}\frac{1}{2}\int_u^v (t-u)(v-t)g''(t)\,dt &= \frac{1}{2}\left[(t-u)(v-t)g'(t)\Big|_u^v - \int_u^v (u+v-2t)g'(t)\,dt\right]\\ &= \frac{1}{2}\left[(2t-u-v)g(t)\Big|_u^v - 2\int_u^v g(t)\,dt\right]\\ &= \frac{g(u)+g(v)}{2}(v-u) - \int_u^v g(t)\,dt.\end{aligned}$$

Thus,

$$\left|\frac{g(u)+g(v)}{2}(v-u)-\int_u^v g(t)\,dt\right| \le \frac{1}{2}\int_u^v (t-u)(v-t)|g''(t)|\,dt$$
$$\le \frac{1}{2}\|g''\|_\infty \int_u^v (t-u)(v-t)\,dt = \frac{1}{12}\|g''\|_\infty (v-u)^3,$$

as desired. □

Utilising (1) we have the approximation for the first moment of a function f.

Lemma 1. *Let $f : [a,b] \to \mathbb{R}$ be an integrable function. We have the following approximation for the first moment of f:*

$$\int_a^b \left(x - \frac{a+b}{2}\right) f(x)\,dx \approx \frac{b-a}{3}\left(\int_{\frac{a+b}{2}}^b f(x)\,dx - \int_a^{\frac{a+b}{2}} f(x)\,dx\right). \quad (6)$$

Proof. Setting $f \equiv g$, $u = \frac{a+b}{2}$ and $v = x$ in (1), and integrate it over $[a,b]$, we have

$$\left|\int_a^b \frac{f(x) + f\left(\frac{a+b}{2}\right)}{2}\left(x - \frac{a+b}{2}\right) dx - \int_a^b \left(\int_{\frac{a+b}{2}}^x f(t)\,dt\right) dx\right|. \quad (7)$$

Now,

$$\int_a^b \frac{f(x) + f\left(\frac{a+b}{2}\right)}{2}\left(x - \frac{a+b}{2}\right) dx = \frac{1}{2}\int_a^b \left(x - \frac{a+b}{2}\right) f(x)\,dx.$$

We also have

$$\int_a^b \left(\int_{\frac{a+b}{2}}^x f(t)\,dt\right) dx$$
$$= \left(x - \frac{a+b}{2}\right)\int_{\frac{a+b}{2}}^x f(t)\,dt\Big|_a^b - \int_a^b \left(x - \frac{a+b}{2}\right) f(x)\,dx$$
$$= \frac{b-a}{2}\int_{\frac{a+b}{2}}^b f(t)dt - \frac{b-a}{2}\int_a^{\frac{a+b}{2}} f(t)dt - \int_a^b \left(x - \frac{a+b}{2}\right) f(x)\,dx.$$

Thus, (7) becomes

$$\left|\frac{3}{2}\int_a^b \left(x - \frac{a+b}{2}\right) f(x)\,dx - \frac{b-a}{2}\left(\int_{\frac{a+b}{2}}^b f(t)dt - \int_a^{\frac{a+b}{2}} f(t)dt\right)\right|.$$

Multiplying the above by $\frac{2}{3}$ completes the proof. □

Let f be an integrable real-valued function defined on $[a,b]$, and set

$$T_1(f) := \int_a^b \left(x - \frac{a+b}{2}\right) f(x)\,dx - \frac{b-a}{3}\left(\int_{\frac{a+b}{2}}^b f(x)\,dx - \int_a^{\frac{a+b}{2}} f(x)\,dx\right).$$

In the next theorem, we give bounds for $|T_1|$, i.e. the error bounds for the approximation in Lemma 1, for different classes of functions.

Theorem 1. *Let $f : [a,b] \to \mathbb{R}$ be an integrable function.*

a. If f of bounded variation on $[a,b]$, then,

$$|T_1(f)| \le \begin{cases} \dfrac{(b-a)^2}{12}\left[\dfrac{1}{2}\bigvee_a^b(f) + \dfrac{1}{2}\left|\bigvee_a^{\frac{a+b}{2}}(f) - \bigvee_{\frac{a+b}{2}}^b(f)\right|\right] \\ \dfrac{b-a}{6}\displaystyle\int_a^b \left|\bigvee_{\frac{a+b}{2}}^x(f)\right| dx \end{cases} \le \frac{1}{6}(b-a)^2\bigvee_a^b(f). \tag{8}$$

b. If f is L-Lipschitz, then

$$|T_1(f)| \le \frac{1}{72}L(b-a)^3. \tag{9}$$

c. If f' is absolutely continuous and $f'' \in L_\infty[a,b]$, then

$$|T_1(f)| \le \frac{1}{576}\|f''\|_\infty (b-a)^4. \tag{10}$$

d. If f' is absolutely continuous, then

$$|T_1(f)| \le \begin{cases} \dfrac{1}{72}\|f'\|_\infty (b-a)^3, & \text{if } f' \in L_\infty[a,b]; \\ \dfrac{q\|f'\|_p (b-a)^{2+\frac{1}{q}}}{3(2q+1)(q+1)^{\frac{1}{q}} 2^{1+\frac{1}{q}}}, & \text{if } f' \in L_p[a,b],\ \frac{1}{p}+\frac{1}{q}=1,\ p>1; \\ \dfrac{1}{12}\|f'\|_1 (b-a)^2, & \text{if } f' \in L_1[a,b]. \end{cases} \tag{11}$$

Proof. Let f be a function of bounded variation on $[a,b]$. Setting $f \equiv g$, $u = x$ and $v = \frac{a+b}{2}$ in (2), we have

$$\left|\frac{f(x)+f\left(\frac{a+b}{2}\right)}{2}\left(x-\frac{a+b}{2}\right)-\int_{\frac{a+b}{2}}^{x} f(t)\,dt\right| \le \frac{1}{2}\left|x-\frac{a+b}{2}\right|\left|\bigvee_{\frac{a+b}{2}}^{x}(f)\right|.$$

Integrating the above on $[a,b]$, we have

$$\begin{aligned}
&\left|\int_a^b \frac{f(x)+f\left(\frac{a+b}{2}\right)}{2}\left(x-\frac{a+b}{2}\right)dx-\int_a^b\left(\int_{\frac{a+b}{2}}^{x} f(t)\,dt\right)dx\right| \\
&\le \int_a^b \left|\frac{f(x)+f\left(\frac{a+b}{2}\right)}{2}\left(x-\frac{a+b}{2}\right)-\int_{\frac{a+b}{2}}^{x} f(t)\,dt\right| dx \\
&\le \frac{1}{2}\int_a^b \left|x-\frac{a+b}{2}\right|\left|\bigvee_{\frac{a+b}{2}}^{x}(f)\right| dx. \qquad (12)
\end{aligned}$$

Following the proof of Lemma 1, the first term of (12) is $\frac{3}{2}|T_1(f)|$. Furthermore, we have

$$\begin{aligned}
\int_a^b \left|x-\frac{a+b}{2}\right|\left|\bigvee_{\frac{a+b}{2}}^{x}(f)\right| dx &\le \begin{cases} \max\limits_{x\in[a,b]}\left|\bigvee_{\frac{a+b}{2}}^{x}(f)\right| \int_a^b \left|x-\frac{a+b}{2}\right| dx \\ \max\limits_{x\in[a,b]}\left|x-\frac{a+b}{2}\right| \int_a^b \left|\bigvee_{\frac{a+b}{2}}^{x}(f)\right| dx \end{cases} \\
&\le \begin{cases} \max\left\{\bigvee_{a}^{\frac{a+b}{2}}(f), \bigvee_{\frac{a+b}{2}}^{b}(f)\right\} \frac{(b-a)^2}{4} \\ \frac{b-a}{2}\int_a^b \left|\bigvee_{\frac{a+b}{2}}^{x}(f)\right| dx \end{cases} \\
&\le \begin{cases} \frac{(b-a)^2}{4}\left[\frac{1}{2}\bigvee_{a}^{b}(f)+\frac{1}{2}\left|\bigvee_{a}^{\frac{a+b}{2}}(f)-\bigvee_{\frac{a+b}{2}}^{b}(f)\right|\right] \\ \frac{b-a}{2}\int_a^b \left|\bigvee_{\frac{a+b}{2}}^{x}(f)\right| dx \end{cases} \\
&=: I.
\end{aligned}$$

Thus, (12) becomes

$$\frac{3}{2}|T_1(f)| \le \frac{1}{2}I \le \frac{1}{4}\bigvee_{a}^{b}(f).$$

Multiplying the above by $\frac{2}{3}$ gives us (8). Let f be L-Lipschitz. We apply similar steps as above and utilise (3) to obtain

$$\frac{3}{2}|T_1(f)| \leq \int_a^b \frac{1}{4}L\left|x-\frac{a+b}{2}\right|^2 dx = \frac{1}{48}L(b-a)^3.$$

Multiplying the above with $\frac{2}{3}$ gives us (9). Let f' be absolutely continuous and $f'' \in L_\infty[a,b]$. We apply similar steps as above and utilise Proposition 2 to obtain

$$\frac{3}{2}|T_1(f)| \leq \int_a^b \frac{1}{12}\|f''\|_\infty\left|x-\frac{a+b}{2}\right|^3 dx = \frac{1}{384}\|f''\|_\infty(b-a)^4.$$

Multiplying the above with $\frac{2}{3}$ gives us (10). Let f' exists and absolutely continuous. We apply similar steps as above and utilise (5). We have

$$\int_a^b \frac{1}{4}\|f'\|_\infty\left(x-\frac{a+b}{2}\right)^2 dx = \frac{1}{48}\|f'\|_\infty(b-a)^3. \tag{13}$$

The second case of the right-hand side of (5) becomes

$$\frac{1}{2(q+1)^{\frac{1}{q}}}\|f'\|_p\int_a^b\left|x-\frac{a+b}{2}\right|^{1+\frac{1}{q}} dx = \frac{q\|f'\|_p(b-a)^{2+\frac{1}{q}}}{(2q+1)(q+1)^{\frac{1}{q}}2^{2+\frac{1}{q}}}. \tag{14}$$

The third case of the right-hand side of (5) becomes

$$\int_a^b \frac{1}{2}\|f'\|_1\left|x-\frac{a+b}{2}\right| dx = \frac{1}{8}\|f'\|_1(b-a)^2. \tag{15}$$

Multiplying (13), (14) and (15) with $\frac{2}{3}$ gives us the right-hand side of (11). □

Proposition 3. *The constant* $\frac{1}{12}$ *in the first case of* (8) *is best possible.*

Proof. Let $f : [a,b] \subset \mathbb{R} \to \mathbb{R}$ be defined as follows: $f(x) = 1$ when $a \leq x < \frac{a+b}{2}$, $f(x) = 0$ when $x = \frac{a+b}{2}$, and $f(x) = -1$ when $\frac{a+b}{2} < x \leq b$. We have $\bigvee_a^{\frac{a+b}{2}}(f) = 1$, $\bigvee_{\frac{a+b}{2}}^b(f) = 1$, and $\bigvee_a^b(f) = 2$. Let us assume that (8) holds for constants $A > 0$ instead $\frac{1}{12}$, i.e.

$$|T_1(f)| \leq A(b-a)^2\left[\frac{1}{2}\bigvee_a^b(f) + \frac{1}{2}\left|\bigvee_a^{\frac{a+b}{2}}(f) - \bigvee_{\frac{a+b}{2}}^b(f)\right|\right]. \tag{16}$$

With the above choice of f, (16) becomes: $\frac{1}{12}(b-a)^2 \leq A(b-a)^2$, which asserts that $A \geq \frac{1}{12}$. □

Proposition 4. *The constant* $\frac{1}{576}$ *in* (10) *is best possible.*

Proof. Let us assume that (10) holds for constants $B > 0$ instead $\frac{1}{576}$, i.e.

$$|T_1(f)| \leq B\|f''\|_\infty (b-a)^4. \tag{17}$$

Let $f : [a,b] \subset \mathbb{R} \to \mathbb{R}$ be defined as follows: $f(x) = \left(x - \frac{a+b}{2}\right)^2$ when $a \leq x \leq \frac{a+b}{2}$ and $f(x) = -\left(x - \frac{a+b}{2}\right)^2$ when $\frac{a+b}{2} < x \leq b$. We note that f'' exists almost everywhere and $\|f''\|_\infty = 2$. With this choice of f, (17) becomes: $\frac{1}{288}(b-a)^4 \leq 2B(b-a)^4$, which asserts that $B \geq \frac{1}{576}$. □

Utilising (1) we have the approximation for the second moment of a function f.

Lemma 2. *Let* $f : [a,b] \to \mathbb{R}$ *be an integrable function. We have the following approximation for* f*:*

$$\int_a^b \left(x - \frac{a+b}{2}\right)^2 f(x)\,dx \approx \frac{1}{8}(b-a)^2 \int_a^b f(t)\,dt - \frac{1}{24} f\left(\frac{a+b}{2}\right)(b-a)^3. \tag{18}$$

Proof. Set $f \equiv g$, $u = \frac{a+b}{2}$ and $v = x$ in (1) and let $F(x) = \int_a^x f(t)\,dt$ to obtain:

$$\left| F(x) - F\left(\frac{a+b}{2}\right) - \frac{f(x) + f\left(\frac{a+b}{2}\right)}{2}\left(x - \frac{a+b}{2}\right)\right|. \tag{19}$$

If we multiply (19) with $\left|x - \frac{a+b}{2}\right|$, and integrate it on $[a,b]$, we have

$$\left|\int_a^b \left[F(x) - F\left(\frac{a+b}{2}\right)\right]\left(x - \frac{a+b}{2}\right) dx \right.$$
$$\left. -\frac{1}{2}\int_a^b \left(x - \frac{a+b}{2}\right)^2 f(x)\,dx - \frac{1}{2} f\left(\frac{a+b}{2}\right)\int_a^b \left(x - \frac{a+b}{2}\right)^2 dx\right|. \tag{20}$$

Now, observe that

$$\begin{aligned}
&\int_a^b \left[F(x) - F\left(\frac{a+b}{2}\right)\right]\left(x - \frac{a+b}{2}\right) dx \\
&= \int_a^b F(x)\left(x - \frac{a+b}{2}\right) dx \\
&= \frac{1}{2}\int_a^b F(x) d\left(\left(x - \frac{a+b}{2}\right)^2\right)
\end{aligned}$$

$$= \frac{1}{2}\left[(F(b) - F(a))\frac{(b-a)^2}{4} - \int_a^b \left(x - \frac{a+b}{2}\right)^2 f(x)\,dx \right]$$

$$= \frac{1}{8}(b-a)^2 \int_a^b f(t)\,dt - \frac{1}{2}\int_a^b \left(x - \frac{a+b}{2}\right)^2 f(x)\,dx;$$

and

$$\int_a^b \left(x - \frac{a+b}{2}\right)^2 dx = \frac{1}{12}(b-a)^3.$$

Then, (20) becomes

$$\left| \int_a^b \left(x-\frac{a+b}{2}\right)^2 f(x)\,dx - \frac{1}{8}(b-a)^2 \int_a^b f(t)\,dt + \frac{1}{24} f\left(\frac{a+b}{2}\right)(b-a)^3 \right|$$

as desired. □

Let $f : [a, b] \to \mathbb{R}$ be an integrable function and set

$$T_2(f) := \int_a^b \left(x - \frac{a+b}{2}\right)^2 f(x)\,dx - \frac{1}{8}(b-a)^2 \int_a^b f(t)\,dt + \frac{1}{24} f\left(\frac{a+b}{2}\right)(b-a)^3.$$

In the next theorem, we have bounds for $|T_2|$, i.e. the error bounds for the approximation in Lemma 2, for different classes of functions.

Theorem 2. *Let $f : [a, b] \to \mathbb{R}$ be an integrable function.*

a. If f is of bounded variation, then

$$|T_2(f)| \le \frac{1}{48} \bigvee_a^b (f)(b-a)^3. \tag{21}$$

b. If f is L-Lipschitz function, then

$$|T_2(f)| \le \frac{1}{128} L(b-a)^4. \tag{22}$$

c. If f' is absolutely continuous and $f'' \in L_\infty[a, b]$, then

$$|T_2(f)| \le \frac{1}{960} \|f''\|_\infty (b-a)^5. \tag{23}$$

d. If f' is absolutely continuous, then

$$|T_2(f)| \le \begin{cases} \frac{1}{128}\|f'\|_\infty (b-a)^4, & \text{if } f' \in L_\infty[a,b]; \\ \frac{q\|f'\|_p (b-a)^{3+\frac{1}{q}}}{(3q+1)(q+1)^{\frac{1}{q}} 2^{3+\frac{1}{q}}}, & \text{if } f' \in L_p[a,b],\ \frac{1}{p}+\frac{1}{q}=1; \\ \frac{1}{24}\|f'\|_1 (b-a)^3 & \text{if } f' \in L_1[a,b]. \end{cases} \tag{24}$$

Proof. Let f be of bounded variation. Let $F(x) = \int_a^x f(t)\,dt$, we have the following by (2):

$$\left| F(x) - F\left(\frac{a+b}{2}\right) - \frac{f(x) + f\left(\frac{a+b}{2}\right)}{2}\left(x - \frac{a+b}{2}\right)\right| \le \frac{1}{2}\left|x - \frac{a+b}{2}\right| \left|\bigvee_x^{\frac{a+b}{2}}(f)\right|. \tag{25}$$

If we multiply (25) with $\left|x - \frac{a+b}{2}\right|$, we get

$$\begin{aligned} &\left|\left[F(x) - F\left(\frac{a+b}{2}\right)\right]\left(x - \frac{a+b}{2}\right) - \frac{1}{2}\left[f(x) + f\left(\frac{a+b}{2}\right)\right]\left(x - \frac{a+b}{2}\right)^2\right| \\ &\le \frac{1}{2}\left|x - \frac{a+b}{2}\right|^2 \left|\bigvee_x^{\frac{a+b}{2}}(f)\right|. \end{aligned} \tag{26}$$

Integrate (26) on $[a,b]$ and follow the proof of Lemma 2 we have:

$$\begin{aligned} |T_2(f)| &\le \frac{1}{2}\int_a^b \left|x - \frac{a+b}{2}\right|^2 \left|\bigvee_x^{\frac{a+b}{2}}(f)\right| dx \\ &= \frac{1}{2}\left[\int_a^{\frac{a+b}{2}} \left(x - \frac{a+b}{2}\right)^2 \bigvee_x^{\frac{a+b}{2}}(f)\,dx + \int_{\frac{a+b}{2}}^b \left(x - \frac{a+b}{2}\right)^2 \bigvee_{\frac{a+b}{2}}^x (f)\,dx\right] \\ &\le \frac{1}{2}\bigvee_a^{\frac{a+b}{2}}(f)\int_a^{\frac{a+b}{2}} \left(x - \frac{a+b}{2}\right)^2 dx + \frac{1}{2}\bigvee_{\frac{a+b}{2}}^b (f)\int_{\frac{a+b}{2}}^b \left(x - \frac{a+b}{2}\right)^2 dx \\ &= \frac{1}{2}\bigvee_a^b (f)\int_a^{\frac{a+b}{2}} \left(x - \frac{a+b}{2}\right)^2 dx = \frac{1}{48}\bigvee_a^b (f)(b-a)^3. \end{aligned}$$

Let f be L-Lipschitz. We apply similar steps as above and utilise (3) to obtain:

$$|T_2(f)| \leq \frac{1}{4}L \int_a^b \left| x - \frac{a+b}{2} \right|^3 dx = \frac{1}{128}L(b-a)^4.$$

Let f be a function such that f' is absolutely continuous and $f'' \in L_\infty[a,b]$. We apply similar steps as above and utilise Proposition 2 to obtain:

$$|T_2(f)| \leq \frac{1}{12}\|f''\|_\infty \int_a^b \left(x - \frac{a+b}{2} \right)^4 dx = \frac{1}{960}\|f''\|_\infty (b-a)^5.$$

Let f be a function such that f' exists and absolutely continuous. We apply similar steps as above and utilise (5) to obtain:

$$\frac{1}{4}\|f'\|_\infty \int_a^b \left| x - \frac{a+b}{2} \right|^3 dx = \frac{1}{128}\|f'\|_\infty (b-a)^4.$$

The second case in the right-hand side of (5) becomes

$$\frac{1}{2(q+1)^{\frac{1}{q}}}\|f'\|_p \int_a^b \left| x - \frac{a+b}{2} \right|^{2+\frac{1}{q}} dx = \frac{q\|f'\|_p (b-a)^{3+\frac{1}{q}}}{(3q+1)(q+1)^{\frac{1}{q}} 2^{3+\frac{1}{q}}}.$$

The last case in the right-hand side of (5) becomes

$$\frac{1}{2}\|f'\|_1 \int_a^b \left(x - \frac{a+b}{2} \right)^2 dx = \frac{1}{24}\|f'\|_1 (b-a)^3.$$

This completes the proof. □

Proposition 5. *The constant $\frac{1}{48}$ in (21) is best possible.*

Proof. We now prove the sharpness of the constant $\frac{1}{48}$. Let $f : [a,b] \subset \mathbb{R} \to \mathbb{R}$ be defined as follows: $f(x) = 0$ when $x = \frac{a+b}{2}$, and $f(x) = 1$ otherwise. Therefore, $\bigvee_a^b(f) = 2$. Let us assume that (21) holds for a constant $C > 0$ instead of $\frac{1}{48}$, i.e.

$$|T_2(f)| \leq C \bigvee_a^b (f)(b-a)^3. \tag{27}$$

With the above choice of f, (27) becomes: $\frac{1}{24}(b-a)^3 \leq 2C(b-a)^3$, which asserts that $C \geq \frac{1}{48}$, hence the constant $\frac{1}{48}$ is best possible. □

Proposition 6. *The constant $\frac{1}{960}$ in (23) is best possible.*

Proof. Assume that (23) holds for a constant K instead of $\frac{1}{960}$, i.e.

$$|T_2(f)| \le K\|f''\|_\infty (b-a)^5. \tag{28}$$

Let $f : [a,b] \to \mathbb{R}$ defined by $f(x) = \frac{1}{2}\left(x - \frac{a+b}{2}\right)^2$. So, $f''(x) = 1$ for all $x \in [a,b]$ and thus $\|f''\|_\infty = 1$. Therefore, (28) becomes: $\frac{1}{960}(b-a)^5 \le K(b-a)^5$, which yields $K \ge \frac{1}{960}$. □

Remark 1. Utilising a similar technique to that of Lemma 2, we are able to obtain the approximations for higher order moments that can be derived from (1). Set $f \equiv g$, $u = \frac{a+b}{2}$ and $v = x$ in (1), let $F(x) = \int_a^x f(t)\,dt$, multiply with $\left|x - \frac{a+b}{2}\right|^n$ $(n \ge 1)$ and integrate with respect to x on $[a,b]$

$$\int_a^b \left(x - \frac{a+b}{2}\right)^{n+1} f(x)\,dx \approx \frac{2(n+1)}{n+3}\left[\frac{1}{n+1}\left(\frac{b-a}{2}\right)^{n+1}\int_a^b f(t)\,dt \right.$$

$$\left. -\int_a^{\frac{a+b}{2}} f(x)\,dx \int_a^b \left(x-\frac{a+b}{2}\right)^n dx - \frac{1}{2} f\left(\frac{a+b}{2}\right)\int_a^b \left(x-\frac{a+b}{2}\right)^{n+1} dx\right].$$

The integral $\int_a^b \left(x - \frac{a+b}{2}\right)^k dx$ vanishes when k is odd; and when k is even,

$$\int_a^b \left(x - \frac{a+b}{2}\right)^k dx = \frac{(b-a)^{k+1}}{2^k(k+1)}.$$

Applications to Mean and Variance

In this section we provide some applications of Theorems 1 and 2 to obtain bounds for expectation and variance of a probability density function.

Let f be a probability density function on $[a,b]$. Let $E_{[a,b]}(f) := \int_a^b x f(x)\,dx$. Thus, T_1 becomes

$$\int_a^b \left(x - \frac{a+b}{2}\right) f(x)\,dx - \frac{b-a}{3}\left(\int_{\frac{a+b}{2}}^b f(x)dx - \int_a^{\frac{a+b}{2}} f(x)dx\right)$$

$$= E_{[a,b]}(f) - \frac{a+b}{2}\int_a^b f(x)\,dx - \frac{b-a}{3}\int_{\frac{a+b}{2}}^b f(x)dx + \frac{b-a}{3}\int_a^{\frac{a+b}{2}} f(x)dx$$

$$= E_{[a,b]}(f) - \int_{\frac{a+b}{2}}^b f(x)dx\left[\frac{a+b}{2} + \frac{b-a}{3}\right] - \int_a^{\frac{a+b}{2}} f(x)dx\left[\frac{a+b}{2} - \frac{b-a}{3}\right]$$

$$= E_{[a,b]}(f) - \frac{a+5b}{6}\int_{\frac{a+b}{2}}^b f(x)dx - \frac{5a+b}{6}\int_a^{\frac{a+b}{2}} f(x)dx =: T_E(f).$$

Then, we have the following results for $f : [a, b] \to \mathbb{R}$:

1. If f is of bounded variation, then

$$|T_E(f)| \leq \frac{1}{6}(b-a)^2 \bigvee_a^b (f). \tag{29}$$

2. If f is L-Lipschitz function, then

$$|T_E(f)| \leq \frac{1}{72} L(b-a)^3. \tag{30}$$

3. If f' is absolutely continuous and $f'' \in L_\infty[a, b]$, then

$$|T_E(f)| \leq \frac{1}{576} \|f''\|_\infty (b-a)^4. \tag{31}$$

4. If f' is absolutely continuous, then

$$|T_E(f)| \leq \begin{cases} \frac{1}{72} \|f'\|_\infty (b-a)^3, & \text{if } f' \in L_\infty[a, b]; \\ \dfrac{q\|f'\|_p (b-a)^{2+\frac{1}{q}}}{3(2q+1)(q+1)^{\frac{1}{q}} 2^{1+\frac{1}{q}}}, & \text{if } f' \in L_p[a, b], \ \frac{1}{p} + \frac{1}{q} = 1; \\ \frac{1}{12} \|f'\|_1 (b-a)^2, & \text{if } f' \in L_1[a, b]. \end{cases} \tag{32}$$

Consider $\delta : a = x_0 < x_1 < x_2 < \cdots < x_{n-1} < x_n = b$ and set

$$T_E^\delta(f) := E_{[a,b]}(f) - \sum_{i=0}^{n-1} \left(\frac{x_i + 5x_{i+1}}{6} \int_{\frac{x_i + x_{i+1}}{2}}^{x_{i+1}} f(x)dx + \frac{5x_i + x_{i+1}}{6} \int_{x_i}^{\frac{x_i + x_{i+1}}{2}} f(x)dx \right).$$

Write (29) for $[x_i, x_{i+1}]$, $i \in \{0, \ldots, n-1\}$ and then use the generalised triangle inequality, we get:

$$|T_E^\delta(f)| \leq \frac{1}{6} \sum_{i=0}^{n-1} h_i^2 \bigvee_{x_i}^{x_{i+1}} (f) \leq \frac{1}{6} \max_{i \in \{0,\ldots,n-1\}} h_i^2 \sum_{i=0}^{n-1} \bigvee_{x_i}^{x_{i+1}} (f) = \frac{1}{2} \Delta^2(\delta) \bigvee_a^b (f)$$

where $\Delta(\delta) := \max_{i \in \{0,\ldots,n-1\}} h_i$ and $h_i = x_{i+1} - x_i$, assuming f is of bounded variation. If f is L-Lipschitz function, then (30) becomes

$$|T_E^\delta(f)| \leq \frac{1}{72} L \sum_{i=0}^{n-1} h_i^3 \leq \frac{1}{72} L \Delta^3(\delta).$$

If f is twice differentiable and f'' is bounded, then (31) becomes

$$|T_E^{\delta}(f)| \le \frac{1}{576}\|f''\|_{\infty}\sum_{i=0}^{n-1} h_i^4 \le \frac{1}{576}\Delta^4(\delta)\|f''\|_{\infty}.$$

If f' is absolutely continuous, then (32) becomes

$$|T_E^{\delta}(f)| \le \begin{cases} \frac{1}{72}\|f'\|_{\infty}\sum_{i=0}^{n-1} h_i^3; \\ \dfrac{q\|f'\|_p \sum_{i=0}^{n-1} h_i^{2+\frac{1}{q}}}{3(2q+1)(q+1)^{\frac{1}{q}}2^{1+\frac{1}{q}}}; \\ \frac{1}{12}\|f'\|_1 \sum_{i=0}^{n-1} h_i^2; \end{cases}$$

$$\le \begin{cases} \frac{1}{72}\|f'\|_{\infty}\Delta^3(\delta), & \text{if } f' \in L_{\infty}[a,b]; \\ \dfrac{q\|f'\|_p \Delta^{2+\frac{1}{q}}(\delta)}{3(2q+1)(q+1)^{\frac{1}{q}}2^{1+\frac{1}{q}}}, & \text{if } f' \in L_p[a,b]; \\ \frac{1}{12}\|f'\|_1\Delta^2(\delta), & \text{if } f' \in L_1[a,b]; \end{cases}$$

where $p > 1$ and $\frac{1}{p} + \frac{1}{q} = 1$.

Remark 2. We note that $T_E(f)$ can be simplified as follows:

$$E_{[a,b]}(f) - \frac{a+5b}{6} + \frac{2(b-a)}{3}\int_a^{\frac{a+b}{2}} f(x)dx, \text{ or,}$$

$$E_{[a,b]}(f) - \frac{5a+b}{6} - \frac{2(b-a)}{3}\int_{\frac{a+b}{2}}^b f(x)dx;$$

in which the partition over $[a,b]$ can be halved.

Let f be a probability density function on $[a,b]$. Let $\mathrm{Var}_{[a,b]}(f) := \int_a^b x^2 f(x)\,dx - [E_{[a,b]}(f)]^2$. Thus,

$$\begin{aligned} &\int_a^b \left(x - \frac{a+b}{2}\right)^2 f(x)\,dx \\ &= \int_a^b x^2 f(x)\,dx - (a+b)\int_a^b x f(x)\,dx + \left(\frac{a+b}{2}\right)^2 \int_a^b f(x)\,dx \\ &= \mathrm{Var}_{[a,b]}(f) + [E_{[a,b]}(f)]^2 - (a+b)E_{[a,b]}(f) + \left(\frac{a+b}{2}\right)^2. \end{aligned}$$

Therefore, T_2 becomes

$$\int_a^b \left(x-\frac{a+b}{2}\right)^2 f(x)\,dx-\frac{1}{8}(b-a)^2\int_a^b f(x)\,dx+\frac{1}{24}f\left(\frac{a+b}{2}\right)(b-a)^3$$
$$= \mathrm{Var}_{[a,b]}(f) + [E_{[a,b]}(f)]^2 - (a+b)E_{[a,b]}(f) + \left(\frac{a+b}{2}\right)^2$$
$$-\frac{1}{8}(b-a)^2 + \frac{1}{24}f\left(\frac{a+b}{2}\right)(b-a)^3$$
$$= \mathrm{Var}_{[a,b]}(f) + \left[E_{[a,b]}(f) - \frac{a+b}{2}\right]^2 - \frac{1}{8}(b-a)^2 + \frac{1}{24}f\left(\frac{a+b}{2}\right)(b-a)^3$$
$$=: T_V(f).$$

Then, we have the following results:

1. If f is of bounded variation, then

$$|T_V(f)| \le \frac{1}{48}\bigvee_a^b(f)(b-a)^3. \tag{33}$$

2. If f is L-Lipschitz function, then

$$|T_V(f)| \le \frac{1}{128}L(b-a)^4. \tag{34}$$

3. If f' is absolutely continuous and $f'' \in L_\infty[a,b]$, then

$$|T_V(f)| \le \frac{1}{960}\|f''\|_\infty(b-a)^5. \tag{35}$$

4. If f' is absolutely continuous, then

$$|T_V(f)| \le \begin{cases} \frac{1}{128}\|f'\|_\infty(b-a)^4, & \text{if } f' \in L_\infty[a,b]; \\ \frac{q\|f'\|_p(b-a)^{3+\frac{1}{q}}}{(3q+1)(q+1)^{\frac{1}{q}}2^{3+\frac{1}{q}}}, & \text{if } f' \in L_p[a,b],\ \frac{1}{p}+\frac{1}{q}=1; \\ \frac{1}{24}\|f'\|_1(b-a)^3, & \text{if } f' \in L_1[a,b]. \end{cases} \tag{36}$$

Consider $\delta : a = x_0 < x_1 < x_2 < \cdots < x_{n-1} < x_n = b$ and set

$$T_V^{\delta}(f) := \mathrm{Var}_{[a,b]}(f) + \sum_{i=1}^{n-1}\left[\left(E_{[x_i,x_{i+1}]}(f) - \frac{x_i + x_{i+1}}{2}\right)^2 - \frac{1}{8}(x_{i+1} - x_i)^2 + \frac{1}{24} f\left(\frac{x_i + x_{i+1}}{2}\right)(x_{i+1} - x_i)^3\right].$$

Write (33) for $[x_i, x_{i+1}]$, $i \in \{0, \ldots, n-1\}$ and then use the generalised triangle inequality, we get:

$$|T_V^{\delta}(f)| \leq \frac{1}{48}\sum_{i=0}^{n-1} h_i^3 \bigvee_{x_i}^{x_{i+1}}(f) \leq \frac{1}{48}\max_{i\in\{0,\ldots,n-1\}} h_i^3 \sum_{i=0}^{n-1}\bigvee_{x_i}^{x_{i+1}}(f) = \frac{1}{2}\Delta^3(\delta)\bigvee_a^b(f)$$

where $\Delta(\delta) := \max_{i\in\{0,\ldots,n-1\}} h_i$ and $h_i = x_{i+1} - x_i$, assuming f is of bounded variation. If f is L-Lipschitz function, then (34) becomes

$$|T_V^{\delta}(f)| \leq \frac{1}{128} L \sum_{i=1}^{n-1} h_i^4 \leq \frac{1}{128} L\Delta^4(\delta).$$

If f is twice differentiable and f'' is bounded, then (35) becomes

$$|T_V^{\delta}(f)| \leq \frac{1}{960}\|f''\|_\infty \sum_{i=1}^{n-1} h_i^5 \leq \frac{1}{960}\|f''\|_\infty \Delta^5(\delta).$$

If f' is absolutely continuous, then (36) becomes

$$|T_V^{\delta}(f)| \leq \begin{cases} \dfrac{1}{128}\|f'\|_\infty \displaystyle\sum_{i=1}^{n-1} h_i^4; \\ \dfrac{q\|f'\|_p \displaystyle\sum_{i=1}^{n-1} h_i^{3+\frac{1}{q}}}{(3q+1)(q+1)^{\frac{1}{q}} 2^{3+\frac{1}{q}}}; \\ \dfrac{1}{24}\|f'\|_1 \displaystyle\sum_{i=1}^{n-1} h_i^3; \end{cases}$$

$$\leq \begin{cases} \dfrac{1}{128}\|f'\|_\infty \Delta^4(\delta), & \text{if } f' \in L_\infty[a,b]; \\ \dfrac{q\|f'\|_p \Delta^{3+\frac{1}{q}}(\delta)}{(3q+1)(q+1)^{\frac{1}{q}} 2^{3+\frac{1}{q}}}, & \text{if } f' \in L_p[a,b]; \\ \dfrac{1}{24}\|f'\|_1 \Delta^3(\delta), & \text{if } f' \in L_1[a,b] \end{cases}$$

where $p > 1$ and $\frac{1}{p} + \frac{1}{q} = 1$.

References

1. Barnett, N.S., Cerone P., and Dragomir, S.S.: Inequalities for distributions on a finite interval. Advances in Mathematical Inequalities Series. Nova Science Publishers Inc., New York (2008)
2. Brnetić, I. and Pečarić, J.: On an Ostrowski type inequality for a random variable. Math. Inequal. Appl. **3**, No. 1, 143–145 (2000)
3. Cacoullos, T. Variance inequalities, characterizations, and a simple proof of the central limit theorem. Data analysis and statistical inference, 27–32 (1992)
4. Cacoullos, T. and Papadatos, N. and Papathanasiou, V.: Variance inequalities for covariance kernels and applications to central limit theorems. Teor. Veroyatnost. i Primenen. **42**, No. 1, 195–201 (1997)
5. Cerone P., and Dragomir S.S.: Trapezoid type rules from an inequalities point of view. Handbook of Analytic Computational Methods in Applied Mathematics. Anastassiou G. (Ed.), pp. 65–134. CRC Press, N.Y. (2000)
6. Chang, W-Y. and Richards, D.St.P.: Variance inequalities for functions of multivariate random variables. Contemp. Math. **234**, 43–67 (1999)
7. Chernoff, H.: A note on an inequality involving the normal distribution. Ann. Probab. **9**, No. 3, 533–535 (1981)
8. Dharmadhikari, S. W. and Joag-Dev, K.: Upper bounds for the variances of certain random variables. Comm. Statist. Theory Methods **18**, No. 9, 3235–3247 (1989)
9. Houdré, C. and Kagan, A.: Variance inequalities for functions of Gaussian variables. J. Theoret. Probab. **8**, No. 1, 23–30 (1995)
10. Prakasa Rao, B. L. S. and Sreehari, M.: Chernoff-type inequality and variance bounds. J. Statist. Plann. Inference **63**, No. 2, 325–335 (1997)

Correlated Phenomena in Wireless Communications: A Copula Approach

S.N. Livieratos, A. Voulkidis, G.E. Chatzarakis, and P.G. Cottis

Abstract Copulas are multivariate joint distributions of random variables with uniform marginal distributions. A quite interesting topic in statistical modelling is how the inefficiencies, appearing when the classical linear (Pearson) correlation coefficient is employed, can be overcome. Copulas are increasingly being involved to address such challenges. In the present article, the concept of copulas is employed in the framework of wireless communications and is related to multivariate correlated fading phenomena as well as to the relevant fade mitigation techniques. The multivariate copula-based models employed in the present work are general and can be customized to any continuous multivariate random variables.

Keywords Wireless fades • Copulas • Fade mitigation techniques • Multipath fading • Rain attenuation

Introduction

Atmospheric phenomena such as reflection, diffraction, and scattering adversely affect the performance of wireless communication systems as they pose severe limitations to wave propagation. As a result, signal transmission may be severely impaired by the existence of various obstacles such as buildings, mountains, or foliage or due to precipitation mechanisms such as rainfall. Moreover, interference

S.N. Livieratos (✉) • G.E. Chatzarakis
Department of Electrical and Electronic Engineering Educators, School of Pedagogical and Technological Education (ASPETE), 14121, N. Heraklio, Athens, Greece
e-mail: slss@otenet.gr; slivieratos@aspete.gr; geaxatz@otenet.gr; gea.xatz@aspete.gr

A. Voulkidis • P.G. Cottis
School of Electrical and Computer Engineering, National Technical University of Athens, 9 Iroon Polytechniou Street, 15780 Zografou, Athens, Greece
e-mail: avoulk@gmail.com; pcottis@central.ntua.gr

N.J. Daras (ed.), *Applications of Mathematics and Informatics in Science and Engineering*, Springer Optimization and Its Applications 91, DOI 10.1007/978-3-319-04720-1_6,

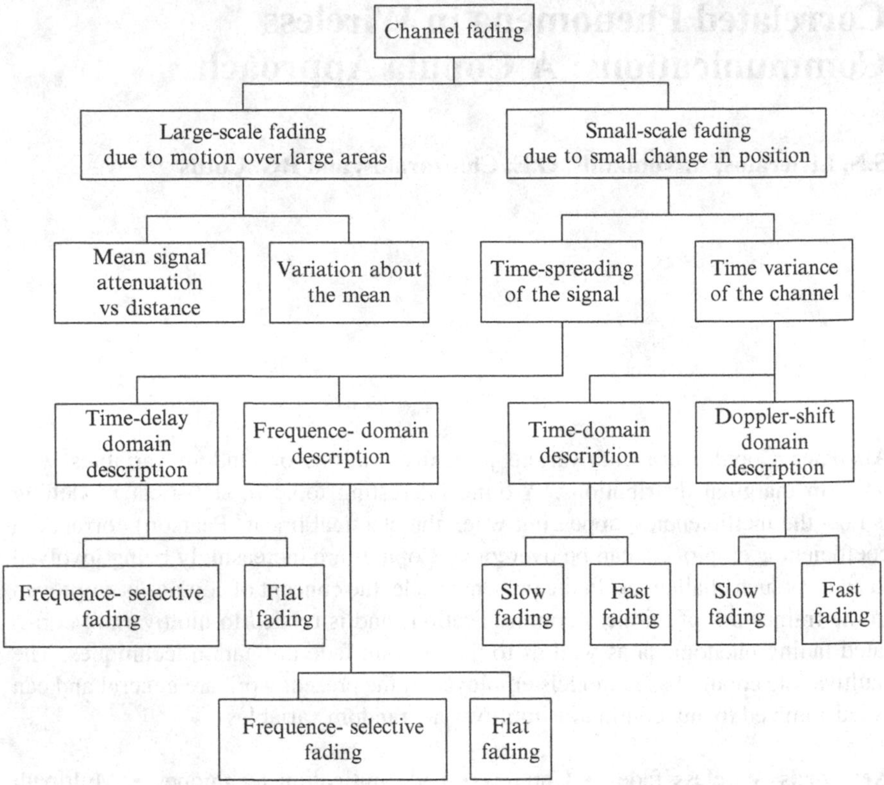

Fig. 1 Indicative fading classification

can further aggravate signal transmission. In general, the wireless channel characteristics are non-stationary and non-predictable and are subjected to fading normally perceived as signal attenuation.

The various types of fading associated with specific fading mechanisms can be classified into two main categories: large-scale fading and small-scale fading [1]. Large-scale fading is dependent on the distance between transmitter and receiver, whereas small-scale fading is caused by small changes in position (of the order of half wavelength) or by changes in the transmission environment (surrounding objects, moving obstacles crossing the line of sight (LOS) between transmitter and receiver, etc.). Likewise, fading can be classified with regard either to its duration or to where it happens (outdoor or indoor). A high-level overview of the various fading types is given in Fig. 1 [2].

To successfully model wireless channels, accurate wave propagation models are required. Such models aim at describing the changes caused to the transmitted waves as they propagate from the transmitter to the receiver and suffer from path loss, interference, noise, and various types of fading. In practical wireless communication systems, the channel state cannot be estimated at the receiver perfectly. Therefore, and regardless of the fading mechanism involved, it is essential to examine the effect of channel estimation errors, on the structure and performance of the receivers by analyzing their performance over correlated fading channels. Indicative examples of fading mechanisms that give rise to correlated microwave terrestrial or satellite transmissions are: (i) fading due to rain attenuation induced on spatially diversified links or (ii) multipath fading. Both the above types of fading are mitigated by well-known diversity combining techniques such as maximal ratio combining and optimal combining. The performance of those diversity techniques deteriorates due to imperfect channel estimation. Hence, to formulate realistic radio channel models, appropriate statistical propagation models are required.

The performance assessment of wireless links in the presence of correlated fading should employ multivariate distributions that represent the joint statistics of different fading mechanisms. In most cases, the relevant distribution is heuristically defined, if possible. The modelling difficulties that arise are related to the nature of the physical phenomena involved and to the algebraic complexity introduced. The commonly adopted performance metrics assessing availability and reliability are the outage probability and the average bit error probability for various transmission rates and quality of service (QoS) levels.

Up to now, various multivariate fading distributions have been employed, such as the Rayleigh, Rice, and Nakagami distributions for short-term fading, caused primarily by multipath propagation, or the lognormal distribution for long-term fading, caused primarily by rain attenuation and shadowing. It should be noted, though, that, in complex propagation environments, more than one type of fadings exist simultaneously [3].

Although correlated fades rarely take large values at the same time, if such an incident happens, it happens in a highly correlated way. The classic linear (Pearson) correlation coefficient fails to appropriately model the interdependence of fading events caused by different mechanisms as their underlying correlation is not linear, particularly as to the tails of the fading distributions. The Rayleigh or the Rice fading models, being stimulated by an inherent Gaussian random process, are basically Gaussian-oriented and do not efficiently model simultaneous deep fades that are affected by an underlying interdependence. Hence, it becomes evident that the benefit expected from diversity in wireless communications, which derives from the assumption that the diversity channels employed should be as de-correlated as possible, must be reconsidered and be modelled via multivariate distributions that can effectively describe the joint statistical behavior between random variables that are not linearly correlated. In this respect, the classic Pearson correlation factor cannot properly represent their underlying interdependence. In particular the n -variate Nakagami distribution describes multipath propagation of relatively large delay time-spread, employing clusters of reflected waves [4]. This distribution incorporates as special cases the Rayleigh distribution and the one-sided Gaussian model distribution, also approximating the Rician fading distribution. Though it performs well for the main part of these distributions, the approximation fails in the tails, which is critical since bit errors or outages occur mainly during deep fades [5].

As for terrestrial or satellite links operating above 10 GHz, where rain attenuation is the primary impairing factor, double site diversity constitutes an effective fade mitigation technique. The basic assumption is that the rain attenuation affecting the (two) diversity links involves two random variables modelled via the bivariate lognormal or, in some cases, the bivariate gamma distribution. The diversity gain determined when deep rain fades occur simultaneously, particularly in tropical zones, does not match the experimental data [6]. Among other reasons, this is attributed to the nonlinear underlying correlation between the two correlated random variables representing the rain attenuation over the two earth-space paths.

Another failure when modelling fading phenomena takes place when short-term fading due to multipath modelled by the Nakagami, Rice, or Rayleigh distribution, coexists with long-term fading due to shadowing modelled by the lognormal distribution [7]. The representation of such complex propagation environments leads to complicated mathematical models inconvenient to accurately evaluate the performance of wireless links. Concluding, the development of alternative mathematical models is necessary to overcome the aforementioned inefficiencies.

Basic Theory of Copulas

One method to model correlated random variables which has recently become quite popular is copula which, in Latin, means "a link, tie or bond." Copulas introduce a bond between correlated random variables and where first employed in mathematics or statistics by Abe Sklar [8, 9]. Copulas are functions that make feasible to obtain a joint distribution having a particular correlation by combining univariate distributions. Let $\boldsymbol{X} = (X_1, X_2, \ldots, X_n)$ be a vector of n random variables modelling n correlated fades having marginal cumulative distribution functions (CDFs) F_i, $i = 1, 2, \ldots, n$, respectively. The relevant multivariate CDF is defined by

$$\boldsymbol{F}(x_1, x_2, \ldots, x_n) = \Pr\{X_1 \leq x_1, \ldots, X_n \leq x_n\} \tag{1}$$

and completely determines the correlation of random variables X_i, $i = 1, 2, \ldots, n$. As pointed out in the previous section, the analytic representation of $\boldsymbol{F}$ might be too complex, making any algebraic evaluations or even numerical estimations practically impossible, thus restricting its practical use. The multivariate Gaussian distribution has become popular because it can easily describe interrelated fades. On the other hand, it has not been proven capable of fitting real data in fading communications channels. The use of copula functions overcomes the issue of estimating multivariate CDFs. This is accomplished by splitting the relevant procedure into two steps:

i. Determine the marginal CDFs F_i, $i = 1, 2, \ldots, n$ of the particular fading phenomena involved; estimate the parameters of CDFs by fitting the available data using well-established statistical methods;
ii. Determine the correlation structure of the random variables X_i, $i = 1, 2, \ldots, n$ and select a suitable copula function.

The goal is twofold: to select the most appropriate marginal CDFs in order to fit the real data of each fading mechanism and the copula function that performs better in properly linking the various marginal CDFs and fitting the joint measurements available.

An n-dimensional copula, hereafter denoted by $\boldsymbol{C}$, is a multivariate CDF with marginals uniformly distributed in [0, 1] that possesses the following properties:

i. $\boldsymbol{C} : [0, 1]^n \to [0, 1]$;
ii. As CDFs are always increasing functions, $\boldsymbol{C}(u_1, \ldots, u_n)$ is increasing with respect to any component u_i, $i = 1, 2, \ldots, n$.
iii. $\boldsymbol{C}$ is grounded, that is $\boldsymbol{C}(u_1, \ldots, u_n) = 0$, if $u_i = 0, i = 1, 2, \ldots, n$.
iv. The marginal with respect to component u_i is obtained by $\boldsymbol{C}(1, \ldots, u_i, \ldots, 1) = u_i$, that is, by setting $u_j = 1$ for any j, $j \neq i$, and as it must be uniformly distributed.

From the above properties, it is deduced that, if $F_1, \ldots, F_n$ are univariate distribution functions, $\boldsymbol{C}(F_1(x_1), \ldots, F_n(x_n))$ is a multivariate CDF with margins $F_1, \ldots, F_n$, since $U_i = F_i(x_i)$, $i = 1, \ldots, n$, are uniformly distributed random variables. Copulas constitute a useful tool to derive and process multivariate distributions. Based on the definitions, the following relation describes the basic properties of copula functions.

$$\begin{aligned}
\boldsymbol{C}(x_1, x_2, \ldots, x_n) &= \Pr\{X_1 \leq x_1, \ldots, X_n \leq x_n\} \\
&= \Pr\{F_1(X_1) \leq F_1(x_1), \ldots, F_n(X_n) \leq F_n(x_n)\} \\
&= \Pr\{U_1 \leq u_1, \ldots, U_n \leq u_n\} = \boldsymbol{C}(u_1, \ldots, u_n) \\
&= \boldsymbol{C}(\Pr\{X_1 < x_1\}, \ldots, \Pr\{X_n < x_n\}) \qquad (2)
\end{aligned}$$

The founding theorem for copulas [8, 9] states that, given a joint multivariate distribution function and the constituent marginal distributions, a copula function exists that relates them. This theorem known as Sklar's Theorem is very important in explaining copula functions because it provides a way to analyze the correlation structure of multivariate distributions without any requirements for setting any specifications concerning the related marginal distributions such as that they must be or have the same parameters. Also, it defines how multivariate CDFs modelled by copulas are used in many practical applications. According to Sklar's Theorem, if $\boldsymbol{F}$ is an n-dimensional CDF with continuous margins F_i, $i = 1, 2, \ldots, n$, then $\boldsymbol{F}$ has the following unique copula representation (canonical decomposition):

$$\boldsymbol{F}(x_1, x_2, \ldots, x_n) = \boldsymbol{C}(F_1(x_1), \ldots, F_n(x_n)) \qquad (3)$$

From (3) it can readily be deduced that, for continuous multivariate distributions, the univariate margins can be separated from the multivariate correlation. The latter can be represented by a suitable copula function which should be formed based on statistically stable experimental data and efficient regression techniques.

An additional property of copula functions follows:

Let $\boldsymbol{F}$ be an n-dimensional CDF with continuous margins $F_1, \ldots, F_n$ and copula $\boldsymbol{C}$. Then, for any $\boldsymbol{u} = (u_1, \ldots, u_n)$ in $[0, 1]^n$ the following relation holds which can be obtained in a straightforward way:

$$\boldsymbol{C}(u_1, \ldots, u_n) = \boldsymbol{F}(F_1^{-1}(u_1), \ldots, F_n^{-1}(u_n)) \tag{4}$$

where F_i^{-1} is the generalized inverse of F_i.

From (4) it is deduced that copulas allow joining together correlated distributions when the constituent marginal distributions are deterministically known. For an n-variate joint distribution $\boldsymbol{F}$, the associated copula is a distribution function $\boldsymbol{C}$ that satisfies

$$\boldsymbol{F}(x_1, x_2, \ldots, x_n) = \boldsymbol{C}(F_1(x_1), \ldots, F_n(x_n); \vartheta) \tag{5}$$

where ϑ is the dependence parameter of the copula measuring the correlation depth between the constituent marginal distributions. The above equation can be the starting point of how copulas can be empirically applied to various problems. Although ϑ is in general, a vector of parameters, for bivariate applications it is a scalar correlation measure to be specified. Thus, a bivarate distribution is expressed in terms of the constituent marginal distributions and a function $\boldsymbol{C}$ that binds them together making use of ϑ. An essential advantage of copula functions is that the various marginal distributions involved may belong to different families. For example, a bivariate distribution might involve the normal distribution representing one random variable and the gamma distribution representing the other. Though in many cases traditional representations of multivariate distributions necessitate that all the random variables involved must have the same marginals, this is not necessary when employing copulas. In this context, the assumption of identically distributed random variables which might be an inefficient simplification in many cases is not necessary when employing copulas.

In general, copulas allow to consider marginal distributions and correlation as two separate though related issues. For many practical applications, the correlation parameter is the main prerequisite for proper estimation. The assumption of linear correlation, which can fully determine elliptic multivariate distributions, must be reconsidered when modelling non-elliptic multivariate distributions. As an example of the weakness of assuming linear correlation, consider two random variables, X which is Gaussian $N(0, 1)$ and X^2. Evidently, the knowledge of X fully determines X^2, that is, the two variables are 100% correlated. However, based on the classic definition of the Pearson correlation coefficient, it is deduced that their covariance is zero, that is

$$Cov(X, X^2) = 0 \tag{6}$$

Consequently their correlation coefficient ρ is also zero which means that they are not correlated.

It should be noted that, as copulas are multivariate distributions of uniformly distributed random variables, they may be expressed in terms of marginal probabilities (CDFs). If a copula is a product of two marginals, independence is deduced allowing the separable estimation of each marginal.

How to Use Copulas in Practice

Based on the previous sections, the question arising is how to select the copula appropriate for a specific problem involving multiple correlated random variables. Often, the choice is based on the usual criteria of familiarity, ease of use, and analytical tractability. The estimation of the marginal distribution and its parameters is not affected by the choice of the copula function used for modeling the dependence of the random variables involved. Hence, any statistical distribution that effectively fits the available experimental data could be adopted to describe the one-dimensional phenomenon whereas the parameters of the distribution can be obtained following well-known fitting/regression methods.

However, there might be some cases where the estimation of conditional measures such as the conditional mean $E(X/Y = y)$ or variance $V(X/Y = y)$ might be affected by the choice of the copula function used to model the dependence between the random variables X and Y. More precisely if X and Y are continuous random variables with distribution functions $F_X(x)$ and $F_Y(y)$ respectively, their CDF satisfies the following expression:

$$\begin{aligned} \boldsymbol{F}_{XY}(x, y) &= \boldsymbol{F}_{XY}\left[F_X^{-1}(u), F_Y^{-1}(v)\right] \\ &= \Pr\left\{X \le F_X^{-1}(u), Y \le F_Y^{-1}(v)\right\} = \boldsymbol{C}(u, v) \end{aligned} \tag{7}$$

Equation (7) shows how the copula function $\boldsymbol{C}$ bridges the marginal and the joint distribution. The existence of $\boldsymbol{C}$ is guaranteed by Sklar [8, 9]. The uniqueness of $\boldsymbol{C}$, once F_X, F_Y, and $\boldsymbol{F}_{XY}$ are defined, is ensured as long as the random variables are continuous. In many instances there may be various options for the marginals and little or no idea about the joint distribution function.

Commonly used copulas are:

(i) the Gumbel copula for extreme distributions:

$$\boldsymbol{C}_{\vartheta}^{\mathrm{Gu}}(u_1, u_2) = \exp\left[-\left((-\ln u_1)^{\vartheta} + (-\ln u_2)^{\vartheta}\right)^{1/\vartheta}\right], \quad \vartheta \in [1, \infty) \tag{8}$$

(ii) the Gaussian copula for linear correlation

$$C_R^{\mathrm{Ga}}(u) = \boldsymbol{\Phi}_R^n\left(\Phi^{-1}(u_1), \ldots, \Phi^{-1}(u_n)\right) \tag{9}$$

$$C_R^{\mathrm{Ga}}(u, v) = \int_{-\infty}^{\Phi^{-1}(u)} \int_{-\infty}^{\Phi^{-1}(v)} \frac{1}{2\pi\sqrt{1 - R_{12}^2}} \exp\left[\frac{s^2 - 2R_{12}st + t^2}{2\left(1 - R_{12}^2\right)}\right] dsdt \tag{10}$$

where R_{12} is the standard linear correlation coefficient of the corresponding bivariate normal distribution

(iii) the $\boldsymbol{t}$-copula for dependence in the tail [10]

$$C_{\nu,R}^t(u) = \boldsymbol{t}_{\nu,R}^n\left(t_\nu^{-1}(u_1), \ldots, t_\nu^{-1}(u_n)\right) \tag{11}$$

$$C_{\nu,R}^t(u, v) = \int_{-\infty}^{t_\nu^{-1}(u)} \int_{-\infty}^{t_\nu^{-1}(v)} \frac{1}{2\pi\sqrt{1 - R_{12}^2}} \left[1 + \frac{s^2 - 2R_{12}st + t^2}{\nu\left(1 - R_{12}^2\right)}\right]^{-(\nu+2)/2} dsdt \tag{12}$$

where R_{12} is simply the usual linear correlation coefficient of the corresponding bivariate $\boldsymbol{t}_\nu$ distribution if $\nu > 2$.

The Gaussian copula derives from the multivariate Gaussian distribution. Other methods of deriving copulas may employ geometry and (4). For example, for two marginal distributions, one following the beta distribution with parameters α and β, and the other following the lognormal distribution with parameters μ and σ, the following copula may be employed

$$C_\vartheta^{\mathrm{Fr}}(u_1, u_2) = -\frac{1}{\vartheta} \ln\left[1 + \frac{\left(e^{-\vartheta u_1} - 1\right)\left(e^{-\vartheta u_2} - 1\right)}{e^{-\vartheta} - 1}\right], \quad \vartheta \in \mathbb{R} \setminus \{0\} \tag{13}$$

which is a member of the Frank's family.

Upon substituting the relevant distribution functions, a new joint distribution comes up. Parameter ϑ determines the depth of correlation between the marginals.

A high-level overview of the fading problems encountered in wireless communications suggests that there are experimental data available for various types of fading (random variables) upon which prediction models should be developed and verified by other similar data, if possible. Copulas allow to build new prediction tools avoiding the use of complicated multivariate distribution functions which are rarely available and too complex. In the attempt to determine the copula function that fits real data, various methods must be followed such as minimizing particular cost functions and identifying the parameter ϑ affecting the correlation depth. The sensitivity of ϑ to the various electric or spatial characteristics of the physical problems involved is examined so that a numerical trend for its determination can be obtained. Consequently, a probabilistic prediction of the fade values is attempted for various engineering configurations of the physical problem as an estimator of the actual system performance.

Methodologically the following procedure is proposed. The first step that has to be taken is to formulate the problem each time in hand. As an example assume a MIMO system with two antennas operating over fading channels. A copula function must be formed to determine the relevant outage probability. If $SNR_i, i = 1, 2$ is the signal-to-noise ratio achieved over the i-th channel and the selection combining criterion is adopted, that is the receiver selects the channel exhibiting the maximum SNR, the statistics of system outage are given by:

$$P_{outage} = \Pr\{SNR_1(dB) \leq SNR_{\text{TH}}(dB), SNR_2(dB) \leq SNR_{\text{TH}}(dB)\} \quad (14)$$

where the threshold SNR level appearing in (14) depends on the application, transmission rate, modulation scheme, etc.

If the CDFs of each $SNR_i, i = 1, 2$, are estimated, only the joint statistics must be modelled to determine the outage probability given by (14). To initiate the copula modelling, measurement data are required for the estimation of the CDFs of the individuals $SNRs$. For this estimation there is no need to derive a closed form distribution function. It must be focused on determining the CDF sample values as accurately as possible. If specific SNR thresholds are employed, the inverse CDFs can correspond to values in [0, 1] related to the uniform random variables $U_i, i = 1, 2$, as in (2). In other words, the one-dimensional datum for each random variable (fade) does not have to be fitted by a closed form statistical distribution since

$$\Pr\{X_1 \leq x_1, \ldots, X_n \leq x_n\} = \boldsymbol{C}(\Pr\{X_1 < x_1\}, \ldots, \Pr\{X_n < x_n\}) \quad (15)$$

However, the joint statistical data have to be represented by various copula functions under specific error criteria. The most widely criterion is the least squares one giving rise to following optimization problem.

$$\text{minimize} \sum_{md} \left[P_{jom} - \boldsymbol{C}\left(P_{som\_1}, P_{som\_2}; \vartheta\right)\right]^2 \quad (16)$$

where

$$P_{som\_1} = \Pr\{SNR_1 < SNR_{\text{TH}}\} \quad (17)$$

$$P_{som\_2} = \Pr\{SNR_2 < SNR_{\text{TH}}\} \quad (18)$$

$$\boldsymbol{C}\left(P_{som\_1}, P_{som\_2}; \vartheta\right) = P_{jop} \quad (19)$$

and the acronyms **md**, **jom**, **som**, and **jop** stand for measurement data, joint outage measured, single outage measured, and joint outage predicted, respectively.

The copula family that minimizes the cost function expressed in (16) performs better and may be the copula selected to describe the phenomenon under consideration and be adopted for prediction purposes. The nature of the regression is to determine the appropriate value of the dependence factor, ϑ, for each copula family employed to minimize the cost function given by (16). Hence, after having selected the copula family, the dependence factor ϑ is also determined. It should be noted that ϑ depends on system parameters such as frequency, polarization, site separation, etc. Therefore, to generalize a copula prediction method, a plethora of measurements are required to generate a statistically stable model for ϑ, involving the spatial and electromagnetic parameters, allowing the use of the same copulas be employed in similar cases under different operational characteristics.

Conclusions

The copula approach needs to specify the marginal distribution of the random variable involved along with the copula function that correlates them. The copula function can be adjusted to take into account the measurements of the correlated constituent random variables. Employing proper correlation parameters can lead to more efficient representations of joint distributions. The copula method being advantageous in capturing correlation regardless of the marginal type is expected to be very useful when dealing with fading in wireless channels.

Acknowledgements This research has been co-financed by the European Union (European Social Fund-ESF) and Greek national funds through the Operational Program "Education and Lifelong Learning" of the National Strategic Reference Framework (NSRF)-Research Funding Program: Thales-Athens University of Economics and Business-New Methods in the Analysis of Market Competition: Oligopoly, Networks and Regulation.

References

1. Parson D., The Mobile Radio Propagation Channel, New York: Halsted Press (John Wiley & Sons, Inc.), 1992.
2. Gibson, J.D., "The Mobile Communications Handbook", 2th ed. CRC Press 2002.
3. I. M. Kostic, "Analytical approach to performance analysis for channel subject to shadowing and fading," IEE Proc., vol. 152, no. 6, pp. 821–827, Dec. 2005.
4. M. Nakagami, "The m-distribution—A general formula for intensity distribution of rapid fading," in Statistical Methods in RadioWave Propagation, W. G. Hoffman, Ed. Oxford, U. K.: Pergamon, 1960.
5. P. J. Crepeau, "Uncoded and coded performance of MFSK and DPSK in Nakagami fading channels," IEEE Trans. Commun., vol. 40, pp. 487–493, Mar. 1992.
6. S.N. Livieratos and P.G. Cottis, "Availability and Performance of single multiple site diversity satellite systems under rain fades", European Transactions on Telecommunications, Vol. 12, No 1, Jan.-Feb., pp. 55–65, 2001.
7. F. Hansen and F. I. Meno, "Mobile fading-Rayleigh and lognormal superimposed," IEEE Trans. Veh. Technol., vol. 26, no. 4, pp. 332–335, Nov. 1977.
8. Sklar A (1959) Fonctions de repartition a n dimensions et leurs marges. Publ Inst Statist Univ Paris 8:229–231.
9. Sklar A (1973) Random variables, joint distributions, and copulas. Kybernetica 9:449–460.
10. Nelsen R. R. 1999. An Introduction to Copulas. Springer, New York.

Stochastic Analysis of Cyber-Attacks

Nicholas J. Daras

Abstract After studying the stochastic process of cyber-attacks against a cyber system, we investigate the set of intermediate times between successive cyber-attacks. Then, we give basic definitions and properties on the expected number of cyber-attacks in a closed time interval and present renewal theorems, as well as a description of exact and asymptotic probability distributions for the main occurrence moments of cyber-attacks. Further, we outline stationary renewal processes of cyber-attacks. The paper concludes with asymptotic properties for the counting function of cyber-attacks.

Keywords Information security • Density function • Distribution • Exponential distribution

Mathematics Subject Classification (2010): 60G07, 60K05

Introduction

Cyber warfare has recently become of increasing importance to the military, the intelligence community, and the business world. The task of securing applications against cyber-attacks is one of the most urgent for now. As an immediate consequence, there are a growing number of scientific papers devoted to cyber-attack's study [1–8].

N.J. Daras (✉)
Department of Mathematics and Engineering Sciences, Hellenic Military Academy, 16673, Vari, Greece
e-mail: darasn@sse.gr; njdaras@gmail.com

N.J. Daras (ed.), *Applications of Mathematics and Informatics in Science and Engineering*, Springer Optimization and Its Applications 91, DOI 10.1007/978-3-319-04720-1_7,

Of course, the main challenge in analysis and identification of cyber-attacks is the asynchronous nature of the problem. However, for obvious reasons of better organization, any cyber system $\mathbb{E}$ must be absolutely aware of *the expected number of cyber-attacks in a closed time interval*, as well as *the most likely occurrence moments for cyber-attacks*. Further, for justifiable reasons of effectiveness, the cyber system must be aware of *the time length between successive cyber-attacks* in order to attempt to achieve its plans within the time which elapses between two successive cyber-attacks.

This paper answers these questions and deals with a short stochastic and renewal analysis of cyber-attacks. Specifically, in section "The Stochastic Process of Cyber-Attacks," we consider the stochastic variable $A(t)$ $(t \geq 0)$ denoting the *total number of cyber-attacks* against a given cyber system $\mathbb{E}$ up to the moment time t from the beginning 0 of a surveillance period. The first main result shows that *the distribution function for the associated cyber-attacks process is a Poisson distribution* (Theorem 1). Next, in section "Intermediate and Waiting Times in a Process of Cyber-Attacks," we consider intermediate times between successive cyber-attacks, and we prove that *if W_ν is the waiting time up to the νth cyber-attack, then the set $(W_\nu : \nu = 1, 2, \ldots)$ forms a stochastic process with probabilities given by an Erlang distribution* (Theorem 2). Moreover, it is showed that *the successive intermediate times between successive cyber-attacks against $\mathbb{E}$ are independent and equidistributed random variables, with common exponential density function* (Theorem 3). The randomness character for a stochastic process of cyber-attacks against the cyber system $\mathbb{E}$ is stated in Theorems 4 and 5 and some reasonable generalizations are given in section "Generalizations for the Stochastic Process of Cyber-Attacks." Section "The Renewal Counting Function of Cyber Attacks: Definitions and Properties" deals with renewal processes of successive cyber-attacks without exponential distribution. Now, $A(t)$ is the *renewal counting function of cyber-attacks against* $\mathbb{E}$. In sections "The Renewal Counting Function of Cyber Attacks: Definitions and Properties" and "Renewal Theorems on Cyber-Attacks," we give basic definitions and renewal properties on the expected number of cyber-attacks in any time interval $[0, t]$. Section "Exact and Asymptotic Distributions for the Occurrence Moments of Cyber-Attacks" describes the exact and asymptotic distributions for the *next cyber-attack occurrence moment*, the *preceding cyber-attack occurrence moment*, and the *successive cyber-attacks' occurrence moments*. Finally, section "Stationary Renewal Processes of Cyber-Attacks" is devoted to *stationary* renewal process of cyber-attacks, while Sect. 10 formulates some asymptotic properties for the counting function $A(t)$.

The Stochastic Process of Cyber-Attacks

Let $\mathbb{E}$ be a cyber system exposed to serious threats of cyber warfare. Let also $A(t)$ be a stochastic variable denoting the total number of *cyber-attacks against* $\mathbb{E}$ up to the moment time t from the beginning $t_0 = 0$ of a surveillance period.

In what follows, we will always assume that

(i) *The number $A(0)$ of $\mathbb{E}$'s cyber-attacks at the beginning $t_0 = 0$ is equal to zero.*
(ii) *For any $h > 0$, the differences $A(t) - A(s)$* and *$A(t+h) - A(s+h)$ are equidistributed events* (we say that *$\mathbb{E}$'s cyber-attacks have stationary increases*).
(iii) *For any $\nu > 2$ and any $t_1 < t_2 < \ldots < t_\nu$, the differences $A(t_2) - A(t_1)$, $A(t_3) - A(t_2), \ldots, A(t_\nu) - A(t_{\nu-1})$ are independent events* (we say that *$\mathbb{E}$'s cyber-attacks have independent increases*).
(iv) *There is at most one attack occurring into any infinitesimal time interval $(t, t+h)$.*

Definition 1. The random variable $A(t)$ with Properties (i), (ii), (iii), and (iv) is said to be a stochastic process of cyber-attacks for $\mathbb{E}$. (*For a concise and lucid introduction to simple stochastic processes, the reader is referred to [9]*).

We will first make some comments.

Remark 1. Property (ii) has the following mathematical interpretation: $P(A(t+s) - A(s) = k) = a_k(t)$ *for any* $k = 0, 1, \ldots$ *and* $t \geq 0, s \geq 0$. As usually, $P(A(s) = k)$ denotes the probability that exactly k cyber-attacks against $\mathbb{E}$ arise from the beginning of a specified time period.

Remark 2. Analogously, Property (iv) has the following mathematical interpretation: *there is a constant $\lambda > 0$ such that $a_1(h) = P(A(t+h) - A(t) = 1) = \lambda h + o(h)$ and $a_0(h) = P(A(t+h) - A(t) = 0) = 1 - \lambda h + o(h)$, where the "remainder" $o(h)$ is sufficiently small, in the sense that* $\lim_{h \to 0}(o(h)/h) = 0$. Notice that the above two relationships on $a_1(h)$ and $a_0(h)$ imply that the *probability of the event "there are more than one cyber-attacks occurring in the time interval $(t, t+h)$" is equal to $a_k(h) = o(h), k > 1$.*

The exact meaning of the second relationship in Remark 2 is $\lim_{h \to 0^+} a_1(h) = \lambda$ and, in this connection, we give the following definition for the parameter λ which can also be viewed as a *speed of cyber-attacks' occurrence*.

Definition 2. The parameter λ is the stochastic process intensity of cyber-attacks against $\mathbb{E}$.

Using the fact that $A(t)$ has independent increases, we obtain the next result.

Theorem 1. *The distribution function for a process of cyber-attacks against $\mathbb{E}$ with intensity λ is a Poisson distribution with parameter λ:*

$$P_k(t) := P(A(t) = k) = e^{-\lambda t}\frac{(\lambda t)^k}{k!}, k = 0, 1, 2, \ldots$$

Proof. From the fact that $A(t)$ has independent increases, and the *total probability theorem*, it follows that

$P_0(t+h) = P_0(t)P_0(h)$ and

$$P_\nu(t+h) = \sum_{k=0}^{\nu} P_{\nu-k}(t)\, P_k(h), \nu = 1, 2, \ldots.$$

So, by Remark 2,

$$\frac{P_0(t+h) - P_0(t)}{h} = -\lambda P_0(t) + o(h)$$

and

$$P_\nu(t+h) - P_\nu(t) = P_\nu(t)\,[P_0(h) - 1] + P_{\nu-1}(t)\, P_1(h) +$$

$$\sum_{j=2}^{\nu} P_{\nu-j}(t)\, P_j(h) = -\lambda P_\nu(t)\, h + P_{\nu-1}(t)\, P_1(h) + o(h),$$

respectively. The last equation is a consequence of the relationship $a_k(h) = \mathrm{o}(h)$ whenever $k > 1$, since

$$\sum_{k=2}^{\nu} P_{\nu-k}(t)\, P_k(h) = \sum_{k=2}^{\nu} P_k(h) = \sum_{k=2}^{\nu} a(h) = o(h).$$

Letting $h \to 0$, and using one more time the relationship $a_k(h) = \mathrm{o}(h)$ whenever $k > 1$, we see that the probabilities $P_\nu(t)$ satisfy the differential equations $(dP_0/dt)(t) = -\lambda P_0(t)$ and $(dP_\nu/dt)(t)(t) = -\lambda P_\nu(t) + P_{\nu-1}(t)$, $\nu = 1, 2, \ldots$, with initial conditions $P_0(0) = 1$ and $P_\nu(0) = 0$, $\nu = 1, 2, \ldots$ Hence, $P_0(t) = e^{-\lambda t}$. Substituting $P_0(t)$ into the equation $(dP_1(t)/dt) = -\lambda P_1(t) + P_0(t)$, and solving with respect to $P_1(t)$, we obtain $P_1(t) = \lambda t e^{-\lambda t}$. Continuing similarly, we prove the desired assertion. □

Remark 3. We are often limited to stochastic processes of cyber-attacks which are *homogeneous* with respect to the time, in the sense that the transition probabilities $p_{i,\,j}(t) := P(A(t+s) = j / A(t) = i)$ for any $s \geq 0$ are *stationary* (i.e., independent of $s = 0$). Since, in such a case, the infinitesimal transition probabilities $p_{i,\,j}(h) := P(A(t+h) - A(t) = j / A(t) = i)$ for "small" $h > 0$ are also stationary, an application of Property (iii) shows that the *conditional probabilities of the two events*:

1. "*(exactly) one cyber-attack occurs in time interval* $(t,\ t+h)$*," and*
2. "*no one cyber-attack occurs in time interval* $(t,\ t+h)$"

coincide with the corresponding unconditional probabilities. So, from Remark 2, it follows that $p_{i,\,1}(h) = a_1(h)$, $p_{i,\,0}(h) = a_0(h)$, and $p_{i,\,j}(h) = \mathrm{o}(h)$ whenever $j > 1$. These equalities reveal a "*forgetfulness*" property for the stochastic processes of cyber-attacks against $\mathbb{E}$: *after any cyber-attack occurrence in time, the process* $\{A(t) : t > \tau\}$ *is independent of the process* $\{A(t) : t = \tau\}$.

Intermediate and Waiting Times in a Process of Cyber-Attacks

In this section we are not only interested for the value $A(t)$, but also for the moments $t_1, t_2, t_3, \ldots$ of occurrence of successive attacks.

Definition 3. The **intermediate times** $T_1, T_2, T_3, \ldots$ between successive occurrences of cyber-attacks against $\mathbb{E}$ can be defined by using the waiting times W_ν up to the νth attack as follows

$$T_1 = W_1,\ T_2 = W_2 - W_1, \ldots,\ T_\nu = W_\nu - W_{\nu-1}, \ldots$$

Conversely, the waiting time W_ν up to the νth cyber-attack occurrence can be defined in terms of the intermediate times:

$$W_1 = T_1,\ W_2 = T_1 + T_2, \ldots,\ W_\nu = T_1 + T_2 + \cdots + T_\nu, \ldots$$

Theorem 2. *The set $(W_\nu : \nu = 1, 2, \ldots)$ is an Erlang distribution with form parameter ν and scalar parameter λ:*

$$F_\nu(t) = 1 - \sum_{k=0}^{\nu-1} e^{-\lambda t} \frac{(\lambda t)^k}{k!}$$

and density

$$f_\nu(t) = \lambda e^{-\lambda t} \frac{(\lambda t)^{\nu-1}}{(\nu-1)!},\ t > 0.$$

Proof. Evidently, for any $\nu = 1, 2, \ldots.,$; we have $W_\nu > t$ if and only if $A(t) < \nu$. Thus

$$P(W_\nu > t) = P(A(t) < v) = \sum_{k=0}^{\nu-1} P_k(t).$$

Combination with Theorem 1 proves the first assertion. As for the proof of the second statement, observe that

$$f_\nu(t)\,dt = P(\text{“}\nu - 1 \text{ cyber–attacks } \textit{occurred in the time interval } (0;;\ t) \textit{ and}$$

$$\textit{the } \nu th \text{ cyber–attack } \textit{occurs in the interval } (t,\ t + dt)\text{”}).$$

Since, by Property (iii), cyber-attacks in time intervals $(0, t)$ and $(t, t + dt)$ are independent events, we write

$$\begin{aligned} f_\nu(t)dt &= P(A(t)) \\ &= (\nu - 1)\, P((\text{“one cyber-attack occurs in the time interval } (t, t + dt)\text{”}). \end{aligned}$$

Hence, by Theorem 1

$$f_\nu(t)\,dt = \lambda e^{-\lambda t}\frac{(\lambda t)^{\nu-1}}{(\nu-1)!}\,dt.$$

□

Remark 4. From Theorem 2, it follows that the time W_1 of waiting up to the first cyber-attack occurrence is exponentially distributed with density $f(t) = \lambda e^{-\lambda t}$, $t > 0$. Further, W_1 is independent of the origin time.

The same exponential law rules the stochastic behavior of the intermediate times $T_1,\ T_2,\ \ldots$ between successive cyber-attacks.

Theorem 3. *The successive cyber-attacks' intermediate occurrence times $T_1,\ T_2, \ldots$ are independent and equidistributed random variables, with common exponential density*

$$f(t) = \lambda e^{-\lambda t}, t > 0.$$

Proof. It suffices to prove that the joint density of $T_1,\ T_2,\ T_3, \ldots, T_\nu$ is given by the product

$$g(t_1,\ t_2,\ t_3, \ldots, t_\nu) = \prod_{i=1}^{\nu}\left(\lambda e^{-\lambda t_i}\right), t_i > 0.$$

To do so, note that if $f(w_1,\ w_2,\ w_3, \ldots, w_\nu)$ is the joint density of the waiting times $w_1 < w_2 < \cdots < w_\nu$, then

$$g(t_1,\ t_2,\ t_3, \ldots, t_\nu) = \frac{f(w_1,\ w_2,\ w_3, \ldots, w_\nu)}{|\det J(w_1,\ w_2,\ w_3, \ldots, w_\nu)|};$$

where $\det J(w_1,\ w_2,\ w_3, \ldots, w_\nu)$ is the Jacobian of the transformation

$$t_i = w_i - w_{i-1}; i = 1, 2, \ldots, \nu\ (w_0 = 0).$$

Since $\det J(w_1,\ w_2,\ w_3, \ldots, w_\nu) = 1$, we get

$$g(t_1,\ t_2,\ t_3, \ldots, t_\nu) = f(w_1,\ w_2,\ w_3, \ldots, w_\nu).$$

Now, to determine f, we consider the following two events:

$$K_i := \text{“}no\ \text{cyber– attack}\ occurs\ into\,(w_{i-1}, w_i)\text{”}$$

and

$$M_i := \text{“}one\ \text{cyber– attack}\ occurs\ into\,(w_i, w_i + dw_i)\text{.”}$$

Since the events $K_1, M_1, \ldots, K_\nu, M_\nu$ are independent, we have

$$f(w_1, w_2, \ldots, w_\nu)\, dw_1 dw_2 \ldots dw_\nu = P(K_1)\, P(M_1) \ldots P(K_\nu)\, P(M_\nu).$$

But, on the other hand, from the fact that $A(t)$ has stationary increases and from Remark 2, it follows that

$$P(K_i) = P(A(w_i) - A(w_{i-1}) = 0) = P(A(w_i - w_{i-1}) = 0)$$

and

$$P(M_i) = \lambda dw_0.$$

By *Theorem 1*, the desired conclusion follows. □

Remark 5. The converse of *Theorem 3* is also valid.

The Randomness Character of a Stochastic Process of Cyber-Attacks

Theorem 4. *The conditional distribution of the occurrence moments $t_1 < t_2 < \cdots < t_\nu$ for the first ν successive cyber-attacks (assuming that $A(t) = \nu$) is exactly the same as the distribution of an ordered random sample of ν uniformly distributed observations on the time interval $[0, t]$. In other words, the corresponding non-ordered moments $t'_1 < t'_2 < \cdots < t'_\nu$ of occurrence of cyber-attacks constitute a uniformly distributed random sample in the time interval $[0, t]$.*

Proof. Let

1. $f(t_1, t_2, \ldots, t_\nu)$ be the joint density of the ordered moments $t_1 < t_2 < \cdots < t_\nu$,
2. $\in_\nu$ be the event: "*(exactly) ν cyber-attacks occur in the time interval* $[0, t]$," and
3. $\notin_\nu$ be the event: "*no cyber-attack occurs in the time interval* $[t_\nu, t]$."

Then

$$f(t_1, t_2, \ldots, t_\nu / \in_\nu) = \begin{cases} \frac{f(t_1, t_2, \ldots, t_\nu) P(\notin_\nu)}{P(\in_\nu)}, & \text{when } 0 < t_1 < t_2 < \cdots < t_\nu < t \\ 0, & \text{otherwise.} \end{cases}$$

Since the moments $t_1, t_2, \ldots, t_\nu$ coincide with the waiting times $w_1, w_2, \ldots, w_\nu$ $(:t_i = w_i)$, we have

$$f(t_1, t_2, \ldots, t_\nu) = \prod_{i=1}^{n} \left(\lambda\, e^{-\lambda\, t_i}\right).$$

Further, $P\left(\notin_{\nu}\right) = P\left(A\left(t - t_{\nu}\right) = 0\right) = e^{-\lambda\,(t-t_{\nu})}$ and

$$P\left(\in_{\nu}\right) = e^{-\lambda\,t}\,\frac{(\lambda\,t)^{\nu}}{\nu\,!}.$$

Hence,

$$f\left(t_1, t_2, \ldots, t_{\nu}/\in_{\nu}\right) = \begin{cases} \frac{\nu!}{t^{\nu}}, \text{ when } 0 < t_1 < t_2 < \cdots < t_{\nu} < t \\ 0, \text{ otherwise} \end{cases}.$$

The corresponding conditional density of $t'_1, t'_2, \ldots, t'_{\nu}$ is

$$\frac{f\left(\ldots/\in_{\nu}\right)}{\nu\,!}.$$

□

Besides Theorem 4, the randomness ruling the cyber-attacks' appearance brings another consequence: *If we know the total number of cyber-attacks against* $\mathbb{E}$ *in a time interval, then the number of cyber-attack occurrences in a subinterval depends only on the length of this subinterval and is unfolded according to the pattern of the Bernoulli distribution.*

Theorem 5. *If* $\{A(t)\ :\ t \geq 0\}$ *is the stochastic process of cyber-attacks against* $\mathbb{E}$, *then, for every* $0 < s \leq t$ *and* $k \leq \nu$, *the conditional distribution of* $A(s)$, *assuming that* $A(t) = \nu$, *is the binomial distribution*

$$b\left(\frac{k}{\nu}, p\right)$$

with $p = s/t$. *In other words, whenever* $0 < s \leq t$ *and* $k \leq \nu$, *it holds*

$$P(A(s) = k/A(t) = \nu) = \binom{\nu}{k}\left(\frac{s}{t}\right)^{k}\left(1 - \frac{s}{t}\right)^{\nu-k}.$$

Proof. It is clear that

$$P\left(A\left(s\right) = k/A\left(t\right) = \nu\right) = \frac{P\left(A\left(s\right) = k,\ A\left(t\right) = \nu\right)}{P\ \left(A\left(t\right) = \nu\right)}$$

$$= \frac{P\left(A\left(s\right) = k,\ A\left(t\right) - A\left(s\right) = \nu - k\right)}{P\ \left(A\left(t\right) = \nu\right)} = \frac{P\left(A\left(s\right) = k\right) P\left(A\left(t - s\right) = \nu - k\right)}{P\ \left(A\left(t\right) = \nu\right)}.$$

The last equality is a consequence of Properties (ii) and (iii) (section "The Stochastic Process of Cyber-Attacks"). The desired assertion follows now from Theorem 2.

□

Remark 6. We have the following generalization. If $\{A(t) : t \geq 0\}$ is the cyber-attacks' stochastic process, then, whenever $0 < s_1 < s_2 < \cdots < s_k < t$ and $\nu_i < \nu$ $(i = 1, 2, \ldots, k)$, the conditional distribution of $A(s_1), \ldots, A(s_k)$, given that $A(t) = \nu$, is the polynomial distribution

$$P\left(A(s_1) = \nu_1, \ldots, A(s_k) = \nu_k / A(t) = \nu\right) = \nu! \prod_{i=1}^{k+1} \frac{p_i^{\nu_i'}}{\nu_i'!}$$

where we have used the notation

$$p_i := \frac{s_i - s_{i-1} - 1}{t}, i = 1, 2, \ldots, k \ (s_0 := 0),$$

$$p_{k+1} := 1 - (p_1 + p_2 + \cdots + p_k)$$

and

$$\nu_i' := \nu_i - \nu_{i-1}, i = 1, 2, \ldots, k \ (\nu_0 := 0)$$

$$\nu_{k+1}' := \nu - \left(\nu_1' + \nu_2' + \cdots + \nu_k'\right).$$

Remark 7. If $\{A_1(t) : t \geq 0\}$ and $\{A_2(t) : t \geq 0\}$ are two independent processes of cyber-attacks against $\mathbb{E}$, then, whenever $k = \nu$, the conditional distribution of $A_1(s)$, assuming that $A_1(t) + A_2(t) = \nu$, is the binomial distribution

$$b\left(\frac{k}{\nu}, p\right),$$

with $p := \lambda_1/(\lambda_1 + \lambda_2)$. λ_1 and λ_2 represent the intensities of the cyber-attacks' processes $\{A_1(t) : t \geq 0\}$ and $\{A_2(t) : t \geq 0\}$, respectively. In other words, for any $k = \nu$ we have

$$P\left(A_1(s) = k / A_1(t) + A_2(t) = \nu\right) = \binom{\nu}{k} \left(\frac{\lambda_1}{\lambda_1 + \lambda_2}\right)^k \left(\frac{\lambda_2}{\lambda_1 + \lambda_2}\right)^{\nu - k}.$$

The reciprocal is also valid: If $\{A_1(t) : t \geq 0\}$ and $\{A_2(t) : t \geq 0\}$ are two independent integer-valued positive stochastic processes, satisfying

$$P\left(A_1(s) = k / A_1(t) + A_2(t) = \nu\right) \sim b\left(\frac{k}{\nu}, p\right)$$

then $\{A_1(t) : t \geq 0\}$ and $\{A_2(t) : t \geq 0\}$ are cyber-attacks' stochastic processes with intensities σ_μ and σ, respectively, where σ is an arbitrary positive constant and $\mu := p/(1 - p)$.

Generalizations for the Stochastic Process of Cyber-Attacks

There is a proper modification on the stochastic process intensity $\lambda > 0$ of cyber-attacks against $\mathbb{E}$. If λ depends on time t, then $\{A(t) : t \geq 0\}$ is referred to be a *non-homogeneous stochastic process of cyber-attacks against* $\mathbb{E}$.

A second generalization is the *decimation*. It results from a modification of the relationship $a_k(h) = o(h)$, in such that a way that *more than one cyber-attack occur in "small" time intervals, given at least one cyber-attack occurrence*. Under this presupposition, Properties (i), (iii), and (iv) of section "The Stochastic Process of Cyber-Attacks" can be used to show that the set $\{A(t) : t \geq 0\}$ is a stochastic process of *cyber-attack*s with stationary and independent increases. The characteristic function of this process is

$$ch_A(u) = exp(\lambda [ch(u) - 1])$$

where $ch(u)$ is the characteristic function of $\{p_k\}$:

$$ch(u) = \sum_{k=1}^{\infty} p_k \, exp(i \, ku).$$

Such a process is said to be a *generalized stochastic process of cyber-attacks against* $\mathbb{E}$. (The usual homogeneous process of cyber-attacks corresponds to the degenerated distribution at the point 1: $p_1 = 1$ and $p_2 = p_3 = \cdots = 0$.) The significance of a generalized stochastic process of cyber-attacks results from his representation as a finite sum $A(t) = \sum_{\nu=1}^{N(t)} \mathscr{X}_\nu$, where $\{N(t) : t \geq 0\}$ is a Poisson process. The $\mathscr{X}_\nu$ are independent and equidistributed random variables, independent of $N(t)$. If $\mathscr{X}_\nu$ are only independent and equidistributed random, then $\{A(t) : t \geq 0\}$ is said to be a *composite stochastic process of cyber-attacks*. In such a case, the characteristic function of $\{A(t) : t \geq 0\}$ is again ch_A, but now $ch(u) = ch_{\mathscr{X}}(u)$, where $ch_{\mathscr{X}}$ is the random characteristic function of $\mathscr{X}_\nu$.

The Renewal Counting Function of Cyber-Attacks: Definitions and Properties

A natural generalization is obtained by considering cyber-attacks against $\mathbb{E}$ with independent and equidistributed intermediate successive times, but without exponential distribution density. Now, the function $A(t)$, $t \geq 0$ will be called the *renewal counting function of cyber-attacks against* $\mathbb{E}$, while the set $\{A(t) : t \geq 0\}$ constitutes the associated *counting function of cyber-attacks*.

Definition 4. i. *The expected number* $M(t) := E(A(t))$ *of cyber-attacks in* $[0, t]$ *is called the renewal counting function of cyber-attacks against* $\mathbb{E}$.

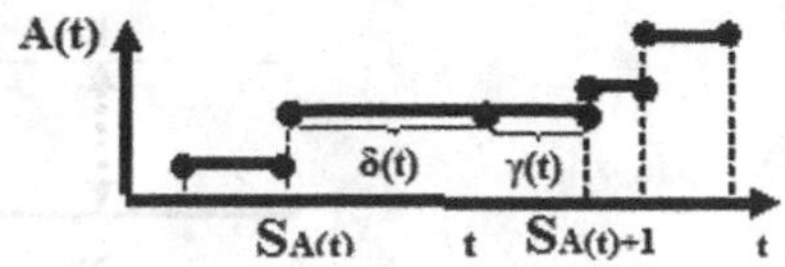

Fig. 1 Successive occurrence moments of cyber-attacks as function of time representing the occurrence moments of the preceding and next cyber-attack

ii. *The density of cyber-attacks against* $\mathbb{E}$ *at occurrence moment t is defined by*

$$\mu(t) := limsup_{\Delta t \to 0} \frac{p\,(there\ are\ cyber-attacks\ in\ (t,\ t+\Delta t))}{\Delta t}.$$

iii. *The* ν*th cyber-attack occurrence moment is the random variable*

$$S_\nu = T_1 + T_2 + \cdots + T_\nu.$$

Of course, $S_\nu = W_\nu$.

iv. *The function of time representing the occurrence moment of the next cyber-attack against* $\mathbb{E}$

$$\gamma\,(t) := \ S_{A(t)+1} - t\ ,\ t > 0\,.$$

It describes the time length between a moment t and the next cyber-attack moment (*see* Fig. 1 above).

v. *The function of time representing the occurrence moment of the preceding cyber-attack against* $\mathbb{E}$ *is the random variable*

$$\delta\,(t) := \ t - \ S_{A(t)}\ ,\ t > \ 0\,.$$

It describes the length of the time interval between a moment t and the preceding *cyber-attack* moment (*see* Fig. 1 above).

vi. *The function of time for the occurrence moments of successive cyber-attacks against* $\mathbb{E}$ *is the random variable*

$$\beta(t)\ := \gamma\,(t) + \delta\,(t)\,(= \ S_{A(t)+1} - S_{A(t)})\ ,\ t\ > \ 0\,.$$

Proposition 1. *Let* $(T_\nu\ :\ \nu = 1,\ 2,\ \ldots)$ *be the renewal process of cyber-attacks against* $\mathbb{E}$ *with distribution* $F(t) := \ P(T_\nu\ = \ t)$*. Then, the renewal counting function* $M(t)$ *of cyber-attacks is given by*

$$M\,(t) = \sum_{k=1}^{\infty} F_k\,(t) < \infty, t > 0,$$

Fig. 2 A $(t) \geq k$ if and only if $S_k \leq t$

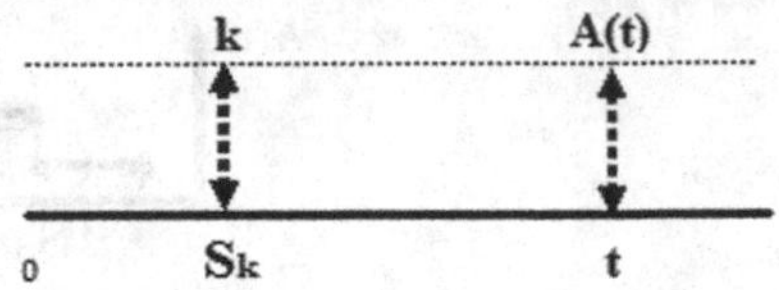

where $F_k\,(t)$ is the kth convolution of F(t) with itself:

$$F_1\,(t) := F(t),$$

$$F_2\,(t) := \int_0^t F_{2-1}\,(t-x)\,dF_1\,(x)$$

$$F_3\,(t) := \int_0^t F_{3-1}\,(t-x)\,dF_1\,(x)\left(= \int_0^t F_{3-2}\,(t-x)\,dF_2\,(x)\right)$$

$$\ldots$$

$$F_k\,(t) := \int_0^t F_{k-i}\,(t-x)\,dF_i\,(x)(1 = i = k-1)$$

$$\ldots$$

Proof. Observe that $A(t) \geq k$ if and only if $S_k \leq t$ (*see* Fig. 2 above).
Thus

$$P\,(A\,(t) = k) = P(S_k = t) = F_k\,(t)$$

for any $t > 0$ and $k = 1,\ 2, \ldots$, and therefore

$$M\,(t) = E\,(A\,(t)) = \sum_{k=0}^{\infty} P\,(A\,(t) = k) = \sum_{k=0}^{\infty} F_k\,(t)\,(t > 0).$$

It remains to show that the series

$$M\,(t) = \sum_{k=1}^{\infty} F_k\,(t)$$

converges whenever $t > 0$. To do so, we note that

$$F_k\,(t) := \int_0^t F_{k-i}\,(t-x)\,dF_i\,(x) =$$

$$F_{k-i}\,(t) \int_0^t dF_i\,(x) = \quad F_{k-i}\,(t) - F_i\,(t)\,, 1 \leq i \leq k-1.$$

It follows that

$$F_{\mu\lambda+k}(t) = [F_\lambda(t)]^\mu F_k(t), 0 \le k \le \lambda - 1,$$

which guarantees that the series $M(t) = \sum_{k=1}^{\infty} F_k(t)$ converges for any $t > 0$ satisfying $F_\lambda(t) < 1$. Since, for any $t > 0$ there is an index λ, such that $F_\lambda(t) < 1$, the proof is complete. □

The principal property of $M(t)$ is stated in the next proposition.

Proposition 2. *The renewal counting function of cyber-attacks against* $\mathbb{E}$ *fulfills the renewal convolutive equation*

$$M(t) = F(t) + F * M(t), t \ge 0.$$

Proof. The conditional value of $M(t)$, given the occurrence moment of the first cyber-attack against $\mathbb{E}$ (after which the cyber-attacks have the same stochastic behavior) is

$$M(t/x) = E(A(t)/T_1 = x) = \begin{cases} 0, \; if \; x > t \\ 1 + M(t-x), \; if \; x \le t \end{cases}.$$

In fact, if the time x of the first cyber-attack exceeds t, then there is no cyber-attack occurrence in $(0, t]$. On the contrary, if the time x is less than or equal to t, then the time interval $(0, t]$ contains surely (at least) one cyber-attack occurrence plus the number $M(t-x)$ of expected cyber-attacks occurrence moments during the time interval $(x, t]$. Hence,

$$M(t) = E(A(t)) = \int_0^\infty M(t/x) dF(x)$$

$$= \int_0^t [1 + M(t-x)] dF(x) = F(x) + \int_0^\infty M(t-x) dF(x).$$

□

We will now discuss the role of the renewal counting function for cyber-attacks in solving most of the general equations.

Definition 5. An integral equation

$$G(t) = g(t) + \int_0^t G(t-x)\; dF(x), t \ge 0$$

where $g(t)$ is a given function, $F(t)$ is a given renewal process distribution of cyber-attacks, and $G(t)$ is the unknown function, is said to be a *renewal equation of cyber-attacks against* $\mathbb{E}$.

Combination with Proposition 2 implies that

$$g(t) = F(t) \ (= \ P(T_\nu = t)).$$

In the next theorem, we shall see that *the solution of any renewal equation of cyber-attacks can be expressed with the aid of the corresponding renewal counting function M (t) of cyber-attacks.*

Theorem 6. *Let* $(T_\nu : \nu = 1, 2, \ldots)$ *be a renewal process of cyber-attacks against* $\mathbb{E}$ *with (common) general density* $f(x)$ *and (common) distribution*

$$F(t) := P(T_\nu = t) = \int_0^t f(x)\, dx, \nu = 1, 2, \ldots.$$

For any piecewise bounded function $g(t)$, *there is only one piecewise bounded function* $G(t)$ *satisfying:*

$$G(t) = g(t) + \int_0^t G(t-x)\, dM(x), t \geq 0.$$

Proof. By Proposition 1 and the convolution associative property, we have $M * g(t) = F * G(t)$. Hence, the function $G(t) = g(t) + M * g(t)$ satisfies the equation $(t) = g(t) + F * G(t)$. It remains to show that $G(t)$ is uniquely defined. To do so, assume that $H(t)$is any function, bounded on every finite time interval and satisfying $H = g + F * H$. Observe that $H = g + \left(\sum_{k=1}^{\nu-1} F_k\right) * g + F_\nu * H$ whenever $\nu = 2, 3, \ldots$ Thus, letting $\nu \to \infty$, we get $H = g + \lim_{\nu\to\infty}\left\{\left(\sum_{k=1}^{\nu-1} F_k\right) * g\right\} + lim_{\nu\to\infty}\{F_\nu * g\}$. But, Proposition 1 and g's piecewise boundedness property imply that $\lim_{\nu\to\infty}\left\{\left(\sum_{k=1}^{\nu-1} F_k\right) * g\right\} = \left(\sum_{k=1}^{\infty} F_k\right) * g = M * g$. Further, Hs piecewise boundedness property guarantees that

$$lim_{\nu\to\infty} |F_\nu * H(t)| = lim_{\nu\to\infty} \left|\int_0^t H(t-x) dF_\nu(x)\right| =$$

$$lim_{\nu\to\infty} [\{sup_{0=x=t} |H(t-x)|\} F_\nu(t)] = 0.$$

Hence $H = g + M * g$, which means that $H = G$. □

Corollary 1. *i. The expected value for the time interval length required to express the number of cyber-attacks against* $\mathbb{E}$ *up to the final moment of this interval as the sum of the number* $A(t)$ *of cyber-attacks up to the moment t plus one more attack is equal to the product of the expected value for the waiting time up to the first cyber-attack occurrence and the expected value for the sum of cyber-attacks number* $A(t)$ *up to the moment t plus one more cyber-attack occurrence:*

$$E(S_{A(t)+1}) = E(T_1) E(A(t) + 1).$$

ii. *The expected value for the time interval length required to express the number of cyber-attacks against* $\mathbb{E}$ *up to the final moment of this interval as the sum of the number* $A(t)$ *of cyber-attacks up to the moment* t *plus one more attack is equal to the product of the expected value for the waiting time up to the first cyber-attack occurrence and the sum of cyber-attacks' expected value in the time interval* $(0, t)$ *plus one more cyber-attack:*

$$E\left(S_{A(t)+1}\right) = E(T_1)\left[M(t) + 1\right].$$

Proof. Applying the renewal reasoning of Proposition 2, one can show that $G(t) = E(S_{A(t)+1})$ satisfies the following renewal equation of cyber-attacks:

$$G(t) = E(T_1) + \int_0^t G(x - t)\, dF(x).$$

If we take $g(t) = E(T_1) = constant$ in *Theorem 6*, then

$$G(t) = E(T_1) + \int_0^t E(T_1)\ dM(x) = E(T_1)\left[1 + M(t)\right].$$ □

Corollary 2. *Given any occurrence moment* t*, the expected value for the interval time length separating this moment* t *and the occurrence moment occurrence of the next cyber-attack against* $\mathbb{E}$ *is*

$$E(\gamma(t)) = E(T_1)\left[1 + M(t)\right] - t.$$

Renewal Theorems on Cyber-Attacks

We will now turn to the so-called renewal theorems on the cyber-attacks. If the expected value for the waiting time T_1 is

$$\mu := E(T_1),$$

then it is reasonable to expect a mean

$$1/\mu$$

of cyber-attacks per unit of time. The next theorem on cyber-attacks gives the asymptotic mean of cyber-attacks against $\mathbb{E}$ per unit of time.

Theorem 7 (The elementary renewal theorem for cyber-attacks). *Let* $(T_\upsilon : \upsilon = 1, 2, \ldots)$ *be a renewal process of cyber-attacks against* $\mathbb{E}$*. Then:*

$$lim_{t\to\infty}\frac{M(t)}{t} = \frac{1}{\mu}.$$

Proof. By the obvious inequality $t < S_{A(t)+1}$ and Corollary 1, we have

$$\liminf_{t\to\infty} \frac{M(t)}{t} = \frac{1}{\mu}.$$

To prove the converse inequality, we consider the *truncated renewal process* of cyber-attacks against $\mathbb{E}$ defined by

$$T_\nu^c := \begin{cases} T_\nu, \text{ if } T_\nu = c \\ c, \text{ if } T_\nu > c \end{cases}, \nu = 0, 1, 2, \ldots$$

where c is an arbitrary positive constant. If W_ν^c and $A^c(t)$ represent the "*waiting time up to the* ν*th cyber-attacks*" and the "*associated counting function of cyber-attacks*," respectively, then the uniform boundedness property of T_ν^c leads to the inequality

$$t + c = E\left(S_{A^c(t)+1}\right) = \mu_c \left[1 + M^c(t)\right],$$

where

$$\mu_c := E\left(T_\nu^c\right) = \int_0^c \left[1 - F(x)\right] dx \text{ and } M^c(t) = E\left(A^c(t)\right).$$

Since $M^c(t) = M(t)$, this inequality gives

$$\limsup_{t\to\infty} \frac{M(t)}{t} = \frac{1}{\mu_c}.$$

Since $\lim_{c\to\infty}\mu_c = \mu$, we obtain

$$\limsup_{t\to\infty} \frac{M(t)}{t} = \lim_{c\to\infty} \frac{1}{\mu_c} = \frac{1}{\mu}.$$

□

Corollary 3. *(The Blackwell's renewal theorem on cyber-attacks) As the cyber warfare's duration grows, the cyber-attacks' mean in an interval time of length h approaches the quantity* h/μ*:*

$$M(t+h) - M(t) \underset{t\to\infty}{\longrightarrow} \frac{h}{\mu}, \textit{ whenever } h > 0.$$

There are two variations of Corollary 3. The first variation will be used in the next section; the second one will be generalized in section "Marginal Theorems for the Counting Function of Cyber-Attacks." We need the following:

Definition 6. *Let $F(t) = P(T_\nu = t)$ be the distribution of a renewal process*

$$(T_\nu : \nu = 1, 2, \ldots)$$

of cyber-attacks against $\mathbb{E}$.

i. *An occurrence moment t_0 is said to be an increase moment of cyber-attacks occurrence if $F(t_0 + \varepsilon) - F(t_0 - \varepsilon) > 0$ $(\varepsilon > 0)$.*
ii. *We say that the distribution F is a graduated distribution if there is a constant $C > 0$ such that every increase occurrence moment of cyber-attacks occurrence lies in the set*

$$\{0,\ C,\ 2C,\ 3C,\ \ldots\}.$$

The largest value for this constant C is the distribution width for the occurrence moments of cyber-attacks occurrence.

Remark 8. If the cyber-attacks' occurrence moments distribution is a (piecewise) continuous function, then this distribution is not graduated. On the contrary, if we agree that the renewal process $(T_\nu\ :\ \nu = 1,\ 2,\ \ldots)$ of cyber-attacks is a discrete process, then it is clear that the corresponding distribution of cyber-attacks' moment times is graduated with integer width.

We are now in the position to state the following

Theorem 8 (The fundamental renewal theorem on cyber-attacks). *Let $F(t) = P(T_\nu = t)$ be the distribution of a renewal process $(T_\nu\ :\ \nu = 1,\ 2,\ \ldots)$ of cyber-attacks against $\mathbb{E}$. Let also $g(t)$ be a monotone absolutely integrable function.*

Let finally $G(t)$ be the unique solution of the renewal equation of cyber-attacks.

1. *If $F(t)$ is not a graduated distribution, then*

$$lim_{t\to\infty} G(t) = \frac{1}{\mu}\int_0^\infty g(x)\,dx.$$

2. *If $F(t)$ is a graduated distribution with width C, then*

$$lim_{\nu\to\infty} G(\alpha + \nu C) = \frac{C}{\mu}\sum_{k=0}^{\infty} g(\alpha + \nu C)$$

whenever $\alpha > 0$.

Let us finally give a second variation of Theorem 8:

Theorem 9. *Let $F(t) = P(T_\nu = t)$ be the distribution for a renewal process $(T_\nu : \nu = 1, 2, \ldots)$ of cyber-attacks against $\mathbb{E}$, with $\mu := E(T_1) < \infty$.*

i. If $F(t)$ is not a graduated distribution, then

$$lim_{t\to\infty}[M(t) - M(t-h)] = \frac{h}{\mu}$$

for any positive constant h.

ii. *If $F(t)$ is a graduated distribution with width C, then the same as above limiting relation holds only for positive constants h which are integer multiples of C.*

Exact and Asymptotic Distributions for the Occurrence Moments of Cyber-Attacks

The Case of the Next Cyber-Attack Occurrence Moment

For a probabilistic description of the random variable $\gamma(t)$, $t > 0$ (*see* Definition 4.iv), it is useful to consider the function $G_T(t) := P(\gamma(t) > T), T > 0$.

Definition 7. *The function $G_T(t)$ is called the exact distribution of the next cyber-attacks occurrence moments. The limit*

$$G_T := lim_{t\to\infty} G_T(t)$$

(if it exists) *is called the asymptotic distribution of the next cyber-attacks occurrence moments during a time length T.*

Remark 9. Since $\gamma(t) > T$ is equivalent to the non-occurrence of cyber-attacks against $\mathbb{E}$ during the time interval $(t, t+T)$, we can write

$$G_T(t) = P(A(t+T) - A(t) = 0).$$

Example 1. Especially for a Poisson renewal process of cyber-attacks with intensity λ and exponentially distributed lengths of intermediate time intervals between successive cyber-attacks according to the rule $F(T) = 1 - e^{-\lambda T}$, we have $P(\gamma(t) > T) = P(A(T) = 0) = e^{-\lambda T}$. This means that *in a Poisson renewal process of cyber-attacks against $\mathbb{E}$, the timing of the following attack occurrence is also an exponentially distributed random variable.* This is due to the "forgetfulness" property of the exponential distribution.

In the sequel, we will determine a renewal equation describing cyber-attacks against $\mathbb{E}$ with unique solution given by the exact distribution $G_T(t)$ of the next cyber-attacks occurrence moments. Then, we will find the asymptotic distribution for the occurrence moments of the next cyber-attacks. For this purpose, let us observe that

$$P([\gamma(t) > T]/[T_1 = x]) = \begin{cases} 1, \ when\ x > t+T \\ 0, \ when\ t+T \geq x > t \\ G_T(t-x), \ when\ t \geq x > 0. \end{cases}$$

By the *total probability theorem*, we have

$$G_T(t) = \int_0^{\infty} P([\gamma(t) > T]/[T_1 = x])\, dF(x) =$$

$$1 - F(t+T) + \int_0^{t} G_T(t-x)\, dF(x).$$

Hence, from Theorem 6, it follows that the solution of the raised renewal equation of cyber-attacks has the representation

$$G_T(t) = 1 - F(t+T) + \int_0^{t} [1 - F(t+T-x)]\, dM(x).$$

If $0 \neq E(T_1) = \mu < \infty$ (in reality this is always valid), the function $1 - F(t+T)$ is absolutely integrable and monotone. So, by Theorem 7, we infer

$$lim_{t\to\infty} G_T(t) = \frac{1}{\mu}\int_0^{\infty} [1 - F(x+T)]\, dx = \frac{1}{\mu}\int_T^{\infty} [1 - F(y)]\, dy < \infty.$$

We have thus proved the

Theorem 10. *Let* $F(t) = P(T_\nu = t)$ *be the distribution function for a renewal process* $(T_\nu : \nu = 1, 2, \ldots)$ *of cyber-attacks against* $\mathbb{E}$.

i. *The exact distribution of the next cyber-attacks occurrence moments is the function*

$$G_T(t) = 1 - F(t+T) + \int_0^{t} [1 - F(t+T-x)]\, dM(x), T > 0.$$

ii. *If* $\mu \neq 0$, *then the asymptotic probability of the next cyber-attacks occurrence moments during a time length* T *exists and is equal to*

$$G_T = \frac{1}{\mu}\int_0^{\infty} [1 - F(y)]\, dy.$$

The Occurrence Moments of Successive Cyber-Attacks

The *asymptotic probability distribution for the occurrence moments of two successive cyber-attacks against* $\mathbb{E}$ *during a time interval of length* T is the limit

$$K_T := lim_{t\to\infty} P(\beta(t) > T)$$

(if it exists). To determine K_T, it suffices to proceed by applying Theorem 8 to the renewal equation of cyber-attacks against $\mathbb{E}$:

$$K_T(t) = 1 - F(\max\{t, T\}) + \int_0^t K_T(t-y)dF(y),$$

which is satisfied by the exact probability distribution function

$$K_T(t) := P(\beta(t) > T).$$

In any case, we are leaded to the following conclusion.

Theorem 11. *If $F(t) = P(T_\nu = t)$ is the distribution function for a renewal process $(T_\nu : \nu = 1, 2, \ldots)$ of cyber-attacks against $\mathbb{E}$, then the asymptotic distribution of two successive cyber-attacks occurrence moments during a time length T equals*

$$K_T = \frac{1}{\mu}\int_T^\infty y\, dF(y).$$

Remark 10. It follows that the distribution of any two successive cyber-attacks occurrence moments during a time length T toward cyber warfare's end approaches the limit

$$L(T) := lim_{t\to\infty} P(\beta(t) = T) = \frac{1}{\mu}\int_0^T y\, dF(y).$$

The last equality guarantees that the expected time length between two successive cyber-attacks toward the end of the cyber warfare is

$$\int_0^T t\, dL(t) = \frac{1}{\mu}\int_0^\infty y^2\, dF(y).$$

Hence

$$\frac{1}{\mu}\int^\infty y^2\, dF(y) = \frac{1}{\mu}\left(\int_0^\infty y\, dF(y)\right)^2 = \mu.$$

In other words, *the expected length for the time interval between two successive cyber-attacks toward cyber warfare's end is not less than the expected time length between two successive cyber-attacks*. This is called "*the sample partiality of time lengths between successive cyber-attacks*."

Example 2. In Example 1, we have seen that, for a Poisson renewal process of cyber-attacks against $\mathbb{E}$ in a cyber warfare with intensity λ, and exponentially distributed lengths of intermediate time intervals between successive cyber-attacks according to the probability distribution $F(t) = 1 - e^{-\lambda T}$, the exact distribution for the occurrence moments of next cyber-attacks during a time length T is given by the exponential function $P(\gamma(t) > T) = e^{-\lambda t}$. In the present example, we will show that, under the same hypotheses, *the preceding cyber-attacks occurrence moments distribution is a mixed distribution with exponential continuous part and discrete part which is concentrated at $\delta(t) = t$ with probability $e^{-\lambda t}$*. In fact, it is clear that $\delta(t) = t$ and $\delta(t) > x$ $(x < t)$ if and only if the time interval $(t - x, t)$ does not contain any cyber-attack occurrence. Hence, the distribution for the occurrence moments $\delta(t)$ of preceding cyber-attacks against $\mathbb{E}$ can be written as

$$P(\delta(t) = x) = \begin{cases} 1 - e^{-\lambda x}, \ when\ 0 = x < t \\ 1, \ when\ x = t. \end{cases}$$

Further, the expected value of $\delta(t)$ is

$$E(\delta(t)) = \int_0^t P(\delta(t) > x)dx = \int_0^t e^{-\lambda x}dx = \frac{1}{\lambda}\left(1 - e^{-\lambda t}\right).$$

Since, by Example 1,

$$E(\gamma(t)) = \frac{1}{\lambda},$$

we conclude that

$$E(\beta(t)) = E(\gamma(t)) + E(\delta(t)) > \frac{1}{\lambda} = \mu.$$

Especially for large t, the value $E(\beta(t))$ becomes the double of the expected time length between successive cyber-attacks.

Remark 11. The random variables $\gamma(t)$ and $\delta(t)$ of occurrence times of the next and preceding cyber-attack, respectively, are independent with joint distribution:

$$P(\gamma(t) > x, \delta(t) > y)$$

$$= \begin{cases} e^{-\lambda(x+y)}, \ when\ x > 0\ and\ 0 < y < t \\ 0, \ when\ y \geq t. \end{cases}$$

Stationary Renewal Processes of Cyber-Attacks

Since the starting moment of a cyber warfare is not uniquely defined, the distribution of the waiting time T_1 up to the first cyber-attack may be different from the distributions of T_2, T_3,.. In such a case, we say that *the process is a modified renewal process of cyber-attacks against* $\mathbb{E}$ (or *a renewal process of cyber-attacks against* $\mathbb{E}$ *with hysteresis*). If, in particular, the density $f_1(t)$ of T_1 is of the form

$$f_1(t) = \frac{1}{\mu}[1 - F(t)],$$

where

1. $\mu = E(T_i)$ $(i = 2, 3, \ldots)$
2. $F(t) =$ the common distribution of T_2, T_3,..,

we say that the renewal process is *a stationary renewal process of* cyber-attacks against $\mathbb{E}$. Then, the distribution of T_1 is

$$F_1(t) = \frac{1}{\mu}\int_0^T [1 - F(t)]\, dt.$$

As it is pointed out in section "The Case of the Next Cyber-Attack Occurrence Moment,"

$$G_T = \frac{1}{\mu}\int_T^{\infty} [1 - F(t)]\, dt.$$

Equivalently, the limit distribution $P(\gamma(t) = T)$ for the next cyber-attacks occurrence moments during a time length T is again

$$G(T) = \frac{1}{\mu}\int_{\mu}^{T} [1 - F(t)]\, dt.$$

Thus, for a *renewal process of* cyber-attacks *against* $\mathbb{E}$, *which began from long ago, the distribution of the waiting time* T_1 *up to the first cyber-attack is equal to the limit distribution for the occurrence moments of the next* cyber-attacks. This justifies the notation "stationary renewal process of cyber-attacks against $\mathbb{E}$."

Theorem 12. *Let* $A_s(t)$ *be the counting function of a stationary renewal process of cyber-attacks against* $\mathbb{E}$.

i. The renewal counting function of cyber-attacks is

$$M_s(t) := E(A_s(t)) = \frac{t}{\mu}.$$

ii. The distribution of the next cyber-attacks occurrence moments is

$$P(\gamma_s(t) \le T) = F_1(t), T > 0.$$

Proof. **i.** As in Proposition 2, one can show that

$$M_s(t) = F_1(t) + \int_0^t F_1(t - x)\, dM(x),$$

where $M(t)$ is the renewal counting function that corresponds to the cyber-attacks' process with distribution $F(t)$. From Theorem 6, it follows that $M_s(t)$ is a solution of the integral equation

$$M_s(t) = F_1(t) + \int_0^t M_s(t-x)\, dF(x).$$

But, since this equation has unique solution, it is sufficient to show that $M_s(t) = t/\mu$ satisfies the above renewal equation. Substitution of $M_s(t)$ and $F_1(t)$ by

$$\frac{t}{\mu} \text{ and } \frac{1}{\mu}\int_0^t [1 - F(x)]\, dx$$

respectively, gives

$$\int_0^t d(xF(x)) = \int_0^t t\, dF(x)$$

which is a tautology.

ii. Similarly, one can verify that the exact distribution $G_T^{(s)}(t) = P(\gamma_s(t) > T)$ of the next cyber-attacks occurrence moments has the integral representation $G_T^{(s)}(t) = 1 - F_1(t+T) + F_1 * G_T(t)$, with $G_T(t) := P(\gamma(t) > T)$ (*see* Definition 7). By the *total probability theorem*, it holds $G_T(t) = g_T(t) + M * g_T(t)$ where $g_T(t) := 1 + F(t+T)$. Substituting into the convolutive equation and taking into account the relationship $M_s(t) = F_1 * M(t)$, we are led to the following equations

$$G_T^{(s)}(t) = 1 - F_1(t+T) + \int_0^t g_T(t-y)dM_s(y).$$

Since $M_s(y) = (y/\mu)$, we conclude that

$$G_T^{(s)}(t) = 1 - F_1(t).$$

□

Remark 12. In a stationary renewal process of cyber-attacks against $\mathbb{E}$ the asymptotic formulae of the elementary renewal theorem on cyber-attacks trivializes to an obvious identity. The same is true for the asymptotic formulae in the fundamental renewal theorem on cyber-attacks against $\mathbb{E}$, as well as for the asymptotic relationship in the variation of fundamental renewal theorem on cyber-attacks against $\mathbb{E}$.

Marginal Theorems for the Counting Function of Cyber-Attacks

We will now formulate asymptotic properties for the counting function $A(t)$ of cyber-attacks against $\mathbb{E}$ in a renewal process $(T_\nu : \nu = 1, 2, \ldots)$ where the intermediate times $T_2, T_3, ..$ between successive cyber-attacks are independent and equidistributed random variables with common distribution $F(t)$. At first such a property is given in the following

Theorem 13 (The central marginal theorem for the counting function of cyber-attacks). *If $\mu = E(T_2) < \infty$ and $\sigma^2 = Var(T_2) < \infty$, then for every $T > 0$, it holds*

$$lim_{t\to\infty} P\left(\frac{A(t) - \frac{t}{\mu}}{\sqrt{\frac{t\,\sigma^2}{\mu^3}}} \leq T\right) \leq \Phi(T),$$

where Φ is the reduced normal distribution function.

Proof. Let $T > 0$ be fixed. If

$$\frac{t - \nu\mu}{\sigma\sqrt{\nu}} \underset{\nu\to\infty,\, t\to\infty}{\longrightarrow} (-T),$$

then, an application of the *central marginal theorem* for the random variable $W_\nu = T_2 + T_3 + \cdots + T_\nu$, representing the waiting time up to the νth cyber-attack, shows that

$$lim_{\nu\to\infty,\, t\to\infty} P(S_\nu > t) = lim_{\nu\to\infty} P\left(\frac{S_\nu - \nu\mu}{\sigma\sqrt{\nu}} > -T\right) =$$

$$1 - \Phi(-T) = \Phi(T).$$

Hence, the desired assertion follows from the equivalence of the events "$S_\nu > t$" and "$A(t) < \nu$." □

Corollary 4 (The strong law of large numbers for the counting function of cyber-attacks). *If $\mu = E(T_2) < \infty$ and $\sigma^2 = Var(T_2) < \infty$, then the following holds.*

i. It is certain that the value of the ration $A(t)/t$ approaches the number $1/\mu$ as the cyber warfare's duration t increases:

$$P\left(lim_{t\to\infty}\frac{A(t)}{t} = \frac{1}{\mu}\right) = 1.$$

ii. As the cyber warfare's duration t increases, the value of the ratio

$$\frac{\Delta\left[A\left(t\right)\right]}{t} := \frac{A\left(t+h\right)-A\left(t\right)}{t}$$

approaches the number σ^2/μ^3*:*

$\lim_{t\to\infty}\frac{\Delta[A(t)]}{t} = \frac{\sigma^2}{\mu^3}$.

Remark 13. According to an equivalent formulation of Theorem 13, the counting function $A\left(t\right)$ of cyber-attacks has asymptotic normal distribution with asymptotic mean and variance as in Corollary .

References

1. S. A. Camtepe and B. Yene, Modeling and detection of complex attacks. In *Proceedings of the Third International Conference on Security and Privacy in Communications Networks and the Workshops,* 2007, IEEE Conference Publications, Nice, France, pp.234–243.
2. N. Cuppens-Boulahia, F. Cuppens, F. Autrel and H. Debar, An ontology-based approach to react to network attacks. *Int. J. of Information and Computer Security*, 3, 3(2009), 280–305.
3. G.C. Dalton, R.F. Mills, J.M. Colombi and R.A. Raines, Analyzing attack trees using generalized stochastic Petri nets. In *IEEE Information Assurance Workshop*, 116–123.
4. K. Ingols, M. Chu, R. Lippmann, S. Webster and S. Boyer, Modeling modern network attacks and countermeasures using attack graphs. In *Proceedings Computer Security Applications Conference*, (*ACSAC*), 117–126.
5. R. Hewett and P. Kijsanayothin, Host-centric model checking for network vulnerability analysis. In *Computer Security Applications Conference*, 225–234.
6. P. Maggi, D. Pozza and R. Sisto, Vulnerability modeling for the analysis of network attacks. In *Dependability of Computer Systems*, 15–22.
7. X. Ou, W. F. Boyer and M. A. McQueen, A scalable approach to attack graph generation. In *ACM Conference on Computer and Communications Security*, 336–345.
8. R. E. Sawilla and X. Ou, Identifying critical attack assets in dependency attack graphs. In the *European Symposium on Research in Computer Security* (*ESORICS*), 18–34.
9. D. Stirzaker, *Stochastic Process and Models*, Oxford University Press.

Numerical Solution of the Defence Force Optimal Positioning Problem

Nicholas J. Daras and Demetrius Triantafyllou

Abstract In this paper we study the positioning of defender forces in order to handle in an efficient way the forces of the attacker. The scope is to determine the minimum amplitude of territories which the invader will occupy. The defender's forces should swoop rapidly to any point of the defence locus in order to protect their territories. The selection of the "optimal" position in which the defender's forces should be placed is a difficult problem and it aims at the minimization of enemy's penetration. The minimization methods result to non-linear equations and there are many classical numerical algorithms for solving such equations. The most known one is Newton's method. Since the selection of a suitable initial point is not a trivial task, we will study the behaviour of these numerical procedures for various initial points and small perturbations of the data in order to present stable procedures which compute efficiently the solution of non-linear equations, leading to the optimal selection of the position, on which the forces of the defender should be placed. All the proposed methods are tested for various sets of data and useful conclusions arise. The algorithms are compared as to the computational complexity and stability through error analysis yielding useful results.

Keywords Optimal positioning of defender forces • Numerical solution of non-linear equations • Newton's algorithm

---

N.J. Daras (✉) • D. Triantafyllou
Department of Mathematics and Engineering Sciences, Hellenic Military Academy, 16673, Vari, Greece
e-mail: njdaras@gmail.com; darasn@sse.gr; dtriant@math.uoa.gr

N.J. Daras (ed.), *Applications of Mathematics and Informatics in Science and Engineering*, Springer Optimization and Its Applications 91, DOI 10.1007/978-3-319-04720-1_8,

Introduction

In military, we need to determine in advance a way that gives the proper placement of the defence forces, so as to effectively protect each critical interval on the line of the defensive front. For this purpose, it must be (entirely or even with a high degree of reliability of information) known all the movements to be performed on the part of the attacking. It is clear that basic priorities of the attacker are the following six objectives.

1. Maximizing the "controlled area of land from attacking (: territory control)".
2. Maximizing the ratio of the "controlled the land area from attacking" to "the multitude of losses of the attacking" (: attack attrition), namely the maximization of territorial gains which will have the attacking while minimizing the number of losses.
3. Maximizing the "duration of the armed conflict (: duration of the armed conflict)", i.e. the possibility of extension of the duration of the armed conflict for so long as is necessary in order to defeat the opposing defense force.
4. Maximizing the "multitude of losses of the defending (: defensive force attrition)".
5. Maximizing the ratio "number of losses of the defending" to "the multitude of losses of the attacking".
6. Maximizing the "duration control of the land from the attacking force (: territory controlled per duration)".

Of course, there are other key objectives—priorities that can be set on the part of the attacking, for example are the control of certain specific areas (: control of specified territory). It goes without saying that there exist further objectives-priorities that can be set in behalf of the attacker; for example one may refer to the attempt for controlling certain specific areas. For each combination of these basic objectives or priorities of the attacking, the defence must devise a corresponding appropriate strategy to minimize the weight of the results obtained by the efforts of the attacking.

Without any doubt, to achieve all these goals, the positioning of the defence forces before attacker's penetration is of crucial importance in military. Since the penetration point is unknown, the computation of the point or points where the defender's forces should be placed is not a trivial problem. The defender should handle attacker's forces efficiently, minimizing the conquested territories. For this purpose, the defender has to select the optimal defence forces positions in order to minimize enemy's penetration. On the other hand, the scope of the invader is the maximization of the territory control, the defensive's force attrition, the duration of the control of the territories and the maximization of the ratios of the territory control over attack attrition and the defensive force attrition over offensive force attrition. For every combination of the basic targets of the attacker, the defender has to find a tactic in order to minimize its loses.

In this paper we present numerical solutions of the defence force optimal positioning problem. To do so, we first study the optimal mobile defence problem when the opponent penetrates from one point or from several different points of the front line [5, 6]. Next, we recall and apply well-known numerical methods solving non-linear equations [1, 2, 4].

Optimal Mobile Defence When the Opponent Penetrates from One Point of the Front Line

There are two basic types of defence. The *distributed static* and the *concentrated mobile defence*. In the first one, the units marshal behind the front line and in the second *a significant percentage of the units are available for immediate intervention to any point where the units attack the invader.*

Throughout this paper the following assumptions are supposed to hold.

i. The opponent forces are moving into positions on either side of one straight line of length M of the defence front line.
ii. The opponent forces have estimated the size of the enemy.
iii. The initial thrusting of the attacking force will be attempted by a single unknown point (0, z) of the segment length M, which may not be known before the outbreak of the first attack(and, therefore, the estimation of the variable z by the defending will be made using stochastic methods).
iv. The invader is moving perpendicularly to the defence front line with velocity v_A.
v. In peacetime, all weapon power of the defending units remain in military base located at the point (x, y) from which these units can pounce at speeds equal to v_D to a point (I, z) located on a defensive arc behind the defensive line in order to cope with the attacking force.

   The pursuit of the defending force is to minimize the penetration of the attacking behind the defensive line. To this end, the Military Base (within which the weapon units of the defending forces were guarded) should be grounded near the defensive line, on a point (x, y) which should be selected in such a way that each defensive weapon unit can pounce quickly to a corresponding point on the defensive arc of which effectively prevents further penetration of the attacking from the defensive line. Thus, it is reasonable to accept that if the penetration attempt by a point $(0, z)$ of the defensive line,
vi. the time that will be required on the part of the attacking force to move an interval of length I is equal to the time needed for the transfer of the defending forces from the point (x, y) of the CBD in (I, z) of the defensive arc. In other words, we assume the following relationship

$$\frac{I}{v_A} = \frac{\sqrt{(x-I)^2 + (y-z)^2}}{v_D}, \tag{1}$$

   where v_A is the velocity of the attacker.

The exact location of the infiltration point $(0, z)$ can only be estimated by the defender. Let $p(z)$ be the probability density distribution determined by the perception of defence forces about the strategic value of the enemy's targets, the morphology of the ground and other similar factors. Because the attacking force aims to break the defensive front attempting a point $(0, z)$ of the critical interval length M on the front line of defence, we have the following obvious relationship

$$\int_0^M p(z)\, dz = 1. \tag{2}$$

We will study the following three cases.

i. $v_A = v_D$
ii. $v_A < v_D$
iii. $v_A > v_D$

The Velocity of the Attacker Is Equal to the Velocity of the Defender

If $v_A = v_D$, then from Eq. (1) the infiltration of the attacker is given by

$$I = I(x, y; z) = \frac{x^2 + (y - z)^2}{2\,x}. \tag{3}$$

Thus, the expected enemy's penetration $\bar{I}\,(x, y)$ is

$$\bar{I}\,(x, y) = \int_0^M p(z)\, I(x, y, z)\, d\,z. \tag{4}$$

Our aim is the selection of an optimal positioning point (x_*, y_*) for the military base in order to minimize the expected enemy's penetration:

$$\bar{I}\;(x_*, y_*) = \min_{x,y} \bar{I}\;(x, y) = \min_{x,y} \left[\int_0^M p\,(z)\; I\;(x, y, z)\; d\,z\right]. \tag{5}$$

The minimization is achieved when the following two conditions hold:

$$\frac{\partial}{\partial x}\left[\int_0^M p\,(z)\; I\;(x, y, z)\; d\,z\right]\bigg|_{x=x_*, y=y_*} = 0 \tag{6}$$

and

$$\frac{\partial}{\partial y}\left[\int_0^M p\,(z)\; I\;(x, y, z)\; d\,z\right]\bigg|_{x=x_*, y=y_*} = 0. \tag{7}$$

From Leibniz's rule we have

$$\int_0^M \frac{\partial}{\partial x}\left[p\;(z)\; I\;(x, y, z)\right]\, d\,z\bigg|_{x=x_*, y=y_*} = 0 \tag{8}$$

and

$$\int_0^M \frac{\partial}{\partial y} [p(z) I(x, y, z)] dz \bigg|_{x=x_*, y=y_*} = 0, \quad (9)$$

respectively. Differentiating Eq. (3) with respect to x and y, we get

$$\frac{\partial I}{\partial x} = \frac{1}{2} - \frac{(y-z)^2}{2x^2} \text{ and } \frac{\partial I}{\partial y} = \frac{y-z}{x}, \quad (10)$$

respectively. Thus,

$$\int_0^M p(z) \frac{\partial I}{\partial x} dz \bigg|_{x=x_*, y=y_*} = \int_0^M p(z) \left(\frac{1}{2} - \frac{(y_* - z)^2}{2x_*^2} \right) dz = 0 \quad (11)$$

and

$$\int_0^M p(z) \frac{\partial I}{\partial y} dz \bigg|_{x=x_*, y=y_*} = \int_0^M p(z) \frac{(y_* - z)}{x_*} dz = 0. \quad (12)$$

Solving (12) with respect to y_*, we have

$$y_* \int_0^M p(z) dz = \int_0^M z p(z) dz.$$

Hence, from (2), it follows

$$y_* = \int_0^M z p(z) dz =: \mu, \quad (13)$$

where μ is the expected (mean) value of the variable z. This expected value represents the point along the front line whereby the attacker is expected to penetrate the defender's line. Thus,

$$y_* = \mu. \quad (14)$$

Substituting y_* into (11) and using (2) we finally obtain

$$x_*^2 = \int_0^M (\mu - z)^2 p(z) dz =: \sigma^2, \quad (15)$$

where σ^2 is the variation of variable z. Therefore, the defender has to place its forces to a point (x^*, y^*) with abscissa

$$x_* = \sigma. \quad (16)$$

The following theorem is proved.

Theorem 1 **($\sigma - \mu$ theorem of Gupta [5, 6]).** *Suppose that the speed of movement of the attacking force is equal to the speed of movement of the defending. The defender commander will be able to minimize in the ground defends the expected aggressive penetration, having had time to send each weapon unit against a unit weapon the attacker, if the commander has previously garnered all available weapon of units on the point* (σ, μ).

Lemma 1. *In particular, if the stochastic variable* z *follows a uniform distribution, i.e. whether* $p(z) = 1/M$, *and if the speed of movement of the attacking is equal to the speed of movement of the defending, then the following applies.*

i. *In order to minimize the penetration, the defender has to place its forces to the point*

$$(\sigma, \mu) = \left(M \middle/ \sqrt{12},\ M/2\right) \tag{17}$$

since, because of Theorem 1 the point where the defender's forces should be placed is (σ, μ), *with*

$$\sigma^2 = \int_0^M \frac{1}{M}\,(\mu - z)^2\, d\,z = \frac{M^2}{12} \quad and \quad \mu = \int_0^M z\,p\,(z)\,d\,z = \int_0^M \frac{z}{M}\,d\,z = \frac{M}{2}. \tag{18}$$

ii. *The penetration* I *of the attacker and the location* z *of the point of the defense locus through which the attacker will penetrate in the defender's ground is given by*

$$I = \frac{\frac{M^2}{12} + \left(\frac{M}{2} - z\right)^2}{\frac{2M}{\sqrt{12}}}. \tag{19}$$

The Velocity of the Attacker Is Less than the Velocity of the Defender

Let $\xi = \frac{v_A}{v_D}$. If $v_A < v_D$ (thus $\xi < 1$) we have the following equation

$$I^2\,\xi^{-2} = (x - I)^2 + (y - z)^2.$$

The positive solution of the previous equation is given by

$$I = \frac{-x + \sqrt{x^2 + (\xi^{-2} - 1)\left(x^2 + [y - z]^2\right)}}{(\xi^{-2} - 1)}. \tag{20}$$

As in section "The Velocity of the Attacker Is Equal with the Velocity of the Defender" our scope is to find an optimal point (x_*, y_*) minimizing enemy's penetration $\bar{I}\,(x,y) = \int_0^M p\,(z)\; I\,(x,y,z)\; d\,z$. As in the preceding case, we will assume that in the beginning, the density $p(z)$ of the stochastic variable z follows the uniform density distribution $p(z) = \frac{1}{M}$. Then, optimization conditions (8) and (9) become

$$\frac{1}{M}\int_0^M \frac{\partial I\,(x,y,z)}{\partial\, x} d\,z\Big|_{x=x_*,y=y_*} = 0 \tag{21}$$

and

$$\frac{1}{M}\int_0^M \frac{\partial I\,(x,y,z)}{\partial\, y} d\,z\Big|_{x=x_*,y=y_*} = 0. \tag{22}$$

Computing the partial derivatives with respect to x and y from Eq. (20), we conclude to

$$\frac{\partial\, I}{\partial\, x} = \frac{1}{\xi^{-2}-1}\left(-1+\frac{\xi^{-2}\,x}{\sqrt{x^2+\left(\xi^{-2}-1\right)\left(x^2+[y-z]^2\right)}}\right) \tag{23}$$

and

$$\frac{\partial\, I}{\partial\, y} = \frac{\xi\,(y-z)}{\sqrt{x^2+[y-z]^{\,2}}}. \tag{24}$$

Substituting (24) into (22) we have

$$\int_0^M \frac{\xi\,(y-z)}{\sqrt{x^2+[y-z]^{\,2}}} d\,z\Big|_{x=x_*,y=y_*} = \int_0^M \frac{\xi\,(y_*-z)}{\sqrt{x_*^2+[y_*-z]^{\,2}}}\, d\,z = 0, \tag{25}$$

which after integration gives

$$y_* = M/2. \tag{26}$$

Similarly, substitution of (23) into (21), with $y_* = M/2$, we obtain

$$x_*\,\xi^{-2}\int_0^M \frac{d\,z}{\sqrt{x_*^2+\left(\xi^{-2}-1\right)\left(\frac{M}{2}-z\right)}} = M. \tag{27}$$

Integrating we conclude to

$$\frac{x_* \xi^{-2}}{\sqrt{\xi^{-2}-1}} \ln \left[\frac{\sqrt{\frac{\xi^{-2} x_*^2}{\xi^{-2}-1} + \frac{M^2}{4}} + \frac{M}{2}}{\sqrt{\frac{\xi^{-2} x_*^2}{\xi^{-2}-1} + \frac{M^2}{4}} - \frac{M}{2}} \right] = M. \tag{28}$$

Equation (28) is transcendental and can be solved using methods of numerical analysis. Under specific assumptions it is possible to find an analytic solution for (28). For example, if $\xi = (v_A/v_D) = 0.5$, then $x_* = 0.0994\, M$.

The Velocity of the Attacker Is Greater than the Velocity of the Defender

If $v_A > v_D$ and $\xi = (v_A/v_D) => 1$ then Eq. (1) gives $I^2 = (x-I)^2 \xi^2 + (y-z)^2 \xi^2$ and therefore

$$I = \frac{x \pm \sqrt{x^2 - (1-\xi^{-2}) \left(x^2 + [y-z]^2\right)}}{(1-\xi^{-2})} = \begin{cases} \frac{x - \sqrt{x^2 - (1-\xi^{-2}) \left(x^2 + [y-z]^2\right)}}{(1-\xi^{-2})} \\ \frac{x + \sqrt{x^2 - (1-\xi^{-2}) \left(x^2 + [y-z]^2\right)}}{(1-\xi^{-2})} . \end{cases} \tag{29}$$

The general solution of (29) has two branches. The first one corresponds to the solution with the positive sign in the numerator of (29) and the second one to that with the negative sign in the numerator. In order to decide which branch has to be chosen, let us assume that $z = y$. Then, Eq. (29) becomes

$$I = I(x, y, z) = \frac{x\ (1 \pm \xi^{-1})}{(1-\xi^{-2})} = \begin{cases} \frac{x\,(1-\xi^{-1})}{(1+\xi^{-1})\,(1-\xi^{-1})} = \frac{x}{1+\xi^{-1}} \\ \frac{x\,(1+\xi^{-1})}{(1+\xi^{-1})\,(1-\xi^{-1})} = \frac{x}{1-\xi^{-1}} . \end{cases} \tag{30}$$

Let t be the time that elapses before the first contact is made between the defence units and the attacker. Then, $I = t\, v_A$ and $x - I = t\, v_D$. Thus, the penetration $I = I(x, y; z)$ of the attacker beginning from the point $(0, y)$ of the defence locus equals to

$$I = I(x, y; z) = \frac{x\,\xi}{1+\xi} = \frac{x}{1+\xi^{-1}} . \tag{31}$$

Comparing Eqs. (30) and (31), with the penetration starting from the point $(0, z) = (0, y)$ of the defence locus, we conclude that we have to choose the branch of the Eq. (29) which corresponds to the negative sign in the numerator of (29). Thus,

$$I = \frac{x - \sqrt{x^2 - (1 - \xi^{-2})\left(x^2 + [y - z]^2\right)}}{(1 - \xi^{-2})}. \tag{32}$$

Using again Eqs. (8) and (9) we will determine an optimal point (x_*, y_*) where the defence forces should be placed in order to minimize the expected attacker's penetration $\bar{I}(x, y)$. In this case, we will assume from the beginning that the density $p(z)$ of the stochastic variable z follows the uniform distribution, thus $p(z) = \frac{1}{M}$. Then, Eqs. (8) and (9) become

$$\frac{1}{M}\int_0^M \frac{\partial I\ (x, y, z)}{\partial x} d\,z|_{x=x_*, y=y_*} = 0 \tag{33}$$

and

$$\frac{1}{M}\int_0^M \frac{\partial I\ (x, y, z)}{\partial y} d\,z|_{x=x_*, y=y_*} = 0. \tag{34}$$

Differentiating Eq. (31) with respect to x and y we get

$$\frac{\partial I}{\partial x} = \frac{1}{1 - \xi^2}\left(1 - \frac{x\,\xi^{-2}}{\sqrt{\xi^{-2}x^2 - (1 - \xi^{-2})\,(y - z)^2}}\right) \tag{35}$$

and

$$\frac{\partial I}{\partial y} = \frac{y - z}{\sqrt{x^2 - (1 - \xi^{-2})\left(x^2 + [y - z]^2\right)}}, \tag{36}$$

respectively. Thus Eq. (34) becomes

$$\int_0^M \frac{y - z}{\sqrt{x^2 - (1 - \xi^{-2})\left(x^2 + [y - z]^2\right)}} d\,z|_{x=x_*, y=y_*}$$

$$= \int_0^M \frac{y_* - z}{\sqrt{x_*^2 - (1 - \xi^{-2})\left(x_*^2 + [y_* - z]^2\right)}} d\,z = 0 \tag{37}$$

and integrating Eq. (37) we conclude

$$y_* = M/2. \tag{38}$$

Similarly, from Eqs. (33) and (35), with $y_* = \frac{M}{2}$, we get

$$\frac{x_* \xi^{-2}}{\sqrt{1-\xi^{-2}}} \int_0^M \frac{d\,z}{\sqrt{\frac{x_*^2 \xi^{-2}}{\xi^{-2}-1} - \left(\frac{M}{2} - z\right)^2}} = M. \tag{39}$$

Integrating Eq. (39) we obtain the following transcendental equation

$$Arc\tan\left(\sqrt{\xi^2-1}\frac{M}{2\,x_*}\right) = \xi\,\sqrt{\xi^2-1}\;\frac{M}{2\,x_*} \tag{40}$$

which can be solved with respect to x_* only using methods of numerical analysis.

Optimal Positioning Defender'S Forces in Mobile Defence when the Attacker Penetrates from Different Points of the Front Line

It is possible for the attacker to penetrate from different points of the front line. Let n be the number of these points and let M be the length of the front line. Then, the defender has to organize its mobile defence forces looking for optimal positioning points

$$X_1 := \sum_{j=1}^{M_1} X_1^{(j)} = X_1^{(1)} + X_1^{(2)} + \ldots + X_1^{(M_1)}$$

of the available units. The actual points $(0, z_1), (0, z_2), \ldots, (0, z_n)$ from which the attacker will penetrate are initially unknown. The defender has to estimate these points. Assume the front line is partitioned into n subintervals. Let $p(z_1), p(z_2), \ldots, p(z_n)$ be the n discrete or continuous densities defined from factors such as the importance that gives the attacker to the targets, the landscaping of surfaces of the front line which may allow easier penetration from that point. Then,

$$\int_0^{M_1} p(z_1)\,d\,z_1 = \int_{M_1}^{M_2} p(z_2)\,d\,z_2 = \cdots = \int_{M_{n-1}}^{M_n} p(z_n)\,d\,z_n = 1.$$

In case that the attacker penetrates from two different points, the defender has to

i. divide the interval of length M of the front line to two equal subintervals, the lower $[0, M_1]$ and the upper $[M_1, M_2]$ each one of length $\frac{M}{2}$
ii. carry the half of its forces $\left(\frac{X_1}{2}\right)$ to every subinterval.

Under these assumptions, the $\sigma - \mu$ theorem of Gupta guarantees the following.

Theorem 2 ([5, 6]). *If the speed of movement of the attacking force is equal to the speed of movement of the defending, then the defending commander will be able to minimize the expected penetration of the attacking behind the defensive line, having had time to send the unit each weapon against each attacking unit if the Governor has previously garnered half $\frac{x_1}{2}$ the number of weapons of units in (σ_1, μ_1) and the other $\frac{x_1}{2}$ weapon of units in (σ_2, μ_2). Here μ_1, μ_2 are the expected points of enemy's penetration at the lower and upper subinterval of length $\frac{M}{2}$ correspondingly and σ_1, σ_2 the standard deviations of μ_1 and μ_2, respectively.*

Lemma 2. *If $v_D = v_A$ and if the two stochastic variables z_1 and z_2 follow the uniform distribution, i.e.*

$$p(z_1) = p(z_2) = (1/(M/2)) \; (= 2/M),$$

the following holds.

i. *The defending commander minimizes the expected enemy's penetration behind the lower subinterval of the defence locus if initially has placed the half $(\frac{X_1}{2})$ of its forces to the point*

$$(\sigma_1, \mu_1) = \left(M \Big/ 2\sqrt{12}, \, M/4\right). \tag{41}$$

Further, the defending commander minimizes the expected enemy's penetration behind the upper subinterval of the defence locus if initially has placed the half $(\frac{X_1}{2})$ of its forces to the point

$$(\sigma_2, \mu_2) = \left(M \Big/ 2\sqrt{12}, \, 3M/4\right). \tag{42}$$

ii. *Under the same assumptions, the following equations between the penetrations I_1 and I_2 and penetration points z_1 and z_2 hold.*

$$I_1 = \frac{\frac{M^2}{24} + \left(\frac{M}{4} - z_1\right)^2}{M \Big/ \sqrt{12}}$$

and

$$I_2 = \frac{\frac{M^2}{24} + \left(\frac{3M}{4} - z_2\right)^2}{M \Big/ \sqrt{12}}. \tag{43}$$

Remark 1. According to the formula (43) of Lemma 2, the positioning of the defence forces to the points (σ_1, μ_1) and (σ_2, μ_2) given by (41) and (42) restricts the expected maximal attacker's penetration by dividing it with a length of order of $M \Big/ \sqrt{12} = 0.2887\,M$ metric units, in contrast with Lemma 1 where the positioning of the defence forces to the point (σ, μ) of formula (17) restricts the expected maximal attacker's penetration by dividing it with a double length of order of $2M \big/ \sqrt{12} = 0.5773\,M$ metric units. The increase of 50% to the value of the expected penetration is caused because in case of Lemma 2 the concentration of the defence forces to the points (σ_1, μ_1) and (σ_2, μ_2) is halved.

In the second simpler case, where again the enemy penetrates from two different points, the defender has to

i. divide the interval of length M of the front line to two equal subintervals such that the lower subinterval $[0, M_1]$ to be of length $\frac{M}{3}$ and the upper one $[M_1, M_2]$ to be of length $\frac{2M}{3}$
ii. carry the half of its forces $\left(\frac{X_1}{2}\right)$ to every subinterval.

Under these assumptions, the $\sigma - \mu$ theorem of Gupta guarantees the following.

Theorem 3 ([5, 6]). *If $v_D = v_A$, then the enemy's penetration is minimized if the half of the defence forces are initially placed at the point (σ_1, μ_1) and the remaining half of the defence forces at the point (σ_2, μ_2), where μ_1, μ_2 are the expected points of enemy's penetration at the lower and upper subinterval of length $\frac{M}{3}$ and $\frac{2M}{3}$ respectively and σ_1, σ_2 are the standard deviations of μ_1 and μ_2 correspondingly.*

Lemma 3. *In case that the two stochastic variables z_1 and z_2 follow the uniform distribution, i.e.*

$$p\,(z_1) = M/3\ (= 3/M)\ and\ p\,(z_2) = 1/\,(2M/3)\ \ (= 3/2M)$$

and if $v_D = v_A$ the following holds.

i. The defending commander minimizes the expected enemy's penetration behind the lower subinterval of the defence locus if initially has placed the half $(X_1/2)$ of its forces to the point

$$(\sigma_1, \mu_1) = \left(2M \big/ 3\sqrt{12}\,,\ M/3\right). \tag{44}$$

Further, the defending commander minimizes the expected enemy's penetration behind the upper subinterval of the defence locus if initially has placed the half $(X_1/2)$ of its forces to the point

$$(\sigma_2, \mu_2) = \left(2M \big/ 3\sqrt{12}\,,\ 2M/3\right). \tag{45}$$

ii. Under the same assumptions, the following equations between the penetrations I_1 and I_2 of the attacker and the penetration points z_1 and z_2 hold.

$$I_1 = \frac{\frac{M^2}{27} + \left(\frac{M}{3} - z_1\right)^2}{4M \big/ 3\sqrt{12}}$$

and

$$I_2 = \frac{\frac{M^2}{27} + \left(\frac{2M}{3} - z_2\right)^2}{4M \Big/ 3\sqrt{12}}. \tag{46}$$

Remark 2. According to the formula (46) of Lemma 2, the expected maximal attacker's penetration is divided by a length of order of $4M \Big/ 3\sqrt{12} = 0.3849\,M$ metric units, in contrast with Lemma 1 where the positioning of the defence forces to the point $(\sigma,\ \mu)$ of formula (17) restricts the expected maximal attacker's penetration by dividing it with a double length of order of $2M \Big/ \sqrt{12} = 0.5773\,M$ metric units.

Numerical Examples

In this section, we present several numerical examples computing the optimal position where the defender's forces should be placed in order to be minimized enemy's penetration. In the following examples we assume that the stochastic variable z follows the uniform distribution.

There are many methods in order to solve a non-linear equation of the form (28) and (40). The most known ones are the bisection method, fixed point methods such as Newton's and secant method and many others [2–4]. Newton's method converges quadratically except in case that there are multiple roots where it converges linearly. In this case, the modified instead of the classical Newton's method can be used in order to have again quadratic convergence.

Quadratic convergence is not appeared very often in practice, since in most cases we have linear or almost linear convergence. Aitken's Δ^2 method [2] accelerates the convergence of a linearly convergent sequence. Steffensen's method is similar with that of Aitken's and under the assumptions mentioned in [2] converges quadratically.

In all methods a small tolerance named tol is used relaxing the notion of zero. Different tolerances may lead to different results and thus the selection of a suitable tolerance is not always a trivial task. In the evaluated examples, a tolerance of order of 10^{-10} is used. The convergence and the stability of the presented methods are also given. The order of the convergence of a method is given by the following definition.

Definition 1. The order of the convergence of a method is p if there is a constant C:

$$\frac{|x_{k+1} - x^*|}{|x_k - x^*|^p}, \ \forall k \in \mathbb{N}.$$

If $p = 1, 2, 3$, the convergence is *linear, quadratic* and *cubic* respectively.

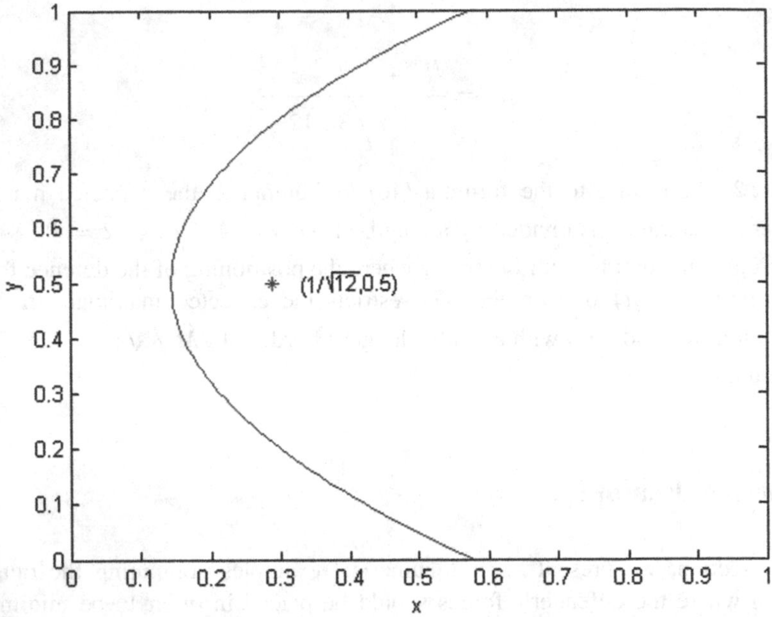

Fig. 1 Optimal distance from the front line

All the methods are tested for various sets of data and tables comparing the above well-known numerical algorithms; in respect of absolute error, convergence and complexity are given.

Example 1. In this example we study the case where the velocities of defender's and attacker's forces are equal, thus $v_D = v_A$ and the attacker penetrates from one point of the defence locus. In Fig. 1 is shown the optimal distance from the front line and an optimal point where the defender's forces should be placed for $M = 1$.

Example 2. In this example we solve the Eq. (28) of section "The Velocity of the Attacker Is Less than the Velocity of the Defender" assuming that $v_D = 2v_A$, thus $\xi = 0.5$ and for $M = 0.5$ computing the optimal position where the defender has to place its forces in order to minimize enemy's penetration. In Fig. 2 is presented the graph of the Eq. (28).

For solving efficiently Eq. (28) we applied all the methods mentioned above using the following initial data: $x_0 = 0.0859$, $tol = 10^{-10}$ and $n_{\max} = 1000$. In Bisection we used as initial interval the $[-0.9141, 1.0859]$ and in Secant as $x_1 = 0.1$.

In Table 1, the approximations of the solution x^* of the Eq. (28), the number of required steps and the required time in seconds of each method are presented.

As it is shown in Table 1, all the methods converge to the solution $0.0994M$ (in this example $M = 0.5$). The required time is of the order of 10^{-4} s for every method. The solution x^* is approached with an accuracy of order 10^{-10}. The most

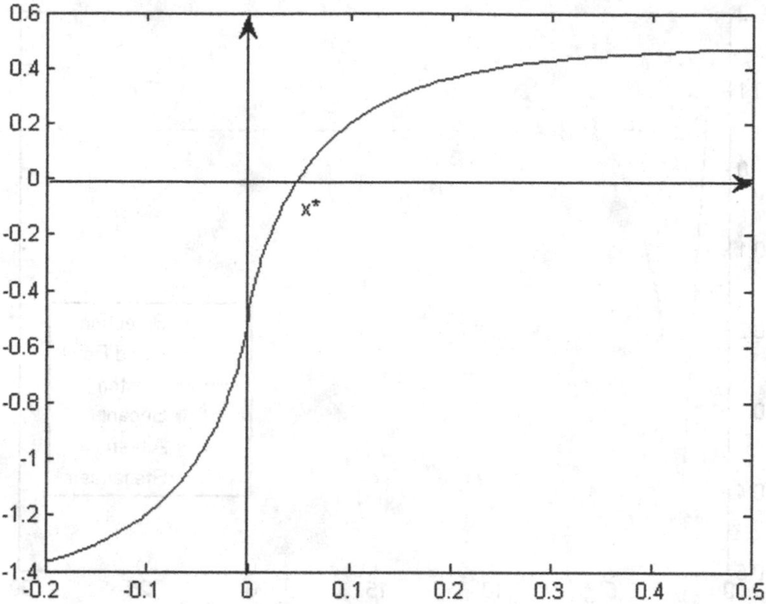

Fig. 2 Graph of Eq. (28)

Table 1 Comparison of the well known numerical methods

Method	Steps k	Time	x_0	x_1	...	x_k
Bisection	27	$6.7889 \cdot 10^{-4}$	0.0859	−0.4141	...	0.049718565011024
Fixed point	24	$5.1831 \cdot 10^{-4}$	0.0859	0.065417753601010	...	0.049718565060335
Newton	5	$6.9059 \cdot 10^{-4}$	0.0859	0.039168049901939	...	0.049718564996934
Secant	8	$2.7506 \cdot 10^{-4}$	0.0859	0.1	...	0.049718564996933
Aitken's	11	$2.6263 \cdot 10^{-4}$	0.0859	0.065417753601010	...	0.049718565010434
Steffensen's	3	$2.1178 \cdot 10^{-4}$	0.0859	0.049984089135137	...	0.049718564996934

steps require the Bisection method and the less Steffensen's. The convergence of Newton's, Aitken's and Steffensen's method is quadratic as

$$\frac{|x_{k+1} - x^*|}{|x_k - x^*|^2} \approx 40.21$$

for all k and

$$\frac{|x_{k+1} - x^*|}{|x_k - x^*|^3} \to +\infty.$$

In Fig. 3 is shown the approximations of the solution for every method in respect of required steps. The solution has been computed with an absolute error of order of 10^{-10} for all methods.

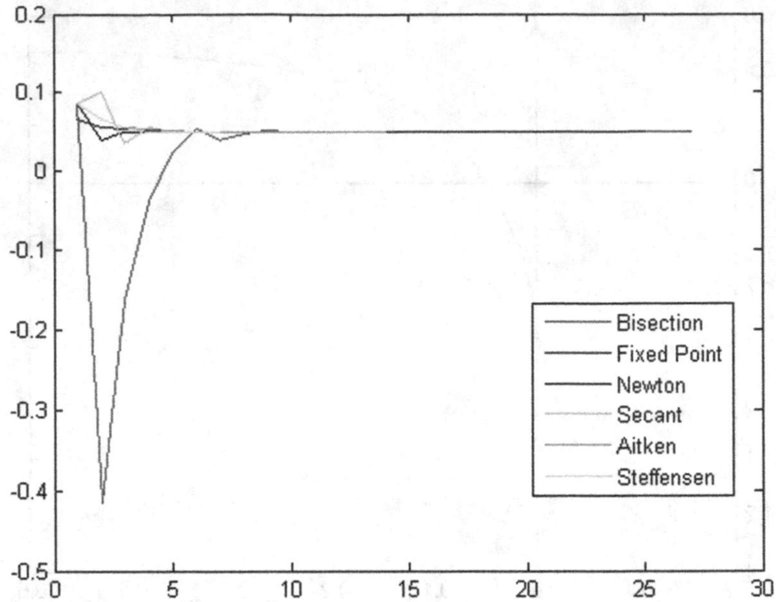

Fig. 3 Approximating the solution of Eq. (28)

In Fig. 4, is shown the point $(x^*, y^*) = \left(0.0994M, \frac{M}{2}\right) = (0.0497, 0.25)$ where the defender should place its forces in order to minimize enemy's penetration. The attacker in this figure penetrates from point$(0, z) = (0, 0.3)$ of the defence locus and its penetration is $I = 0.027381341895084$.

Example 3. In this example we solve the problem of optimal positioning when the velocity of the attacker is greater than the velocity of the defender. We study the cases when $\frac{v_A}{v_D} = \xi = 1.5, 2, 2.5, 3, 3.5, 4$, for $M = 1$ and $M = 2$. We solve the Eq. (40) of section "The Velocity of the Attacker Is Greater than the Velocity of the Defender" for every combination of M and ξ in order to find an optimal position where the defender should place its forces minimizing enemy's penetration applying some steps of Bisection in order to approximate the solution and continuing with Newton's method for computing the solution with an absolute errors of order of 10^{-10}. In Figs. 5 and 6 is presented the penetration $I(z)$ [given by Eq. (29)] of the attacker as a function of the penetration point $(0, z)$ for $\xi = 1.5, 2, 2.5, 3, 3.5, 4$ and $M = 1$ and $M = 2$, respectively.

The values of z for which the value under the root of Eq. (29) is negative are not presented in Figs. 5 and 6.

Example 4. In this example we compute the optimal positions where the defender should place its forces in order to minimize enemy's penetration, when the attacker

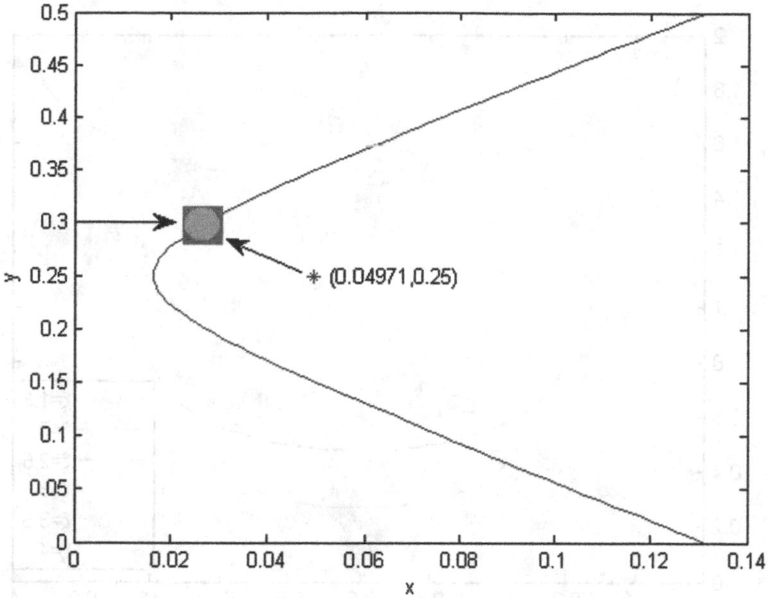

Fig. 4 Optimal positioning of defender's forces.
red square: attacker's forces
green circle: defender's forces (Colour figure online)

penetrates from two different points. As previously, we assume that the stochastic variable z follows the uniform distribution. The defender has to partition the initial interval $[0, M]$ to two subintervals, the lower one $[0, M_1]$ and the upper one $[M_1, M_2]$ and place the half of its forces to the point (σ_1, μ_1) of the lower and (σ_2, μ_2) of the upper subinterval. In this example we have studied two cases. In the fist one we have partitioned the initial interval to two subintervals of equal length $\frac{M}{2}$ and in the second one into two subintervals of length $\frac{M}{3}$ and $\frac{2M}{3}$, respectively. The computed points, where the defender should place its forces are shown in Table 2.

In Fig. 7 is shown the optimal distance from the front line in the case that the defender's and attacker's velocities are equal ($v_D = v_A$), the attacker penetrates from two different points and the defender has partitioned the initial interval of length $M = 1$ to two subintervals of lengths $\frac{1}{3}$ and $\frac{2}{3}$, respectively (see Theorem 3). From Lemma 3 the penetrations of the attacker's forces I_1 and I_2 are given from Eq. (46).

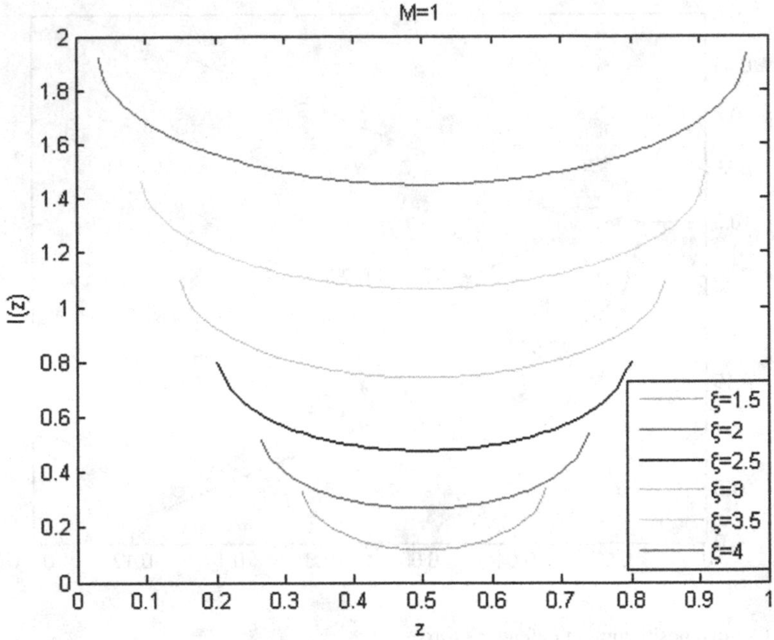

Fig. 5 Attacker's penetration for various values of $\xi = \frac{v_A}{v_D}$

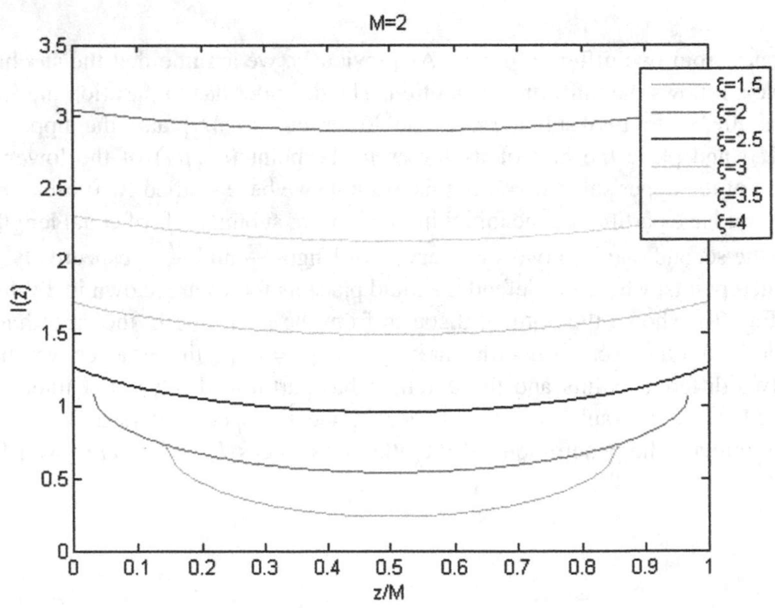

Fig. 6 Attacker's penetration for various values of $\xi = \frac{v_A}{v_D}$

Table 2 Optimal positioning of defence forces

Intervals	$[0, M_1]$	$[M_1, M_2]$	Lower subinterval	Upper subinterval
Length	$\frac{M}{2}$	$\frac{M}{2}$	$(\sigma_1, \mu_1) = \left(\frac{M}{2}\sqrt{12}, \frac{M}{4}\right)$	$(\sigma_2, \mu_2) = \left(\frac{M}{2}\sqrt{12}, \frac{3M}{4}\right)$
Length	$\frac{M}{3}$	$\frac{2M}{3}$	$(\sigma_1, \mu_1) = \left(\frac{2M}{3}\sqrt{12}, \frac{M}{3}\right)$	$(\sigma_2, \mu_2) = \left(\frac{2M}{3}\sqrt{12}, \frac{2M}{3}\right)$

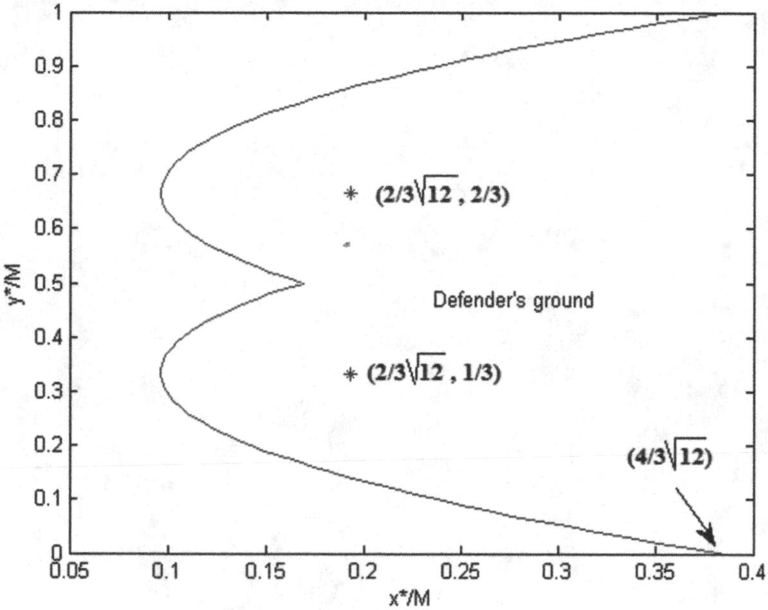

Fig. 7 Optimal distance from the front line

References

1. B. Brandie, A Friendly Introduction to Numerical Analysis, Pearson Education Inc., New Jersey, 2006.
2. R. L. Burden and J. D. Faires, Numerical Analysis, 8th edition, Thomson Brooks Cole, USA, 2005.
3. L. V. Fausett, Applied Numerical Analysis Using Matlab, Second Edition, Pearson Education Inc., New Jersey, 2008.
4. G.E. Forsythe, M.A. Malcolm and C.B. Möller, Computer methods for mathematical computations, Prentice-Hall, Englewood Cliffs, N. J., 1977.
5. R. Gupta, Defense Positioning and Geometry, the Brookings Inst., Washington, DC, 1993.
6. J. Przemieniecki, Mathematical Methods in Defense Analyses, Third Edition, Education Series, American Institute of Aeronautics and Astronautics (AIAA), Inc., Reston, VA 20191-4344, 2000.

Invisibility Regions and Regular Metamaterials

George Dassios

Abstract For more than half a century people are trying to develop effective methods of remote sensing. RADAR and SONAR systems are the most advertised techniques toward this end, and the contemporary level of sophistication of both of these modalities is really amazing. In fact, so many things have been achieved with these identification methods that the inverse question has been naturally raised: Is it possible to isolate a region of space where nothing can be detected via scattering techniques? Much to our surprise the answer to this question is "yes" and the way to achieve it is knowing as "cloaking." This is possible through the construction of a material, called "metamaterial" surrounding the cloaked region, which has particular preassigned properties. Cloaking has a history of less than a decade and almost all realistic cloaking regions share the shape of a sphere. However, spherically cloaked regions demand metamaterials with singular conductivity tensors, a consequence of the highly focusing effects of the spherical system as it collapses down to its center. We will demonstrate an ellipsoidal cloaking region, which, as a consequence of the fact that the ellipsoidal system springs from its characteristic focal ellipse, the necessary metamaterial that creates the invisibility region is regular throughout, leaving this way its realization at the level of engineering construction.

Introduction

The fact that there exist hidden primary or secondary (such as scatterers) sources goes back to Helmholtz, who, as far back as 1853 [17], demonstrated that it is possible to have currents within a conductive medium which generate a null field in its exterior. Electroencephalography, Magnetoencephalography, and Electric

G. Dassios (✉)
Department of Chemical Engineering, Division of Applied Mathematics,
University of Patras and ICE-HT/FORTH, 265 04 Patras, Greece
e-mail: gdassios@otenet.g

N.J. Daras (ed.), *Applications of Mathematics and Informatics in Science and Engineering*, Springer Optimization and Its Applications 91, DOI 10.1007/978-3-319-04720-1_9,

Impedance Tomography are intimately related with this observation. Devaney and Wolf [8], Friedlander [9], Kerker [19] and Bleistein and Cohen [1] have investigated further this idea of hidden primary sources at the level of radiation problems. Nevertheless, it is the last decay that the idea of cloaking, as a technique of making objects "invisible," has gained a lot of attention [2–6, 10, 12–16, 20–23]. There are actually two ways to prove the existence of cloaked regions. One is based on spectral methods, where one proves that it is possible to choose the coefficients of an appropriate eigenfunction expansion in such a way that the excitation field does not enter the region we want to hide [5]. The other one utilizes the invariants of the governing equations with respect to coordinate transformations in such a way that a single point singularity in the domain of the transformation blows up to a three-dimensional cloaked region in its range [11]. A fairly extended literature on the subject is reported in [12].

Cloaking at the highly symmetric environment of the spherical geometry causes singularities which are due to focusing effects. The situation though is much more regular if we extend our investigation to the more realistic and much more general geometry of the ellipsoidal system, which essentially governs the geometry of the anisotropic space. Since, as far as the mathematics is concerned, the fundamental part of the cloaking problem focuses on the control of the differential operator, and since the basic differential operator appearing in the spectral form of the wave equation is Laplacian, we restrict our attention to the problem of the conductivity profile that secures electrostatic cloaking. This is actually the problem of Electric Impedance Tomography. Therefore, the present work is focused on the construction of a particular three-dimensional cloak which can be adapted to almost any shape. In contrast to the isotropic behavior of a spherical cloak the cloak we propose here has complete anisotropy which can be chosen at will. In fact, in utilizing the invariants of the differential equations in spherical geometry, we arrive at two possibilities. The first one is to blow-up a point, in which case we obtain perfect cloaking but singular metamaterial tensor, and the second one is to blow-up a small sphere, in which case we obtain almost perfect cloaking but regular metamaterial tensor. On the other hand, as we demonstrate in the present report using ellipsoidal geometry, it is possible to avoid both of these difficulties and achieve at the same time perfect cloaking and regular metamaterial tensor.

This report is organized as follows. In section "The Case of Spherical Symmetry" we discuss in brief, the corresponding spherical problem in order to indicate the mathematical difficulties that generated. In section "The Ellipsoidal System" we provide a short introduction to the ellipsoidal coordinate system which makes the paper self readable. Then the transformation and its inverse is introduced and discussed in section "The Ellipsoidal Transformation." Finally the material tensor, which guides the field to avoid the cloaking region, is calculated in section "The Material Tensor."

The Case of Spherical Symmetry

We demonstrate the problem of electrostatic cloaking in spherical geometry with the following well-known example [12]. Suppose that a sphere centered at the origin with radius 2 is a homogeneous conductor with conductivity equal to one. That is

$$\tilde{\boldsymbol{\sigma}} = \tilde{\mathbf{I}} \tag{1}$$

where $\tilde{\boldsymbol{\sigma}}$ denotes the conductivity tensor, in its dyadic form, and $\tilde{\mathbf{I}}$ is the identity dyadic. Applying the diffeomorphism

$$\mathbf{F}(\mathbf{x}) = \mathbf{y} = \left(\frac{|\mathbf{x}|}{2} + 1\right)\hat{\mathbf{x}} \tag{2}$$

where $\hat{\mathbf{x}}$ stands for the unit vector along $\mathbf{x}$, it follows that the initial sphere is mapped into a spherical shell. In particular the boundary of the sphere $|\mathbf{x}| = 2$ stays invariant, while the origin is blown up to the sphere of radius one, forming the interior boundary of the shell. The metric dyadic of the inverse transformation,

$$\mathbf{x} = 2(|\mathbf{y}| - 1)\hat{\mathbf{y}} \tag{3}$$

where $\mathbf{y}$ assumes the spherical representation (r, ϑ, φ), leads to

$$\tilde{\mathbf{g}} = 4\hat{\mathbf{r}} \otimes \hat{\mathbf{r}} + 4(r-1)^2 \hat{\boldsymbol{\vartheta}} \otimes \hat{\boldsymbol{\vartheta}} + 4(r-1)^2 \sin^2\vartheta \hat{\boldsymbol{\varphi}} \otimes \hat{\boldsymbol{\varphi}} \tag{4}$$

which has the determinant

$$|\tilde{\mathbf{g}}| = 64(r-1)^4 \sin^2\vartheta. \tag{5}$$

In view of the form

$$\Delta = \frac{1}{\sqrt{|g|}} \sum_{i=1}^{3} \sum_{j=1}^{3} \frac{\partial}{\partial x_i} \left(\sqrt{|g|} g^{ij} \frac{\partial}{\partial x_j}\right) \tag{6}$$

that the Laplacian assumes in the general coordinate system

$$x_i = x_i(q_1, q_2, q_3), \quad i = 1, 2, 3 \tag{7}$$

we calculate the conductivity dyadic in the spherical shell to be equal to

$$\tilde{\boldsymbol{\sigma}} = \sqrt{|\tilde{\mathbf{g}}|}\tilde{\mathbf{g}}^{-1} = 2(r-1)^2 \sin\vartheta \hat{\mathbf{r}} \otimes \hat{\mathbf{r}} + 2\sin\vartheta \hat{\boldsymbol{\vartheta}} \otimes \hat{\boldsymbol{\vartheta}} + \frac{2}{\sin\vartheta} \hat{\boldsymbol{\varphi}} \otimes \hat{\boldsymbol{\varphi}}. \tag{8}$$

Since the determinant of the conductivity dyadic is given by

$$|\tilde{\boldsymbol{\sigma}}| = 8(r-1)^2 \sin\vartheta \tag{9}$$

it follows that the conductivity becomes singular in the inner boundary $r - 1$ of the shell.

One way to avoid this singular behavior is to start with an initial manifold that instead of a single point is a small sphere of radius ε, and prove that the conductivity dyadic does not become singular anymore, but the field in the cloaked region is of the order of ε^n, where n is the dimensionality of the space [12]. This corresponds to a weak interior field which vanishes as the radius ε approaches zero.

The Ellipsoidal System

We start with the essentials of the ellipsoidal system in order to fix the notation [18, 24]. The definition of an ellipsoidal system demands the determination of a reference ellipsoid

$$\frac{x_1^2}{a_1^2} + \frac{x_2^2}{a_2^2} + \frac{x_3^2}{a_3^2} = 1 \tag{10}$$

where $0 < a_3 < a_2 < a_1 < +\infty$, which fixes the foci of the system and establishes the standards of every spatial direction. The reference ellipsoid (10) plays the role of the unit sphere in the case of the spherical system. The six foci of the ellipsoidal system are located at the points $(\pm h_2, 0, 0)$, $(\pm h_3, 0, 0)$, and $(0, \pm h_1, 0)$, where

$$h_1^2 = a_2^2 - a_3^2 \tag{11}$$

$$h_2^2 = a_1^2 - a_3^2 \tag{12}$$

$$h_3^2 = a_1^2 - a_2^2 \tag{13}$$

and they are related by the equation

$$h_1^2 - h_2^2 + h_3^2 = 0. \tag{14}$$

The back bone of the ellipsoidal system is given by the focal ellipse

$$\frac{x_1^2}{h_2^2} + \frac{x_2^2}{h_1^2} = 1, x_3 = 0 \tag{15}$$

and the focal hyperbola

$$\frac{x_1^2}{h_3^2} - \frac{x_3^2}{h_1^2} = 1, x_2 = 0. \tag{16}$$

In the first octant, the ellipsoidal coordinates (ρ, μ, ν) are related to the Cartesian coordinates by

$$x_1 = \frac{\rho\mu\nu}{h_2 h_3}, \quad h_2 < \rho < +\infty \tag{17}$$

$$x_2 = \frac{\sqrt{\rho^2 - h_3^2}\sqrt{\mu^2 - h_3^2}\sqrt{h_3^2 - \nu^2}}{h_1 h_3}, \quad h_3 < \mu < h_2 \tag{18}$$

$$x_3 = \frac{\sqrt{\rho^2 - h_2^2}\sqrt{h_2^2 - \mu^2}\sqrt{h_2^2 - \nu^2}}{h_1 h_2}, \quad 0 < \nu < h_3 \tag{19}$$

while the other seven octants are specified by considering the appropriate signs of the x_i's. The variable ρ =constant specifies an ellipsoid and therefore it corresponds to the radial variable of the spherical system. In particular, the focal ellipse (6) corresponds to the value $\rho = h_2$. The pair (μ, ν) identifies a point on the ellipsoid ρ =constant and therefore it can be considered as the (θ, φ) pair of the spherical orientation system.

By varying the variable ρ we obtain a family of ellipsoids

$$\frac{x_1^2}{\rho^2} + \frac{x_2^2}{\rho^2 - h_3^2} + \frac{x_3^2}{\rho^2 - h_2^2} = 1, \quad \rho^2 \in (h_2^2, +\infty). \tag{20}$$

Similarly, the variation of the variable μ defines the family of hyperboloids of one sheet

$$\frac{x_1^2}{\mu^2} + \frac{x_2^2}{\mu^2 - h_3^2} + \frac{x_3^2}{\mu^2 - h_2^2} = 1, \quad \mu^2 \in (h_3^2, h_2^2) \tag{21}$$

and the variation of the variable ν defines the family of hyperboloids of two sheets

$$\frac{x_1^2}{\nu^2} + \frac{x_2^2}{\nu^2 - h_3^2} + \frac{x_3^2}{\nu^2 - h_2^2} = 1, \quad \nu^2 \in (0, h_3^2). \tag{22}$$

For a detailed analysis of the ellipsoidal system we refer to [7].

The Ellipsoidal Transformation

Suppose we want to cloak an ellipsoidal region that is given by (10), or, in view of (20), by $\rho = a_1$. Then we choose an ellipsoidal system (ρ, μ, ν) having (1) as reference ellipsoid and define the transformation

$$\mathbf{f}(\rho,\mu,\nu) = R(\rho)\frac{\mu\nu}{h_2h_3}\hat{\mathbf{x}}_1 + \sqrt{R^2(\rho)-h_3^2}\frac{\sqrt{\mu^2-h_3^2}\sqrt{h_3^2-\nu^2}}{h_1h_3}\hat{\mathbf{x}}_2$$
$$+\sqrt{R^2(\rho)-h_2^2}\frac{\sqrt{h_2^2-\mu^2}\sqrt{h_2^2-\nu^2}}{h_1h_2}\hat{\mathbf{x}}_3 \tag{23}$$

where

$$R(\rho) = \sqrt{\frac{\rho_1^2-a_1^2}{\rho_1^2-h_2^2}(\rho^2-h_2^2)+a_1^2} \tag{24}$$

with $\rho_1 > a_1$ and $0 \leqslant \nu^2 \leqslant h_3^2 \leqslant \mu^2 \leqslant h_2^2 \leqslant \rho^2$. Transformation (23) maps the focal ellipse $\rho = h_2$ to the reference ellipsoid $\rho = a_1$ and leaves the outer ellipsoid $\rho = \rho_1$ invariant. Hence, **f** restricts to a one to one map from the interior of the ellipsoid $\rho = \rho_1$ except the focal ellipse, to the ellipsoidal shell $a_1 < \rho < \rho_1$. Note that **f** is singular on the focal ellipse. This mapping connects the two points ρ and R on the coordinate curve $(\mu,\nu) =$ constant and therefore, it is completely determined by the scalar transformation (24). Consequently,

$$\rho(R) = \sqrt{\frac{\rho_1^2-h_2^2}{\rho_1^2-a_1^2}(R^2-a_1^2)+h_2^2} \tag{25}$$

and the inverse of **f** is given by

$$\mathbf{f}^{-1}(R,\mu,\nu) = \rho(R)\frac{\mu\nu}{h_2h_3}\hat{\mathbf{x}}_1 + \sqrt{\rho^2(R)-h_3^2}\frac{\sqrt{\mu^2-h_3^2}\sqrt{h_3^2-\nu^2}}{h_1h_3}\hat{\mathbf{x}}_2$$
$$+\sqrt{\rho^2(R)-h_2^2}\frac{\sqrt{h_2^2-\mu^2}\sqrt{h_2^2-\nu^2}}{h_1h_2}\hat{\mathbf{x}}_3. \tag{26}$$

The value of a_1 determines the size and the shape of the cloaked region, the value of ρ_1 determines the invariant exterior boundary, and the ellipsoidal distance $\rho_1 - a_1$ controls the thickness of the cloak.

The Material Tensor

The Dirichlet problem for the Laplace equation in a domain V refers to finding a harmonic function u in V which takes the preassigned values h on the boundary ∂V. If the domain V is equipped with the Riemannian metric (g_{ij}), then

$$\Delta u(x_1, x_2, x_3) = \frac{1}{\sqrt{|g|}} \sum_{i=1}^{3} \sum_{j=1}^{3} \frac{\partial}{\partial x_i} \left(\sqrt{|g|} g^{ij} \frac{\partial}{\partial x_j} \right) u(x_1, x_2, x_3) \quad (27)$$

where (g^{ij}) is the inverse of (g_{ij}) and $|g|$ denotes the determinant of the metric tensor. The map

$$\Lambda_g(h) = \sum_{i=1}^{3} \sum_{j=1}^{3} \left(\sqrt{|g|} n_i g^{ij} \frac{\partial}{\partial x_j} \right) u, \quad \mathbf{r} \in \partial V \quad (28)$$

known as the Dirichlet-to-Neumann map remains invariant under any diffeomorphic transformation that reduces to the identity on the boundary [10]. This observation allows to interpret the effect of the transformation as a change of the material tensor that characterizes the medium in V. In other words, the mathematical transformation is absorbed by the physical characteristics of the medium. The material properties of the medium are represented by the symmetric tensor

$$\sigma^{ij} = \sqrt{|g|} g^{ij} . \quad (29)$$

Therefore, we have to calculate the metric that corresponds to the inverted transformation mapping $\mathbf{f}^{-1}$. This will lead us to the metric $\tilde{g}^{ij}$ and to the material tensor

$$\tilde{\sigma}^{ij} \rho = \sqrt{|\tilde{g}|} \tilde{g}^{ij} . \quad (30)$$

After long and tedious calculations with the inverted ellipsoidal transformation $\mathbf{f}^{-1}$, we obtain the following expression for the inverted metric

$$\tilde{g}_{\mathrm{RR}} = \frac{R^2}{\rho^2(R)} \left(\frac{\rho_1^2 - h_2^2}{\rho_1^2 - a_1^2} \right)^2 \frac{(\rho^2(R) - \mu^2)(\rho^2(R) - \nu^2)}{(\rho^2(R) - h_3^2)(\rho^2(R) - h_2^2)} \quad (31)$$

$$\tilde{g}_{\mu\mu} = \frac{(\rho^2(R) - \mu^2)(\mu^2 - \nu^2)}{(\mu^2 - h_3^2)(h_2^2 - \mu^2)} \quad (32)$$

$$\tilde{g}_{\nu\nu} = \frac{(\rho^2(R) - \nu^2)(\mu^2 - \nu^2)}{(h_3^2 - \nu^2)(h_2^2 - \nu^2)} \quad (33)$$

where, due to the orthogonality of the ellipsoidal system, every other component of the metric tensor vanishes. Furthermore,

$$\sqrt{|\tilde{g}|} = \frac{\rho_1^2 - h_2^2}{\rho_1^2 - a_1^2} \frac{R(\rho^2(R) - \mu^2)(\rho^2(R) - \nu^2)(\mu^2 - \nu^2)}{\rho(R)\sqrt{\rho^2(R) - h_3^2}\sqrt{\rho^2(R) - h_2^2}\sqrt{\mu^2 - h_3^2}\sqrt{h_2^2 - \mu^2}\sqrt{h_3^2 - \nu^2}\sqrt{h_2^2 - \nu^2}}$$

$$= \frac{\rho_1^2 - h_2^2}{\rho_1^2 - a_1^2} \frac{R\mu\nu}{x_1 x_2 x_3} \frac{(\rho^2(R) - \mu^2)(\rho^2(R) - \nu^2)(\mu^2 - \nu^2)}{h_1^2 h_2^2 h_3^2} \tag{34}$$

which finally implies the material tensor

$$\tilde{\sigma} = \begin{pmatrix} \tilde{\sigma}_{\rho\rho} & 0 & 0 \\ 0 & \tilde{\sigma}_{\mu\mu} & 0 \\ 0 & 0 & \tilde{\sigma}_{\nu\nu} \end{pmatrix} \tag{35}$$

where

$$\tilde{\sigma}_{\rho\rho} = \frac{\rho_1^2 - a_1^2}{\rho_1^2 - h_2^2} \frac{\rho^2\mu\nu}{x_1 x_2 x_3} \frac{(\rho^2 - h_3^2)(\rho^2 - h_2^2)(\mu^2 - \nu^2)}{h_1^2 h_2^2 h_3^2} \times \left[\frac{\rho_1^2 - a_1^2}{\rho_1^2 - h_2^2}(\rho^2 - h_2^2) + a_1^2 \right]^{-\frac{1}{2}} \tag{36}$$

$$\tilde{\sigma}_{\mu\mu} = \frac{\rho_1^2 - h_2^2}{\rho_1^2 - a_1^2} \frac{\mu\nu}{x_1 x_2 x_3} \frac{(\mu^2 - h_3^2)(h_2^2 - \mu^2)(\rho^2 - \nu^2)}{h_1^2 h_2^2 h_3^2} \times \left[\frac{\rho_1^2 - a_1^2}{\rho_1^2 - h_2^2}(\rho^2 - h_2^2) + a_1^2 \right]^{\frac{1}{2}} \tag{37}$$

$$\tilde{\sigma}_{\nu\nu} = \frac{\rho_1^2 - h_2^2}{\rho_1^2 - a_1^2} \frac{\mu\nu}{x_1 x_2 x_3} \frac{(h_3^2 - \nu^2)(h_2^2 - \nu^2)(\rho^2 - \mu^2)}{h_1^2 h_2^2 h_3^2} \times \left[\frac{\rho_1^2 - a_1^2}{\rho_1^2 - h_2^2}(\rho^2 - h_2^2) + a_1^2 \right]^{\frac{1}{2}}. \tag{38}$$

It is of interest to note that the material tensor (36)–(38) above remains bounded away from zero, as well as from infinity, on both boundaries of the cloak, i.e., on $\rho = a_1$ and on $\rho = \rho_1$. The relative problem for the case of a spherical cloak leads to a material tensor that vanishes in the inner boundary of the cloak [12]. It seems that the singular behavior of the spherical case is due to the fact that the transformation map blows up the origin to a full sphere, while in the ellipsoidal case a point on the focal ellipse is mapped just to two symmetric points on the inner boundary of the cloak. In other words, the inverse map exhibits a strong focusing effect in the spherical case, sending a two-dimensional manifold to a single point, while in the ellipsoidal case, the inverse map sends just two points to one. On the other hand, in both the spherical and the ellipsoidal case, there are submanifolds in the interior of the cloak where the material tensor vanishes because the determinant of the metric vanishes there, but this is due to the particular system.

References

1. N. Bleistein and J.K. Cohen. Nonuniqueness in the inverse source problem in acoustics and electromagnetics. *Journal of Mathematical Physics*, 18:194–201, 1977.
2. H. Chen and C.T. Chan. Acoustic cloaking in three dimensions using acoustic metamaterials. *Applied Physics Letters*, 91, 183518:1–3, 2007.
3. H. Chen and G. Uhlmann. Cloaking a sensor for three-dimensional Maxwell's equations:transformation optics approach. *Optics Express*, 19:20518–20530, 2011.
4. H. Chen, B-I. Wu, B. Zhang, and J.A. Kong. Electromagnetic wave interactions with a metamaterial cloak. *Physical Review Letters*, 99, 063903:1–4, 2007.
5. A.S. Cummer, B-I. Popa, D. Schurig, D.R. Smith, J. Pendry, M. Rahm, and A. Starr. Scattering theory derivation of a 3D acoustic cloaking shell. *Physical Review Letters*, PRL100,024301:1–4, 2008.
6. A.S. Cummer and D. Schurig. On path to acoustic cloaking. *New Journal of Physics*, **9**(45):1–8, 2007.
7. G. Dassios. *Ellipsoidal Harmonics. Theory and Applications*. Cambridge University Press, Cambridge, 2012.
8. A.J. Devaney and E. Wolf. Radiating and nonradiating classical current distributions and the fields they generate. *Physical Review D*, 8:1044–1047, 1973.
9. F.G. Friedlander. An inverse problem for radiation fields. *Proceedings of the London Mathematical Society*, 27:551–576, 1973.
10. A. Greenleaf, Y. Kurylev, M. Lassas, and G. Uhlmann. Full-wave invisibility of active devices at all frequencies. *Communications in Mathematical Physics*, 275:749–789, 2007.
11. A. Greenleaf, Y. Kurylev, M. Lassas, and G. Uhlmann. Cloaking devises, electromagnetic wormholes, and transformation optics. *SIAM Review*, 51:3–33, 2009.
12. A. Greenleaf, Y. Kurylev, M. Lassas, and G. Uhlmann. Invisibility and inverse problems. *Bulletin of the Americal Mathematical Society*, 46:55–97, 2009.
13. A. Greenleaf, Y. Kurylev, M. Lassas, and G. Uhlmann. Cloaking a sensor via transformation optics. *Physical Review E*, 83, 016603:1–6, 2011.
14. A. Greenleaf, M. Lassas, and G. Uhlmann. Anisotropic conductivities that cannot detected in EIT. *Physilogical Measurement (special issue on Impedance Tomography)*, 24:413–420, 2003.
15. A. Greenleaf, M. Lassas, and G. Uhlmann. On nonuniqueness for the Calderón's inverse problem. *Mathematical Research Letters*, 10:685–693, 2003.
16. A. Greenleaf, M. Lassas, and G. Uhlmann. The Calderón problem for conormal potentials, I: global uniqueness and reconstraction. *Communications in Pure and Applied Mathematics*, 56:328–352, 2003.
17. H. Helmholtz. Ueber einige Gesetze der Vertheilung elektrischer Ströme in körperlichen Leitern mit Anwendung auf die thierisch-elektrischen Versuche;. *Annalen der Physik und Chemie*, 89:211–233 and 353–377, 1853.
18. E.W. Hobson. *The Theory of Spherical and Ellipsoidal Harmonics*. Cambridge University Press, U.K., Cambridge, first edition, 1931.
19. M. Kerker. Invisible bodies. *Journal of the Optical Society of America*, 65:376–379, 1975.
20. U. Leonhardt. Optical conformal mapping. *Science*, 312:1777–1780, 2006.
21. A. Norris. Acoustic cloaking theory. *Proceedings of the Royal Society A*, 464:2411–2434, 2008.
22. J.B. Pendry, D. Schurig, and D.R. Smith. Controlling electromagnetic fields. *Science*, 312:1780–1782, 2006.
23. R. Weder. A rigorous analysis of high-order electromagnetic invisibility cloaks. *Journal of Physics A: Mathematical and Theoretical*, 41, 065207:1–21, 2008.
24. E.T. Whittaker and G.N. Watson. *A Course of Modern Analysis*. Cambridge University Press, 3rd edition, 1920.

Evaluating UAV Impact in the Tactical Context of a Mechanized Infantry Scout Platoon Through Military Simulation Software

E. Mavratzotis, G. Drakopoulos, A. Voulodimos, A. Vatikalos, K. Kouvelis, S. Papadopoulos, and M. Sakelariou

Abstract Military simulation has been established as a computational and scientific tool for assessing the performance in combat of equipment, ranging from land mines to aircraft, and the suitability of tactics, ranging from platoon to division level. To this end, specialized software has been developed, replacing the traditional Prussian dice-throwing, turn-based war games. JANUS is such a suite, allowing human-in-the-loop simulations from squadron to battalion scale based on realistic combat models based on historical conflict data. This paper presents the initial results of a recent large scale campaign of experiments aimed to assess the effects of incorporating a UAV to a typical Hellenic mechanized infantry scout platoon. To the authors best knowledge, this is the first campaign of experiments undertaken by the Hellenic Military Academy. Therefore, there have been key contributions in a number of levels. On the software side, there are the development of a realistic mechanized infantry platoon advance scenario for JANUS, the creation and insertion of an appropriate UAV to JANUS unit database, the assessment of JANUS strengths and limitations for simulations of this scale, and the software development

---

E. Mavratzotis (✉)
KEPYES - Hellenic Army IT Center, Papagou Camp, Mesogeion 227–231, Holargos 15451, Greece

Division of Mathematics and Engineering Studies, Department of Military Science, Hellenic Military Academy, Vari 16673, Greece
e-mail: elisseosmav@gmail.com; e.k.mavratzotis@army.gr

G. Drakopoulos
Computer Engineering and Informatics Department, University of Patras, Patras 26500, Greece

A. Voulodimos • A. Vatikalos
School of Electrical and Computer Engineering, National Technical University of Athens, Iroon Polytecniou 9, Zografou Campus, 15780 Athens, Greece

K. Kouvelis • S. Papadopoulos • M. Sakelariou
Division of Mathematics and Engineering Studies, Department of Military Science, Hellenic Military Academy, Vari 16673, Greece

N.J. Daras (ed.), *Applications of Mathematics and Informatics in Science and Engineering*, Springer Optimization and Its Applications 91, DOI 10.1007/978-3-319-04720-1_10,

for parsing JANUS voluminous output report text files. From a mathematical perspective, there is the statistical analysis and interpretation of simulation results. Finally, there is the experimental aspect, where heavy emphasis has been placed on selecting the experiment independent, dependent, and control variables. Of equal importance was the skill level evaluation of the class III Cadets which have been volunteered as JANUS operators as well as their subsequent training. Ultimately, the software, mathematical, and experimentation aspects combined yield a framework for conducting large-scale defense experiments. As for the simulation per se, results indicate a considerable advantage to UAV possession as a reconnaissance asset, as scout platoons equipped with a UAV were able on average to fire more rounds over longer distances and inflict more losses to enemy forces. Ultimately, these factors enabled friendly units to accomplish their objectives.

Keywords Constructive simulation • Combat model • Defense experimentation • UAV • Military simulation • Wargames • Scout platoon • Descriptive statistics • Termination criteria • JANUS

Introduction

During recent years a rapid increase of the worldwide military conflicts complexity in tactical, operational, and strategic level has been observed. Said complexity combined with requirements regarding operational cost curbing, environmental protection, as well as increased force protection measures, led to the realization that Western armed forces need to focus on modernizing personnel training, on developing and testing doctrines reflecting the context within current conflict occurs, and on implementing new systems or upgrading existing ones in order to satisfy any set of realistic operational constraints and to furthermore achieve the expected goals in the best possible way [1]. Within the typical framework of a nations' armed forces the set of requirements along with the associated expected outcome is determined by the General Staff whereas the assessment of a proposed operational solution is carried out among others by a dedicated military simulation group tasked with defense experimentation. Given the highly technical nature of military simulation and the complex nature of conflict itself, in order to deliver meaningful answers to upper echelons this simulation group should be fully familiar with both the underlying mathematical model [2] and the actual simulation system, typically a combination of specialized software and hardware.

Overview and Goals

KEPYES, the Hellenic Army IT center, has a Military Simulation and Wargames staff section which operates within the general framework of developing, evaluating, and deploying defense experimentation tools and systems in order to assist the Hellenic Army General Staff select the best possible alternative among Hellenic Army operational options.

During spring of 2012 KEPYES Military Simulation and Wargames designed and oversaw the first stage of a larger campaign of experiments [2–4]. In these experiments a group of class III Cadet Officers from the Hellenic Military Academy initially were trained in the use of the latest edition of JANUS version 7.3 war game. Subsequently, they assumed the role of a typical in size and composition mechanized infantry platoon commander in the JANUS digital battlefield. An extensive series of simulations evaluated the impact of providing a suitably equipped UAV as a reconnaissance asset to the scout platoon. Besides this obvious objective, two longer-term goals were to create a defense experimentation framework and a statistical analysis framework in order for reaching conclusions quickly and accurately.

Methodology

In this section the experimentation methodology is outlined. The experiment setup is explained, followed by a detailed outline of the simulation scenario as it was played in JANUS platform, and finally the actual JANUS parameters are listed.

Simulation Setup and Operator Selection

At the early experiment design stages, given that JANUS was the simulation platform of choice, human-in-the-loop simulation methodology [2] has been selected. The latter implies that human operators with continuous, real-time interaction with simulated forces and/or equipment were to be part of the simulation process. In general, human operators must be selected in such a way as not to interfere in any conceivable way with the simulation outcome, unless of course their performance or another aspect of their behavior is to be observed.

As it was stated in the goals section, the immediate experimentation objective was to quantify the effects of adding a UAV to a scout platoon assets. Therefore, any human operator interference had to be isolated. To this end, the original pool of 60 class III Cadet Officers was divided into two equally sized groups A and B in a way that the distributions of the military science course weighted averages were almost identical within the two groups. Cadets of group A were operating a UAV whether those of group B were relying on existing target acquiring systems (Fig. 1).

For the experiment purposes, two personal figures of merit for each Cadet have been computed. $Tmil_i$ is based only on military science courses and is an indicator of each Cadet military skill level, whereas G_i is based on all courses and is an indicator of each Cadet overall skill level. The subscript i ranges over the N Cadets that have been volunteered for this experiment.

$Tmil_i$ is calculated in the following manner. Each applied military training course a had a weight of 0.75, each staff course b had a weight of 0.15, and each leadership course c had a weight of 0.1. Let a_i^A, b_i^A, and c_i^A be the course grades for cadet i for

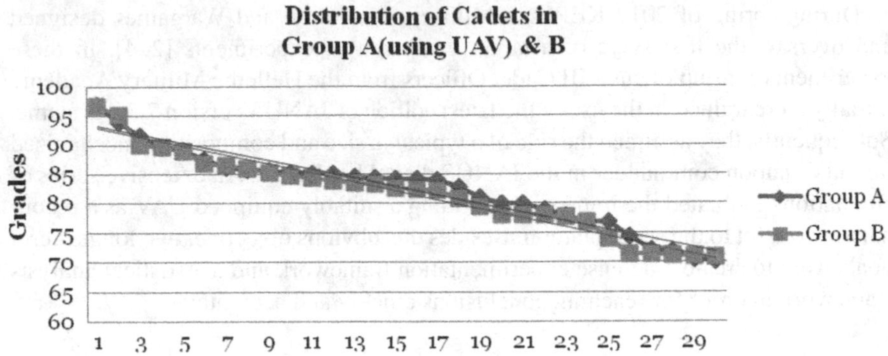

Fig. 1 Weighted course grade average of the Cadets

the first year of studies and a_i^B, b_i^B, c_i^B for the second. Then, the weighted average of the first- and second-year courses yield $Tmil_i^A$ and $Tmil_i^B$, respectively according to the formulae

$$Tmil_i^A = 0.75 \sum_{a_i^A} a_i^A + 0.15 \sum_{b_i^A} b_i^A + 0.1 \sum_{c_i^A} c_i^A, \qquad 1 \le i \le N$$

$$Tmil_i^B = 0.75 \sum_{a_i^B} a_i^B + 0.15 \sum_{b_i^B} b_i^B + 0.1 \sum_{c_i^B} c_i^B, \qquad 1 \le i \le N$$

and finally

$$Tmil = 0.6\, Tmil_i^B + 0.4\, Tmil_i^A, \qquad 1 \le i \le N$$

Let p_i^A and p_i^B be the course grade for Cadet i, where p^A and p^B ranges over courses taught in first- and second-year of study, respectively. First, p_i^A and p_i^B are normalized as percentiles of the respective maximum grades

$$g_i^A = \frac{p_i^A}{\max\{g_i^A\}} 100, \qquad 1 \le i \le N$$

$$g_i^B = \frac{p_i^B}{\max\{g_i^B\}} 100, \qquad 1 \le i \le N$$

Then

$$G_i^A = \frac{1}{|g_i^A|} \sum_{g_i^A} g_i^A, \qquad 1 \le i \le N$$

$$G_i^B = \frac{1}{|g_i^B|} \sum_{g_i^B} g_i^B, \qquad 1 \le i \le N$$

and finally

$$G_i = 0.6\,G_i^B + 0.4\,G_i^A, \qquad 1 \leq i \leq N$$

The final ranking W_i for each Cadet was the weight linear combination of the military ranking $Tmil_i$ and the general ranking G_i as follows

$$W_i = 0.6\,Tmil_i + 0.4\,G_i, \qquad 1 \leq i \leq N$$

Once the rankings were available, they have been sorted and every other Cadet has been assigned to group B. This simple scheme yielded two groups whose skill distributions were very similar. Class III Cadet groups A and B can be considered homogeneous for our statistical analysis purposes for the following reasons.

- They are of relatively young age and, thus, their overall experiences are still limited.
- They have quite similar social background and they have received formal primary and secondary education under a centralized education system.
- They have received the same level of training with the same intensity. This is particularly true both for the infantry tactical training at the team and platoon levels and for the JANUS software suite itself.

As a final notice, although the original Cadet pool was of mixed gender, creating separate two subgroups, one for male and one for female Cadets, within each group A and B served no apparent purpose, as human influence to the experiment outcome had to be isolated. Instead, a gender-neutral policy was deemed appropriate for the purposes of this experiment.

Scenario

The scenario is built around blue force alpha, a reinforced mechanized infantry scout platoon operated by a Cadet Officer comprised of 2 Leopard 2A4 and 2 VBL armored scout vehicles. Blue force alpha spearheads the advance of its parent company, blue force beta, which has been ordered to capture an objective point code named P3. On the other side, there is red force, a reinforced tank platoon consisting of 4 M-48A3 MOLF, 1 mechanized infantry squadron, 2 M901 ITV, and 2 armored scout vehicles. Red force, operated as combat outpost, has been assigned the triple task to prevent blue forces from observing the defensive actions undertaken by other red forces in the area, to make blue forces assume battle formation, and to inflict casualties on the blue forces. To this end, red force has deployed its elements in four areas A, B, C, and D, all strategically located close to P3.

The weather conditions correspond to a clear summer day in with excellent visibility with both sides equally lit as the scenario starts at seven am local time. The terrain is mostly plains surrounded by hills with lots of vegetation, providing

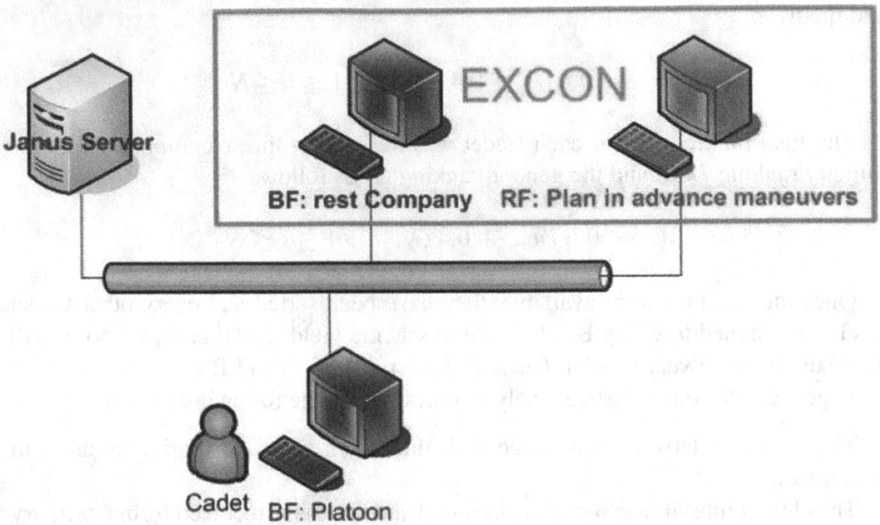

Fig. 2 Simulation network architecture

sufficient cover for both the attacker and the defender. Neither side has access to indirect fire or non-conventional weapons. Finally, all vehicles are in their respective basic configuration and personnel is carrying only standard issued equipment and weaponry. In short, this scenario represents a very common case of infantry combat (Fig. 2).

Part of the scenario were also the following simulation termination criteria, which have been derived from both national infantry military regulations and from NATO guidelines.

- T1: Blue force beta occupies and maintains a garrison to P3 in less than 3 h (victory).
- T2: Blue force beta fails to capture P3 within 3 h (defeat-unacceptable advance delay).
- T3: Blue force alpha losses exceed 30% (defeat-scout platoon in need of replacements).

The actual combat scenario proceeds as follows. As blue force alpha advances, it inevitably encounters the red force elements. At each of these contact points, the blue force alpha commander must assess the current tactical situation and judge as accurately as possible whether blue force alpha can by itself engage and overcome the opposition or not. In the former case, blue force alpha attacks, while, in the latter case, blue force alpha goes into defilade mode and waits for the arrival of blue force beta. In each of these cases, there are four possible outcomes (Fig. 3).

- Blue force alpha can overcome the opposition and its commander decides to attack. In this case, blue force alpha clears the advance path in minimal time with

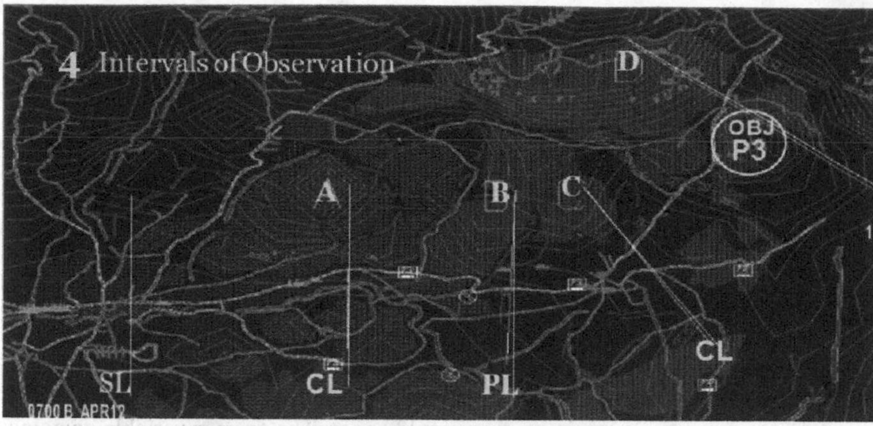

Fig. 3 Main points of target overlay (EXCON view in JANUS)

zero or few casualties. It is the optimal case from the blue perspective, but this case in the "run times" of the scenario appears rarely.

- Blue force alpha can overcome the opposition but its commander decides to wait for blue force beta, the latter being strong enough to overcome any red force element. However, local victory comes at the expense of time. This may endanger the entire mission if termination criterion T2 is met and, thus, commanders who wait too often are penalized.
- Blue force alpha cannot overcome the opposition and its commander decides to wait for blue force beta. As in the previous case, the overall advance is delayed and the termination condition T2 may hold true. However, blue force alpha remains intact and, hence, there is no risk for triggering termination condition T3.
- Blue force alpha cannot overcome the opposition but its commander decides to attack. In this case, blue force alpha either destroys the opposing forces after having suffered heavy losses or it is itself destroyed. Therefore, termination criterion T3 may be satisfied with high probability, which is the toll for an excessively aggressive tactic.

The above implies that the termination conditions T2 and T3 are complementary in the sense that the former excludes an overtly quiet tactic with no or few blue force alpha engagements, while the latter prohibits too risky tactics with blue force alpha engaging opposition at each opportunity. Between these two extrema there is middle ground for a number of tactics as well as limited margin for judgement errors on behalf of blue force alpha commander.

Blue force alpha commander judgement is formulated by skill, accounting for training and experience combined, as well as the reconnaissance data collected by blue force alpha vehicles and UAV, wherever applicable. Given that Cadet skill level distributions within groups A and B are identical as explained earlier, any difference in blue force alpha combat power should be attributed to reconnaissance data quality difference. In turn, since blue force alpha has identical composition across all scenario runs, it follows that any observed blue force alpha performance change should be ultimately attributed to UAV availability or lack thereof.

Experiment Variables and Experiment Hypotheses

Based on combat power definition [6] "*[...] elements are maneuver, firepower, protection, leadership and information guide the employment of all infantry forces [...]*," the dependent variables of interest in this campaign of experiments were the number of blue casualties and the occupation (or not) of the final objective by the friendly forces. The latter variable is qualitative in nature and outlines the general scenario outcome as a strategic planner would see it. The former variable contains information regarding the actual engagement and is mostly of interest to tactical planners. Other dependent variables to be assessed were the engagement average distance, the time taken to reach each objective, the percentage of missed shots, the number of objectives captured, and the kill exchange ratio of between blue and red forces.

The only control variable was whether blue force alpha had a UAV (or not) at its disposal. In order to ensure that there were no other hidden control variables affecting the experiment outcome, every parameter taken into account by operation planners, such as red force strength, terrain, and weather, has been the same across all scenario runs. Identifying all such possible control variables and the way JANUS handles them required familiarization with the combat simulation software suite. The independent variables were the strength and deployment pattern of the red force, the weather, the terrain, and the strength of blue force alpha and beta.

Finally, the null hypothesis was that the UAVs had no effect to the overall platoon combat power.

JANUS Run-Time Settings and UAV Settings

The weather conditions were typical of a summer Mediterranean day with a clear sky and full sunlight, namely Ambient Light Level 3, with a visibility of 12 Km, 90 degrees wind direction, 11 Km/h wind velocity, 6 Km cloud ceiling, 30% relative humidity, and 29 degrees of Celsius. Blue and red forces were equally lit and neither of them could exploit weather conditions to their advantage.

Defilade time, namely the time required for a stationary unit status to switch from exposed status to partial defilade, was 30 s. Detection cycle, the time required for a unit to complete a target detection cycle, was 3 s. Target list cycle, defined as the longest time for direct fire units to acquire targets and update their target list, was 50 s. Return to duty time, the amount of time a soldier system will be inactive while performing first aid, was 15 min. The hit and kill probability has been the

Table 1 JANUS database parameters for the new UAV

Parameter	Value
Speed (Km/h)	48–96
Flight time (min)	60–90
Altitude (m)	152
Sensor type	Classified
Machine type	Electrical rotor
Range of control	Classified

Table 2 Simulation results

Performance indicator	Group A	Group B
Percentage of victorious runs	10	0
Average shots (standard deviation)	90 (36.77)	74 (30.41)
Average engage distance	1,581 m	1,369 m
Average casualties	16	20

same in each of the 30 runs for each of the two groups. Finally, in each scenario execution JANUS linear congruential pseudo-number generator seed was always the same integer value.

Although JANUS database has a UAV type, early trial scenario runs have shown that its specifications and operating mode are unsuitable for a scout platoon. Thus, a new UAV type had to be created. The specifications used for the creation of a T-mini UAV model in JANUS database are summarized in Table 1. A run speed factor of 4.00 will run at approximately 10 min of simulated time for every minute of real time.

Results

Using the JANUS Analyst Workstation and its Post Processing capabilities, a large report file in text format has been generated for each scenario run from the JANUS binary event recording files. These ASCII report files were in turn parsed using a custom shell script in order for data of interest to the specific experiment to be isolated and extracted among the detailed account of each combat event.

In total 8 scenario run from both Cadet groups that were deemed as outliers and were not examined further. Simulation data from the remaining 52 runs were processed using descriptive statistics and survival analysis in MATLAB and STATA and yielded the results of Table 2. Points A, B, C, and D have been used as references in Figs. 4 and 5.

As it is shown in Table 2, 10% of group A commanders accomplished the mission in contrast to group B, where no commander managed to do so. Additionally, blue forces in group A on average fired 20% more rounds over a 14% longer distance and, in principle at least, had better chances to inflict losses to the red force elements

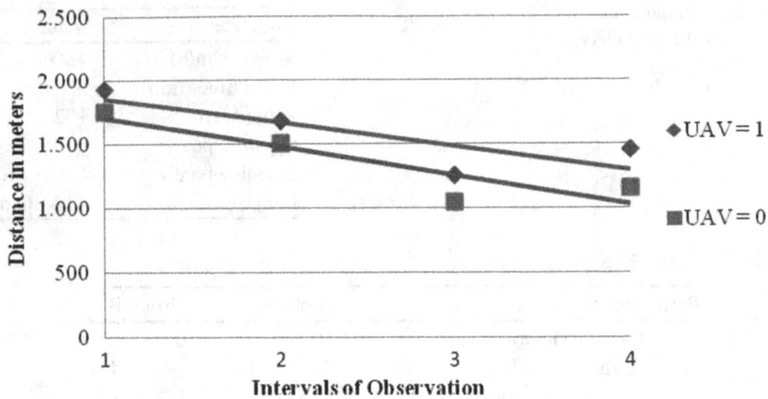

Fig. 4 Average distance of shots for blue force alpha

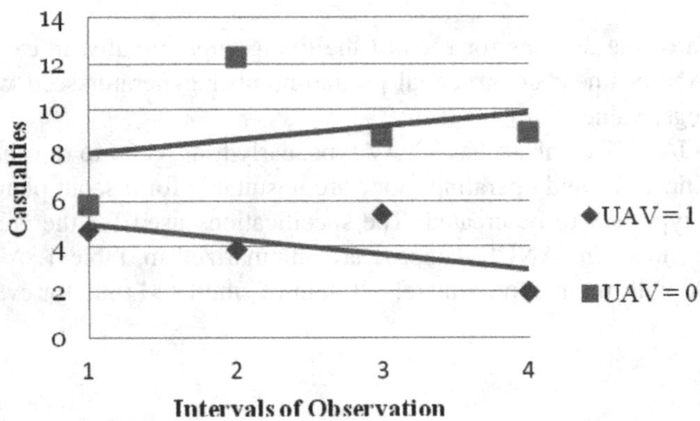

Fig. 5 Average number of casualties for blue force alpha

from a safer distance. At the same time, blue forces in group A suffered on average 20% less casualties. Notice that the above results are statistically significant as p was less than 0.05.

One of JANUS known limitations [5], circumvented in modern simulation platforms, is its lack of human behavior modeling. In any scenario execution friendly and enemy units carry out their orders with a varying degree of efficiency independent of combat events and general troop morale and affected primarily by unit operational status and type—for instance a sniper is better in target acquiring than a regular infantryman. Simulation results should be interpreted therefore under this light.

Conclusions

This paper outlines a summary of the results obtained from of a recent large scale campaign of experiments, which took place at the premises of the Hellenic Military Academy using the JANUS military simulation suite. These experiments have been conducted under the supervision of KEPYES Military Simulation and Wargames staff section with the assistance of class III Cadet volunteers acting as JANUS operators. To the best knowledge of the authors, this is the first time that a campaign of experiments of this scale took place at the Hellenic Military Academy. This implied that KEPYES Military Simulation and Wargames staff section had to develop most analytical and software tools from start, including a bash script for parsing JANUS long output, STATA and MATLAB functions to compute statistical quantities of interest, and to determine the dependent, independent, and control variables. Though not a trivial task, an initial Hellenic experimentation framework, covering software, mathematical, and experimentation aspects, has been established to serve as a future guideline.

In parallel to developing this framework, the actual simulation runs took place. The simulation scenario entails the advance of a mechanized infantry scout platoon, a common and realistic situation in ground combat, and aims at evaluating the effects of adding a UAV to this platoon as a reconnaissance asset. Results obtained through statistical analysis revealed that UAV possession considerably improves battlefield survival probability. A quantitative review indicated that UAVs compared to existing observation and target acquisition systems extended effective visibility, which in turn translated to more accurate shots over 14% longer distances on average as well as to 20% less casualties on average.

Major efforts have been placed on ensuring that human factor did not influence the simulation outcome. To this end, it was imperative that the JANUS volunteer operator pool be divided into two equally sized groups A and B of comparable capabilities. This has been achieved by ranking the Cadets according to a weighted average of their course grades, sorting the result, and assigning every other Cadet to group B. This practice resulted in two groups whose ranking distributions were extremely similar. Group B operators relied only on reconnaissance data from their scout platoon vehicles, whereas group A operators could additionally deploy a UAV.

To obtain more accurate conclusions besides the general trend, more experiments need to be conducted regarding the UAV integration into army tactical echelons. Moreover, it should be underlined that a detailed knowledge of combat model used by the simulation software is a prerequisite for defense experimentation.

References

1. NATO WG CD and E, 2011
2. Code of Best Practice “Experimentation”, DoD Command and Control Research Program (CCRP) Publication Series, Alberts, David S., Hayes, Richard E, July 2002

3. Code of Best Practice “Campaigns of Experimentation”, DoD Command and Control Research Program (CCRP) Publication Series, Alberts, David S., Hayes, Richard E, March 2005
4. Guide for Understanding and Implementing Defense Experimentation (GUIDEx), The Technical Cooperation Program (TTCP), Version 1.1 February 2006
5. JANUS Model Documentation, KEPYES 2012.
6. FM 3-21-71 Mechanized infantry platoon and squad (Bradely fighting vehicle), US Army 2002.

Balanced Integer Solutions of Linear Equations

Konstantinos A. Draziotis

Abstract We use lattice-based methods in order to get an integer solution of the linear equation $a_1x_1 + \cdots + a_nx_n = a_0$, which satisfies the bound constraints $|x_j| \leq X_j$. Further we study the corresponding homogeneous linear equation under constraints and finally we apply our method to Knapsack problem.

Keywords Linear diophantine equation • Lattice • LLL • CVP

2000 Mathematics Subject Classification: Primary 11D04; Secondary 11Y50.

Introduction-Statement of Results

Let $a_j \in \mathbb{Z} - \{0\}$, $(0 \leq j \leq n)$. We consider the linear equation

$$f(x_1, \ldots, x_n) = \sum_{j=1}^{n} a_j x_j = a_0. \tag{1}$$

We are interested in the integer solutions of (1) under the constraints $|x_j| \leq X_j$, for some $X_j \in \mathbb{Z}_{>0}$. This is an NP-complete problem, but without the bound constraints is solved in polynomial time. This problem has some important applications in discrete optimization, in designing integrated circuits [1] and is also applied in Merkle-Hellman and the Chor-Rivest knapsack cryptography systems [10, 12]. Further we shall apply our method to the knapsack problem of density 1 and dimension ≤ 40.

K.A. Draziotis (✉)
Department of Informatics, Aristotle University of Thessaloniki, 54124, Thessaloniki, Greece
e-mail: drazioti@csd.auth.gr; drazioti@gmail.com

N.J. Daras (ed.), *Applications of Mathematics and Informatics in Science and Engineering*, 173
Springer Optimization and Its Applications 91, DOI 10.1007/978-3-319-04720-1_11,

In [1] the authors found a method for the solutions of Eq. (1) under the bound constraints $0 < x_j < X_j$. Their method contains two parts, one deterministic (application of LLL) having polynomial time complexity and the other heuristic. Further, in [15] the method of Rosser starts with the matrix $M = (I_n, \mathbf{a}^T)$, $\mathbf{a} = (a_1, \ldots, a_n)$ and a new matrix M' is obtained (using linear integer transformations) with $M' = (U, \mathbf{d}^T)$, $\mathbf{d} = (0, 0, \ldots, \gcd(a_1, \ldots, a_n))$ and $U \in SL_n(\mathbb{Z})$. Then the general solution can be expressed in terms of the row vectors of M'. Many recent methods are based on the original idea of Rosser.

Also there is another approach to this problem which is based to the Closest Vector Problem (CVP). Let $\mathbf{y} = (y_1, y_2, \ldots, y_n)$ (the target vector) be a solution of (1), not necessarily small (for instance, one can use Euclidean algorithm). Let L be the lattice generated by the solutions of the homogeneous equation $\sum_{j=1}^{n} a_j x_j = 0$. We solve the CVP instance $CVP(L, \mathbf{y})$ and we get the solution, say $\mathbf{t} = (t_1, t_2, .., t_n)$ (cvp vector). Then $\mathbf{x} = \{\text{target vector}\} - \{\text{cvp vector}\} = \mathbf{y} - \mathbf{t}$ is an integer solution of (1) and has small absolute value $||\mathbf{x}||$. For this approach for instance see [13]. Implementations of CVP can be found in fplll [14] and in Magma [2]. Here we use fplll. We shall provide some examples which compare the CVP approach with our algorithm and we shall conclude (based on experiments) that the better strategy is to combine the two methods (for $n \leq 60$). For large values of n, say $n \geq 80$, the CVP-solver of fplll (and Magma) is (relatively) slow. So in this case our strategy is to apply our algorithm which is fast for large values of n. For instance, we made some tests for $n = 100$ and our algorithm terminated (with a small solution) in less than 5 s in all the examples while fplll needed at least 3 h of running time in a 3Ghz Dual Core Pc (without getting any solution). Further, the algorithm given in [1] is also fast for large values of n (since it uses LLL) but in general it gives longest vectors than ours (see example (ii) in section "Examples").

Before we state our basic result we must set up some notation. Let $\{\mathbf{e}_1, \ldots, \mathbf{e}_{n+2}\}$ be the standard basis of $\mathbf{R}^{n+2}$, that is $\mathbf{e}_j = (\ldots, \delta_{ij}, \ldots)$, where δ_{ij} is the delta of Kronecker and is located in the jth entry of the vector $\mathbf{e}_j$. We define the vectors

$$\mathbf{b}_j = \frac{1}{X_j}\mathbf{e}_j + a_j \mathbf{e}_{n+2} \quad (1 \leq j \leq n).$$

Let L be the lattice of $\mathbf{R}^{n+2}$ spanned by the vectors $\{\mathbf{b}_1, \ldots, \mathbf{b}_n\}$. In matrix form L is spanned by the rows of the $n \times (n+2)$ matrix

$$B = [\mathbf{b}_1, .., \mathbf{b}_n] = \begin{bmatrix} \frac{1}{X_1} & 0 & \cdots & 0 & 0 & a_1 \\ 0 & \frac{1}{X_2} & \cdots & 0 & 0 & a_2 \\ \cdots & \cdots & \cdots & \cdots & \cdots & \cdots \\ 0 & 0 & \cdots & \frac{1}{X_n} & 0 & a_n \end{bmatrix}.$$

(For reasons that later will become clear, we added the column with zeros). Let $\{\mathbf{b}'_1, \ldots, \mathbf{b}'_n\}$ be its LLL-reduced basis. We set $\mathbf{b}'_j = (b'_{j1}, \ldots, b'_{j,n+2})$. We consider the Gram-Schmidt orthogonalization process

$$\mathbf{b}_1'^* = \mathbf{b}_1', \mathbf{b}_i'^* = \mathbf{b}_i' - \sum_{j=1}^{i-1} \mu_{ij} \mathbf{b}_j'^*, \tag{2}$$

where

$$\mu_{ij} = \frac{\mathbf{b}_i' \cdot \mathbf{b}_j'^*}{B_j^2}, \quad B_j = ||\mathbf{b}_j'^*||.$$

Also if $\lfloor \cdot \rfloor$ is the floor function, we define $\lceil x \rfloor = \lfloor x + 0.5 \rfloor$ with $x \in \mathbf{R}$. That is the closest integer to x.

We shall prove the following Theorem.

Theorem 1. *Let* $\gcd(a_1, \ldots, a_n) = 1$. *If the following two assumptions hold*

A_1. $(a_n X_n)^2 + (a_j X_j)^2 < \dfrac{1}{2^{n+1}} (X_n X_j)^2, \ j = 1, 2, \ldots, n-1, \ X_j \in \mathbb{Z}_{>0}$,

A_2. $\left\lceil \dfrac{a_0}{B_n^2} \right\rfloor = a_0$,

then there is an integer solution $(x_1, \ldots, x_n)$ *of Eq. (1), such that*

$$|x_j| < c(n) \prod_{i=1}^{n} X_i, \ \ j = 1, 2, \ldots, n, \ c(n) = \sqrt{3}(1.25)^{(n-1)/2}. \tag{3}$$

Further, we obtain this solution in polynomial time.

Assumption A_1 guarantees that LLL-reduction will generate (some) solutions of the homogeneous linear Eq. (thus balanced multipliers for the $\gcd(a_1, \ldots, a_n)$). Further, this assumption is crucial for the computation of integer solutions of Eq. (1). Assumption A_2 can be rewritten

$$a_0 \leq \frac{a_0}{B_n^2} + \frac{1}{2} < a_0 + 1,$$

thus,

$$B_n^2 \left(a_0 - \frac{1}{2}\right) \leq a_0 < B_n^2 (a_0 + \frac{1}{2}). \tag{4}$$

This assumption guarantees that our procedure will end up with a solution to Eq. (1). The type of lattice we used here looks like the one used by Coppersmith in [3], which is used in order to attack the RSA cryptosystem. Although the bound is seemingly theoretical, in practice we get the desired solution (if there is any) $|x_j| \leq X_j$, without the strong assumption A_1. As far as I know, the methods given in the bibliography for the solution of the problem (1) under $|x_j| \leq X_j$ provide us with a small solution, but there is not any theoretical result that guarantees that the method will work. Our method provides a theoretical result, in the sense that if assumptions A_1, A_2 are fulfilled, then the method shall work and also experiments show that the method will end up with solutions satisfying $|x_j| < X_j$, instead of inequality (3).

If $x_j \in \mathbb{N}$ and $a_0 = 0$ (i.e., the homogeneous version of our problem) then the problem of deciding if there is any integer solution is NP-complete [11]. We study a variant of this problem [see Lemma 1 and Proposition 2]. In fact we prove that if there is a Shortest Vector Problem (SVP) oracle then we can find an integer solution of $\sum_{j=1}^{n} a_j x_j = 0$, with $|x_j| < \sqrt{2} \max_{1 \le i \le n}\{|a_i|\}$ $(1 \le j \le n)$. Finally, If $a_j > 0$ and we restrict the solutions $x_j \in \{0, 1\}$, then we have the $0 - 1$ Knapsack or subset sum problem. The decisional version is NP-complete [6]. This has many applications in public key cryptography. We shall apply our method in this problem.

We give a brief outline of the paper. In the next section we give some preliminaries propositions about LLL and in section "Auxiliary Results" some basic auxiliary results which we shall use for the proof of our theorem in section "Proof of the Theorem". In section "The Homogeneous Case" we study the homogeneous linear equation and in section "Examples" we provide some examples. In the last section we use our approach to knapsack problem and give some examples.

Preliminaries on LLL

Our method uses LLL-reduction algorithm, but with a different lattice than the one used in [1]. Lattices have many applications in cryptanalysis, for example [3, 9]. Here we shall not provide analytically the theory of lattices and LLL-algorithm. For instance, the reader can study [7, 17].

Definition 1. A subset $L \subset \mathbf{R}^n$ is called a lattice if there exists linearly independent vectors $\mathbf{b}_1, \mathbf{b}_2, \ldots, \mathbf{b}_k$ of $\mathbf{R}^n$ such that

$$L = \left\{ \sum_{j=1}^{k} \alpha_j \mathbf{b}_j : \alpha_j \in \mathbb{Z}, 1 \le j \le k \right\} := L(\mathbf{b}_1, \mathbf{b}_2, \ldots, \mathbf{b}_k).$$

The vectors $\mathbf{b}_1, \mathbf{b}_2, \ldots, \mathbf{b}_k$ are called a lattice basis of L.

All the bases have the same number of elements, and this common number is called the *dimension* or *rank* of the lattice.

Lemma 1. *Let* $\mathbf{b}_1, \mathbf{b}_2, \ldots, \mathbf{b}_k$ *be an* LLL*-reduced basis of the lattice* $L \subset \mathbf{R}^n$ *and* $\mathbf{x}_1, \mathbf{x}_2, \ldots, \mathbf{x}_t$ *linearly independent vectors in* L. *Then for all* $j \le t$ *we have*

$$||\mathbf{b}_j||^2 \le 2^{n-1} \max\{||\mathbf{x}_1||^2, \ldots, ||\mathbf{x}_t||^2\}.$$

Proof. [17, Theorem 7.10]. □

Auxiliary Results

Let L be the lattice of $\mathbf{R}^{n+2}$ spanned by the vectors $\{\mathbf{b}_1, \ldots, \mathbf{b}_n\}$ and let $\{\mathbf{b}'_1, \ldots, \mathbf{b}'_n\}$ be its LLL-reduced basis. We set

$$\mathbf{b}'_j = (b'_{j1}, \ldots, b'_{jn}, b'_{j,n+1}, b'_{j,n+2}).$$

Further, from Gram-Schimdt orthogonalization process we get the vectors,

$$\mathbf{b}'^*_i = (\hat{b}^*_{i1}, \ldots, \hat{b}^*_{in}, \hat{b}^*_{i,n+1}, \hat{b}^*_{i,n+2}),\ (1 \le i \le n),$$

where $\hat{b}^*_{ij} = \dfrac{b^*_{ij}}{X_j}$ and $\hat{A} = [\hat{b}^*_{ij}]_{1 \le i,j \le n}$. We shall prove the following Lemma.

Lemma 1.

$$\prod_{j=1}^{n} X_j = \frac{1}{|\det \hat{A}|}.$$

Proof. Let B, B' be the matrices $[\mathbf{b}_1, \ldots, \mathbf{b}_n]^T$, $[\mathbf{b}'_1, \ldots, \mathbf{b}'_n]^T$, respectively (we work row-wise). Then there is a matrix $U = [\lambda_{ij}]_{1 \le i,j \le n} \in SL_n(\mathbb{Z})$ such that $B' = UB$. Indeed, we consider the $n \times n$ matrix $U_1 = \mathrm{diag}(\frac{1}{X_1}, \ldots, \frac{1}{X_n})$. We apply LLL to the rows of B, thus we get B'. The linear changes and swaps of rows made by LLL in B also applied to U_1, since $B = [U_1, \mathbf{a}^T]$ where, $\mathbf{a} = (a_1, \ldots, a_n)$. Then we get a new $n \times n$ matrix U_2 which equals to $\left[\frac{\lambda_{ij}}{X_j}\right]_{1 \le i,j, \le n}$. So

$$|\det U_1| = |\det U_2| = \frac{1}{\prod_{j=1}^{n} X_j}.$$

Also,

$$|\det U_2| = \frac{1}{\prod_{j=1}^{n} X_j} |\det U|,$$

thus $|\det U| = 1$. Now we shall prove that

$$|\det[b^*_{ij}]| = |\det[\lambda_{ij}]| = 1. \tag{5}$$

Indeed, from $B' = UB$ we get

$$\mathbf{b}'_i = \sum_{j=1}^{n} \lambda_{ij} \mathbf{b}_j = \left(b'_{i1}, \ldots, b'_{in}, *, *\right),$$

where λ_{ij}, b'_{ij} are related by $\lambda_{ij} = X_j b'_{ij}$. From relations (2) we get

$$\hat{b}^*_{1j} = b'_{1j},\ \hat{b}^*_{2j} = b'_{2j} - \mu_{21} b'_{1j}, \ldots (1 \leq j \leq n)$$

thus, multiplying by X_j we get

$$b^*_{1j} = \lambda_{1j},\ b^*_{2j} = \lambda_{2j} - \mu_{21}\lambda_{1j}, \ldots (1 \leq j \leq n),$$

so (5) is proved. We conclude that

$$|\det \hat{A}| = \frac{1}{\prod_{j=1}^n X_j} |\det[b^*_{ij}]| = \frac{1}{\prod_{j=1}^n X_j}.$$

□

Also we get the following Proposition.

Proposition 2. *Assume that*

$$(a_n X_n)^2 + (a_j X_j)^2 < \frac{1}{2^{n+1}} (X_n X_j)^2, \tag{6}$$

for all j such that $1 \leq j \leq n-1$. Then

$$b'_{j,n+1} = b'_{j,n+2} = 0 \text{ for } 1 \leq j \leq n-1.$$

Further,

$$b'_{n,n+1} = 0,\ |b'_{n,n+2}| = 1.$$

Proof. We set $\mathbf{r}_j = a_n \mathbf{b}_j - a_j \mathbf{b}_n$. These are independent in L, so from Lemma 1 we get

$$||\mathbf{b}'_j||^2 \leq 2^{n+1} \max_{1 \leq j \leq n-1} \{||\mathbf{r}_j||^2\} \leq 2^{n+1} \max_{1 \leq j \leq n-1} \left\{ \frac{a_n^2}{X_j^2} + \frac{a_j^2}{X_n^2} \right\}.$$

Since

$$(a_n X_n)^2 + (a_j X_j)^2 < \frac{1}{2^{n+1}} (X_n X_j)^2$$

we get

$$2^{n+1} \left(\frac{a_n^2}{X_j^2} + \frac{a_j^2}{X_n^2} \right) < 1\ (1 \leq j \leq n-1).$$

Thus $||\mathbf{b}'_j||^2 < 1$. Now we assume that $b'_{j,r} \neq 0$ for $r \in \{n+1, n+2\}$ and $1 \leq j \leq n-1$. Since $b'_{j,n}, b'_{j,n+1} \in \mathbb{Z}$ we get $||\mathbf{b}'_j||^2 \geq 1$, contradicting to the previous inequality. The first part of the Proposition follows.

Since LLL-algorithm makes only linear transformation of the form

$$\mathbf{b}_j \leftrightarrow \mathbf{b}_i \text{ or } \mathbf{b}_j \leftarrow \mathbf{b}_j - r\mathbf{b}_i,\ r \in \mathbb{Z},\ j > i$$

and the $(n+1)$th column consists only from zeros, we get $b'_{n,n+1} = 0$. Further, the gcd of the last column remain the same in every step of the LLL-process. Since we assumed that $\gcd(a_1, \ldots, a_n) = 1$ we get

$$\begin{aligned} 1 = \gcd(a_1, \ldots, a_n) &= \gcd(b'_{1,n+2}, b'_{2,n+2}, \ldots, b'_{n,n+2}) = \gcd(0, 0, \ldots, 0, b'_{n,n+2}) \\ &= |b'_{n,n+2}|. \end{aligned}$$

□

Remark 3. From assumption (6) we get

$$|a_n| < \frac{1}{2^{(n+1)/2}} X_j,\ (1 \leq j \leq n-1) \text{ and } |a_j| < \frac{1}{2^{(n+1)/2}} X_n.$$

We shall show that, under the assumption (6), we have $\hat{b}^*_{i,n+1} = \hat{b}^*_{i,n+2} = 0$ for $1 \leq i \leq n-1$ and $\hat{b}^*_{n,n+1} = 0,\ \hat{b}^*_{n,n+2} = \pm 1$. Indeed, it is easy for $\mathbf{b}_1'^*$ since it is equal to $\mathbf{b}'_1 = (\ldots, 0, 0)$. Also for

$$\mathbf{b}_2'^* = \mathbf{b}'_2 - \mu_{21}\mathbf{b}_1'^* = (*, *, \ldots, 0, 0) - \mu_{21}(*, *, \ldots, 0, 0) = (*, *, \ldots, 0, 0).$$

Inductively we can show that $\mathbf{b}_i'^* = (*, *, \ldots, 0, 0)$, for $1 \leq i \leq n-1$. For

$$\begin{aligned} \mathbf{b}_n'^* &= (*, *, \ldots, 0, \pm 1) - \mu_{n2}(*, *, \ldots, 0, 0) - \cdots - \mu_{n,n-1}(*, *, \ldots, 0, 0) \\ &= (*, *, \ldots, 0, \pm 1). \end{aligned}$$

So we proved the following.

Corollary 4. *Under the assumption (6) we get*

$$\mathbf{b}_i'^* = (\hat{b}^*_{i1}, \ldots, \hat{b}^*_{in}, 0, 0)\ (1 \leq i \leq n-1) \textit{ and } \hat{b}^*_{n,n+1} = 0,\ |\hat{b}^*_{n,n+2}| = 1.$$

The Homogeneous Case

We get the following Lemma concerning the integer solutions of the corresponding homogeneous linear equation.

Lemma 1. *Under the assumption* A_1*, we can find an integer solution* $(x_1, \ldots, x_n)$ *of the homogeneous linear equation* $\sum_{j=1}^{n} a_j x_j = 0$*, with* $|x_j| < X_j$*, in polynomial time.*

Proof. Every vector of the LLL-reduced basis of the lattice L is of the form

$$\Big(\frac{\lambda_1}{X_1}, \ldots, \frac{\lambda_n}{X_n}, \beta, \sum_{j=1}^{n} \lambda_j a_j - \beta a_0\Big),$$

with $\lambda_j, \beta \in \mathbb{Z}$ (note also that the two last coordinates are integers). From the previous Proposition we get $\beta = 0$ and $\sum_{j=1}^{n} \lambda_j a_j = 0$ for the first $n-1$ vectors of the LLL-reduced basis. Thus, the n-tuples

$$\{\lambda_{ij} : 1 \leq j \leq n\}$$

for $i = 1, 2, \ldots, n-1$ are solutions of the equation $a_1 x_1 + \cdots + a_n x_n = 0$. Where λ_{ij} are as in Lemma 5. Moreover, since $||\mathbf{b}_i'||^2 < 1$ for each $i = 1, 2, \ldots, n-1$, we conclude that $|\lambda_{ij}| < X_j$ for each $j \in \{1, 2, \ldots, n-1\}$. □

In the previous Lemma we used LLL reduction. It is known that LLL runs in polynomial time and it provides vectors that are exponentially longer than the shortest vectors. Assume now that we have a SVP oracle. That is a probabilistic algorithm which computes with high probability, a shortest vector of a lattice in polynomial time. LLL-algorithm behaves as a SVP oracle for small dimensions (≤ 40) and further $BKZ - 20$ reduction algorithm for dimensions ≤ 50 (see [5], Fig. 1) (remark that we do not have any proof that BKZ runs in polynomial time, but in practice for dimensions ≤ 50 is fast) . Assuming the existence of a SVP oracle, let $\mathbf{b}$, a shortest vector of our lattice, which has the form

$$\Big(\frac{\lambda_1}{X_1}, \ldots, \frac{\lambda_n}{X_n}, \beta, \sum_{j=1}^{n} \lambda_j a_j - \beta a_0\Big), \ \lambda_j \in \mathbb{Z}.$$

Assume that there is some j_0 with $|a_{j_0}| \neq |a_n|$ and set

$$X = X_1 = \cdots = X_n = \sqrt{2} \max_{1 \leq j \leq n} |a_j|.$$

Since $||\mathbf{b}|| \leq ||\mathbf{r}_j||$, $(1 \leq j \leq n-1)$ we get

$$||\mathbf{b}||^2 \leq ||\mathbf{r}_{j_0}||^2 = \frac{a_n^2 + a_{j_0}^2}{X^2} < 1.$$

Thus necessarily, $\beta = 0$, since if not $||\mathbf{b}|| \geq 1$. So $\beta = 0$ and

$$\Big|\frac{\lambda_j}{X}\Big| < 1, \ \sum_{j=1}^{n} \lambda_j a_j = 0.$$

So we proved the following.

Proposition 2. *Assume that there is a SVP oracle and at least one* $|a_j| \neq |a_n|$. *Then we can compute in polynomial time an integer solution* $(x_1, \ldots, x_n)$ *of the homogeneous equation* $\sum_{j=1}^{n} a_j x_j = 0$, *with* $|x_j| < \sqrt{2} \max_{1 \le i \le n} |a_i|$.

Proof of the Theorem

We set $\mathbf{b}_{n+1} = \mathbf{e}_{n+1} - a_0 \mathbf{e}_{n+2}$. As usual $L = L(\mathbf{b}_1, \ldots, \mathbf{b}_n)$ and $L' = L'(\mathbf{b}'_1, \ldots, \mathbf{b}'_n)$ be the LLL-reduced basis of L. We consider the lattice $\hat{L}$ generated by the set $\{\mathbf{b}'_1, \ldots, \mathbf{b}'_n, \mathbf{b}_{n+1}\}$. We apply size reduction to $\hat{L}$ that is

$$row(n+1) \leftarrow row(n+1) - \lceil \mu_{n+1,n} \rfloor row(n). \tag{7}$$

Note that at this stage $\mu_{n+1,j} = 0$, for $1 \le j \le n-1$. Also, from Proposition 2 we get

$$\mu_{n+1,n} = \frac{1}{B_n^2}(\mathbf{b}_{n+1} \cdot \mathbf{b}'_n) = \frac{1}{B_n^2}\big((0, 0, \ldots, 1, -a_0) \cdot (*, *, \ldots, 0, \pm 1)\big) = \frac{\pm a_0}{B_n^2}.$$

From assumption A_2, we get $\lceil \mu_{n+1,n} \rfloor = a_0$ if $b'_{n,n+2} = -1$ and $\lceil \mu_{n+1,n} \rfloor = -a_0$ if $b'_{n,n+2} = 1$. That is always

$$b'_{n+1,n+2} = -a_0 - \lceil \mu_{n+1,n} \rfloor \cdot b'_{n,n+2} = 0.$$

We continue with

$$row(n+1) \leftarrow row(n+1) - \lceil \mu_{n+1,j} \rfloor row(j) \text{ for } j = 1, 2, \ldots, n-1.$$

Let $\mathbf{b}'_{n+1}$ the size reduced (last) row. Then $\mathbf{b}'_{n+1}$ has the form $(\hat{x}_1, \ldots, \hat{x}_n, 1, 0)$. The new basis $\{\mathbf{b}'_1, \ldots, \mathbf{b}'_n, \mathbf{b}'_{n+1}\}$ has the property $\mu_{n+1,j} < 1/2$ for $1 \le j \le n$. Since

$$\mu_{n+1,j} = \frac{\mathbf{b}'_{n+1} \cdot \mathbf{b}'^*_j}{B_j^2}, \quad B_j = ||\mathbf{b}'^*_j||$$

we get (using Corollary (4)) the system,

$$\hat{x}_1 \hat{b}^*_{j1} + \cdots + \hat{x}_n \hat{b}^*_{jn} = \varepsilon_j, \ 1 \le j \le n \tag{8}$$

where

$$\hat{b}^*_{ij} = \frac{b^*_{ij}}{X_j} \text{ and } |\varepsilon_j| = |\mu_{n+1,j}|\, B_j^2 < \frac{B_j^2}{2}.$$

Since $B_j^2 = ||\mathbf{b}_j'^*||^2 < ||\mathbf{b}_j'||^2$, we get

$$|\varepsilon_j| < \frac{||\mathbf{b}_j'||^2}{2} < \frac{1}{2} \ (1 \le j \le n-1).$$

For the case $j = n$ we get

$$|\varepsilon_n| = |\mu_{n+1,n}|\ B_n^2 < \frac{1}{2}B_n^2 < \frac{1}{2}2 = 1. \tag{9}$$

We used that $B_n^2 < 2$. Indeed, from the first part of inequality (4) we get

$$B_n^2 < \frac{2a_0}{2a_0 - 1}.$$

Since $a_0 \in \mathbb{Z} - \{0\}$, we get

$$0 < \frac{2a_0}{2a_0 - 1} < 2,$$

thus $B_n^2 < 2$. Let $\hat{A}$ be the matrix of the system, that is $\hat{A} = [\hat{b}_{ij}^*]_{1 \le i,j \le n}$. If we substitute the jth column of $\hat{A}$ with the column vector $(\varepsilon_1, \ldots, \varepsilon_n)^T$ we get the matrix $\hat{A}_j$. Then

$$|\hat{x}_j| = \left|\frac{\det \hat{A}_j}{\det \hat{A}}\right|.$$

From Lemma (1), the system (8) has determinant $|\det \hat{A}| = \dfrac{1}{\prod_{i=1}^n X_i}$. Thus

$$|\hat{x}_j| = |\det \hat{A}_j| \prod_{i=1}^n X_i. \tag{10}$$

We apply Hadamard inequality to $\det \hat{A}_j$. We shall get

$$|\det \hat{A}_j|^2 < \prod_{i=1}^n ||row[\hat{A}_j]_i||^2. \tag{11}$$

The ith row of the matrix $\hat{A}_j$ is

$$row[\hat{A}_j]_i = \left(\frac{b_{i1}^*}{X_1}, \ldots, \varepsilon_i, \ldots, \frac{b_{in}^*}{X_n}\right),$$

where ε_i is in the j-entry. The square of its length for $j \neq n$ is

$$||row[\hat{A}_j]_i||^2 < B_i^2 + \varepsilon_i^2 < 1 + \varepsilon_i^2 < 1 + \frac{1}{4} = \frac{5}{4} = 1.25.$$

For the case $j = n$ we get

$$||row[\hat{A}_j]_n||^2 = \left(\frac{b_{n1}^*}{X_1}\right)^2 + \cdots + \left(\frac{b_{n,n-1}^*}{X_{n-1}}\right)^2 + \varepsilon_n^2 < B_n^2 + 1 < 3.$$

Thus, from inequality (11) we get

$$|\det \hat{A}_j| < \sqrt{3}(1.25)^{(n-1)/2} = c(n).$$

So from relation (10) we get

$$|\hat{x}_j| < c(n) \prod_{i=1}^{n} X_i, \quad j = 1, 2, \ldots, n.$$

Since all the previous computations can be done in polynomial time (LLL and size reduction), the Theorem follows.

Remark 1. In the case where $a_0 < 0$, instead of relation (9) we have the better inequality $|\varepsilon_n| < 1/2$, since $B_n^2 < 1$. Thus $c(n) = (1.25)^{n/2}$.

Examples

The first example is given only to explicitly show how we apply our method.

(i). Let

$$84 \cdot 10^5 x_1 + 4 \cdot 10^6 x_2 + 15688 x_3 + 6720 x_4 + 15 x_5 = 371065262.$$

This is example 1 of [1] and they get the solution $\mathbf{x} = (36, 17, 39, 8, -22)$, with $||\mathbf{x}|| \simeq 60.44$. Assumption A_1 is fulfilled if

$$\max_{1 \le j \le 4} |a_j| < \frac{1}{8} X_5, \; |a_5| < \frac{1}{8} \max_{1 \le j \le 5} |X_j|.$$

So it is enough to choose

$$X = X_1 = \cdots = X_5 = 8 \cdot 84 \cdot 10^5 + 1.$$

We consider the matrix

$$M = \begin{bmatrix} \frac{1}{67200001} & 0 & 0 & 0 & 0 & 0 & 8400000 \\ 0 & \frac{1}{67200001} & 0 & 0 & 0 & 0 & 4000000 \\ 0 & 0 & \frac{1}{67200001} & 0 & 0 & 0 & 15688 \\ 0 & 0 & 0 & \frac{1}{67200001} & 0 & 0 & 6720 \\ 0 & 0 & 0 & 0 & \frac{1}{67200001} & 0 & 15 \end{bmatrix}$$

Applying LLL to the rows of M we get

$$M_{\text{LLL}} = \begin{bmatrix} -\frac{10}{67200001} & \frac{21}{67200001} & 0 & 0 & 0 & 0 & 0 \\ 0 & 0 & \frac{15}{67200001} & -\frac{35}{67200001} & -\frac{8}{67200001} & 0 & 0 \\ \frac{1}{67200001} & -\frac{2}{67200001} & -\frac{25}{67200001} & -\frac{1}{67200001} & -\frac{72}{67200001} & 0 & 0 \\ \frac{5}{67200001} & -\frac{10}{67200001} & -\frac{95}{67200001} & -\frac{76}{67200001} & \frac{72}{67200001} & 0 & 0 \\ \frac{2}{67200001} & -\frac{4}{67200001} & -\frac{42}{67200001} & -\frac{21}{67200001} & \frac{1}{67200001} & 0 & -1 \end{bmatrix}$$

Then applying size reduction to the lattice $\hat{L}$ generated by the set $\{\mathbf{b}'_1, \ldots, \mathbf{b}'_5, \mathbf{b}_6\}$, where $\mathbf{b}_6 = (0, 0, 0, 0, 0, 1, -a_0)$, we get

$$\hat{M}_{\text{LLL}} = \begin{bmatrix} -\frac{10}{67200001} & \frac{21}{67200001} & 0 & 0 & 0 & 0 & 0 \\ 0 & 0 & \frac{15}{67200001} & -\frac{35}{67200001} & -\frac{8}{67200001} & 0 & 0 \\ \frac{1}{67200001} & -\frac{2}{67200001} & -\frac{25}{67200001} & -\frac{1}{67200001} & -\frac{72}{67200001} & 0 & 0 \\ \frac{5}{67200001} & -\frac{10}{67200001} & -\frac{95}{67200001} & -\frac{76}{67200001} & \frac{72}{67200001} & 0 & 0 \\ \frac{2}{67200001} & -\frac{4}{67200001} & -\frac{42}{67200001} & -\frac{21}{67200001} & \frac{1}{67200001} & 0 & -1 \\ \frac{36}{67200001} & \frac{17}{67200001} & \frac{39}{67200001} & \frac{8}{67200001} & -\frac{2}{6109091} & 1 & 0 \end{bmatrix}$$

We take the 6th row and multiply each entry with X, then we shall get the vector $\mathbf{x}$. If we want the coordinates of the solution vector to satisfy

$$|x_1| < 30, |x_j| < 50 \ (2 \leq j \leq 5),$$

then taking $X_1 = 30, X_2 = X_3 = X_4 = X_5 = 50$ we get the following solution $\mathbf{x} = (26, 38, 39, 8, -22)$. This solution has absolute value 64.72. Notice that it is larger (with respect to Euclidean length) than the previous solution, but it has the advantage that satisfies our constraints.

(ii). We consider now $n = 50$.

$$\begin{aligned}\{a_j\}_j = [&872934629013064, 362643350651979, 231593889792433, 1084529488472651, 152647947850799,\\ &739407904067188, 1078361055147110, 522287723336618, 1048278073142822, 71464720981315,\\ &1026144865997912, 401128969656441, 1104125375426692, 223040948030783, 259134135114376,\\ &477165086702863, 693696459173357, 956101007737750, 1076391779531258, 887808907972169,\\ &154289043341408, 1123813906929138, 1006407849303380, 1028038257417354, 126747913149526,\\ &345001039716371, 173180910604612, 376756743710801, 462057825850822, 105084485099476,\\ &193285152829384, 663950233902816, 1005024177016821, 350981819196027, 1049577315489835,\\ &455051495653072, 1014366278972062, 905067265314795, 972603957926899, 1110054606397627,\\ &768533772552959, 798515502008744, 705587377794293, 64248048456242, 771519628719865,\\ &190006526706907, 481482852515889, 916067763534188, 768875611228651, 666640039086558]\end{aligned}$$

$$a_0 = 17297404087862459$$

Using our algorithm with $X_j = 3$ $(1 \leq j \leq n)$ we get (using Sage [16])

$$\begin{aligned}\mathbf{x}_1 = [&1, 0, 1, 2, 0, 0, 1, -1, 0, 0, 1, 2, -1, 0, 0, 0, 0, 0, 1, 1, 1, 0, 0, 2, 2, 0, 0, -1,\\ &1, 1, -1, -1, 0, 0, 1, 0, 1, 1, 1, 1, 1, 0, 1, 0, 1, 0, 0, 1, 1, 0]\end{aligned}$$

with Euclidean length 6.324. It took less than 5 s in order to find it. This solution satisfies our bound $|x_j| \leq X_j = 3$. Using the algorithm of [1] we get

$$\begin{aligned}\mathbf{x}_2 = [&0, 0, 0, 1, 1, 2, -1, 3, 0, 1, 2, 1, 2, 2, 0, 0, 0, 0, 1, -1, 0, -1, 0, 2, 1, 0, 0, 1,\\ &2, 0, -1, 0, 1, 0, 0, 0, 1, 0, 1, 1, 0, 0, 0, 0, 1, 0, 0, 1, 1, 0]\end{aligned}$$

which has Euclidean length 7.141. Now using CVP method (as implemented in fplll) we manage to get the vector

$$\begin{aligned}\mathbf{x}_3 = [&0, 0, 0, 1, 0, 1, 1, 0, 1, 0, 2, 0, 1, 0, 0, 0, 1, 1, 1, 0, 0, 0, 0, 0, 1,\\ &0, 0, 0, 1, 0, 0, 0, 0, 1, -1, 1, 0, 0, 1, 0, 0, 0, 2, 1, 0, 0, 0, 1, 0, 0, 1]\end{aligned}$$

which has norm 5. It took almost 4 min in order to compute it. We worked as follows. Let L be the lattice generated by the homogeneous equation $\sum_{j=1}^{n} a_j x_j = 0$. Then we choose as target vector, a vector $\mathbf{t}$ which is a solution of $\sum_{j=1}^{n} a_j x_j = a_0$. If $\mathbf{c}$ is the output of the CVP instance $CVP(L, \mathbf{t})$, then a small solution is given by $\mathbf{x} = \mathbf{t} - \mathbf{c}$. We noticed that, if the target vector has large length then fplll provide us with a large solution $\mathbf{x}$ (but smaller than the length of target vector). That is the CVP solvers are "sensitive" to the choice of the target vector. In order to get the previous solution $\mathbf{x}_3$ we used as target vector, the vector $\mathbf{x}_1$ from the application of our method. So it seems that a nice strategy in order to get a small solution is to combine the CVP solvers and our algorithm (which shall give us the target vector). In case we have large n, say $n \geq 80$ then the CVP solver of fplll(and Magma)[1] is slow. Thus, we conclude (at least experimentally) that for large dimensions it is better to use our algorithm and for smaller say $n < 80$ a combination of CVP and our method.

---

[1]For a detailed account how these solvers work see [8].

An Application to Knapsack Problem

The subset sum or knapsack problem is the following. Given a list of n positive integers $\{a_1, \ldots, a_n\}$ and an integer s such that $\max\{a_i\}_i \le s \le \sum_{i=1}^n a_i$ find a binary vector $\mathbf{x} = (x_i)_i$ such that $\sum_{i=1}^n x_i a_i = s$. The decisional version is known to be NP-complete [6]. The variant, multiple knapsack problem is used in many loading and scheduling problems in operational research.

We use the following lattice

$$B = [\mathbf{b}_1, .., \mathbf{b}_{n+1}] = \begin{bmatrix} \frac{N_1}{X_1} & 0 & \cdots & 0 & 0 & N_1 a_1 \\ 0 & \frac{N_1}{X_2} & \cdots & 0 & 0 & N_1 a_2 \\ \cdots & \cdots & \cdots & \cdots & \cdots & \\ 0 & 0 & \cdots & \frac{N_1}{X_n} & 0 & N_1 a_n \\ 0 & 0 & \cdots & 0 & N_2 & -N_1 s \end{bmatrix},$$

for some positive integers N_1, N_2. When we apply LLL-algorithm to the previous lattice we get vectors of the form

$$\Big(\frac{\lambda_1 N_1}{X_1}, \ldots, \frac{\lambda_n N_1}{X_n}, \beta N_2, N_1\big(\sum_{j=1}^n \lambda_j a_j - \beta s\big)\Big).$$

This is because LLL-algorithm uses transformations of the form

$$\mathbf{b}_j \leftrightarrow \mathbf{b}_i \text{ or } \mathbf{b}_j \leftarrow \mathbf{b}_j - r\mathbf{b}_i,\ r \in \mathbb{Z},\ j > i.$$

Since we expect small vectors from LLL, we probably get vectors of the form

$$\Big(\frac{\lambda_1 N_1}{X_1}, \ldots, \frac{\lambda_n N_1}{X_n}, N_2, 0\Big)$$

with $\lambda_j \in \{0, 1\}$ (in fact for small dimensions $n \le 40$ is very probable, at least experimentally, to get vectors of the previous form). Then a solution of the knapsack is $(\lambda_1, \ldots, \lambda_n)$. We shall fix N_1, N_2 and we randomly choose $X_1, \ldots, X_n \in \{1, 2, \ldots, k\}$ with k say ≤ 10.

After making many experiments with dimension $n \le 40$ we concluded that whenever the algorithm of Coster et al. [4] is working is faster than ours. But there are cases where the algorithm of [4] is not working as we will see below. For large values of n say $n > 40$ and density very close to 1 our algorithm is slow. All the examples below have density very close to 1.

Example 1. ($n = 20$). Let

$$\mathbf{a} = [231578, 90066, 426782, 989541, 428396, 861588, 366246, 430412, 329226, 299869,$$

$$179689, 288142, 916676, 447222, 1040519, 271141, 652751, 132316, 548527, 907547]$$

and $s = 6507929$. Using the algorithm of Coster et al. with $N = 15$ we did not manage to get a solution. Our algorithm with $N_1 = 90$, $N_2 = 80$ and using random denominators from the set $\{1, 2, 3, 4, 5\}$, in the $38th$ round we got the solution $\mathbf{x} = (0, 1, 1, 1, 1, 1, 1, 0, 0, 0, 0, 1, 1, 1, 1, 0, 1, 0, 0, 0)$.

Example 2. ($n = 30$). We set

$$\mathbf{a} = [257957069, 211449890, 453588748, 460393904, 806269638, 965676997, 722998227,$$

$$557173347, 544414881, 605777707, 308224438, 609694552, 614806334, 86201849,$$

$$3033849, 54567875, 749134183, 136657534, 339166263, 622170807, 339856371, 565613209,$$

$$66643022, 732672773, 874884984, 522967114, 168924289, 405266804, 946333809, 879669424]$$

and $s = 6835888107$. Again using Coster et al, with $N = 10$ we did not manage to get any solution. Using our algorithm with $N_1 = 250$, $N_2 = 230$ and using random denominators from the set $\{1, 2, 3\}$, we got

$$\mathbf{x} = (1, 0, 1, 1, 0, 1, 0, 1, 0, 0, 0, 0, 1, 1, 0, 0, 0, 1, 1, 1, 0, 1, 0, 1, 1, 0, 1, 0, 0, 0)$$

in the 439th round. We got these results in some minutes.

Example 3. ($n = 35$). We set

$$[25757712619, 1703301249, 29787913497, 12224812308, 10842851796,$$

$$12515371588, 32028450775, 34098294238, 3343156310, 27995252025, 8010200960,$$

$$15769634246, 23243451953, 18423819032, 4905368619, 18951710032, 18461896729,$$

$$31018788743, 33944716414, 30577978749, 19433865371, 21833994553, 16822791334,$$

$$9873829642, 32574703247, 16993191260, 34144724289, 6412642125, 15206763392,$$

$$17781019093, 29173151234, 25267831499, 32387438669, 18801581598, 19492385639].$$

and $s = 206027365036$. Again using Coster et al, with various values of $N \in \{5, 10, 15\}$ we did not get any solution. Using our algorithm with $N_1 = 25000$, $N_2 = 20000$ and using random denominators from the set $\{1, 2, 3, 4\}$, we got

$$\mathbf{x} = (1, 1, 0, 1, 0, 0, 0, 0, 1, 0, 1, 0, 1, 1, 1, 0, 0, 0, 0, 0, 0, 1, 0, 0, 1, 0, 0, 1, 1, 0, 0, 0, 1, 0, 0)$$

after 163.2 min.

References

1. Aardal, Karen; Hurkens, Cor.; Lenstra, Arjen K.; Solving a linear Diophantine equation with lower and upper bounds on the variables. Integer programming and combinatorial optimization p.229–242, Lecture Notes in Comput. Sci., **1412**, Springer, Berlin, 1998.
2. Bosma, Wieb, Cannon, John and Playoust, Catherine; The Magma algebra system. I. The user language, J. Symbolic Comput.Vol. **24** (1997).
3. Coppersmith, D.: Finding small solutions to small degree polynomials. Lecture Notes in Computer Science **2146** (2001) p.20–31.
4. Coster J.M., Joux A., LaMacchia B.A., Odlyzko A.M., Schnorr C-P., and Stern J., Improved low-density subset sum algorithms. Computational Complexity, 2:111–128, 1992.
5. Gama N. and Nguyen P.Q., Predicting Lattice reduction, Eurocrypt 2008, LNCS **4965**, p.31–51 (2008)
6. Garey, M.R.;Johnson, D.S.; Computers and intractability : A guide to the theory of NP-completeness. W.H.Freeman and Company, NY (1979).
7. Galbraith, S.; Mathematics of Public Key Cryptography, Version 1.1, December 1, 2011, http://www.math.auckland.ac.nz/~sgal018/crypto-book/crypto-book.html.
8. Hanrot G., Pujol X. and D.Stehle. Algorithms for the Shortest and Closest Lattice Vector Problems. IWCC'11.
9. Lagarias, J.C., Odlyzko, A.M.: Solving low-density subset sum problems. J. Assoc. Comput. Mach. **32(1)** (1985) 229–246.
10. Lenstra, H.W.; On the Chor-Rivest knapsack cryptosystem; Journal of Cryptology, Volume **3**, Number 3 (1991), p.149–155.
11. Lueker, G.S; Two NP-complete problems in nonnegative integer programming, report **178** CSL, Princeton University (1975).
12. Merkle, R., Hellman, M.; Hiding information and Signatures in trapdoor cryptosystem, IEEE Trans.Inf.Theory **IT-24**(1978), p.525–530.
13. Nguyen, P.Q., Stern, J., The two faces of Lattices in Cryptography, Cryptography and lattices. 1st international conference, CaLC 2001, Providence, RI, USA, March 29–30, 2001. Revised papers. Berlin: Springer. Lect. Notes Comput. Sci. **2146**, 146–180 (2001).
14. Pujol X., Stehle D., fplll mathematic software (version 4.0), http://xpujol.net/fplll.
15. Rosser, J. B.; A note on the linear Diophantine Equation, American Maths. Monthly **48**(1941).
16. Stein, W.A. et al. Sage Mathematics Software (Version 4.5.1), The Sage Development Team, 2012, http://www.sagemath.org.
17. Vasilenko, O.N.; Number-Theoretic Algorithms in Cryptography, Translations of Mathematical Monographs (AMS), Volume **232** ,Translated by Alex Martsinkovsky.

A Short Exposition of Topological Applications to Security Systems

D. Panagopoulos and S. Hassapis

Abstract In this article several practical applications of algebraic topology are presented. After a short technical review of the necessary theory applications to sensor networks are presented. A very short reference of applications to data analysis follows.

Introduction

For many years algebraic topology has been considered as an abstract mathematical field with none or few practical applications. However many abstract mathematical ideas have found unexpected applications in real-life problems. This is also the case with algebraic topology during the past couple of years. Theories and techniques created by mathematicians in order to answer abstract problems are being used to answer problems such as protein docking, image analysis, data analysis and space coverage by sensor networks.

This article focuses, mainly, on the last two cases, i.e., applications of algebraic topology to sensor networks and data analysis. A brief introduction to the necessary mathematical background is given in the beginning of the article. In the second section applications to sensor networks are presented while in the third section approaches related to data analysis are discussed.

D. Panagopoulos (✉)
Eureka Module, Pelopa 3, 15344, Gerakas, Attiki, Greece
e-mail: dpanagop@yahoo.com; dpanagop@gmail.com
S. Hassapis
Evangeliki Model School of Smyrna, Lesvou 4, 17123, N. Smirni, Greece
e-mail: shasapis@gmail.com

N.J. Daras (ed.), *Applications of Mathematics and Informatics in Science and Engineering*, Springer Optimization and Its Applications 91, DOI 10.1007/978-3-319-04720-1_12,

Mathematical Background

Algebraic Homology

In this section we give a short introduction to Homology theory. We begin with the definition of a (simplicial) complex.

Definition 1. Let V be a finite set. A collection $\mathcal{K}$ of subsets of V is called a complex if $\alpha \in \mathcal{K}$ and $\beta \subseteq \alpha$ implies $\beta \in \mathcal{K}$.

A set in $\mathcal{C}$ with $k+1$ elements is called a k-simplex and we define its dimension to be k. If $\beta \subseteq \alpha$ then we call β a face of α. We call β a proper face if in addition $\beta \neq \alpha$.

In this article complexes will, mostly, refer to geometric objects. A 0-simplex will be a point, a 1-simplex an arc, a 2-simplex a triangle and so on. In Fig. 1 there is an example of a complex which is consisted of a triangle $\{u_0, u_1, u_2\}$ and a line $\{u_2, u_3\}$ (and of course all their faces).

Given a complex $\mathcal{K}$, we define $C_n(K)$ to be the vector space whose base is the set of n-simplices of $\mathcal{K}$ with coefficients over a field. The base element that corresponds to the n-simplex $\{u_0, \ldots, u_n\}$ is denoted by $[u_0, \ldots, u_n]$.

For example for the complex of Fig. 1 we have that $C_2(K)$ is a one dimensional vector space with base $[u_0, u_1, u_2]$ while $C_1(K)$ is a four dimensional vector space with base $[u_0, u_1], [u_0, u_2], [u_1, u_2], [u_3, u_4]$. For each n we define the boundary operator $\vartheta_n : C_n(K) \to C_{n-1}(K)$ to be the linear map with

$$\vartheta([u_0, \ldots, u_n]) = \sum_{i=0}^{n} (-1)^i [u_0, \ldots, \widehat{u_i}, \ldots, u_n]$$

where $\widehat{u_i}$ means that u_i is deleted. A straight forward calculation [7, p. 105] verifies that $\vartheta_{n-1} \circ \vartheta_n = 0$. Hence we have a chain complex

$$C_\bullet(K) : \cdots \xrightarrow{\vartheta_{n+2}} C_{n+1}(K) \xrightarrow{\vartheta_{n+1}} C_n(K) \xrightarrow{\vartheta_n} C_{n-1}(K) \xrightarrow{\vartheta_{n-1}} \cdots$$

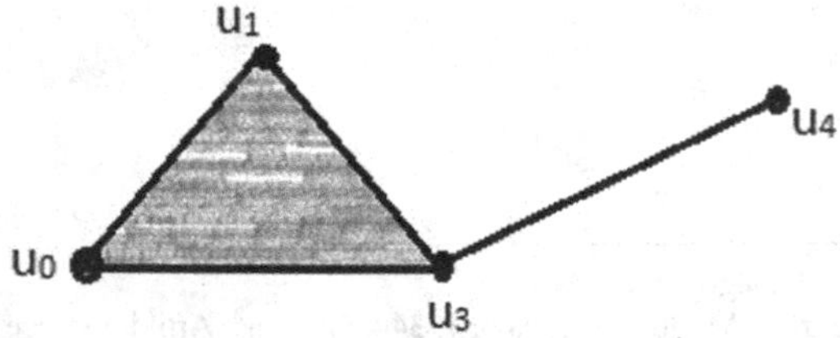

Fig. 1 A simple example of a simplicial complex

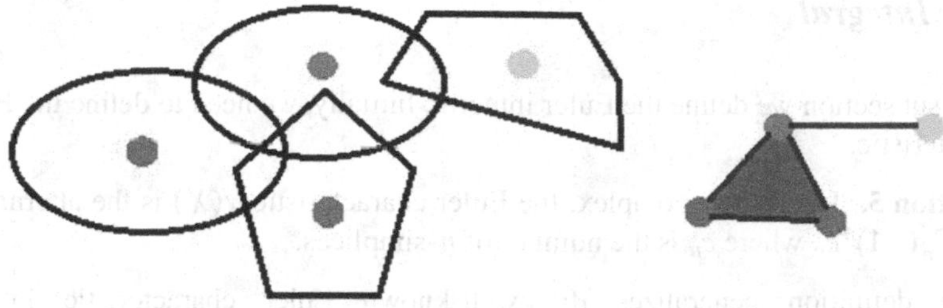

Fig. 2 Four sets in the plane and their corresponding nerve

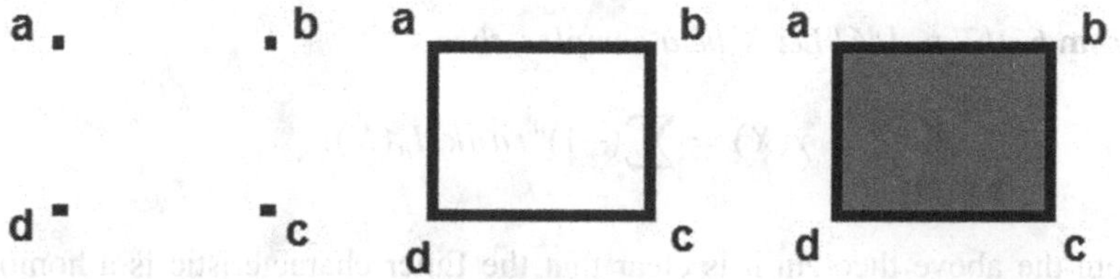

Fig. 3 The Rips complexes of four points on the vertices of a unit square. On the *left* for ϵ in $[0, 1)$, in the *middle* for values in $[1, \sqrt{2})$ and on the *right* for values in $[\sqrt{2}, +\infty)$

Definition 2. Let $C_\bullet(K)$ be a chain complex as above. The n-th Homology group $H_n(K)$ is the group quotient $Ker\vartheta_n / Im\vartheta_{n+1}$. The rank of the n-th homology group is called the n-th betti number.

Let K be a complex and A a subcomplex contained in it. Then we can construct the quotient K/A. Geometrically K/A is obtained from K by collapsing A to a single point. The homology of K/A is called the relative homology of K with respect to A and is symbolized by $H_{\bullet(K,A)}$.

Finally, for our purposes we will require the following two constructions.

Definition 3. [3, 5] Let $\mathcal{S}$ be a set of sets. We define the nerve $Nrv\mathcal{S}$ of $\mathcal{S}$ to be the set of subsets of $\mathcal{S}$ with no empty intersection:

$$Nrv\mathcal{S} = \{X \subseteq \mathcal{S} : \bigcap X \neq \emptyset\}$$

The nerve is always a complex. In Fig. 2 an example with four sets in $\mathbb{R}^2$ is given and on the right-hand side the corresponding complex is depicted. If $\mathcal{S}$ contains disks, then the corresponding complex is called the Čech complex.

Definition 4. [3] Let $V = \{v_1, \ldots, v_n\}$ be a set of points of a metric space (X, d). Then for a real number $\epsilon > 0$ we define the Rips (or Rips-Vietoris) complex R_ϵ to be the complex whose k-simplices are $\{v_{i_0}, \ldots, v_{i_k}\}$ with $d(v_{i_j}, v_{i_l}) \leq \epsilon$.

Euler Integral

In this subsection we define the Euler integral. Initially, we need to define the Euler characteristic.

Definition 5. Let X be a complex, the Euler characteristic $\chi(X)$ is the alternating sum $\sum_n (-1)^n c_n$ where c_n is the number of n-simplices.

The definition generalizes the well-known Euler characteristic in the two-dimensional complexes. The relation of the Euler characteristic to the n-th homology group is given by the following theorem.

Theorem 6. *[7, p. 146] Let X be a complex, then*

$$\chi(X) = \sum_n (-1)^n rank H_n(X).$$

From the above theorem it is clear that the Euler characteristic is a homotopy invariant. Hence it can be defined for many topological spaces. For example:

1. for a finite set X the Euler characteristic equals the number of points in X,
2. $\chi(X)$ =vertices−edges+faces=2 for plane graphs,
3. for a subset X of $\mathbb{R}^2$ with n holes $\chi(X) = 1 - n$,
4. $\chi(X) = 2 - 2g$ for an orientable surface X of genus g.

A simple argument based on the Mayer-Vietoris sequence [7] gives that for two complexes A, B

$$\chi(A \cup B) = \chi(A) + \chi(B) - \chi(A \cap B).$$

This fact allows us to set up an integration theory using Euler characteristic.

Definition 7. [1, 2] Let X be a complex and $CF(X)$ the abelian group of functions from X to $\mathbb{Z}$ with generators the characteristic functions 1_σ, where σ is a closed simplex of X. Then for a function $h = \sum_\alpha c_\alpha 1_\alpha \in CF(X)$ the Euler integral with respect to the Euler characteristic is defined to be

$$\int_X h d\chi = \sum_\alpha c_\alpha \chi(\alpha).$$

Homology and Sensor Networks

Hole Detection

Perhaps the most important question about a sensor network is whether it covers an entire area or not. Furthermore, if there exists a hole in the coverage we should have ways to detect it. This problem is easily solved if the location of each sensor of the network is known. On the other hand there exist scenarios where our sensors, or at least most of them, do not have any information about their location. This might be due to the fact that our sensors are too small to carry positioning systems or because these systems are too expensive. Maybe in the near future swarms of low-cost sensors will be spread in an area for collecting data. For example, Smart Dust was a research proposal [10] to DARPA to build wireless sensor nodes with a volume of one cubic millimeter. The project led to a working mote smaller than a grain of rice (Fig. 4).

The first step in using homology is to create a complex from the sensor network. Assuming that each sensor covers a disk of radius r_c, and of course that it can detect the presence and the identity of any other sensors in that disk, we can create the Čech complex C of the coverage disks (see Fig. 5). For the Čech complex of a set $\{U_\alpha\}$ of sets the following theorem holds:

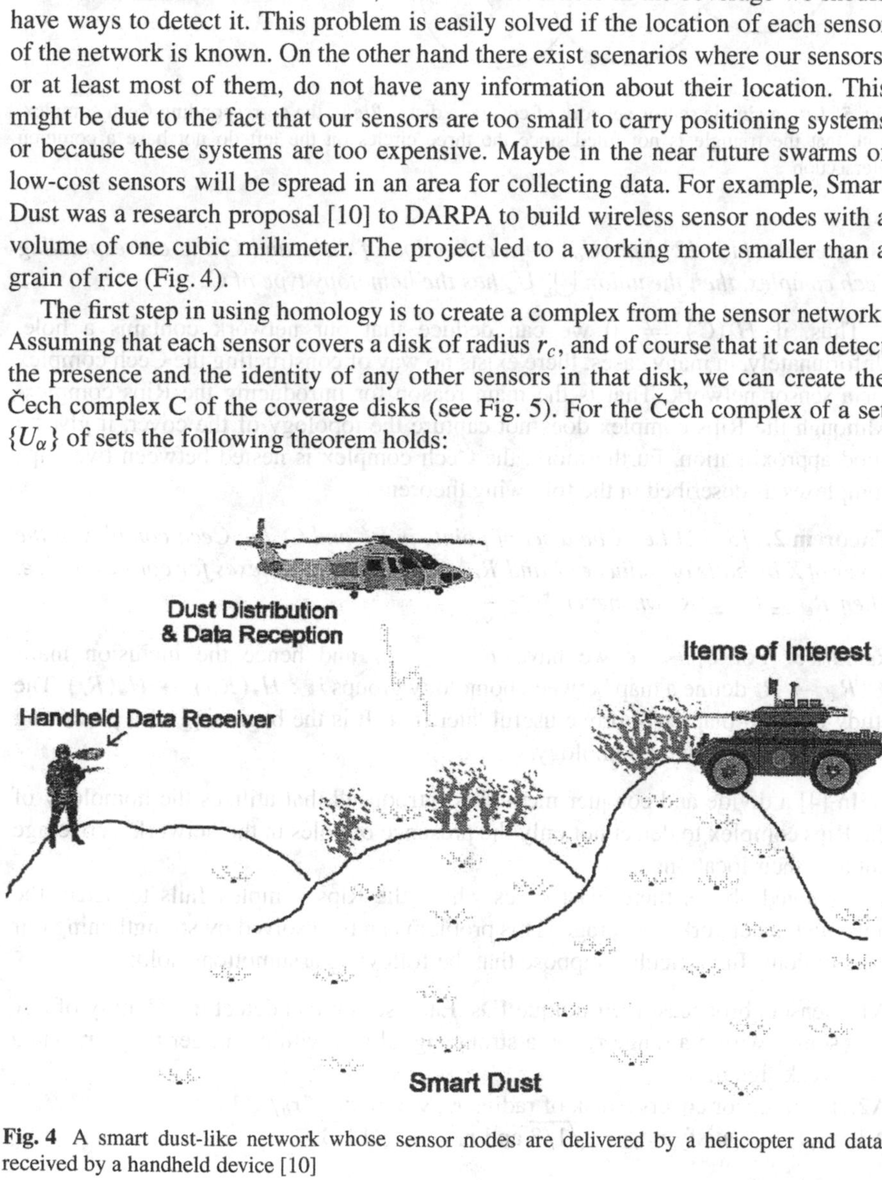

Fig. 4 A smart dust-like network whose sensor nodes are delivered by a helicopter and data received by a handheld device [10]

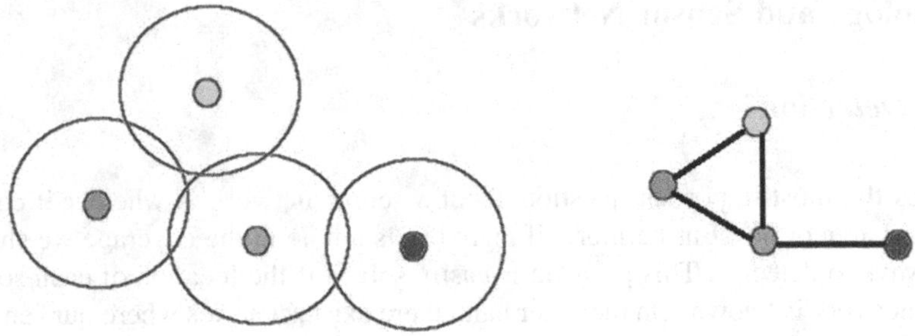

Fig. 5 *Left*: a simple sensor network of coverage disks. *Right*: the corresponding Čech complex. Note that the triangle is not filled since the three circles on the left do not have a common intersection

Theorem 1. *[11, 12] Let $\{U_\alpha\}$ be a collection of disks and C the corresponding Čech complex, then the union $\bigcup_\alpha U_\alpha$ has the homotopy type of C.*

Thus, if $H_1(C) \neq 0$ we can deduce that our network contains a hole. Unfortunately, in many cases, there exists no way of constructing the Čech complex for a sensor network. That is the main reason for introducing the Rips complex. Although the Rips complex does not capture the topology of the cover, it gives a good approximation. Furthermore, the Čech complex is nested between two Rips complexes as described in the following theorem :

Theorem 2. *[3, 12] Let X be a set of points in $\mathbb{R}^2$ and C_ϵ the Čech complex of the cover of X by balls of radius $\epsilon/2$ and $R_{\epsilon'}$, R_ϵ the Rips complexes for constants ϵ', ϵ. Then $R_{\epsilon'} \subseteq C_\epsilon \subseteq R_\epsilon$ whenever $\frac{\epsilon}{\epsilon'} \geq \frac{2}{\sqrt{3}}$.*

Remark 3. For $\epsilon' < \epsilon$ we have $R_{\epsilon'} \subseteq R_\epsilon$ and hence the inclusion maps $i : R_{\epsilon'} \to R_\epsilon$ define a map between homology groups $i_* : H_\bullet(R_{\epsilon'}) \to H_\bullet(R_\epsilon)$. The study of these maps will prove useful later in 4. It is the beginning of a promising theory called persistent homology.

In [4] a divide and conquer method is introduced that utilizes the homology of the Rips complex to detect not only the presence of holes in the network's coverage but also their location.

As noted above, there exist cases where the Rips complex fails to detect the holes in the network's coverage. This problem can be resolved by strengthening our assumptions. In particular, suppose that the following assumptions hold:

A1. Sensors broadcast their unique IDs. Each sensor can detect the identity of any sensor within a range r_s via a strong signal and within a larger range r_w via a weak signal.
A2. Each sensor covers a disk of radius r_c, where $r_c \geq r_b/\sqrt{3}$.
A3. r_c, r_s, r_w satisfy $r_c \geq r_s\sqrt{1/3}$ and $r_w \geq r_s\sqrt{13/3}$.

A4. The sensors are placed on a bounded subset D of the plane. Moreover sensors can detect the presence (but not the location or direction) of the boundary within a fixed fence detection radius $r_f \geq 0$. ϑD is connected and piecewise linear.

A5. The restricted domain $D - C$ is connected, where $C = \{x \in D : ||x - \vartheta D|| \leq r_f + \frac{1}{2} r_s\}$.

A6. The curve(s) $\{x \in D : ||x - \vartheta D|| = r_f\}$ have internal injectivity radius at least $r_s/\sqrt{2}$ and external injectivity radius at least r_s.

In [12] the following theorem is proved:

Theorem 4. *For a network satisfying A1-A6 let F_s, F_w be the subcomplexes of R_s, R_w respectively whose vertices correspond to the sensors that lie within the fence detection radius. The region $D - C$ is contained in the cover if there is a homology class in $H_2(R_s, F_s)$ which is nonzero in $H_2(R_w, F_w)$.*

Target Enumeration

In this section we outline the use of the Euler integral to the target enumeration problem of a sensor network. In particular, given a network of sensors which are only able to detect the presence of an other sensor or of certain "targets" within a disk of radius r_c and where relaying of messages is done between sensors within distance r_c, we want to enumerate the targets in the network's coverage. It should be emphasized that each sensor counts only the number of targets in it's covering disk. It cannot identify the targets nor it can detect the direction of the other sensors or the targets. Hence, if a target is contained in the intersection of the covers of two sensors it is counted twice.

Assuming that the sensors cover an entire subspace X of $\mathbb{R}^2$ and that for every target α the subset U_α of X is the set containing all the sensors which detect the target, then we can define a function $h : X \to \mathbb{N}$, where $h(x)$ is the number of targets α which can be detected by the sensor located at x ($h(x) = \#\{\alpha : x \in U_\alpha\}$). The following theorem states that we can use the Euler integral to enumerate the targets in X.

Theorem 5. *[1, 2] Given $h : X \to \mathbb{N}$ the counting function of compact target supports in X satisfying $\chi(U_\alpha) = n \neq 0$ for all targets α. Then $\#\alpha = \frac{1}{n} \int_X h \, d\chi$.*

Naturally, the assumption that at every point of X a sensor is located is not realistic. In practice, we assume that the sensors are located on the vertices of a triangulation of X. In that case the values of h are known only for the vertices and we can integrate the piecewise linear interpolation of h [2].

Further Applications

We should point out that there exist more applications of algebraic homology to sensor networks. For example, let the following assumptions hold:

B1. the sensors broadcast their unique IDs and each node can detect the ID of any node within radius r_b,
B2. sensors cover a disk of radius $r_c \geq r_b\sqrt{3}$,
B3. sensors lie in a compact connected subset of the plane whose boundary is connected and piecewise linear,
B4. every sensor on the boundary knows the IDs of its two adjacent boundary nodes which both lie within distance r_b.

Then the sensor cover contains D if there exists $[a] \in H_2(R, \vartheta D)$ such that $\vartheta a \neq 0$ [11]. This fact enables us to detect reductant sensors. (The non-reductant sensors are part of a minimal generator of $H_2(R, \vartheta D)$). Thus, the network can for, example, conserve energy by placing in sleeping mode the reductant sensors.

Another example mentioned in [11] has to do with the problem of determining whether an evader can avoid detection in a network whose sensors are on the move or come online and offline at various times. Under some reasonable assumptions there exists an affirmative answer to this problem.

Finally, in [4] a method for detecting wormhole attacks on a network is presented. In this kind of attack involving two malicious/infected sensors, a signal received by one sensor is transmitted over a low-latency link and replayed by the other creating the analogous of a wormhole. Thus, the network is tricked into believing that the sensors in the vicinity of the wormhole are in close distance. If the two malicious/infected sensors are located far apart, this can cause several problems to the network. For example, the network might choose to channel signals through the wormhole something that can cause congestion and deplete the power reserves of the sensors near the wormhole [8]. The basic idea in [4] is that a wormhole creates a hole in the network and hence a non-trivial homology element. To distinguish holes created by wormholes from ordinary coverage holes we can use the fact that removing a cycle created by an ordinary hole divides the coverage into two components while a cycle created by a wormhole does not.

Data Analysis

General Methods: A First Approach

Another unexpected application of topology is to data analysis. The main advantage of using topological methods is that these methods are noise tolerant.

Quite naturally the first application is to use topology to classify different datasets. A dataset may be viewed as a set of points of some n-dimensional real space $\mathbb{R}^n$ (this set is often called cloud set). From the dataset we can construct a complex (i.e., a Rips complex) and calculate it's homology groups and the corresponding betti numbers. It is then possible to try to classify different datasets by examining their respective betti numbers. Several researchers have conducted tests attempting to classify data coming from a great variety of sources.

A method proposed in [3, 13] is given a cloud set X to use a continuous function $f : X \to \mathbb{R}$ and a covering $\cup_\alpha A_\alpha$ for the image A of f to obtain a covering $\cup X_\alpha$, where $X_\alpha = f^{-1}(A_\alpha)$, of X. Finally, a clustering algorithm is used on the set $\{X_\alpha\}$ and the result is used to create a complex. In [3, 13] examples of the technique applied to data coming from a diabetes study, hand-drawn copies of the digit "two" and on a library of 3D models can be found.

Persistent Homology

The study of the evolution of the homology of a Rips complex R_ϵ obtained from a dataset for various values of ϵ (c.f. 4) can be used in various applications. For example, it can be used for identifying the main topological characteristics of an object that is sampled. The main topological characteristics are those that are persistent while ϵ changes.

In [15] a Rips complex R_ϵ from a set of text documents is constructed. Each vertex represents a document and the distance between two documents is calculated by an appropriate function. By selecting various values for the constant ϵ, we can construct a sequence of Rips complexes R_{ϵ_i}. From this sequence the persistent homology is calculated. It is hopped that in this way different corpora can be identified.

Another possible application of persistence is to fine tune the Lazy Learning machine learning algorithm. Given a dataset Lazy Learning selects a small subset and performs regression to predict the outcome for a given input. In [9] it is suggested that there is a correlation between the size of the subsets used by the Lazy learning algorithm and the barcode diagrams obtained by persistent homology. In that paper six different datasets were tested. For each dataset a series of Rips complexes R_{ϵ_i} were constructed for a sequence of constants $\epsilon_0, \ldots, \epsilon_m$. The complexes were used to calculate the persistence homology. The resulting barcode diagrams were compared with the mean regression error of the Lazy Learning method where discs of radius ϵ_i were used for regression.

More general, the persistence of homology has several practical applications from optical character recognition [6] protein docking and image analysis [5, 14]. The basic idea is to use a Morse function on a given dataset to create a filtered complex which can be used to calculate the persistent homology.

Conclusion

The authors would like to point out that this is a very short exposition. Many details are omitted while a large number of applications is left out. We only hope that this article will motivate the readers to search for more detailed information. The field is relatively new and will certainly welcome researchers from diverse backgrounds ranging from mathematics to software engineering.

References

1. Y. Baryshnikov, R. Ghrist, *Target enumeration via Euler characteristic integrals*, SIAM J. Appl. Math. 70(3) (2009), 825–844.
2. Y. Baryshnikov, R. Ghrist, *Target enumeration via integration over planar sensor networks*, Proc. Robotics: Science and Systems, 2008.
3. G. Carlsson, *Topology and Data*, Bull. of the AMS Vol. 46 N. 2 (2009), 255–308.
4. H. Chintakunta, H. Krim, *Topological Fidelity in Sensor Networks*, arXiv:1106.6069.
5. H. Edelsbrunner, J. L. Harer, *Computational Homology*, AMS, 2010.
6. D. Freedman, C. Chen, *Algebraic topology for computer vision*, HP Laboratories HPL-2009-375, 2009.
7. A. Hatcher, *Algebraic Topology*, Cambridge Un. Press, 2002, also available at http://www.math.cornell.edu/~hatcher/AT/ATpage.html.
8. M. Y. Malik, *An Outline of Security in Wireless Sensor Networks: Threats, Countermeasures and Implementations*, arXiv:1301.3022.
9. D. Panagopoulos, *Homology and Lazy Learning*, 1st National conference of HMS and HSOR 2011, 2011.
10. K. S. J. Pister, *Smart Dust*, BAA97-43 Proposal Abstract
11. V. de Silva, R. Ghrist, *Homological Sensor Networks*, Notices of the AMS Vol. 54 N. 1 (2007), 10–17.
12. V. de Silva, R. Ghrist,A. Muhammad, *Blind Swarms for Coverage in 2-D*, Conference: Robotics: Science and Systems - RSS , 2005, 335–342.
13. G. Singh, F. Mémoli, G. Carlsson, *Topological Methods for the analysis of high dimensional data sets and 3D object recognition*, Point Based Graphics 2007, Prague - Slides. Sep 2007.
14. A. J. Zomorodian, *Topology for computing*, Cambridge Un. Press, 2005.
15. H. Wagner, *Computational topology in text mining*, ATMCS 2012, 2012.

SAR Imaging: An Autofocusing Method for Improving Image Quality and MFS Image Classification Technique

A. Malamou, C. Pandis, A. Karakasiliotis, P. Stefaneas, E. Kallitsis, and P. Frangos

Abstract In the first part of this paper several aspects of the SAR imaging are presented. Firstly, the mathematical theory and methodology for generating SAR synthetic backscattered data are developed. The simulated target is a ship, which is located on the sea surface. A two-dimensional and a three-dimensional target (ship) implementations are included in the simulations. Both cases of airborne and spaceborne SAR are simulated. Furthermore, the case of varying target scattering intensity is presented. In addition an application of an autofocusing algorithm, previously developed by the authors for the case of Inverse Synthetic Aperture Radar (ISAR) and Synthetic Aperture Radar (SAR) geometry for simulated data, is presented here for the case of real-field radar data, provided to us by SET 163 Working Group. This algorithm is named "CPI-split-algorithm", where CPI stands for "Coherent Processing Interval". Numerical results presented in this paper show the effectiveness of the proposed autofocusing algorithm for SAR image enhancement.

In the second part of this paper the Modified Fractal Signature (MFS) method is presented. This method uses the "blanket" technique to provide useful information for SAR image classification. It is based on the calculation of the volume of a "blanket", corresponding to the image to be classified, and then on the calculation of the corresponding fractal signature (MFS) of the image. We present here some results concerning the application of MFS method to the classification of SAR images. The MFS method is applied both in simulated data (comparison of a focused and an unfocused image) and in real-field data provided to us by SET 163 Working Group (comparison of a "town" area, "suburban" area and "sea" area). In these results it is clearly seen that the focusing of the SAR radar image clearly correlates with the value of MFS signature for the simulated data, and that the type of area can be distinguished by the value of MFS signature for the real data.

A. Malamou • C. Pandis • A. Karakasiliotis • P. Stefaneas (✉) • E. Kallitsis • P. Frangos
National Technical University of Athens, Greece
e-mail: pfrangos@central.ntua.gr; petrosstefaneas@gmail.com

N.J. Daras (ed.), *Applications of Mathematics and Informatics in Science and Engineering*, Springer Optimization and Its Applications 91, DOI 10.1007/978-3-319-04720-1_13,

Keywords Autofocusing • Post processing algorithm • Synthetic aperture radar (SAR) imaging • MFS method • SAR image classification

Part 1: SAR Imaging Techniques: Application of an Autofocusing Algorithm for Improving Image Quality in the Case of Simulated and Real Radar Data

Introduction

Synthetic aperture radar (SAR) has been widely used not only for military but also for non-military purposes. It is a radio frequency (RF) sensor that can be used in a wide variety of applications such as long-range imaging, remote sensing and global positioning. The radar signal has the ability to penetrate through clouds, haze, rain, fog and precipitation with very little attenuation, thus it can perform with high image resolution at long range, regardless the weather conditions. Moreover the radar can illuminate with variable look angle and can select a wide area of coverage. One of its main uses is in target detection and recognition for civilian or military applications. The extended range of SAR applications led to the development of a number of airborne and spaceborne SAR systems. The range-Doppler information collected by the SAR antenna leads to the synthesis of the SAR image of the target with high resolution [1, 2].

The purpose of this paper is to examine several aspects of the SAR imaging. Both cases of airborne and spaceborne SAR are simulated. Also a two-dimensional and a three-dimensional target (ship) implementations are included in the simulations. Furthermore the case of varying target scattering intensity is presented. Hence, the simulations contribute to analyse, clarify and understand better the cases encountered in real-field radar data.

In addition, due to target movement, the SAR image is usually degraded by defocus, distortion or displacement. In this paper the post processing CPI-split autofocusing algorithm [3] is also applied to the case of real-field data of a moving ship (airborne SAR), provided to us by SET 163 Working Group (see acknowledgement below for more details), in order to obtain a focused SAR image of a moving target [5, 6].

Simulated Target Geometry and Mathematical Formulations

The SAR geometry which is used in our SAR imaging simulations of a ship target is presented in Fig. 1. In general, in SAR geometry, the antenna of the radar can be mounted either on an aircraft (case of an airborne platform) or on a satellite (case of a spaceborne platform). In both cases, the radar illuminates the target during

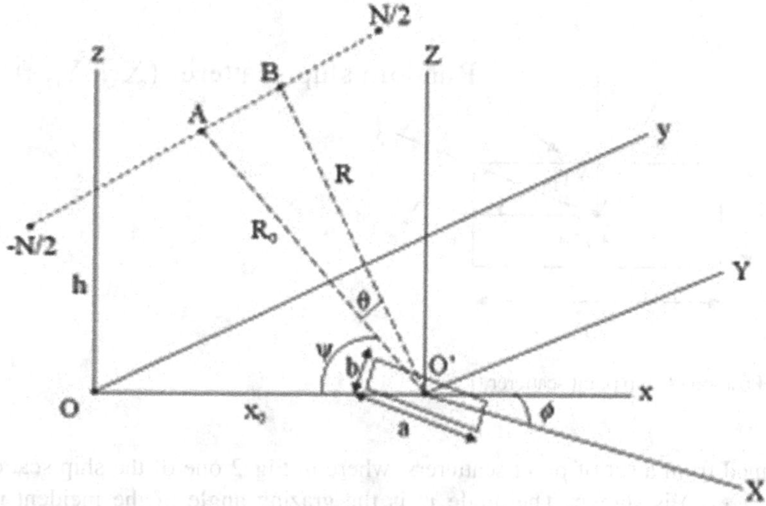

Fig. 1 SAR geometry

the flight [1]. According to our SAR geometry simulation presented in Fig. 1, the aircraft or satellite travels along positive y-axis, with constant velocity v, along a flight path from $-N/2$ to $N/2$, where N is the number of bursts during one CPI. As it is shown in Fig. 1, the centre of the flight path is considered to be point A. Moreover, the radar antenna is assumed to emit Stepped Frequency (SF) pulses, where M stepped frequencies are emitted per burst ($m = 1$ to M) and N bursts per CPI [1, 2].

The target of observation is a ship located on the sea surface. A two-dimensional (2-D) geometry as well as a three-dimensional (3-D) geometry of the ship is simulated below. Regarding both 2-D and 3-D simulations the basic dimensions of the ship are: length a and width b. In Fig. 1 two coordinate systems are presented. The coordinate system of the target (ship) to be imaged is called "local" coordinate system O'XYZ. The other coordinate system is called "earth" coordinate system Oxyz. The origin O' of the "local" coordinate system is placed in the mass centre of the ship. The distance R_0 is the distance between the centre of the flight path and the origin O' of the "local" coordinate system. According to the above geometry the vector of the distance R is given by the following formula:

$$\mathbf{R} = x_0\hat{x} - vnT_b\hat{y} - h\hat{z} \tag{1}$$

where h is the radar platform altitude, $(\hat{x}, \hat{y}, \hat{z})$ are the unit vectors along the (x, y, z) axes, respectively, $T_b = M \cdot PRI$ is the burst duration (PRI is the Pulse Repetition Interval of the transmitted radar waveform) and n is the burst index ($n = -N/2, \cdots, -1, 0, 1, \cdots, N/2$). The simulated target (ship) is considered to

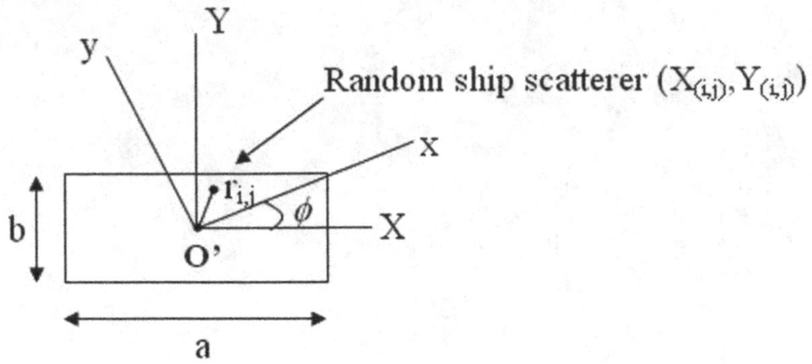

Fig. 2 Position of ship point scatterer (i, j)

be formed from a set of point scatterers, where in Fig. 2 one of the ship scatterers $(X_{(i,j)}, Y_{(i,j)})$ is shown. The angle ψ is the grazing angle of the incident radar electromagnetic (EM) wave and the angle θ is the azimuthal angle of observation of the target within the CPI. The angle θ is determined by the following equation:

$$\cos\theta = \left(\frac{x_0^2 + h^2}{x_0^2 + h^2 + v^2 n^2 T_b^2}\right)^{\frac{1}{2}} \tag{2}$$

The total distance from the SAR antenna to an arbitrary ship scatterer is given by the following equation:

$$\mathbf{R}_{i,j} = \mathbf{R} + [A] \cdot \mathbf{r}_{i,j} \tag{3}$$

where $[A]$ is the transformation matrix from the "local" coordinate system to the "earth" coordinate system:

$$A = \begin{pmatrix} \cos\phi & \sin\phi \\ -\sin\phi & \cos\phi \end{pmatrix} \tag{4}$$

and ϕ is the angle between the "local" and the "earth" coordinate systems. The angle ϕ also determines the orientation of the ship with respect to the "earth" coordinate system Oxyz (axis Ox, in particular).

The distance $r_{i,j}$ is the distance between the origin O' of the "local" coordinate system and an arbitrary ship scatterer and is given by the formula:

$$\mathbf{r}_{i,j} = (X_{i,j}, Y_{i,j}) = X_{i,j}\hat{X} + Y_{i,j}\hat{Y} \tag{5}$$

As a result, combining the above equations, the total distance from the SAR antenna to an arbitrary ship scatterer is given by the formula:

$$\mathbf{R}_{i,j} = \mathbf{R} + [(X_{i,j}\cos\phi + Y_{i,j}\sin\phi)\hat{x} + (-X_{i,j}\sin\phi + Y_{i,j}\cos\phi)\hat{y}] \tag{6}$$

It can be easily seen that the incident wavevector **k** is given by:

$$\mathbf{k} = k\cos\theta\cos\psi\cdot\hat{x} - k\sin\theta\cdot\hat{y} - k\cos\theta\sin\psi\cdot\hat{z} \tag{7}$$

where θ is given by (2) and k is the wavenumber for the emitted SF waveform:

$$k = \frac{\omega_m}{c} = \frac{2\pi f_m}{c} \tag{8}$$

where $f_m = f_0 + (m-1)\Delta f, (m = 1, 2, \cdots, M)$ is the emitted stepped frequency (SF) [1, 2] and ω_m is the corresponding emitted angular frequency.

Hence, the phase $\phi_{i,j}$ for the (i, j) scatterer of the target is calculated from analytic (geometric) calculation of the distance $R_{i,j}$ between the radar and the (i, j) scatterer of (6), as well as from the analytic expression for the incident wavevector **k** of (7), as follows:

$$\phi_{i,j} = 2\cdot\mathbf{k}\cdot\mathbf{R}_{i,j} \tag{9}$$

Then, from (6), (7) and (9) above, the following formula for the "local scattering phase" is obtained, assuming that the target is stationary:

$$\phi_{i,j}^m = \frac{4\pi f_m}{c}[\cos\theta\cos\psi(X_{i,j}\cos\phi + Y_{i,j}\sin\phi) + \sin\theta(X_{i,j}\sin\phi - Y_{i,j}\cos\phi)] \tag{10}$$

whereas the phase corresponding to distance from position B of the platform to the centre of the target (ship) O' is compensated by the radar processor during SAR image synthesis [see also (1)]:

$$\phi_{i,j}^m = \frac{4\pi f_m}{c}(x_0\cos\theta\cos\psi + h\cos\theta\sin\psi + vnT_b\sin\theta) \tag{11}$$

Furthermore, m is the stepped frequency index ($m = 1, 2, \cdots, M$); n is the burst index ($n = 1, \cdots, N\cdot N_{\text{CPI}}$) for a number of simulated CPI's (N_{CPI}); N is the number of bursts during one CPI and $(X_{(i,j)}, Y_{(i,j)})$ are the local coordinates of the ship scatterers. In this simulation one CPI is simulated ($N_{\text{CPI}} = 1$) and as a result N bursts are simulated for this particular time period.

In the case of the 3-D target (ship) simulation, the mathematical expression for the phase of the backscattered signal at SAR receiver changes in order to simulate the 3-D target. An extra term is added to include the contribution of the z-axis scatterers in the signal.

$$\begin{aligned}\phi_{i,j}^m =& \frac{4\pi f_m}{c}(x_0\cos\theta\cos\psi + h\cos\theta\sin\psi + vnT_b\sin\theta)\\ &+ \frac{4\pi f_m}{c}[\cos\theta\cos\psi(X_{i,j,k}\cos\phi + Y_{i,j,k}\sin\phi)\\ &+ \sin\theta(X_{i,j,k}\sin\phi - Y_{i,j,k}\cos\phi)] - Z_{i,j,k}\cos\theta\sin\psi\end{aligned} \tag{12}$$

The backscattered radar data are simulated through the following formula:

$$x(m,n) = \sum_{d} s_{i,j} exp[j\phi_{i,j}^{m}] + u(m,n) \tag{13}$$

where d is the number of the scatterers of the target and $s_{i,j}$ is the scattering intensity for the (i, j) scatterer. In the simulations below we examine not also the case where, without loss of generality, all scatterers have the same strength in amplitude ($s_{i,j} = 1$ for all i, j), but also the case where some scatterers have greater intensities than other and as a result they have greater contribution in the signal. The term $\phi_{i,j}$ is the phase of the backscattered signal (10) or (12), while $u(m, n)$ is the two-dimensional additive white Gaussian noise component.

In the numerical simulations below, the raw data matrices are formed through (13). It is worthy to mention that the dependence on "slow-time" index n in (10) and (12) becomes effective through the aspect angle θ, see (2).

The SAR images are constructed from the raw data matrices through the traditional "Range-Doppler" imaging technique, involving FFT processing in both range and Doppler directions [2].

Numerical Simulations Regarding the Application of an Autofocusing Algorithm for Improving SAR Image Quality

SAR Imaging simulation of a 2-D ship target (airborne scenario)

The simulated ship geometry is shown in Fig. 3. It is a point scatterer model which consists of 233 scatterers. The corresponding radar and geometry parameters are shown in Table 1. Note here that through suitable selection of these parameters "square resolution" of SAR images is obtained [4].

In Fig. 4 the produced SAR image regarding the SAR geometry and simulation parameters described above is presented.

SAR Imaging simulation of a 2-D ship target with varying scattering intensity

In this simulation scenario, five scatterers have greater intensities than the others and as a result they have greater contribution in the signal. This can be easily seen in Fig. 5.

SAR Imaging simulation of a 3-D ship target

This simulation scenario examines the SAR imaging of a three-dimensional target. The target is the point scatterer model of a ship presented above, with the addition of some scatterers along the Z axis. These scatterers represent a mast in the centre of the ship. In Fig. 6 the produced SAR image (two-dimensional imaging) regarding the 3-D ship target is presented.

SAR Imaging simulation of a 2-D ship target (spaceborne scenario)

In the case of the spaceborne simulation scenario the antenna of the radar is placed on a satellite. The basic SAR geometry stays the same (Fig. 1) but the

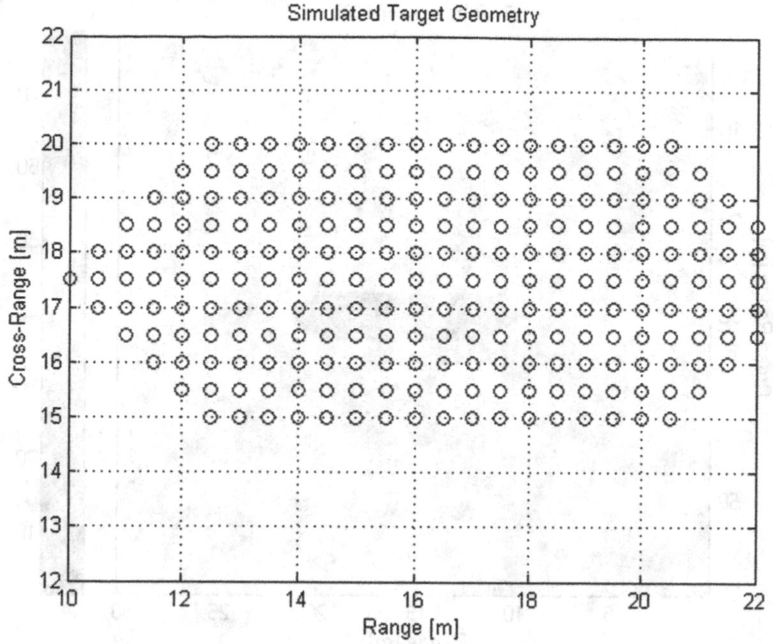

Fig. 3 Geometry of the simulated ship target

Table 1 SAR simulation parameters

Parameter	Value (units)
Carrier frequency, f_0	10 (GHz)
Radar bandwidth, B	300 (MHz)
Number of frequencies, M	64
Pulse repetition frequency, PRF	2.74 (KHz)
Burst duration, T_b	0.0234 (s)
Coherent processing interval, CPI	3 (s)
Number of bursts, N	128
Number of CPIs, N_{CPI}	1
Range distance to center of target, R_0	10 (km)
Height of SAR platform, h	2 (km)
Position angle of the ship, ϕ	0
Velocity of platform, v	100 (m/s)

parameters of the simulation such as the distance between the radar and the target and the velocity of the radar platform, alter. The corresponding parameters regarding the airborne and spaceborne scenarios are shown in Table 2.

In Fig. 7 the produced SAR image for the simulated spaceborne scenario is presented.

Numerical results of the application of the proposed autofocusing algorithm for improving image quality in the case of real radar data

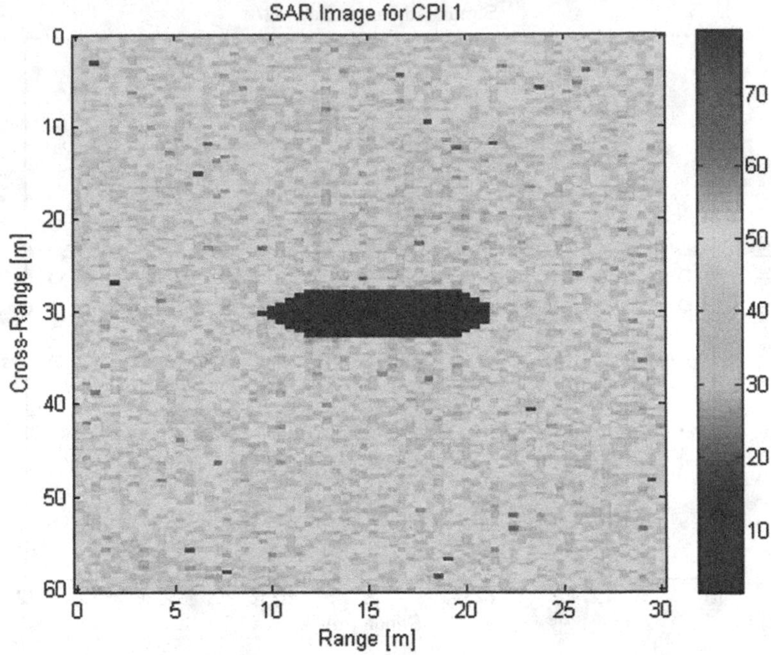

Fig. 4 SAR image for one CPI (airborne scenario/two dimensional target)

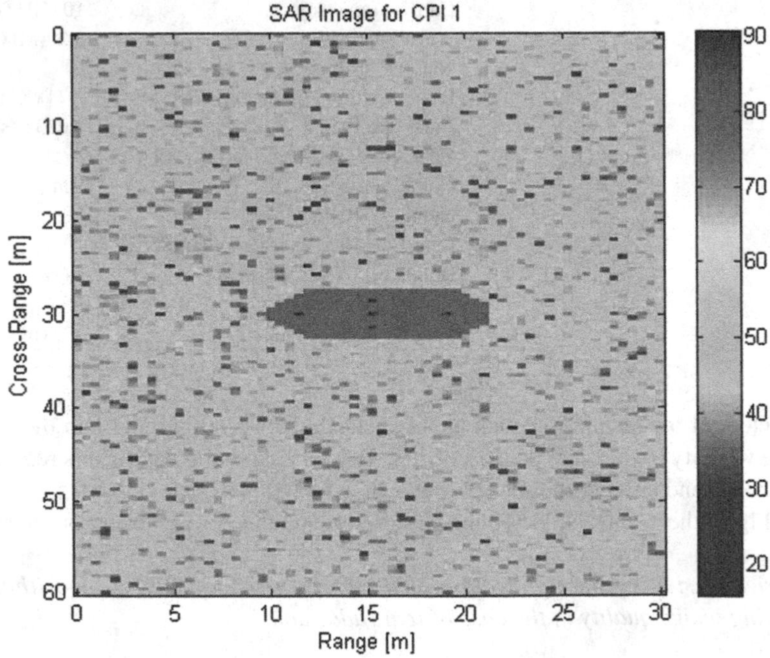

Fig. 5 SAR image for one CPI with varying scattering intensity (airborne scenario)

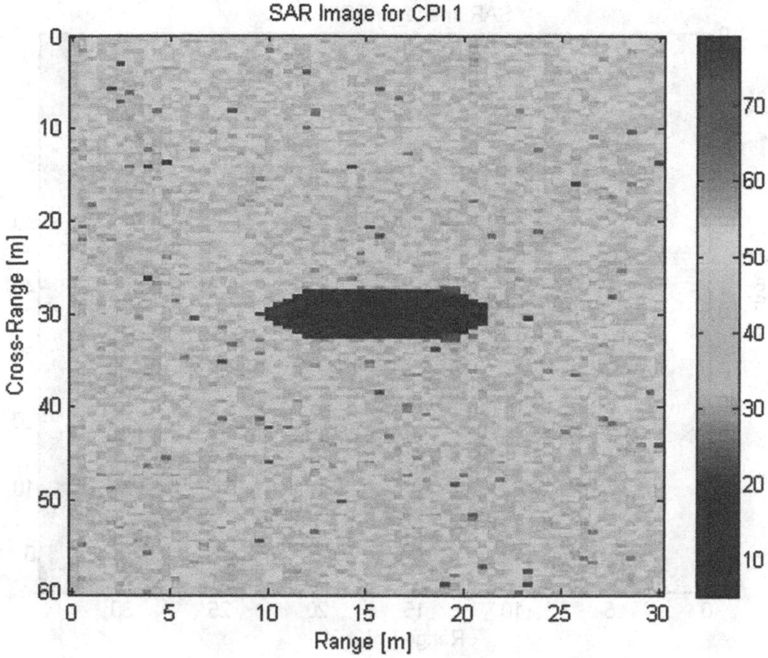

Fig. 6 SAR image for one CPI (airborne scenario/three dimensional target)

Table 2 SAR simulation parameters airborne–spaceborne scenario

Parameter	Airborne value (units)	Speceborne value (units)
Carrier frequency, f_0	10 (GHz)	10 (GHz)
Radar bandwidth, B	300 (MHz)	300 (MHz)
Number of frequencies, M	64	64
Burst duration, T_b	23.4 (ms)	0.085 (ms)
Number of bursts, N	128	128
Range distance to center of target, R_0	10 (km)	1,000 (km)
Height of SAR platform, h	2 (km)	$2 \cdot 10^5$ (m)
Velocity of platform, v	100 (m/s)	2,900 (m/s)

In this section we incorporate the post processing CPI-split autofocusing algorithm recently introduced by our research group [3, 8] in the case of real-field radar data. This algorithm has already been tested for simulated data in the cases of SAR and ISAR geometry [3, 7, 8]. The application of our proposed algorithm produced excellent focusing of SAR and ISAR images for several cases of moving targets [3, 7, 8].

The real-field radar data, which are examined here, were provided to us by SET 163 Working Group (see the "Acknowledgement" below for more details). This radar transmits linear frequency modulated waveform (LFM), whereas in our simulation scenarios the radar antenna is assumed to emit Stepped Frequency

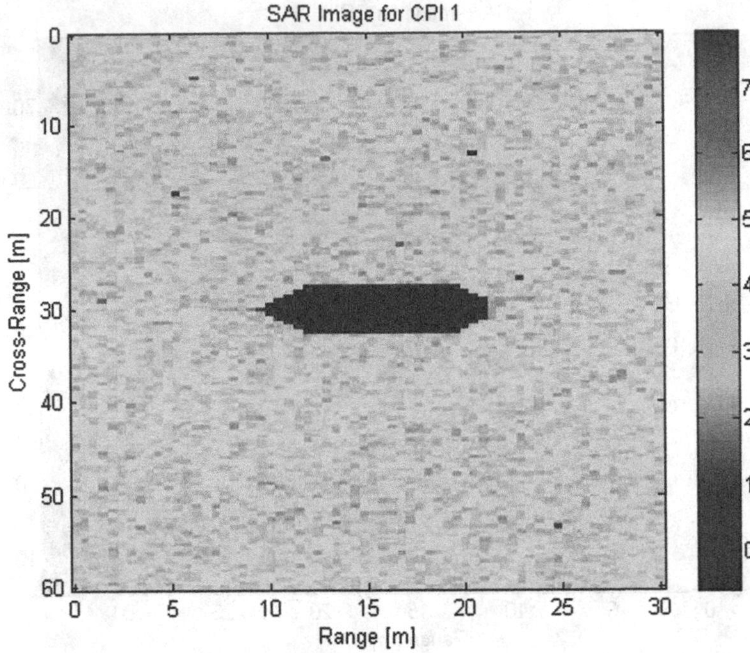

Fig. 7 SAR image for one CPI (spaceborne scenario)

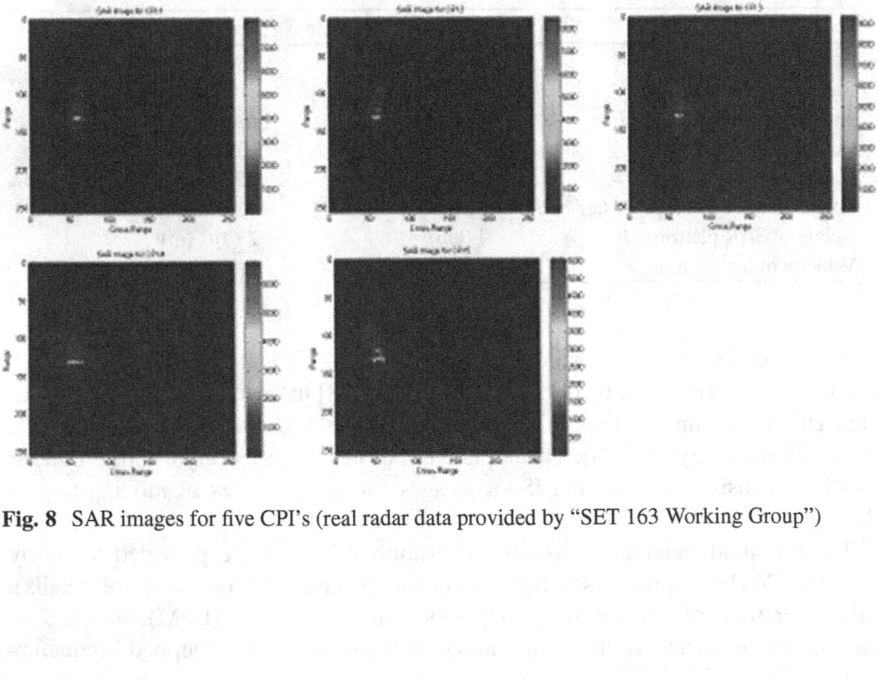

Fig. 8 SAR images for five CPI's (real radar data provided by "SET 163 Working Group")

Table 3 Entropy values

SAR Image	Entropy	Minimum entropy combination
1st CPI	6.4684	
2nd CPI	6.2411	
3rd CPI	6.1885	
4th CPI, unfocused	6.7877	
5th CPI	6.7082	
4th CPI, focused	*6.5991*	*Stage 4, segment 2, combination 7 [3]*

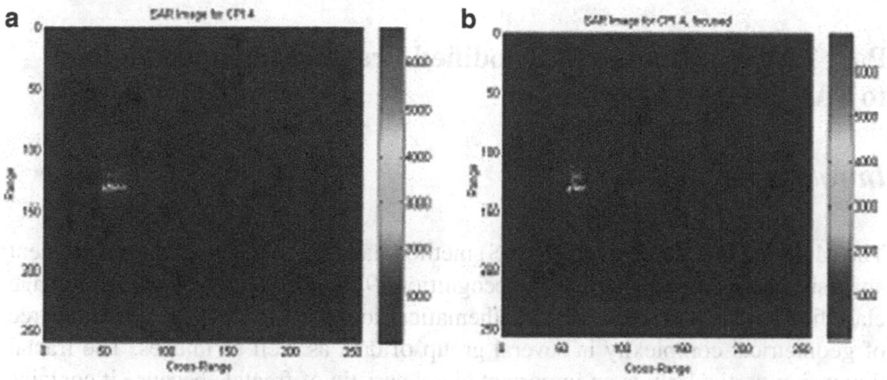

Fig. 9 SAR image before (**a**) and after (**b**) the application of the proposed autofocusing algorithm

(SF) pulses. These raw radar data yield ultimately (after appropriate SAR signal processing) a moving ship target, which is being imaged by the SAR radar. In Fig. 8 five SAR images of consecutive CPI's are presented.

In order to compare the quality of the above SAR images, the entropy value of each image is computed [3, 4]. In Table 3 the entropy values for the five SAR images presented in Fig. 8 are listed. The CPI-split autofocusing algorithm is employed in the CPI which entropy value exceeds a threshold that represents an acceptable SAR image quality [3]. The images with entropy values below the entropy threshold are called "focused" images, while the images with entropy values over the threshold are called "unfocused". We have applied the autofocusing algorithm to the previously presented SAR images for 5 CPIs. As seen in Table 3, the 4th CPI has the greater entropy value and therefore is the "unfocused" CPI. The entropy threshold was set to 6.7.

In Fig. 9 the SAR images for the 4th CPI are presented, before and after the application of the autofocusing algorithm.

The SAR image for the 4th CPI after the application of the autofocusing algorithm is clearly more focused than the SAR image before the application of the autofocusing algorithm. This result is also validated by the entropy values presented in Table 3. The entropy value of the SAR image (4th CPI) before the application of the autofocusing algorithm is greater than the entropy value of the

SAR image (4th CPI) after the application of the autofocusing algorithm. Moreover the entropy value of the "focused" image is within the acceptable entropy values (below the entropy threshold). The real-field data results presented above show that the proposed algorithm is effective in producing focused SAR images. Based on SAR image entropy minimization criterion, the proposed algorithm neglects data leading to ISAR images of poor quality and uses only data leading to ISAR images of superior quality. The simulations' results verify the adaptiveness of the autofocusing procedure to different SAR imaging conditions.

Part 2: Application of the Modified Fractal Signature Method to SAR Image Classification

Introduction

The Modified Fractal Signature (MFS) method has already been used for document analysis, classification and pattern recognition [9] as well as for biomedical image classification [10]. Fractals are a mathematical tool used to describe a high degree of geometrical complexity in several group of data as well as images. The fractal dimension particularly is an important characteristic of fractals because it contains information about their geometric structure. As a result the computation firstly of fractal dimension and consequently of fractal signature is of great importance with regard to radar image classification.

This method includes fractal analysis [11, 12] of surfaces derived from radar images, both from simulated data and from real radar data, provided to us by SET 163 Working Group, using a "blanket" technique [9, 10] described below. The main idea concerning this technique is the fact that different classes of images yield different values of fractal signature (FS) and fractal dimension (FD), upon which classification of different types of images is possible. In this paper we are interested in the classification of SAR radar images, and, in particular, in the discrimination of "focused" or "unfocused" SAR images [8]. In addition, preliminary results regarding the classification of real SAR radar images ("Oslo Fjord") to "town" area, "suburban" area and "sea" area are presented.

Mathematical Formulation of the MFS Method

In this section the implementation of the "blanket" (MFS) method [9, 10] is described. Initially, the SAR image is converted to a grey-level function $g(x, y)$. Subsequently the whole SAR image is divided into several non-overlapping sub-images, and the fractal signature is calculated for each sub-image. The overall fractal signature of the initial image is calculated ultimately by summation of the corresponding values of the sub-images [9]. In addition in order to compute the fractal dimension, we need to measure the area of the grey level surface.

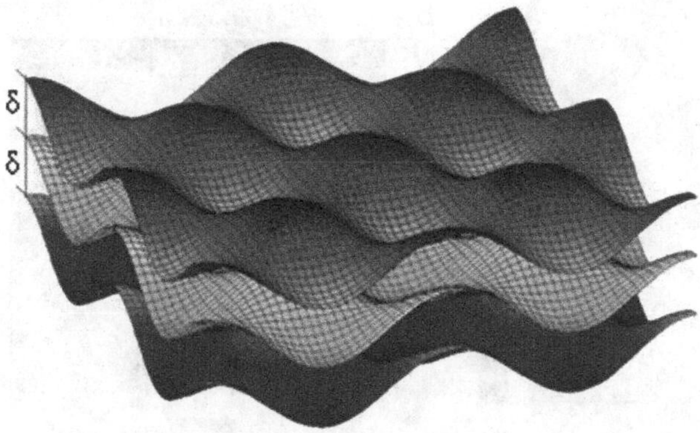

Fig. 10 "Blanket" of thickness 2δ defined by its upper $u_\delta(x, y)$ and lower $b_\delta(x, y)$ surface

In the blanket technique, all points of the three-dimensional space at distance δ from the grey level surface $g(x, y)$ are considered. These points construct a "blanket" of thickness 2δ covering the initial surface. The covering blanket is defined by its upper surface $u_\delta(x, y)$ and its lower surface $b_\delta(x, y)$ as it is presented in Fig. 10.

The algorithm used to compute the upper and lower surface includes the following steps. Initially, the iteration δ equals zero ($\delta = 0$), the grey-level function equals the upper and lower surfaces, namely: $u_0(x, y) = b_0(x, y) = g(x, y)$. For iteration $\delta = 1, 2, \cdots$ the blanket surfaces are calculated through the following iterative formulae :

$$u_{\delta(x,y)} = \max[u_{\delta-1}(x, y) + 1, \max_{|(m,n)-(x,y)|\leq 1} u_{\delta-1}(m, n)] \tag{14}$$

$$b_{\delta(x,y)} = \min[b_{\delta-1}(x, y) - 1, \min_{|(m,n)-(x,y)|\leq 1} b_{\delta-1}(m, n)] \tag{15}$$

Subsequently, the volume of the "blanket" is calculated from $u_\delta(x, y)$ and $b_\delta(x, y)$ by:

$$Vol_\delta = \sum_{(x,y)} [u_\delta(x, y) - b_\delta(x, y)] \tag{16}$$

Furthermore, the fractal signature A_δ is calculated by

$$A_\delta = \frac{Vol_\delta}{2\delta} \quad or \quad A_\delta = \frac{Vol_\delta - Vol_{\delta-1}}{2} \tag{17}$$

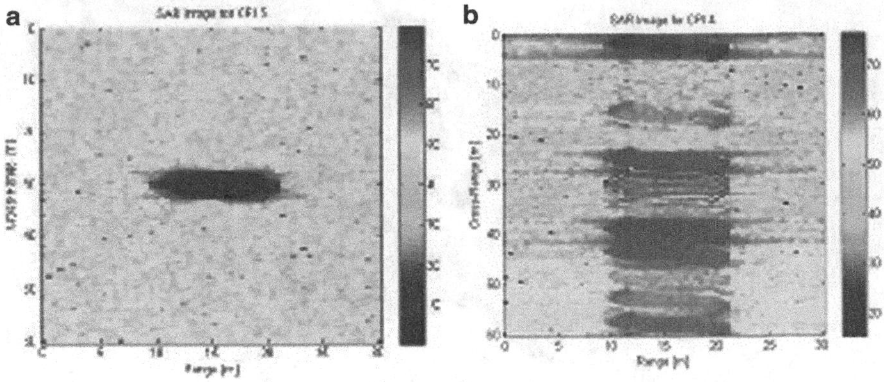

Fig. 11 SAR images (**a**) focused image of ship, (**b**) unfocused image of ship

Finally, concerning the calculation of the corresponding fractal dimension D, the following formula is used [9]:

$$A_\delta \simeq \beta \delta^{2-D} \tag{18}$$

where β is a constant, from which the fractal dimension can be calculated from successive fractal signature values as follows

$$D \simeq 2 - \frac{\log_2 A_{\delta_1} - \log_2 A_{\delta_2}}{\log_2 \delta_1 - \log_2 \delta_2}, \quad where \quad \delta = 1, 2, \cdots \tag{19}$$

Numerical Results

SAR radar images from simulated data

Firstly, the application of the MFS method on Synthetic Aperture Radar (SAR) images from simulated data is presented. For classification purposes, two SAR images are examined: the first is a "focused" image of a ship and the second is an "unfocused" image of a ship (Fig. 11). The case of the "focused" and "unfocused" ISAR image (the target was a rapidly manoeuvring aircraft) has already been examined in [13]. The main idea is whether the two images can be discriminated using the fractal signature and fractal dimension values that occur after the application of the MFS method.

In Fig. 12, the fractal signature A_δ as a function of iteration δ and the fractal dimension D as a function of iteration δ for the cases of the "focused" and the "unfocused" image are presented.

It is apparent from the results in Fig. 12 that the criteria of "fractal signature" and "fractal dimension" for SAR image characterization work in a satisfactory way for the above simulations. It appears that the conclusions about image characterization follow for small values of iteration δ. It can be easily understood, that for large

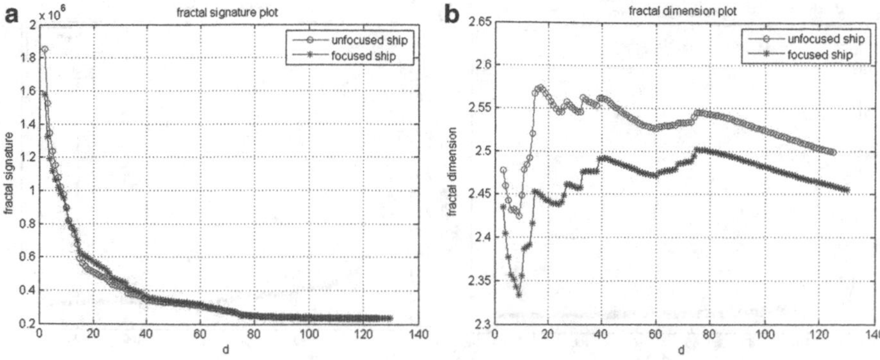

Fig. 12 SAR images (**a**) fractal signature A_δ as a function of iteration δ, (**b**) fractal dimension D as a function of iteration δ for the cases of the "focused" and the "unfocused" ship

Fig. 13 SAR image of "Oslo fjord" (provided by "SET 163 Working Group")

number of iterations (δ), the proposed algorithm of Eqs. (14) and (15) appears to select the ± 1 values of the previous iteration, so that the value of fractal signature equals, in this limit, the number of the pixels of the image.

SAR radar images from real radar data ("Oslo fjord").

Subsequently, the application of the MFS method to Synthetic Aperture Radar (SAR) images from real radar data (SAR image of "Oslo fjord") provided to us by SET 163 Working Group is presented. The SAR image examined here is shown in Fig. 13. Three sub-images were obtained from the initial SAR image: the first includes a "town" area, the second a "suburban" area and the third a "sea" area.

In Fig. 14, the fractal signature A_δ as a function of iteration δ and the fractal dimension D as a function of iteration δ for the cases of three sub-images: "town" area, "suburban" area and "sea" area are presented.

It appears that the proposed algorithm provides interesting characterization results for the cases of "town" area, "suburban" area and "sea" area.

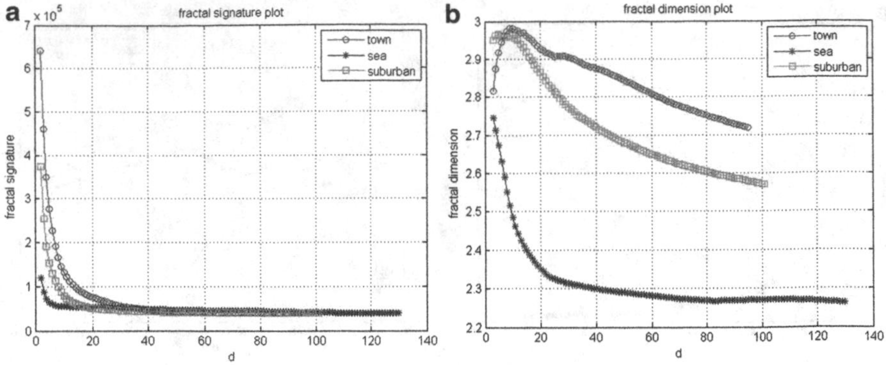

Fig. 14 (**a**) fractal signature A_δ as a function of iteration δ, (**b**) fractal dimension D as a function of iteration δ for the cases of "town" area, "suburban" area and "sea" area

Conclusions

In this paper several aspects of the SAR imaging are examined. The cases of airborne and spaceborne SAR are simulated and also a two-dimensional and a three-dimensional targets (ship) are implemented as well as the case of varying target scattering intensity. Hence, the simulations contribute to better analyse, clarify and understand the cases of real-field radar data. In addition the proposed "CPI-split autofocusing algorithm" is incorporated for the case of real-field radar data. The real-field data results presented above show that the proposed algorithm is effective in producing focused SAR images. The simulations' results verify the adaptiveness of the autofocusing procedure to different SAR imaging conditions.

Furthermore, an iterative MFS technique [9, 10] is applied aiming in SAR radar image characterization (both simulated and real data). As confirmed by the results presented above, the proposed algorithm provides interesting characterization results. It is apparent that the criteria of "fractal signature" and "fractal dimension" for SAR image characterization work in a satisfactory way for the above simulations. In these results it is clearly seen that the focusing of the SAR radar image clearly correlates with the value of "fractal signature" and "fractal dimension" for the simulated data. Also the type of area can be distinguished with the value of "fractal signature" and "fractal dimension" for the real-field radar data.

Acknowledgements The authors (AM, AK, EK, PF) would like to acknowledge SET 163 Working Group, and its Chairman Dr. Luc Vignaud (ONERA, France) in particular, for providing us with the real-field Synthetic Aperture Radar (SAR) data, which were used to reconstruct the SAR images of moving ship shown in Figs. 8 and 9 above, as well as the SAR image of "Oslo fjord" shown in Fig. 13. In particular, the radar data concerning the moving ship of Figs. 8 and 9 were provided to SET 111 and SET 163 Working Groups by Dr. William Miceli (ONR) and their origin is from a radar developed by "Radar Branch of the Naval Command Control and Ocean Surveillance Center", Research Development Test and Evaluation Division (NRaD), San Diego, CA, USA. Furthermore, the "Oslo fjord" image of Fig. 13 was produced by DLR, Germany (spaceborne image). To all the above institutes and involved scientists we express our sincere thanks for providing these real-field radar data to us, in the framework of SET 163 Working Group.

This research has been co-financed by the European Union (European Social Fund) and the Greek National Funds through the Operational Program "Education and Lifelong Learning" of the National Strategic Reference Framework (NSRF)—Research Funding Program: THALIS.

References

1. D. Wehner, High-Resolution Radar,Artech House, 2nd edition, London, 1995.
2. V. Chen, H. Ling, Time-Frequency Transforms for Radar Imaging and Signal Analysis,Artech House,London, 2002.
3. E. Kallitsis, A. Karakasiliotis, G. Boultadakis, P. Frangos, "A Fully Automatic Autofocusing Algorithm for Post-processing ISAR Imaging based on Image Entropy Minimization", Electronics and Electrical Engineering Journal, No. 4 (110), pp. 125–130, April 2011.
4. A. Lazarov, C. Minchev, "ISAR Signal Modeling and Image Reconstruction with Entropy Minimization Autofocusing", Proc. DASC, Portland, USA, Oct 2006, pp. 3E5-1–3E5-11.
5. J. Li, R. Wu, V. Chen, "Robust autofocus algorithm for ISAR imaging of moving", IEEE Trans. Aerosp. Electron. Syst., Vol. 37, No 3, pp. 1056–1069, 2001.
6. D. Pastina, A. Farina, J. Gunning, P. Lombardo, "Two-dimensional super resolution spectral analysis applied to SAR images", IEE Proc., Radar Sonar Navig., Vol. 145, No 5, pp. 281–290, 1998.
7. A. Malamou, A. Karakasiliotis, E. Kallitsis, G. Boultadakis and P. Frangos, "An autofocusing algorithm for post-processing of synthetic aperture radar (SAR) images based on image entropy minimization", CEMA'12 International Conference, National Technical University of Athens (NTUA), Athens, Greece, 8-10/11/2012, pp. 53–56.
8. A. Malamou, A. Karakasiliotis, E. Kallitsis, G. Boultadakis, P. Frangos, 'Application of a Fully Automatic Autofocusing Algorithm for Post Processing of Synthetic Aperture Radar Images based on Image Entropy Minimization', 'Electronics and Electrical Engineering' Journal, Vol. 19, No.6, June 2013, pp. 95–98.
9. Y. Tang, H. Ma, D. Xi, X. Mao, C. Suen, 'Modified Fractal Signature (MFS): A New Approach to Document Analysis for Automatic Knowledge Acquisition', IEEE Transactions on knowledge and data engineering, Vol. 9, No. 5, Sept. – Oct. 1997, pp. 747–762.
10. N.B. Ampilova, E.Y. Gurevich, I.P. Soloviev, 'Application of modified fractal signature and Regny spectrum methods to the analysis of biomedical preparation images, CEMA'11 Conference, Sofia, Bulgaria, October 2011, pp. 96–100.
11. K. J. Falconer, 'Fractal Geometry : Mathematical Foundations and Applications', J. Wiley and Sons, 1990.
12. D. Jaggard, A. Jaggard and P. Frangos, 'Fractal electrodynamics : surfaces and superlattices', in 'Frontiers in Electromagnetics', IEEE Press, Edited by D. Werner and R. Mittra, 2000.
13. C. Pandis, A. Malamou, P. Stefaneas and P. Frangos, 'Applying the modified fractal signature method to image classification : some preliminary results for ISAR radar images', CEMA'12 International Conference, National Technical University of Athens (NTUA), Athens, Greece, 8-10/11/2012.

Optimal Preventive Maintenance of a Production-Inventory System When the Action of "Idling" Is Permissible

Constantinos C. Karamatsoukis and Epaminondas G. Kyriakidis

Abstract In this paper we consider a manufacturing system in which an input generating installation (I) supplies a buffer (B) with a raw material, and a production unit (PU) pulls the raw material from the buffer with constant rate $d > 0$. The capacity of the buffer is equal to K units of raw material. The input rate P is assumed to be a discrete random variable whose possible values belong to the set $\{d, d+1, \ldots, d+K-x\}$ where $x \in \{0, \ldots, K\}$ is the content of the buffer. The installation deteriorates as time evolves and the problem of its preventive maintenance is considered. There are three possible decisions when the installation is at operative condition: (i) the action of allowing the installation to operate, (ii) the action of leaving the installation idle, and (iii) the action of initiating a preventive maintenance of the installation. The objective is to find a policy (i.e., a rule for choosing actions) that minimizes the expected long-run average cost per unit time. The cost structure includes operating costs of the installation, maintenance costs of the installation, storage costs, and costs due to the lost production when a maintenance is performed on the installation and the buffer is empty.

Using the dynamic programming equations that correspond to the problem and some results from the theory of Markov decision processes we prove that the average-cost optimal policy initiates a preventive maintenance of the installation if and only if, for some fixed buffer content x, the degree of deterioration of the installation is greater or equal to a critical level $i^*(x)$ that depends on x. The optimal policy and the minimum average cost can be computed numerically using the value iteration algorithm. For fixed buffer content x, extensive numerical results provide strong evidence that there exists another critical level $\tilde{i}(x) \leq i^*(x)$ such

---

C.C. Karamatsoukis (✉)
Department of Mathematics and Engineering, Hellenic Army Academy, Vari, Attiki, Greece
e-mail: k.karamatsoukis@fme.aegean.gr

E.G. Kyriakidis
Department of Statistics, Athens University of Economics and Business, Athens, Greece
e-mail: ekyriak@aueb.gr

N.J. Daras (ed.), *Applications of Mathematics and Informatics in Science and Engineering*, Springer Optimization and Its Applications 91, DOI 10.1007/978-3-319-04720-1_14,

that the average-cost optimal policy allows the installation to operate if its degree of deterioration is smaller than $\tilde{i}(x)$ and leaves the installation idle if its degree of deterioration is greater or equal to $\tilde{i}(x)$ and smaller than $i^*(x)$. A proof of this conjecture seems to be difficult.

Keywords Dynamic programming • Maintenance • Production-inventory system

Introduction

The preventive maintenance of a production-inventory system is an effective way to reduce operating costs and the occurrence of failures. The reliability of a production system is a crucial issue in cases such as public utilities (e.g., electric power plants) and defense systems (e.g., missile shield or airplane engine compressor blades see [4]). The growing interest in preventive maintenance has led many researchers to develop mathematical models which propose several kinds of maintenance policies.

The paper of Wang [13] is a survey of different kinds of maintenance policies in deteriorating systems. Dekker [1] provides a review of various maintenance models and many applications of these models in real life. Markov Decision Models have been proved a powerful tool for the description and the solution of problems which are related to the maintenance of a system (or a component of a system). In the papers of Douer and Yechiali [3], Van der Duyn Schouten and Vanneste [12], Sloan (2004), the maintenance problems are modeled and analyzed by suitable Markov Decision Models.

In maintenance literature, it is generally assumed that a system is inspected in discrete time epochs, and the deteriorating components of the system are classified into conditions. In the papers of [7], Karamatsoukis and Kyriakidis (2010) and [5] the preventive maintenance depends not only on the working condition of a deteriorating installation of the system but also on the content of a buffer in which a raw material is transferred from the installation. The capacity of the buffer is assumed to be fixed and the raw material is transferred to the buffer at a constant rate p. A production unit pulls the raw material with a constant rate d $(p > d)$. The state of the system is represented by the pair (i, x), where i is the degree of the deterioration of the installation and x is the buffer content. If the installation is at an operative condition, there are two possible actions: to initiate a preventive maintenance on the installation or to allow the installation to operate. The preventive and corrective maintenance are nonpreemptive, i.e., they cannot be interrupted, and they bring the installation to a perfect condition. It is assumed that the preventive and corrective repair times (expressed in time units) are geometrically distributed.

In the present paper we modify a model (see Kyriakidis and Dimitrakos [7]) in which a deteriorating installation transfers raw material to a buffer and a production unit pulls the raw material from the buffer. The production unit is always in an operative condition. If the installation is found to be at a failed condition, a corrective maintenance must be commenced. We assume that the input rate P is a discrete random variable, whereas the rate d at which the production unit pulls raw material from the buffer is constant. It is assumed that the possible values of P belong to the set $d, d+1, ..., d+K-x$ where $x \in 0, ..., K$. Furthermore, if the installation is found to be at an operative condition, the possible actions are different from Kyriakidis's and Dimitrakos's [7] model (there is another possible action which is to leave the installation idle). It is proved that the average-cost optimal policy initiates a preventive maintenance of the installation if and only if, for some fixed buffer content, the degree of deterioration of the installation is greater or equal to a critical level $i^*(x)$ that depends on x. Extensive numerical results provide strong evidence that there is another critical level $\tilde{i}(x) \leq i^*(x)$, such that the average-cost optimal policy, for some fixed buffer content x, allows the installation to operate if its degree of deterioration is smaller than $\tilde{i}(x)$ and leaves it idle if its degree of deterioration is greater or equal to $\tilde{i}(x)$ and smaller than $i^*(x)$. A proof of this conjecture seems to be difficult.

The rest of the paper is organized as follows. The description of the model is given in section "The Model." In section "The Form of the Optimal Policy" it is shown that, for fixed buffer content, the optimal policy is of control-limit type. In section "Numerical Results" two numerical examples are presented and in section "Conclusions" the main conclusions of the paper are summarized.

The Model

In this paper we generalize the results obtained in the paper of [7] that are concerned with the preventive maintenance of a production-inventory system which consists of three components. We consider a manufacturing system in which an input-generating installation (I) supplies a buffer (B) with raw material and a production unit (PU) pulls the raw material from the buffer with constant rate $d > 0$. The three components of the system are depicted in Fig. 1.

The capacity of the buffer is equal to K units of raw material. The input rate P is assumed to be a discrete random variable whose possible values belong to the set $\{d, d+1, \ldots, d+K-x\}$ where $x \in \{0, \ldots, K\}$. When the buffer is full, the rate P at which the installation supplies the buffer can take only the value d.

An example of this manufacturing system (see [8]) could be an automobile general assembly where the production unit represents the assembly line and the installation represents one of many parallel operations that directly supply the line. In the military industry we can assume that this system produces tanks or light weapons.

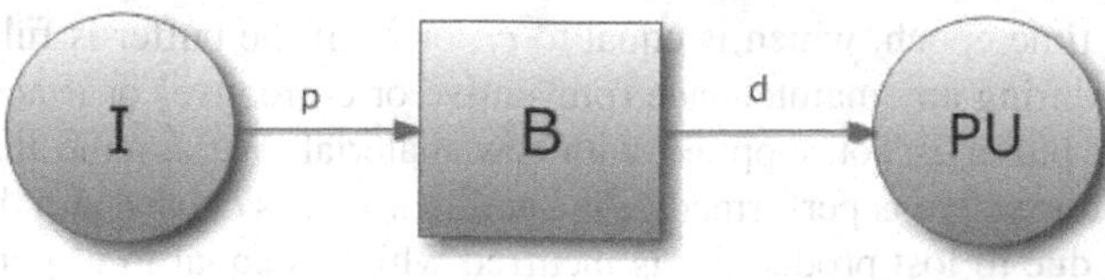

Fig. 1 The three components of the system

The system is inspected at discrete, equidistant time epochs $\tau = 0, 1, \ldots$ (say every hour), and the installation is classified into $m+2$ conditions $0, 1, \ldots, m+1$ which describe increasing degrees of deterioration. State 0 represents a new installation before any deterioration occurs whereas state $m+1$ represents the failure state of the installation. The intermediate states $1, \ldots, m$ are operative. If at a time epoch τ the state of the installation is $i < m+1$ and the content of the buffer is $x < K$ then the content of the buffer at the epoch $\tau + 1$ will be $\min(x + p - d, K)$. This increase of the buffer content will happen even if the state of the installation at the time epoch $\tau + 1$ is the failure state $m + 1$. If the action of allowing the installation to operate is chosen, the transition probability of moving from condition i at time epoch τ to working condition j at time $\tau + 1$ is equal to p_{ij}. If the action of leaving the installation idle is chosen, the transition probability of moving from condition i at time epoch τ to working condition j is equal to q_{ij}. It is assumed that $p_{ij} \leq q_{ij}, 0 \leq j \leq i$ and $q_{ij} \leq p_{ij}, i < j \leq m+1$ because the operation of the installation causes, possibly, more serious deterioration. We assume that the probability of eventually reaching the condition $m + 1$ from any initial state i is nonzero.

There are four possible actions $u \in \{0, 1, 2, 3\}$, which are selected at each time epoch. The possible actions are: (i) the action of allowing the installation to operate ($u = 0$), (ii) the action of leaving the installation idle ($u = 1$), (iii) the action of starting a preventive maintenance of the installation ($u = 2$), and the (iv) the action of starting a corrective maintenance of the installation ($u = 3$). If at a time epoch the installation is found to be at working condition $m + 1$ (failure state), the action ($u = 3$) is compulsory. If at a time epoch the installation is found to be at an operative condition $i \in \{0, \ldots, m\}$, then we may choose the action $u = 0$ or action $u = 1$ or action $u = 2$. Both preventive and corrective maintenance are nonpreemptive, i.e., they cannot be interrupted and they bring the installation to a perfect condition 0. It is assumed that the preventive and corrective maintenance repair are geometrically distributed with the probability of success a and b, i.e. the probability that they last $t \geq 1$ time units is equal to $(1-a)^{t-1}a$ and $(1-b)^{t-1}b$, respectively. This assumption can be considered as realistic for situations in which the repair consists of successive efforts to bring the installation to a perfect condition. For example, the repair may have the form of successive attempts to replace a part of the deteriorating machine or to assemble two parts that are disconnected or restore a setting of the machine. If the probability of success is fixed we can assume that times are geometrically distributed (see [5, 9, 12]).

The cost rates during a preventive and a corrective maintenance are equal to c_p and c_f, respectively. If at a time epoch the installation is found to be at an operative state $i \in \{0, \ldots, m\}$, and action $u = 0$ is chosen, an operating cost is incurred until the next time epoch, which is equal to c_i, or $\tilde{c}_i$, if the buffer is full. We also, suppose, that during any maintenance (preventive or corrective) or if action $u = 1$ is chosen, the buffer is not supplied with raw material. We assume that, when a preventive or corrective is performed or the action $u = 1$ is chosen and the buffer is empty, a cost due to lost production is incurred which is equal to C per unit time.

We also suppose that the cost of holding a unit of the raw material in the buffer for one unit of time is equal to $h > 0$.

We introduce the state PM to denote the situation that a preventive maintenance is performed. Then the state of the system is:

$$S = \{0, \ldots, m+1, PM\} \times \{0, \ldots, K\}$$

where $(i, x) \in S$ is the state in which i is the working condition of the installation and x is the content of the buffer.

The following conditions on the cost structure, on the transition probabilities, and on the expected maintenance times are assumed to be valid:

Condition 1. $c_0 \leq c_1 \leq \cdots \leq c_m, \tilde{c}_0 \leq \tilde{c}_1 \leq \cdots \leq \tilde{c}_m$. That is, the working condition of the installation deteriorates, the operating cost increases.

Condition 2. $\tilde{c}_i \leq c_i, 0 \leq i \leq m$. That is, when the buffer is full the installation slows down and that causes smaller operating costs.

Condition 3. $0 < b < a \leq 1$. That is, the expected time required for a preventive maintenance is smaller than the expected time required for a corrective maintenance.

Condition 4. $c_p \leq c_f$. That is, the cost rate of a preventive maintenance does not exceed the cost rate of a corrective maintenance.

Condition 5. (Increasing Failure Rate Assumptions). For each $k = 0, \ldots, m+1$, the function $D_k(i) = \sum_{j=k}^{m+1} p_{ij}$ and $G_k(i) = \sum_{j=k}^{m+1} q_{ij}$ are nondecreasing in $i, 0 \leq i \leq m$.

It can be shown (see [2], pp. 122–123) that this condition is equivalent to the following one:

Condition 6. For any nondecreasing functions $h(j), h'(j), 0 \leq j \leq m+1$ the quantities $\sum_{j=k}^{m+1} p_{ij} h(j)$ and $\sum_{j=k}^{m+1} q_{ij} h'(j)$ are nondecreasing in $i, 0 \leq i \leq m$.

We consider a Markov Decision Model in which we aim to find a stationary policy which minimizes the long-run expected average cost per unit time. In the next section we prove that the optimal policy has a particular structure.

The Form of the Optimal Policy

Let $\alpha (0 < \alpha < 1)$ be a discount factor. The minimum n-step expected discounted cost $V_n^{\alpha}(i, x)$, where (i, x) is the initial state, can found for all $n = 1, 2, \ldots$ recursively, from the following equations:

$$V_n^{\alpha}(i, x) = \min \left\{ c_i + hx + \alpha \sum_{p=d}^{d+K-x} \sum_{j=0}^{m+1} Pr(P=p) p_{ij} V_{n-1}^{\alpha}(j, \min(x+p-d, K)), \right.$$

$$hx + C(d-x)^+ + \alpha \sum_{j=0}^{m+1} q_{ij} V_{n-1}^{\alpha}(j, (x-d)^+),$$

$$\left. V_n^{\alpha}(PM, x) \right\}, 0 \le i \le m, 0 \le x \le K. \quad (1)$$

$$V_n^{\alpha}(PM, x) = c_p + hx + C(d-x)^+ + \alpha a V_{n-1}^{\alpha}(0, (x-d)^+)$$
$$+\alpha(1-a) V_{n-1}^{\alpha}(PM, (x-d)^+),$$
$$0 \le x \le K,$$

$$V_n^{\alpha}(m+1, x) = c_p + hx + C(d-x)^+ + \alpha b V_{n-1}^{\alpha}(0, (x-d)^+)$$
$$+\alpha(1-b) V_{n-1}^{\alpha}(m+1, (x-d)^+),$$
$$0 \le x \le K,$$

with initial value

$$V_0^{\alpha}(i, x) = 0, \ (i, x) \in S.$$

Note that $(d-x)^+ = \max(d-x, 0)$ and $(x-d)^+ = \max(x-d, 0)$ represent the one-period demand lost and the buffer content, respectively, when the buffer content equals x at the beginning of that period and if action 1 or 2 or 3 is chosen. In Eq. (1), the first term corresponds to action $u = 0$, the second term corresponds to action $u = 1$, and the third term to action $u = 2$. The quantity $V_n^{\alpha}(i, K)$ coinsides with the right side of (1) with $x = K$ if, in the first term inside the curly brackets we replace c_i with $\tilde{c}_i$.

The following lemma is needed to prove that the optimal policy is of control-limit type.

Lemma 1. *For each $n = 0, 1, \ldots$ we have that*

$$(i) \ V_n^{\alpha}(i, x) \le V_n^{\alpha}(i+1, x), \ 0 \le i \le m, 0 \le x \le K$$
$$(ii) \ V_n^{\alpha}(PM, x) \le V_n^{\alpha}(m+1, x), 0 \le x \le K$$

Proof. We prove the lemma by induction on n. The lemma holds for $n = 0$ since $V_0^{\alpha}(i, x) = 0$. We assume that the lemma holds for $n - 1$. We will first prove the part (ii) holds for n and then we will prove the part (i) holds for n.
Part (ii):

$$V_n^{\alpha}(PM, x)$$
$$= c_p + hx + C(d-x)^+ + \alpha a V_{n-1}^{\alpha}(0, (x-d)^+) + \alpha(1-a) V_{n-1}^{\alpha}(PM, (x-d)^+)$$
$$\le c_f + hx + C(d-x)^+ + \alpha a V_{n-1}^{\alpha}(0, (x-d)^+) + \alpha(1-a) V_{n-1}^{\alpha}(m+1, (x-d)^+)$$

$$= c_f + hx + C(d-x)^+ + aV^\alpha_{n-1}(m+1,(x-d)^+) - \alpha a D$$
$$\le c_f + hx + C(d-x)^+ + aV^\alpha_{n-1}(m+1,(x-d)^+) - \alpha b D = V^\alpha_n(m+1,x)$$

where $D = V^\alpha_{n-1}(m+1,(x-d)^+) - V^\alpha_{n-1}(0,(x-d)^+)$. The first inequality follows from the Condition 4 and induction hypothesis of part (ii). The second inequality follows from the Condition 3 and from inequality $D \ge 0$ which is a consequence of part (i) of the induction hypothesis.

Part (i): We have to show that

$$V^\alpha_n(i,x) \le V^\alpha_n(i+1,x),\ 0 \le i \le m-1,\ 0 \le x \le K, \tag{2}$$

and

$$V^\alpha_n(m,x) \le V^\alpha_n(m+1,x),\ 0 \le x \le K. \tag{3}$$

The inequality (3) is easily verified for $x \in \{0,\dots,K-1\}$, using part (ii) above:

$$V^\alpha_n(m,x) = \Bigg\{ c_m + hx + \alpha \sum_{p=d}^{d+K-x}\sum_{j=0}^{m+1} Pr(P=p)p_{mj}V^\alpha_{n-1}(j,\min(x+p-d,K)),$$
$$hx + C(d-x)^+ + \alpha\sum_{j=0}^{m+1} q_{mj}V^\alpha_{n-1}(j,(x-d)^+),$$
$$V^\alpha_n(PM,x)\Bigg\} \le V^\alpha_n(PM,x) \le V^\alpha_n(m+1,x).$$

Similarly, we obtain the inequality for $x = K$. For $x \in \{0,1,\dots,K-1\}$ and $i \in \{0,\dots,m-1\}$ we obtain

$$V^\alpha_n(i,x) = \Bigg\{ c_i + hx + \alpha \sum_{p=d}^{d+K-x}\sum_{j=0}^{m+1} Pr(P=p)p_{ij}V^\alpha_{n-1}(j,\min(x+p-d,K)),$$
$$hx + C(d-x)^+ + \alpha\sum_{j=0}^{m+1} q_{ij}V^\alpha_{n-1}(j,(x-d)^+), V^\alpha_n(PM,x)\Bigg\}$$
$$\le \Bigg\{ c_{i+1} + hx + \alpha \sum_{p=d}^{d+K-x}\sum_{j=0}^{m+1} Pr(P=p)p_{i+1,j}V^\alpha_{n-1}(j,\min(x+p-d,K)),$$
$$hx + C(d-x)^+ + \alpha\sum_{j=0}^{m+1} q_{i+1,j}V^\alpha_{n-1}(j,(x-d)^+), V^\alpha_n(PM,x)\Bigg\}$$
$$= V^\alpha_n(i+1,x).$$

The above inequality follows from Condition 1 and the inequalities:

$$\sum_{p=d}^{d+K-x}\sum_{j=0}^{m+1} Pr(P=p)p_{ij}V_{n-1}^{\alpha}(j,\min(x+p-d,K))$$
$$\leq \sum_{p=d}^{d+K-x}\sum_{j=0}^{m+1} Pr(P=p)p_{i+1,j}V_{n-1}^{\alpha}(j,\min(x+p-d,K)),$$

and

$$\sum_{j=0}^{m+1} q_{ij}V_{n-1}^{\alpha}(j,(x-d)^+) \leq \sum_{j=0}^{m+1} q_{i+1,j}V_{n-1}^{\alpha}(j,(x-d)^+)$$

which are implied by part (i) induction hypothesis and Condition 6. Hence, the inequality (2) has been proved for $x \in \{0,\ldots,K-1\}$. Similarly, we obtain the inequality (2) for $x = K$. □

Since, the state space of the system is finite and the state $(0,0)$ is accessible from every other state under any stationary policy it follows that there exist numbers $v(s), s \in S$ and a constant g such that the following average-cost optimality equations hold (see Corollary 2.5 in [10] p. 98):

$$v(i,x) = \min\Big\{c_i+hx-g+\sum_{p=d}^{d+K-x}\sum_{j=0}^{m+1} Pr(P=p)p_{ij}v(j,\min(x+p-d,K)),$$
$$hx+C(d-x)^+-g+\sum_{j=0}^{m+1} q_{ij}v(j,(x-d)^+), v(PM,x)\Big\},$$
$$0 \leq i \leq m, 0 \leq x \leq K, \tag{4}$$

$$v(PM,x) = c_p+hx+C(d-x)^+-g+av(0,(x-d)^+)+(1-a)v(PM,(x-d)^+),$$
$$0 \leq x \leq K,$$

$$v(m+1,x) = c_p+hx+C(d-x)^+-g+bv(0,(x-d)^+)$$
$$+(1-b)v(m+1,(x-d)^+),$$
$$0 \leq x \leq K.$$

The first term in curly brackets in Eq. (4) corresponds to action $u = 0$ (allow to operate), the second term corresponds to $u = 1$ (leave the installation idle), and the third term to action $u = 2$ (initiate a preventive maintenance).

In view of part (i) of Lemma 1 and Theorem 2.2 in [10], we have the following result:

Corollary 1.

$$v(i,x) \leq v(i+1,x),\ 0 \leq i \leq m,\ 0 \leq x \leq K.$$

The following proposition gives that the optimal policy is of control-limit type.

Proposition 1. *For fixed buffer content x, $0 \leq x \leq K$ there exists a critical working condition $i^*(= i^*(x)) \in \{0,\ldots,m+1\}$ such that the optimal policy initiates a preventive maintenance of the installation if and only if the working condition of the installation is greater than or equal to i^*.*

Proof. Suppose that, for $0 \leq x < K$ the optimal policy initiates a preventive maintenance of the installation at state (i,x), where $i \in \{0,\ldots,m\}$. This implies that:

$$v(PM,x) \leq \min\left\{c_i + hx - g + \sum_{p=d}^{d+K-x}\sum_{j=0}^{m+1} Pr(P=p)p_{ij}\,v(j,\min(x+p-d,K)),\right.$$
$$\left. hx + C(d-x)^+ - g + \sum_{j=0}^{m+1} q_{ij}\,v(j,(x-d)^+)\right\}. \tag{5}$$

To show that the optimal policy prescribes action 1 at state $(i+1,x)$ it is enough to verify that

$$v(PM,x)$$
$$\leq \min\left\{c_{i+1} + hx - g + \sum_{p=d}^{d+K-x}\sum_{j=0}^{m+1} Pr(P=p)p_{i+1,j}\,v(j,\min(x+p-d,K)),\right.$$
$$\left. hx + C(d-x)^+ - g + \sum_{j=0}^{m+1} q_{i+1,j}\,v(j,(x-d)^+)\right\}. \tag{6}$$

From Conditions 1 and 6 and Corollary 1 it follows that the right side of (6) is greater or equal to the right-side of (5). Hence (5) implies (6). The same result is obtained similarly when $x = K$. □

There is a strong numerical evidence that there exists, for fixed buffer content, another critical level $\tilde{i}(x) \leq i^*(x)$ such that the average-cost optimal policy allows the installation to operate if its degree of deterioration is smaller than $\tilde{i}(x)$ and leaves the installation idle if its degree of deterioration is greater or equal to $\tilde{i}(x)$ and smaller than $i^*(x)$. A proof of this conjecture seems to be difficult.

Table 1 The critical numbers $i^*(x)$ and $\tilde{i}(x)$

x	$\tilde{i}(x)$	$i^*(x)$
0	6	8
1	6	8
2	6	8
3	6	8
4	7	8
5	8	8

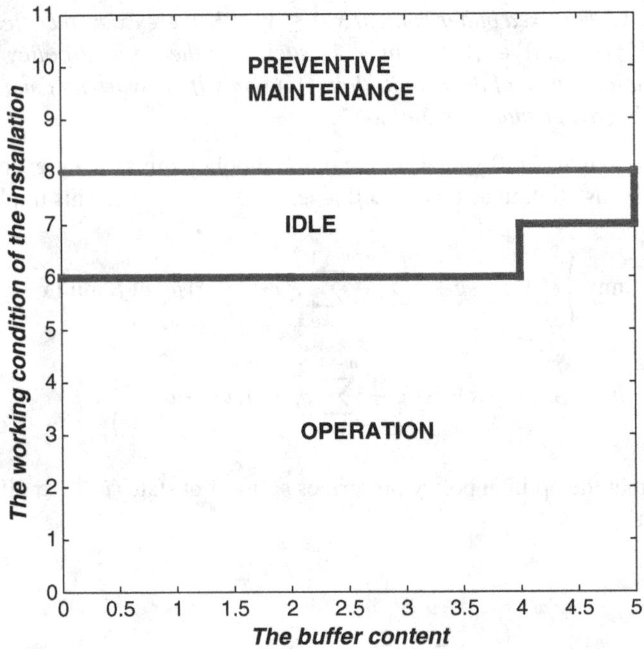

Fig. 2 The average-cost optimal policy generated by value iteration algorithm

Numerical Results

As illustration we consider the following two examples:

Example 1. We suppose that $m = 10, K = 5, d = 20, a = 0.5, b = 0.3, c_p = 25, c_f = 30, h = 1, C = 1, c_i = 3(i+1), \tilde{c}_i = 1.5(i+1), 0 \le i \le 10, q_{ij} = (12-i)^{-1}, 0 \le i \le j \le 11$, and $p_{ij} = (11-i)^{-1}, 0 \le i < j < 11, p_{10,11} = 1$. The random variable P is uniformly distributed in $\{20, 21, \ldots, 25 - x\}$ for each value of $x \in \{0, \ldots, 5\}$.

The iterations of value iteration algorithm are 16. The average cost is equal to 28.58 (Table 1). The optimal control-limit policy is characterized by the following critical numbers:

The average-cost optimal maintenance policy is depicted in Fig. 2. We observe that the optimal policy is of control-limit type. For $x = 5$ the critical numbers $i^*(5)$

Table 2 The critical numbers $i^*(x)$ and $\tilde{i}(x)$

x	$\tilde{i}(x)$	$i^*(x)$
0	10	10
1	10	10
2	10	10
3	10	10
4	9	9
5	9	9
6	9	9
7	8	8
8	8	8
9	8	8
10	8	8
11	7	7
12	6	8
13	5	8
14	4	8
15	3	8
16	3	8
17	3	8
18	2	8
19	2	8
20	2	8

and $\tilde{i}(5)$ are equal ($\tilde{i}(5) = i^*(5) = 8$) which means that the optimal policy never prescribes the action of leaving the installation idle.

Example 2. We suppose that $m = 10, K = 20, d = 5, a = 0.5, b = 0.3, c_p = 50, c_f = 60, h = 0.5, C = 10, c_i = 3(i+1), \tilde{c}_i = 2.5(i+1), 0 \leq i \leq 10, q_{ij} = (12-i)^{-1}, 0 \leq i \leq j \leq 11$, and $p_{ij} = (11-i)^{-1}, 0 \leq i < j < 11, p_{10,11} = 1$. The random variable P is uniformly distributed in $\{5, 6, \ldots, 25-x\}$ for each value of $x \in \{0, \ldots, 20\}$. In the following table we present the critical numbers $i^*(x)$ and $\tilde{i}(x)$, $0 \leq x \leq K$ (Table 2).

The iterations of value iteration algorithm are 18. The average cost is equal to 43.14. We also observe that for $x = 1, \ldots, 11$ the critical numbers $i^*(x)$ and $\tilde{i}(x)$ are equal which means that the optimal policy never prescribes the action of leaving the installation idle.

Conclusions

In this paper we modified a Markov Decision Model, which is proposed by [7], for the optimal preventive maintenance of an installation that supplies a raw material to an intermediate buffer and a production unit that pulls the raw material from the buffer. We assumed that the input rate is a discrete random variable and that there are other possible actions when the installation is in an operative condition. If the maintenance times are geometrically distributed, we proved that the policy that minimizes the long-run average cost per unit time is characterized by critical numbers $(i^*(x), 0 \leq x \leq K)$ such that the preventive maintenance of the machine is initiated only if its degree of deterioration is greater or equal to these numbers. Extensive numerical results provide strong evidence that there are another critical numbers $(\tilde{i}(x), 0 \leq x \leq K)$ such that the optimal policy prescribe to leave the installation idle if its degree of deterioration is smaller than these numbers and allowing to operate if its degree of deterioration is greater or equal to these numbers. It holds for any $x \in \{0, \ldots, K\}$ that $\tilde{i}(x) \leq i^*(x)$. A proof of this conjecture could be the subject of future research.

References

1. Dekker R. (1996) Applications of maintenance optimization models: a review and analysis, *Reliability Engineering and System Safety* **51**, 229–240.
2. Derman C. (1970), Finite State Markovian Decision Processes, Academic Press, New York 1970.
3. Douer N., Yechiali U.(1994) Optimal repair and replacement in Markovian systems, *Stochastic Models* **10**, 253–270.
4. Hopp W., Kuo U.L. (1998) An optimal structured policy for maintenance of partially observable aircraft engine components, *Naval Research Logistics* **45**, 335–352.
5. Karamatsoukis C.C., Kyriakidis E.G. (2009) Optimal preventive maintenance of a production-inventory system with idle periods, *European Journal of Operational Research* **196**, 744–751.
6. Karamatsoukis C.C., Kyriakidis E.G. (2010) Optimal preventive maintenance of two stochastically deteriorating machines with an indermediate buffer, *European Journal of Operational Research* **207**, 297–308
7. Kyriakidis E.G., Dimitrakos T.D. (2006) Optimal preventive maintenance of a production system with an indermediate buffer, *European Journal of Operational Research* **168**, 75–84.
8. Meller R.D., (1996) The impact of preventive maintenance on system cost and buffer size, *European Journal of Operational Research* **95**, 577–591.
9. Pavitsos A., Kyriakidis E.G. (2009) Markov decision models for the optimal maintenance of a production unit with an upstream buffer, *Computers and Operations Research* **36**, 1993–2006.
10. Ross S.M. (1983), Introduction to Stochastic Dynamic Programming,Academic Press, New York, 1983.
11. Sloan T.W. (2004) A periodic review production and maintenance model with random demand, deteriorating system, deteriorating equipment, and binomial yield, *Journal of Operational Research Society* **55**, 647–656.
12. Van Der Duyn Schouten F.A., Vanneste S.G.(1995) Maintenance optimization of a production system with buffer capacity, *European Journal of Operational Research* **82**, 323–338.
13. Wang H. (2002) A survey of maintenance of deteriorating systems, *European Journal of Operational Research* **139**, 469–489.

Operational Planning for Military Demolitions: An Integrated Approach

Matthew G. Karlaftis, Konstantinos Kepaptsoglou, and Antonios Spanakis

Abstract Rapid and efficient implementation of demolition along invasion routes is critical for ground defence operations. The related process includes a series of interrelated actions that have to be carefully planned in order to successfully complete demolition operations; units have to estimate and allocate human and material resources and establish appropriate access to demolition sites. This paper provides efficient operational models that can be applied by army engineering units in order to estimate resources and plan logistic activities for demolitions. Further, a decision support tool developed for real-time military operations is developed and presented.

Keywords Military operations • Demolitions • Logistic activities • Mathematical programming

Introduction

Demolition operations are military defence activities whose aim is to deny enemy forces from the use of invasion routes; demolitions are detonated along those invasion routes most likely to be used [1, 2]. While preparation and coordination of such operations and corresponding tactical plans are the responsibility of division

M.G. Karlaftis
School of Civil Engineering, National Technical University of Athens, Greece
e-mail: mgk@central.ntua.gr

K. Kepaptsoglou (✉)
School of Rural and Surveying Engineering, National Technical University of Athens, Greece
e-mail: kkepap@central.ntua.gr

A. Spanakis
Hellenic Army Staff, Army Corps of Engineers, Greece
e-mail: aspanakis@yahoo.gr

N.J. Daras (ed.), *Applications of Mathematics and Informatics in Science and Engineering*, Springer Optimization and Its Applications 91, DOI 10.1007/978-3-319-04720-1_15,

or higher-echelon commanders and their staff, technical advice and supervision are provided by engineering units [4]. Engineering units have to estimate and allocate human and material resources needed to implement demolition operations as well as establish the appropriate routes to access demolition sites [1, 3].

Demolitions implemented to create obstacles may be either prepared in advance as a part of tactical defence plans (deliberate, either charged or uncharged demolitions) or decided and performed during defensive operations (hasty demolitions) [5]. In deliberate demolitions, infrastructures along with necessary equipment, material and labor resources have already been estimated, reserved and partially prepared, while tactical plans define in each case those demolitions that have to be implemented. In this case, given potential tactical conditions and plan limitations, personnel availability and so on, successful implementation relies upon rapid and effective transportation of personnel, material and equipment to the demolition sites. Hasty demolitions on the other hand are decided upon tactical conditions and their objective is to maximize the effects of implementing some of them, given available resources.

The objective of this paper is to provide efficient operational models that can be applied by engineering units to estimate resources and plan logistics activities for both deliberate and hasty demolition implementation. Models are presented, along with a decision support tool, developed for supporting hasty demolition planning. The rest of the paper is organized as follows: The second section presents an overview of the problem. In the third and fourth section, models for planning deliberate and hasty demolitions are presented while the development of a decision support system for hasty demolitions is discussed in the fifth section. The paper's conclusions are included in the sixth section.

Overview

As mentioned earlier, demolitions may be either deliberate or hasty. For implementing deliberate demolitions, personnel (squads) and equipment must be transported to demolition sites to make final preparations and detonate demolitions. Squads are assembled in camps and move to warehouses where equipments (explosives and so on) are kept. Necessary equipment are loaded and then transported to demolition sites where demolitions are prepared and detonated so that the obstacles can be created. Usually, obstacles are positioned along potential invasion routes. Figure 1 is a typical sketch depicting the positions of demolition sites, invasion routes, warehouses and so on.

Invasion routes are usually known a priori since tactical plans define such possible routes. Each route includes a number of demolition sites that are already partially prepared (depending upon whether explosives are already there or not, these are characterized as charged or uncharged). Squads are dispatched from larger formations (companies, divisions, etc.) located at different camps and are assigned to demolitions. Given that part of the routes to demolition sites is already defined

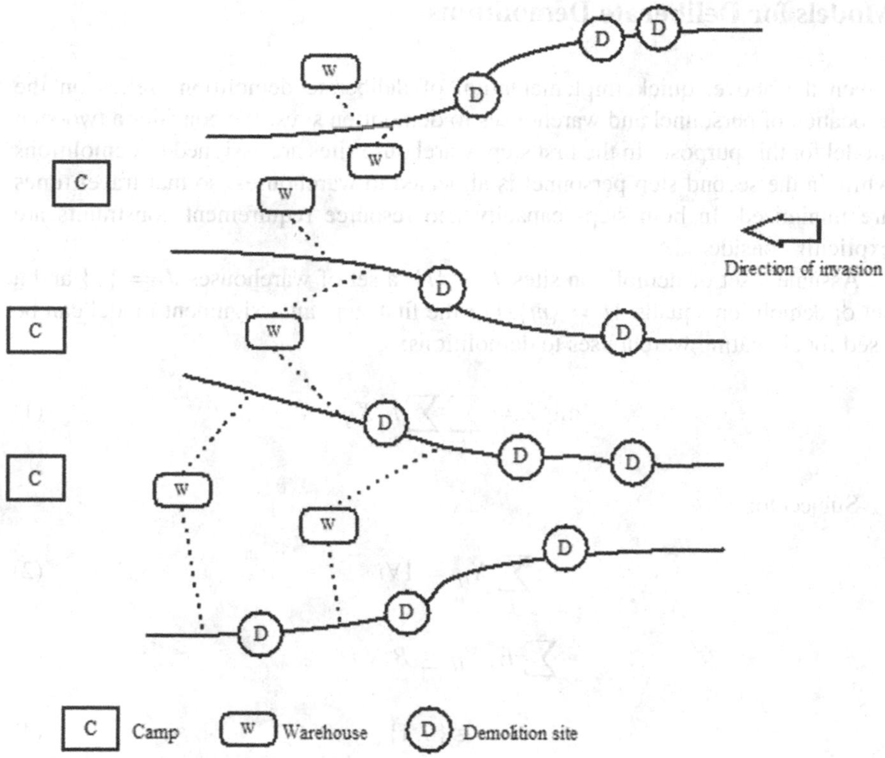

Fig. 1 Sketch depicting demolition sites and support infrastructure

we are interested in (a) the selection of potential warehouses to carry that equipment (usually military warehouses carry various equipment for different operations, e.g. weapons, ammunitions, explosives) and (b) the assignment of squads (initially located at different camps) to demolition sites.

Ad-hoc demolition operations rely on commander decisions depending upon tactical conditions. Such a decision arising from tactical conditions is the selection of potential demolitions to be rapidly prepared and detonated in order to cause a maximum of the effect to invading forces. In such cases, personnel is dispatched so that as many obstacles as possible are created within a limited time window.

In general, demolition planning includes those courses of action that must be undertaken (US Army 1993). Transportation of personnel and resources along with manoeuver schemes are such actions and must be successfully completed for the process to be effective.

Models for Deliberate Demolitions

Given the above, quick implementation of deliberate demolitions relies on the allocation of personnel and warehouses to demolition sites. We consider a two-step model for this purpose: In the first step, warehouse sites are assigned to demolitions while in the second step personnel is allocated to warehouses, so that travel times are minimized. In both steps capacity and resource requirement constraints are explicitly considered.

Assume a set of demolition sites $I = \{i\}$, a set of warehouses $J = \{j\}$ and a set of demolition squads $M = \{m\}$. For the first step, an assignment model can be used for allocating warehouses to demolitions:

$$\min Z = \sum_i \sum_j t_{ij} X_{ij} \tag{1}$$

Subject to:

$$\sum_j X_{ij} = 1 \forall i \tag{2}$$

$$\sum_j E_j X_{ij} \leq B_j \forall j \tag{3}$$

$$X_{ij} \in \{0, 1\} \tag{4}$$

where

i : Demolition
j : Warehouse site
t_{ij} : Travel time between i and j
E_i : Volume of equipment (explosives etc) needed for demolition i B_i : Capacity of warehouse j

$$X_{ij} = \begin{cases} 1, & \text{if warehouse i is assigned to demolition j.} \\ 0, & \text{otherwise.} \end{cases}$$

According to the objective function (Eq. (1)), total travel time between warehouses and demolition sites is minimized. Equation (2) ensures that each demolition is assigned to one warehouse only while Eq. (3) ensures that warehouse capacity is not exceeded.

The allocation of personnel to warehouses (second step model) is constrained by the carrying capacity of each squad. The model can be formulated as follows:

$$\min W = \sum_{mj} t_{mj} \cdot x_{mj} \tag{5}$$

Subject to

$$\sum_{j} x_{mj} = 1 \forall m \quad (6)$$

$$\sum_{m} x_{mj} \cdot q_m \geq B_j \forall j \quad (7)$$

$$w_{ij} = \begin{cases} 0, & \forall \text{ i} \rightarrow \text{j.} \\ 1, & \text{otherwise.} \end{cases} \quad (8)$$

$$y_{mi} \leq \max\{w_{ij} \cdot x_{mj}\}_j \forall m, i \quad (9)$$

$$\sum_{i} y_{mi} \cdot E_i \leq q_m \forall m \quad (10)$$

$$\sum_{i} y_{mi} = 1 \forall i \quad (11)$$

$$x_{mj} \in \{0, 1\} \forall m, j \quad (12)$$

$$y_{mi} \in \{0, 1\} \forall m, i \quad (13)$$

where

j : Warehouse site
m : Squad
i : Demolition site i
t_{mj}: Travel time between m and j
q_m: Quantity of equipment (explosives etc) that can be carried by a squad m
B_j: Quantity of equipment stored in warehouse j to be used for demolitions.
E_i: Quantity of equipment needed to detonate demolition i

$$x_{mj} = \begin{cases} \text{1, if squad m is assigned to warehouse j.} \\ \text{0, otherwise.} \end{cases}$$

$$y_{mi} = \begin{cases} \text{1, if squad m is assigned to warehouse j.} \\ \text{0, otherwise.} \end{cases}$$

Objective function (Eq. (5)) seeks to minimize total travel time between squads and warehouses. Constraints in Eq. (6) indicate that a squad may be assigned to only one warehouse, and constraints of Eq. (7) ensure that enough squads are assigned to demolitions so that all equipment is carried. Constraints in Eq. (8) assign demolitions to warehouses (according to the results of the previous model), constraints in Eq. (9) make sure that squads are assigned to corresponding demolition sites and warehouses, constraints in Eq. (10) ensure that each squad does not carry more equipment than its capabilities and according to constraints in Eq. (11) each demolition site is assigned to a single squad.

Model for Hasty Demolitions

Depending upon tactical conditions, it may be necessary to create more obstacles to invading forces by rapidly implementing and detonating additional (hasty) demolitions. Squads are simultaneously dispatched, each one carrying equipment to prepare and detonate a single demolition and return to camp. Therefore, assuming a set $\mathrm{I} = i$ of potential demolitions to be implemented along a set of routes $\mathrm{R} = r$, demolitions must be selected so that their effect is maximized in terms of estimated time delays. The model may be formulated as follows:

$$\max z^n = \sum_i t_i^k x_i \tag{14}$$

Subject to

$$\sum_i E_i \cdot x_i \leq Y \tag{15}$$

$$\sum_i O_i \cdot x_i \leq P \tag{16}$$

$$\max\{t_i x_i\} \leq T \tag{17}$$

$$\sum_i \left(\left[\frac{E_i}{q} \right] + 1 \right) \cdot x_i \leq N \tag{18}$$

$$x_i \in \{0, 1\} \forall i \tag{19}$$

where

i : Demolition
t_j: Estimated delay to the enemy caused by demolition i if implemented
E_i: Necessary quantity of explosives to implement demolition i
Y: Available quantity of explosives
O_i: Necessary personnel to implement demolition i
P: Available personnel
t_i: Time to implement demolition i (includes loading/unloading of equipment, transportation to demolition site, preparation, detonation and return).
T: Maximum time to complete all demolitions
q: Vehicle capacity
N: Available vehicles

$$x_j = \begin{cases} 1, & \text{if hasty demolition i is selected.} \\ 0, & \text{otherwise.} \end{cases}$$

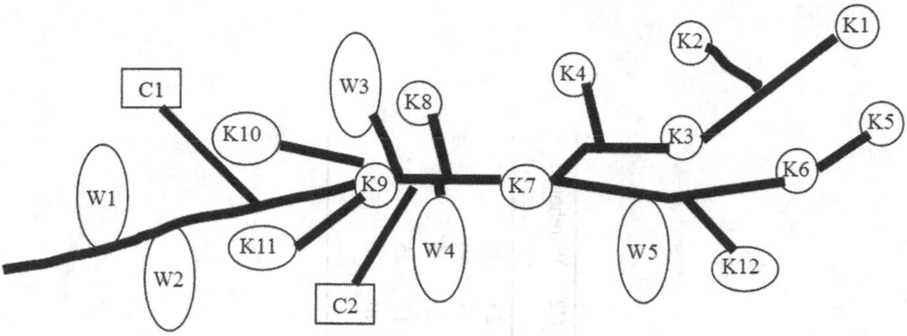

Fig. 2 Example area for P/R demolitions model application

According to the objective function (Eq. (13)) the total time to complete all demolitions must be minimized. Constraints in Eqs. (14) and (15) are equipment and personnel availability constraints, constraint in Eq. (16) ensures that the maximum time available to complete all actions is not exceeded and constraint in Eq. (17) is a maximum vehicle constraint.

Application

In order to demonstrate the application of the models described previously, two illustrative examples are presented.

Example for Deliberate Demolitions

Figure 2 shows a sketch of the testing area: the demolition sites are denoted by K_i ($i = 1, 2, 12$), the warehouses by W_j and the camps by Cr ($r = 1, 2$).

Table 1 summarizes all the data needed to apply the models.

Additionally, seven squads are assigned to prepare and detonate the demolitions (1–4 in camp C1 and 5–7 in camp C2), each one being able to carry explosives of $5\,m^3$. After solving the first step model (using a widely available solver such as Microsoft Excel Solver), the results are shown in Table 2:

As expected, most demolitions are assigned to their closest warehouses which are W4 and W5. The results after the second step model are shown in Tables 3 and 5:

By applying both models, squads are assigned to warehouses and corresponding demolitions, so that travel times are minimized and constraints are met. Demolitions assigned to the same warehouse are the responsibility of the same squad as long as its carrying capabilities are not exceeded. Additionally, results show that squad 5 is not assigned to a demolition since more squads than necessary are available to perform the operation (Table 4).

Table 1 Application data

Time between Sites (min)	1	2	3	4	5	6	7	8	9	10	11	12	C1	C2	Warehouse capacity (m^3)
W1	45	45	35	35	55	45	25	25	15	20	20	40	20	25	30
W2	40	40	30	30	50	40	20	20	10	15	15	35	20	25	40
W3	35	35	25	25	45	35	10	10	5	10	10	30	20	20	23
W4	30	30	20	20	40	30	5	5	5	10	10	25	35	17	25
W5	30	30	20	20	20	10	10	25	25	30	30	10	45	40	20
Needs (m^3)	1	2	3	1	2	1	3	1	4	1	2	1			

Table 2 First step model results

	K1	K2	K3	K4	K5	K6	K7	K8	K9	K10	K11	K12
W1							X					
W2			X							X		
W3								X	X		X	
W4	X			X								X
W5		X			X	X						

Table 3 Second step model results—assignment of personnel to warehouses

	W1	W2	W3	W4	W5
C1			X		
C2		X			
C3	X				
C4			X		
C5				X	
C6					X
C7				X	

Table 4 Second step model results—assignment of personnel to warehouses

	1	2	3	4	5	6	7	8	9	10	11	12
C1											X	
C2			X							X		
C3							X					
C4								X	X			
C5												
C6		X			X	X						
C7	X			X								X

Table 5 Candidate hasty demolition data

Demolition ID	Estimated delay (min)	Time to implement (min)	Explosives (m^3)	Personnel
1	40	25	2	4
2	32	38	1.5	3
3	25	95	1	2
4	40	35	2	4
5	35	62	2	3
6	22	47	1	2
7	49	77	2.5	5
8	55	48	3	5
9	35	25	4	3
10	74	39	3.5	5

Table 6 Hasty demolitions to be implemented

Demolition ID	
1	YES
2	No
3	No
4	YES
5	No
6	No
7	No
8	YES
9	No
10	YES
11	YES
12	No
13	No
14	No
15	No
16	No
17	YES
18	YES
19	No
20	YES

Example for Hasty Demolitions

A set of 20 potential Rapid demolitions, some of which must be selected during defence operations, is provided in Table 5:

The engineering unit has 40 men to dispatch along with 30 m^3 of explosives and 10 vehicles, each being able to carry 8 m^3 of explosives (Table 6). Given the prevailing tactical conditions, all Rapid demolitions must be completed within 60 min. By applying the model, the following demolitions must be implemented:

Out of the 20 potential demolitions, only 9 are selected for implementation with a total gain of 420 min (value of the objective function).

Decision Support System for Hasty Demolitions

Decision support systems (DSS) can be valuable during military operations as they can provide the right information at the right time and level of detail, while keeping the user confident about their results [8] Goodman and Pohl [6] provide a general framework for developing military logistics decision support systems and outline the characteristics of such systems. Extensive reviews on DSS applications in military operations can be found in Keefer et al. [7] and Tonfoni and Jain [9]. A simple and effective DSS is presented, aimed at aiding decisions on performing Rapid demolitions at the medium and lower command levels. The DSS is developed on

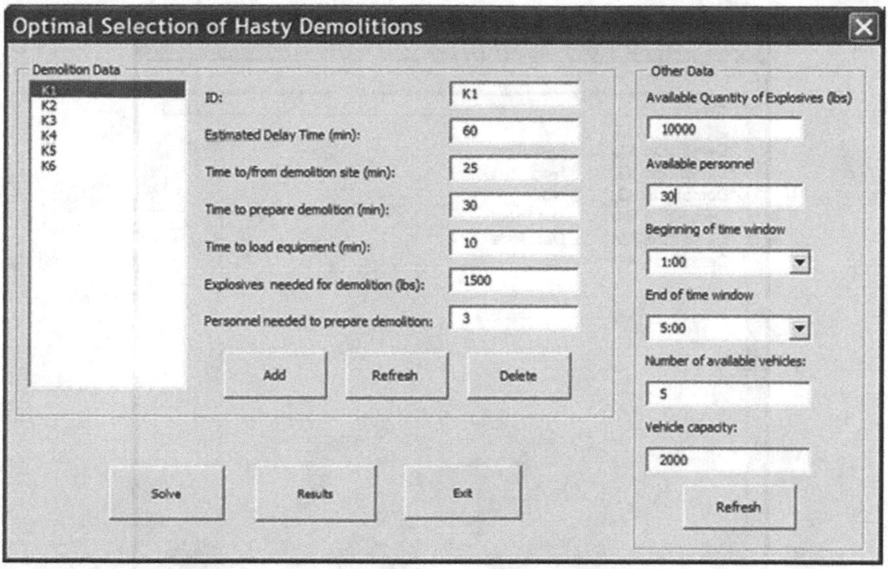

Fig. 3 DSS main window

a Microsoft Excel and VBA platform and supported by the SOLVER application. The objective is to have the hasty demolitions model incorporated in a user-friendly tool based on a widespread software package. This way, the DSS can be used in a personal computer or portable device during defensive operations, to select among candidate demolitions to be implemented. Figure 3 shows the DSS main window:

The user should enter the following data and solve the model, using the Solver™ module:

- the expected impact by detonating a single demolition in terms of time,
- the time to approach/clear the demolition site,
- the time to load/unload the equipment,
- the time to prepare and detonate the demolition,
- the necessary equipment and personnel to prepare and detonate each single demolition,
- the available personnel and the number of vehicles,
- the available quantity of explosives and,
- the time window to complete the operation.

The output consists of the demolition sites that will be finally selected, to achieve the largest impact under prevailing conditions. Given the typical size of the models to be solved (up to 20 variables at most), the results are provided in a few seconds. Figure 4 shows the DSS results window:

The medium and low-level commanders have an easy-to-use tool that can quickly assist in deciding upon performing hasty demolitions through the previously described optimization process.

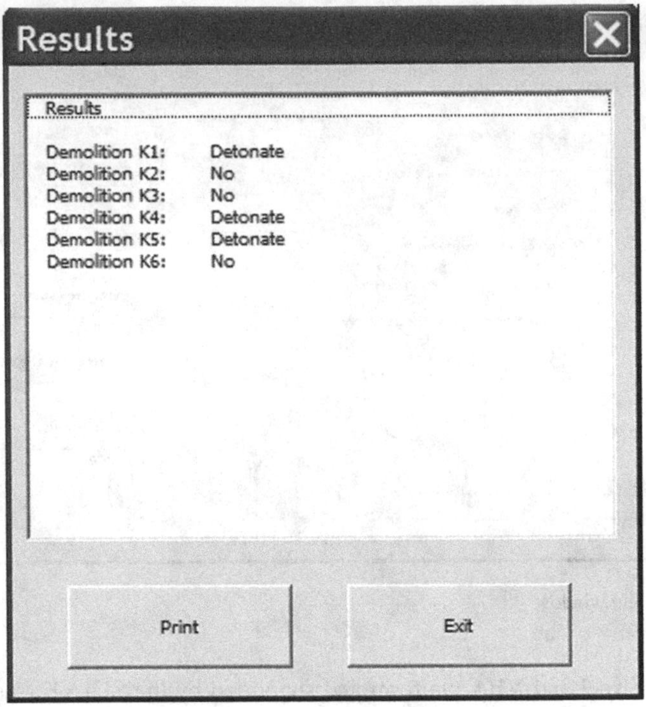

Fig. 4 DSS results window

Conclusions

Military logistics are an area where operational models have traditionally been implemented. In this paper, efficient models are developed for planning defensive operations in the tactical level. Specifically, models for planning obstacle emplacement operations are presented. These models can be solved using popular software packages and provide insights and aid in planning the logistics of defensive procedures. Moreover, a simple yet effective DSS was presented to incorporate such models as well as support command decisions during operations.

References

1. US Army (1993). Field Manual 5-71-100: Division Engineer Combat Operations, http://www.globalsecurity.org/military/library/policy/army/fm/5-71-100/index.html, USA.
2. US Army (1994). Field Manual 90-7: Combined Arms Obstacle Integration, http://www.globalsecurity.org/military/library/policy/army/fm/90-7/index.html, USA.
3. US Army (1995). Field Manual 5-100-15: Corps Engineer Operations, http://www.globalsecurity.org/military/library/policy/army/fm/5-100-15/index.html, USA.

4. US Army (1996). Field Manual 5-100: Engineer Operations http://www.globalsecurity.org/military/library/policy/army/fm/5-100/index.html, USA.
5. US Army (1990). Field Manual 7-10: The Infantry Rifle Company. http://www.globalsecurity.org/military/library/policy/army/fm/7-10/index.html, USA.
6. Goodman, S. and Pohl, S., *ISMIS: A Military Transportation Decision Support Framework. Proceedings of the 1999 International Conference on Systems Research, Informatics and Cybernetics*, Baden-Baden, Germany, 1999.
7. Keefer, D.L., Kirkwood, C.W. and Corner J.L. Summary of Decision Analysis Applications in the Operations Research Literature, 1990–2001, Technical Paper, Arizona State University, Arizona, USA, 2002
8. Pedersen, D., Van Zandt, J.R., Vogel, A.L. and Williamson, M.R. (1999) Decision Support System Engineering for Time Critical Targeting. MITRE Technical Paper, http://www.mitre.org/work/tech_papers/tech_papers_99/pedersen_decision/index.html., USA.
9. Tonfoni, G. and Jain, L. (2003) Innovations in Decision Support Systems, Advanced Knowledge International Pty Ltd, Australia.

Satellite Telecommunications in the Military: Advantages, Limitations and the Networking Challenge

Georgios Katsoulis

Abstract This paper analyzes the impact and benefits resulting from using modern satellite telecommunications in the military. During the last two decades, satellite telecommunications have been considered as the most significant and prioritized tool of information superiority in any operational theatre.
As it will be explained, their extensive use encompasses not only the traditional strategic military telecommunications but tactical networks as well. However, their effective and efficient use necessitates the proper evaluation of their capabilities and their limitations. Starting with a brief historical reference, we proceed by considering their advantages, together with modern modulation and networking technologies used to limit their drawbacks.
Finally, it will be shown that the most effective approach to exploit their capabilities is to include in a common network architecture all available means of wideband telecommunications, terrestrial and wireless. The idea can be further applied to provide high performance connectivity for a networked sensors C4ISR infrastructure.

Introduction

During the last 20 years, satellite telecommunications have been recognized as the most significant and prioritized tool of information superiority in any operational theatre, by continuously supplying information services to all military users, irrespective of their position, movement, weather and climate conditions [3]. To accomplish his mission, every military commander needs effective means of exercising command and control. Command is concerned with the distribution of force, e.g., the allocation of combat power, while control is concerned with properly

G. Katsoulis (✉)
Commander, Hellenic Navy General Staff Greek Ministry of Defence, Mesogeion 227-231, Athens, Greece
e-mail: georges.katsoulis@gmail.com

N.J. Daras (ed.), *Applications of Mathematics and Informatics in Science and Engineering*, Springer Optimization and Its Applications 91, DOI 10.1007/978-3-319-04720-1_16,

executing the decisions that have been made. Both command and control necessitate the capability of gathering, processing, analyzing and disseminating a vast amount of information, data and orders. This is where the satellite provides the best high bandwidth-high data rate means of achieving information transmission among a vast number of headquarters, decision makers, fast-moving vehicles and units of the various armed forces. Initially used for strategic telecommunication purposes, the satellite is more and more exploited on mobile networks for tactical use. This is also the case during a two-sided coastal campaign, where HF-VHF-UHF telecommunications can be easily intercepted, vessels, vehicles and aircraft can be located and indications about operational modes, units advancing or deploying and intentions can be deduced. However, the extensive and sometimes exclusive application of satellite telecommunications to achieve this goal has almost completely replaced to the mind of both the tacticians and operational planners, the term "bandwidth" by the term "satellite." This is why for every telecommunications operational tasking, the need for a satellite is first coming into mind, while in fact the need should refer to the availability of a high bandwidth telecommunications medium. This approach fails to recognize that a thorough evaluation of both satellite telecommunication capabilities and limitations should be considered, before deciding what data network infrastructure best fits our telecommunication needs in both strategic and tactical levels.

Satellite Technology in General

The space adventure of satellite communications started back in 1957 with the former USSR launch of SPUTNIK I. This operation verified the idea of Arthur C. Clarke that a satellite can virtually remain over the same spot on the earth's surface, if it executes a 36,000 km radius equatorial orbit.

Historically, the first satellites like ECHO and WESTFORD acted as communication relays by reflecting incoming signals [5]. The TELSTAR 1 and 2 satellites, launched by NASA for AT&T Bell Laboratories in 1962 and 1963 respectively, were the first active wideband communications satellites. The first satellites were operating on C-Band and first programs were exclusively military, with next generations including commercial applications. The technology of the 1970s and the rapid cost reduction of satellite equipment, allowed the shift from a single-country to regional satellite communications. The 1990s introduced the move to higher RF frequencies and the implementation of advanced techniques like on-board digital signal processing. The new millennium has seen extensive video and audio services, together with low orbit satellites and internetworking capabilities, converting the satellite to a network node in the sky. In satellite telecommunications, the satellite functions as a relay that joins users located in great distances among them. The next table explains four different categories of satellites, concerning their distance from the earth and their orbit characteristics.

Orbit	Type	Altitude	Delay	Remarks
Geosynchronous Orbit (GSO)	Circular	35,000 km	250 ms	Since the satellite is at a fixed position in the sky, the ground antenna does not need to track it
Medium Earth Orbit (MEO)	Circular	1,600–4,200 km	100 ms	Satellite tracking needed. Popular for GPS navigation satellites
Low Earth Orbit (LEO)	Circular	160–640 km	10 ms	Satellite tracking needed
High Earth Orbit (HEO)	Elliptical	From LEO to MEO	10–260 ms	Used for coverage of high-altitude locations

One revolution's period is a function of the altitude and varies between 100 min for the case of the low orbit satellites and 24 h at the distance of 35,786 km.

Advantages of Satellite Communications

Geographical Coverage

Satellite communication links provide coverage over extended geographical areas (Fig. 1). For a geo-stationary satellite to achieve maximum possible earth coverage, a 17.5 degrees antenna beamwidth is necessary. For the case of a parabolic antenna reflector, the 3-dB antenna beamwidth is given by [4]:

$$\phi = \frac{70 \times \lambda}{d} = \frac{70 \times C}{f \times d} \qquad (1)$$

where C is the speed of light, λ the wavelength of the transmitted signal, f its frequency and d the parabolic dish diameter. Substituting we get that for a transmission frequency of 12GHz, a 10cm dish diameter is enough to achieve maximum possible earth coverage. This property is extensively exploited by media companies that can reach millions of users by using a single satellite transponder.

Remote Areas and the Open Sea Reach

Moreover, satellite telecommunication can reach anybody anywhere on earth [3]. Consider that other popular broadband telecommunication means such as optical fibers or cellular networks, still keep low density population areas, remote rural or mountain areas and the open sea out of their reach. However, these areas constitute

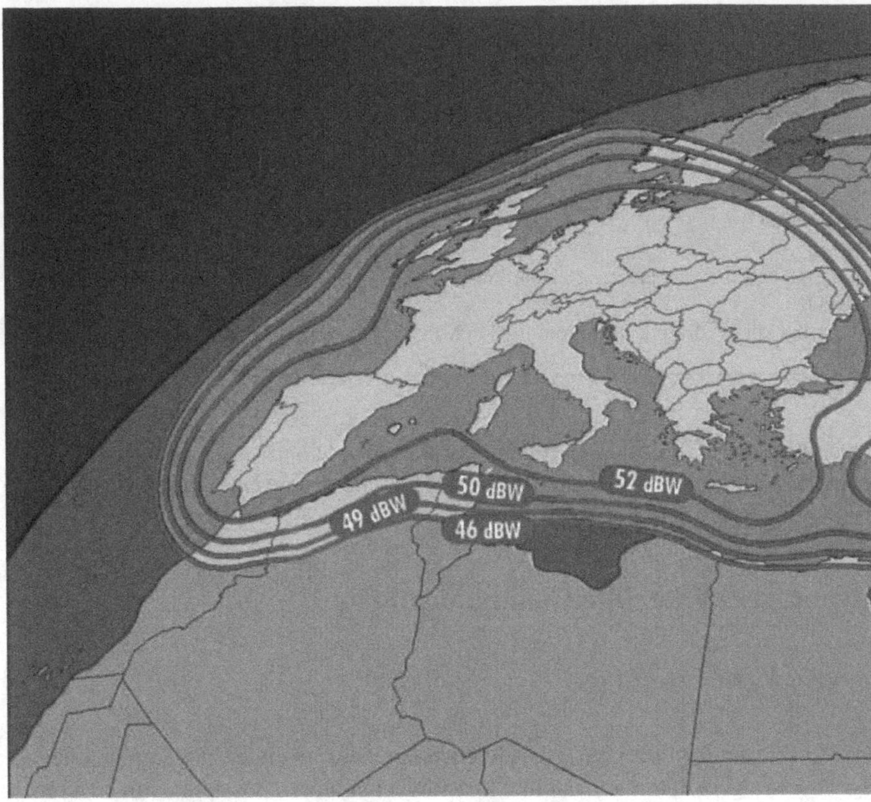

Fig. 1 Hellas satellite footprint

the common operational territory of the armed forces. This is why, for the users operating in these areas, satellite communication is the exclusive solution. Among all the cases mentioned above, satellite communication has been operationally proved to be the most effective means of providing network access in maritime environments.

High Capacity Traffic

The frequency bands used in satellite transmissions facilitate the achievement of high data rates, which can accommodate the data transfer needs of even the most demanding battlefield. This capacity also pairs with the perfect signal quality, inherent in satellite telecommunication, to guarantee to the operational commander, multiple channels of very high capacity traffic potential.

Communication Cost Independent of Distance

The capability of satellite communication to reach anybody anywhere is combined with the advantage that the communication cost is independent of the connection distance. This is not the case of either optical fiber or cellular telecommunication, where more infrastructure is needed to reach remote users. This fact drastically increases their deployment cost. For satellite communication, this cost is irrelevant.

Excellent Signal Quality

The distance from the satellite to the earth being high, however, the signal propagation path is almost vertical from the satellite antenna to the earth's surface. The vertical travelling through the atmosphere, especially through the ionosphere, which is responsible for considerable losses and attenuations to electromagnetic emissions, permits an excellent signal quality for satellite channel subscribers.

Suitability for Multi and Broadcast Services

The combination of a satellite's footprint which provides extended area coverage together with its information carrying capacity, make it suitable for multi and broadcast services, such as audio, video and internet, transmitted simultaneously to a large number of users.

Satellite Communications' Limitations

Geographical Coverage

The extended geographical coverage of satellite communications that was previously recognized as a great advantage for media transmissions, however, constitutes a considerable drawback for military communications. This is due to the fact that the military user would prefer a more directive transmission to friendly users, while leaving out of reach the third-party observers. This is impossible for the satellite downlink transmission. Thus, the interception and monitoring of the overall downlink transmission activity is very easy. This is why satellite communication is equipped with advanced modulation, coding and cyphering techniques, to avoid interception and jamming of friendly transmissions [1].

Latency

Satellite transmission is responsible for considerable latencies, which are very unpleasant to network communications. Latency is defined as the amount of delay, measured in milliseconds, which occurs in a round-trip data transmission. For geostationary satellites, the time required to traverse the distance from the satellite to the earth station, is in the order of 250ms. This is equivalent to a round-trip delay of $2 \times 250 = 500ms$ [5].
These propagation times are much more considerable than those encountered in terrestrial systems. This latency is responsible for the resulting echo on telephone circuits. It is also the cause of a number of problems in networking applications, such as buffer overflows, high queuing delays and significant packet drops. Low earth orbit satellites and advanced networking solutions constitute one way to mitigate the latency problem.

Fading

While the transmission signal path is almost vertical through the ionosphere, travelling through this area is not completely free of problems. The ionosphere cannot damage the satellite signal by refracting, reflecting or absorbing it, but it induces considerable time and phase delays. The result is that the receiver on earth not only receives the proper satellite signal, but also delayed and out of phase versions of it. This phenomenon is known as multipath fading. Again, advanced modulation techniques and Forward Error Correction Coding can be used to limit the effects of fading.

Security of Infrastructure

For a satellite communication network to operate, several earth stations are necessary. First of all, we need the Tracking, Telemetry and Command Stations (TT&C). We also need a satellite control center, where all decisions related to the maintenance of the satellite in operational condition are made and where all vital functions of the satellite are monitored and verified. If those stations are for any reason neutralized, e.g., from any physical damage or enemy action, the entire communication is gone. This is why these stations are highly protected. Neighboring nations should also consider other alternatives, such as maintaining mobile stations' capabilities or even emergency installations abroad to allied countries.

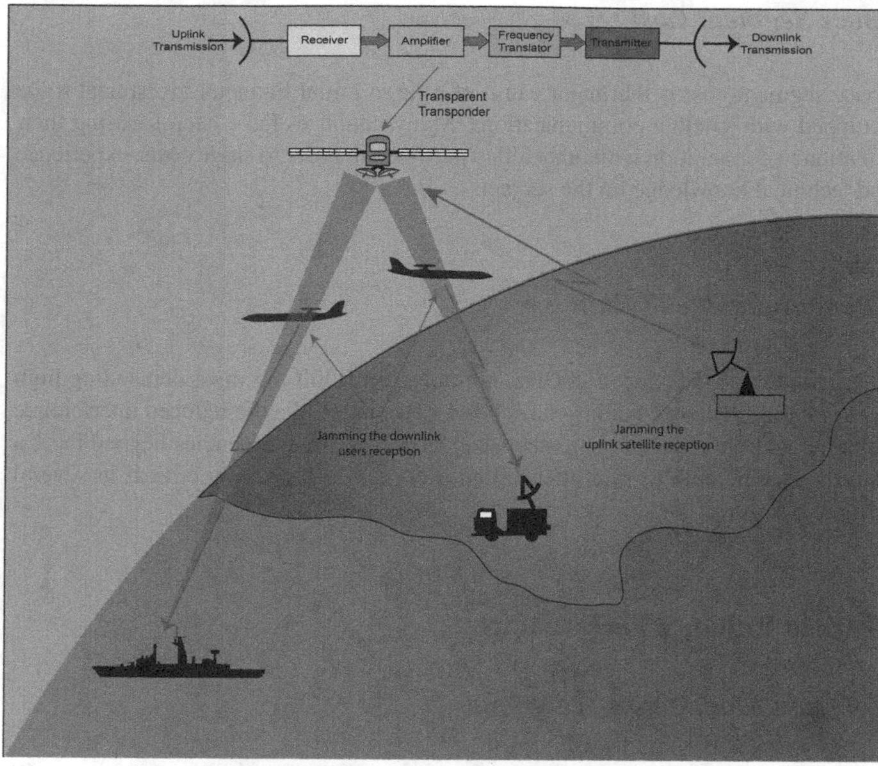

Fig. 2 While jamming the downlink user's reception requires operating into the narrow beamwidth of the receiver's antenna, jamming the uplink satellite reception can be easily implemented by using a high intensity signal in conjunction with a high directivity antenna. The jammer's uplink transmission will thus saturate all available satellite's resources, denying the friendly communications

Transponder Transparency and Anti-Jamming Capabilities

The most common satellite transponder is called a transparent transponder. This kind of transponder merely acts like a focal communications point that receives the uplink transmission at a certain frequency, amplifies it and then translates it to the downlink transmission frequency. This simple structure can permit a third-party user to just transmit a powerful signal over the uplink frequency towards the satellite and completely saturate the satellite resources, thus denying the communication to the friendly users. This condition can be avoided if the satellite possesses transponders with on-board processing capabilities that properly filter the uplink transmissions by recognizing the friendly signals and rejecting all the others. These transponders perform a series of operations including digital signal processing (DSP), regeneration and base band signal processing [1] (Fig. 2).

Space Segment Cost

Space segment cost is the major cause of a large initial financial investment to be equipped with satellite communications. Many countries have been focusing their attention to cooperation with other allied partners, in order to share costs, experience and technical knowledge on the subject.

Attenuation Due to Rain

The frequency bands above 20GHz, not only can fulfill the most demanding high data rare requirements, but also have other advantages like the reduced interference potential and the reduced equipment size. However, the frequencies beyond the Ka Band are vulnerable to rain attenuation, a condition especially critical in several parts of the world.

Ways to Enhance Performance

Common Modulation Techniques

Satellite communications are being used for the transmission of both analog and digital signals. However, modern technology focuses almost entirely on the transmission of digital signals by the use of digital modulation techniques. This also accounts for the case of analog signals, which are first converted to digital and then transmitted using digital modulation techniques. For this reason we generally refer to data modulation, since any digital signal, audio, video, still image etc. is merely a collection of digital data. So, the usual choice of modulation used in satellite communication is either BPSK (Binary Phase Shift Keying) or QPSK (Quadrature Phase Shift Keying). This kind of modulation is chosen for two reasons. First, due to its robustness since the most common interference is amplitude interference that does not affect the phase modulation. Second, because this modulation technique provides an excellent spectral efficiency, e.g. a perfect management of the available bandwidth resources.

Both BPSK and QPSK are general cases of M-PSK modulation, where $M = 2^k$. M represents the number of different symbols-signal waveforms that can be transmitted and k represents the number of bits per symbol. The transmission signal of the ith bit is given by [6]:

$$s(t) = \sqrt{2} \times A_c \times \cos(2\pi f_c t + \theta_d(t)) \tag{2}$$

where $\theta_d(t)$ represents the data. For the BPSK case, $M = 2, k = 1$, so each transmitted symbol represents one bit and $\theta_d(t)$ can assume two possible values, $\theta_1 = 0$ representing bit "1" and $\theta_2 = \pi$ representing bit "0." For these choices, the transmitted signal may also be represented by:

$$s(t) = d(t) \times \sqrt{2} \times A_c \times \cos(2\pi f_c t) \tag{3}$$

where $d(t) = \pm 1$. For the QPSK case, $M = 4, k = 2$, so each transmitted symbol represents two bits, one transmitted by the in-phase component of the carrier and the other by the quadrature component of the carrier. For QPSK, we generally choose the four possible transmitted phases $\theta_d(t)$ representing the symbols $\theta_1 = \frac{\pi}{4}, \theta_2 = \frac{3\pi}{4}, \theta_3 = -\frac{\pi}{4}, \theta_4 = -\frac{3\pi}{4}$. Consequently,

$$s_{\text{QPSK}}(t) = s_i(t) + s_q(t) = \sqrt{2} \times A_c \times [\cos(\theta_m)\cos(2\pi f_c t) - \sin(\theta_m)\sin(2\pi f_c t)] \tag{4}$$

That gives:

$$\sqrt{2} \times A_c \times [d_i(t)\cos(2\pi f_c t) - d_q(t)\sin(2\pi f_c t)] \tag{5}$$

where $d_i(t) = \pm 1, d_q(t) = \pm 1$ during each symbol interval and represent the data bits transmitted on the in-phase and the quadrature components of the carrier respectively. For a fixed bit rate, QPSK requires half the transmission bandwidth of BPSK. Conversely, for the same transmission bandwidth, a QPSK system can transmit twice as many bits per unit time as a BPSK system. For this reason, QPSK is often preferred to BPSK.

Advanced Modulation Techniques

Direct Sequence Spread Spectrum (DS-SS) Modulation

The power spectral density of a BPSK signal as in (2) is given by [6]:

$$S_{\text{BPSK}}(f) = \frac{A_c^2 T_b}{2} \times \{\text{sinc}^2[T_b \times (f - f_c)] + \text{sinc}^2[T_b \times (f + f_c)]\} \tag{6}$$

The null-to-null bandwidth of this signal is equal to $B_{\text{nn}} = 2R_b$. We now define the waveform $c(t) = \pm 1$, called the chipping signal, to be a polar random binary wave like d(t), with the only difference that now the duration between possible transitions is $T_c < T_b$. In fact $T_b = k \times T_c$, meaning that there are k chips per bit. Assuming that the chipping signal $c(t)$ is synchronized with $d(t)$, the direct sequence spread spectrum signal is generated by multiplying $c(t)$ with (3) to get:

$$s(t) = c(t) \times d(t) \times \sqrt{2} \times A_c \times \cos(2\pi f_c t) \tag{7}$$

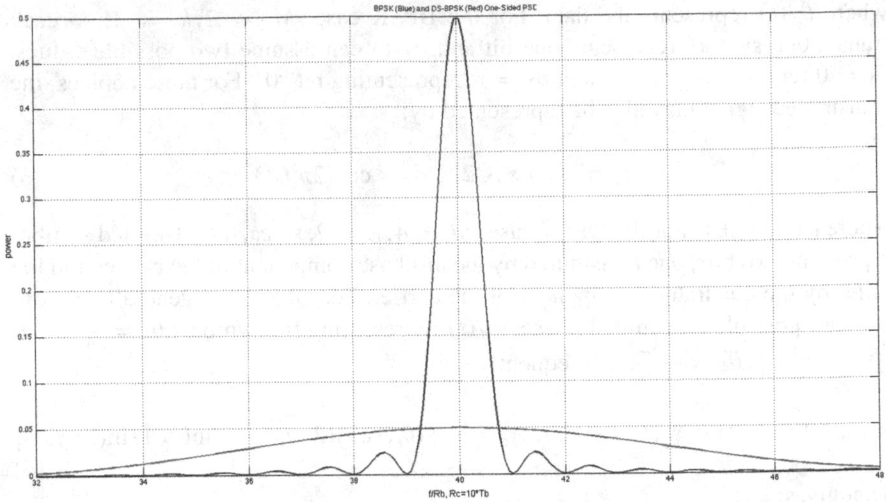

Fig. 3 One-sided PSD of a simple BPSK (*Blue*) and a DS-BPSK (*Red*) signal. Both curves have the same area but the DS-BPSK curve is spread in frequency by a factor of k. For plotting purposes, $k = 10$ was used, e.g. $T_b = 10 \times T_c$. For k much greater than 1, the spreading effect is so high that the signal is understood as white noise by the third-party observer

Since the signals $c(t)$ and $d(t)$ are synchronized, their multiplication results in a new signal $c'(t) = c(t) \times d(t)$, which is just another polar random binary wave with duration between possible transitions equal to T_c. The power spectral density of this new signal is identical to (6), with the exception that T_b is replaced by T_c.

$$S_{\text{DS-BPSK}}(f) = \frac{A_c^2 T_c}{2} \times \{\text{sinc}^2[T_c \times (f - f_c)] + \text{sinc}^2[T_c \times (f + f_c)]\} \quad (8)$$

So, the overall magnitude of the power spectral density of the DS-SS signal is a factor of k less than that of the original BPSK signal, while its transmission bandwidth is a factor of k greater. If k is much greater than 1, we get a DS-SS signal with a power spectral density of such a small magnitude that it is extremely difficult for a hostile observer to even detect its presence. The receiver possesses and locally reproduces the same chipping sequence $c(t)$ that was used to spread the transmitted signal (Fig. 4, below). After synchronizing the received DS-SS signal (7) with $c(t)$, it multiplies it by $c(t)$ to get:

$$\begin{aligned} &c^2(t) \times b(t) \times \sqrt{2} \times A_c \times \cos(2\pi f_c t) \\ &= (\pm 1)^2 \times b(t) \times \cos(2\pi f_c t) \\ &= b(t) \times \sqrt{2} \times A_c \times \cos(2\pi f_c t) \end{aligned} \quad (9)$$

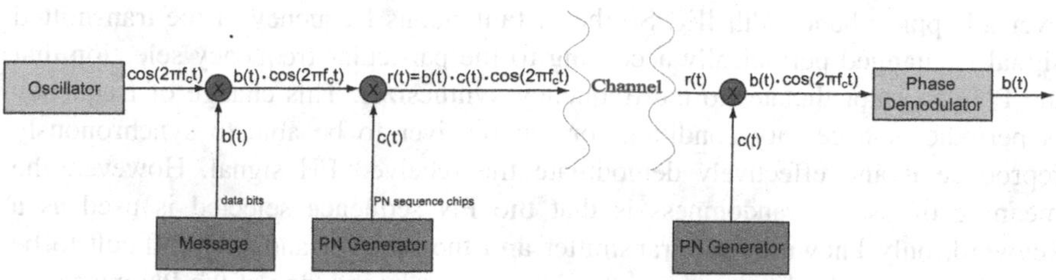

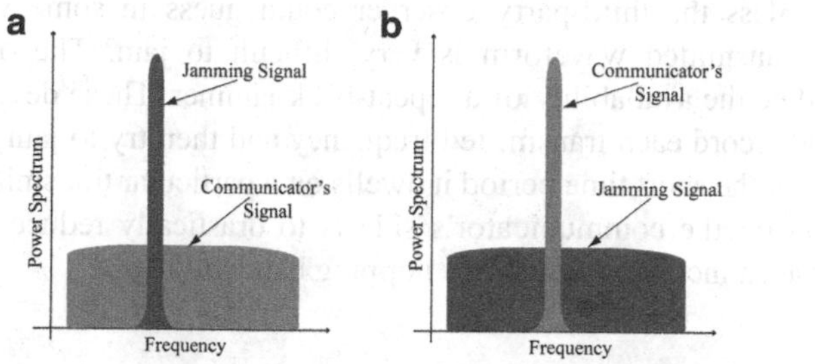

Fig. 4 Multiplication at the receiver by the chipping sequence $c(t)$, despreads the communication signal while simultaneously spreading a potential jamming signal

Frequency Hopping Modulation

The main idea of frequency-hopping spread spectrum communications is to provide protection against hostile jamming by increasing the bandwidth that the information signal occupies much more than the minimum required for proper signal transmission. By technically increasing the overall operational frequency range, we force the jammer to spread its power resources over a wider frequency band, and thus be less effective compared to a "per-frequency" signal corruption jamming capability point of view. For this purpose, the signal prior to be transmitted is multiplied by an intermediate frequency, which is selected pseudo-randomly by a frequency synthesizer. The frequency selection is performed by the use of a pseudo-noise (PN) generator that feeds the frequency synthesizer with a 2^l-length sequence word. The intermediate frequency is of the form:

$$f_i = f_1 + (i-1) \times \Delta f_{\text{fh}}, i = 1, 2, \ldots N \tag{10}$$

Where $N = 2^l - 1$ is the maximum number of possible frequency hop bins, Δf_{fh} is the separation between the carrier frequencies of adjacent bins and i changes pseudo-randomly every T_c seconds. This time is equal to the period that the hopping signal dwells on each particular hop bin. This parameter and its reciprocal $R_c = \frac{1}{T_c}$, the frequency hopping rate, is very important. The goal is to make the frequency-hopping rate as fast as possible, to deny the enemy the capability to locate the friendly emitter. By applying this technique, the entire spectrum of the transmitted signal is shifted from the initial carrier frequency f_c to the new carrier frequencies:

$$f_{\text{ci}} = f_c + (i-1) \times \Delta f_{\text{fh}} \tag{11}$$

over a hopping bandwidth W_{ss}. So the instantaneous frequency of the transmitted signal is changed periodically according to the particular frequency selection that the PN-generator dictates to the frequency synthesizer. This change of frequency is periodic, a necessary condition for the receiver to be able to synchronously reproduce it and effectively demodulate the received FH signal. However, the meaning of pseudo-randomness is that the PN sequence selected is used as a keyword, only known by the transmitter and the receiver and very difficult to be reproduced by a third-party observer without prior knowledge of the PN-generator coding structure. So, unless the third-party observer could guess in some way the PN-sequence, the transmitted waveform is very difficult to jam. The only possible solution would be the availability of a repeat-back jammer. These devices effectively intercept and record each transmitted frequency and then try to jam the communicator's signal for the short time period it dwells on a particular transmitted frequency. The solution for the communicator's side is to drastically reduce the dwelling time T_c and use an increased frequency hopping bandwidth W_{ss}.

Error Correction Coding Techniques

The idea is to use a coding mechanism in order to provide automatic error detection and correction capabilities. To implement it, redundant data is added to each portion of the message data, in such a way that the new encoded data sequence provided can sustain one or more bits being corrupted, by still providing the correct data sequence at the receiver's side. More precisely, for a (n, k) code, every k-length data bit sequence is mapped to a n-length ($n > k$) coded bit sequence, by adding $n-k$ redundant bits. The new coded bit sequence provided is algebraically related to the message bit sequence by the use of an encoding mechanism. That way, we accept to sacrifice somehow the data rate, with the benefit of increasing the robustness of the transmission and the message's sustainability against the errors resulting from the transmission channel interference. This technique is effective against any noise source, including the action of an enemy jammer. The more effective the code, the more redundancy is needed concerning the data bits added, with the drawback of further reducing the data rate. The most commonly used coding technique is named linear block coding. To give an idea of how the encoding mechanism is implemented, consider a simple $(n, k) = (7, 4)$ linear block coding example, described by the following modulo-2 table multiplication:

$$\boldsymbol{w} = \boldsymbol{u} \cdot \boldsymbol{G} \tag{12}$$

Figure 5 explains how the coded sequences are derived.

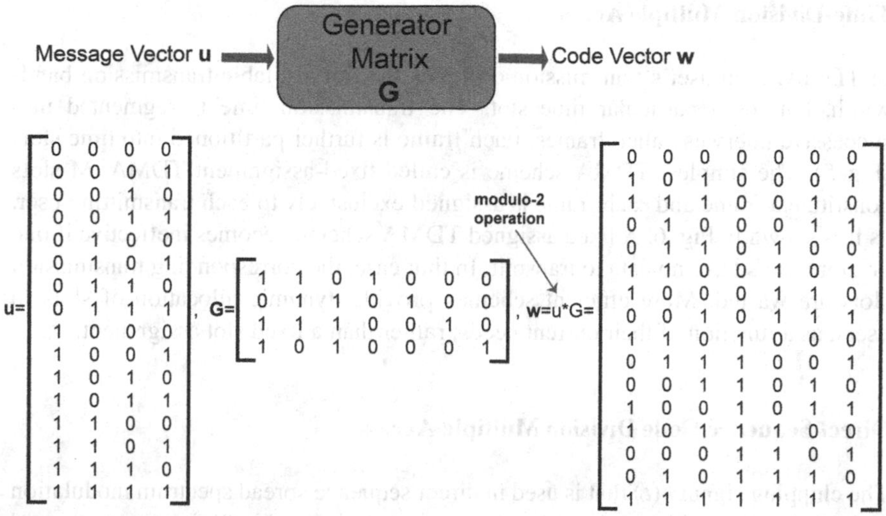

Fig. 5 Seven-length coded sequences generated out of the four-length data sequences using the generator matrix G

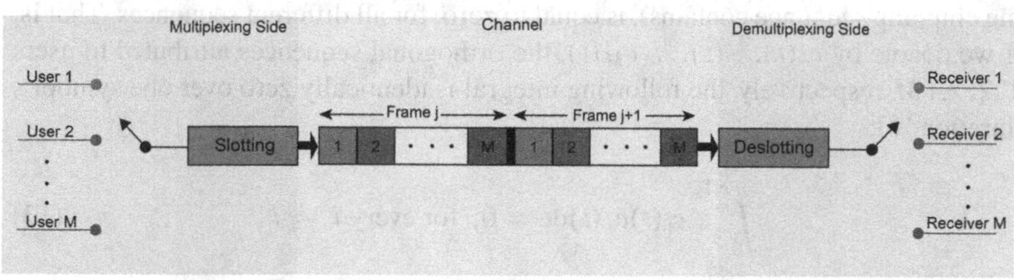

Fig. 6 Typical TDMA configuration

Multiple Access Techniques

Multiple access schemes refer to the support of a multitude of users over the same transmission bandwidth, to transmit their signals without interfering with one another [5]. The classic approach for civil telecommunications is Time Division Multiple Access (TDMA), which effectively provides very good bandwidth management. For military communications, the technique of Code Division Multiple Access (CDMA) is extensively applied, also providing Low Probability of Detection (LPD) and Low Probability of Intercept (LPI) characteristics. This technique is further divided into two different types, Frequency-Hopped CDMA (FH-CDMA) and Direct Sequence CDMA (DS-CDMA) (Fig. 6).

Time-Division Multiple Access

In TDMA, each user's transmission occupies the full available transmission bandwidth, but for a particular time slot. The transmission time is segmented into successive intervals called frames. Each frame is further partitioned into time slots (Fig. 5). The simplest TDMA scheme is called fixed-assignment TDMA. M slots constitute a frame and each frame is assigned exclusively to each transmitting user, as it is shown in Fig. 6. A fixed assigned TDMA scheme becomes ineffective if one or more users have no data to transmit. In that case, the corresponding transmission slots are wasted. More efficient schemes provide dynamic allocation of slots to users, as a function of their current needs, rather than a fixed slot assignment.

Direct Sequence Code Division Multiple Access

The chipping signal $c(t)$ that is used in direct sequence spread spectrum modulation in both the transmitter and receiver is generated from a kind of sequence named pseudo-noise (PN) sequences. PN sequences display the property that their correlation integral for a sequence's period $N \times T_s$ (where N is the number of chips that the chipping sequence contains), is equal to zero, for all different sequences. That is, if we denote by $c_1(t), c_2(t), \ldots c_M(t)$, the orthogonal sequences attributed to users $1, 2, \ldots M$ respectively, the following integral is identically zero over one symbol's duration [6]:

$$\int_0^{\mathrm{NT}_s} c_i(t)c_j(t)\mathrm{d}t = 0, \text{ for every } i \neq j \tag{13}$$

For instance, consider the sequences appearing below which display this property. Those sequences are periodic, since both the transmitter and the receiver must generate them for a spread spectrum system to work. However, their characteristics are much like those of a true random binary sequence (Fig. 7). We now consider that each group of users is assigned a different chipping sequence code, $c_i(t), i = 1, 2, \ldots N$ for N users, generated by the corresponding PN code. As described in the DS-SS paragraph, the first stage of a DS receiver multiplies the incoming signal by the chipping sequence $c_i(t)$. This operation dispreads the intended transmission, while operating a correlation-equivalent process to all other transmissions and rejecting it. This again permits an almost perfect co-existence of different groups of users over the same transmission bandwidth. In FH-CDMA, many different users can hop over the hopping bandwidth W_{ss}, with the orthogonality of the used PN-sequences guaranteeing that no two users simultaneously transmit on the same frequency hop bin. This condition can be achieved by exploiting the orthogonality properties of the PN sequences. The effect of feeding each group of users frequency synthesizer with its own PN sequence, guarantees that independent hopping successions can co-exist on the same hopping bandwidth. This technique is effective both as a multiple access and anti-jamming technique, since any interceptor

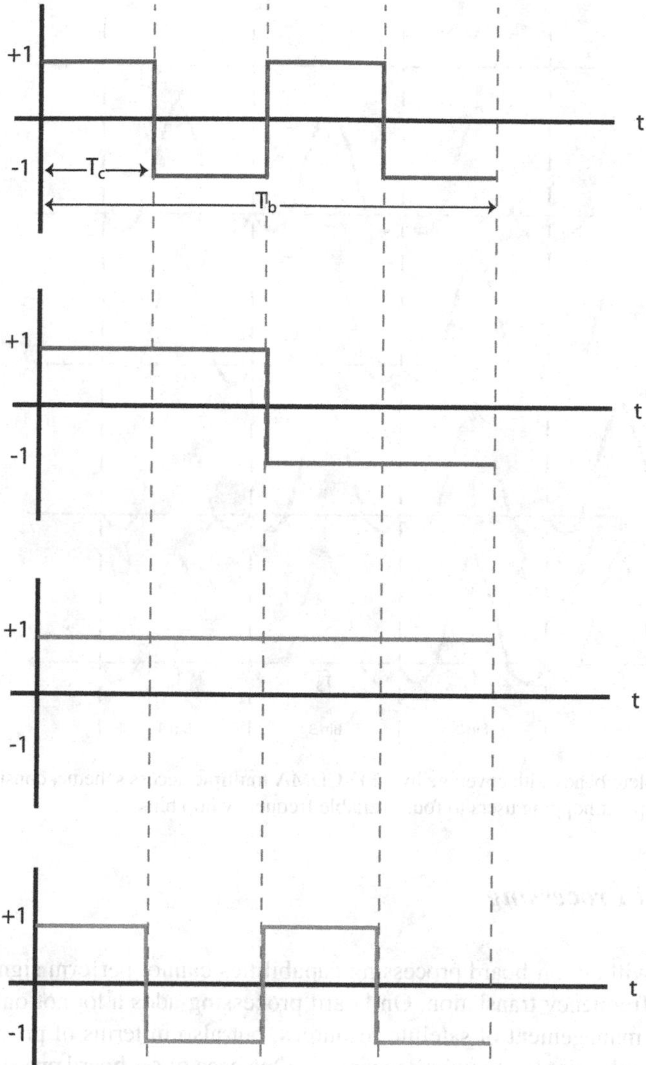

Fig. 7 A simple example of orthogonal PN sequences

would observe the majority of the hopping bandwidth W_{ss} occupied, without being able to attribute a particular emission to a particular hopping network. Figure 8 illustrates a simple example of complete hopping bandwidth coverage, executed by the simultaneous hopping of four different hopping groups of users to four available frequency hop bins:

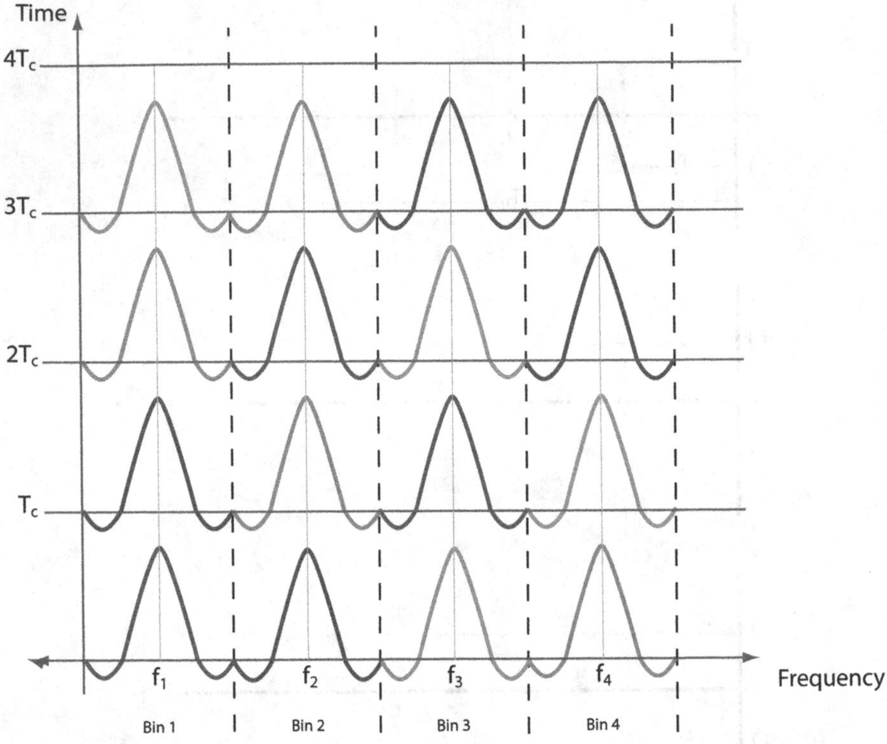

Fig. 8 Complete bandwidth coverage by a FH-CDMA multiple access scheme, consisting of four different groups of hopping users to four available frequency hop bins

On-Board Processing

A satellite without on-board processing capabilities cannot perform signal amplification and frequency translation. On-board processing adds a lot not only in terms of efficient management of satellite resources, but also in terms of providing anti-jamming capabilities to the uplink reception. One case of on-board processing is the regenerative transponder. This technology provides complete separation of uplink and downlink transmissions, which in turn increases the bit error rate performance. The uplink transmission is received, the channel interference is removed and this enhanced signal version is further sifted to the downlink channel. This operation prevents the direct addition of uplink noise to the downlink noise. Further on, if spread spectrum modulation techniques are applied, the on-board processor can be equipped with a receiver-like dispreading mechanism, which dispreads the uplink transmission before regenerating it and retransmitting it toward the downlink channel. The dispreading function also limits the impact of a potential noise jamming transmission. Moreover, the overall network efficiency is drastically improved if on-board processing includes routing capabilities. In that case, the satellite on-board

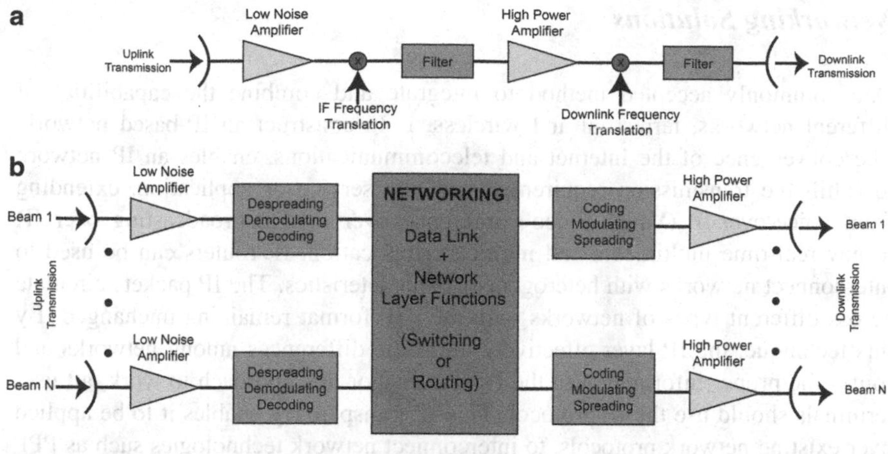

Fig. 9 Part (**a**) displays the configuration of a transparent satellite transponder. Part (**b**) refers to a satellite with full on-board processing capabilities

router acts as an IP-compliant space-based asset, which transforms the satellite into a network node in the sky [2, 3]. This technology can be further applied in inter-satellite links. On-board switching can also be used in conjunction with spot beams. With this technique, different uplink channels are switched on-board to their destination earth stations through the proper spot beams. More agile beams can also be provided on demand and adjusted by an operator, if an on-board processing mechanism possesses the capability to modify the beam's width and channel capacity (Fig. 9).

Spot Beams and Frequency Reuse

Using the spot beams approach, the directivity of the transmitting and receiving antennas is drastically increased, to the extent that the antenna beamwidth is to the order of 0.4 degrees. Each satellite is equipped with many such beams. Several antenna beams provide coverage over different service zones, thus effectively splitting the earth's surface to multiple coverage areas of relatively small space, compared to the satellite's footprint resulting by the use of traditional antennas. This technique also permits to reuse the same transmission frequency in different spot beams. This capability constitutes a bandwidth efficient method, since the network's total capacity is increased without sacrificing the available bandwidth resources. On-board switching can also be used in conjunction with spot beams. With this technique, different channels can be transmitted simultaneously to the satellite and further switched on-board to their destination earth stations through the proper spot beams. Other efficient beamforming techniques can also be used to dynamically change the transponders' earth coverage, like the phased array antenna technology.

Networking Solutions

One commonly accepted method to integrate and combine the capabilities of different networks, terrestrial and wireless, is to construct an IP-based network. The convergence of the Internet and telecommunications, enables an IP network to fulfill the transmission requirements of any service or application, extending from voice-over-IP (VoIP), video conference over IP and broadcasting over IP, to any real-time multimedia and multicast application. IP routers can be used to interconnect networks with heterogeneous characteristics. The IP packets circulate across different types of networks with their IP format remaining unchanged. By this technique, the IP layer effectively hides the differences among networks and routes the proper information to the right user. For this approach to work, all user terminals should use the IP protocol. This IP transparency enables it to be applied over existing network protocols, to interconnect network technologies such as PPP dialup, ATM and DVB-S/DBV-RCS, which support Internet protocols or interwork with the Internet. One commonly used technique to achieve IP compatibility over different networks, is the IP packet encapsulation. At each protocol level of the OSI model, a packet is divided into its header and payload parts. Given that different network technologies use different frame formats, frame sizes and bit rates for transporting IP packets, the encapsulation technique arranges that the whole packet of each protocol, becomes the payload part of the next layer's down protocol, as explained in Fig. 10.

Future Perspectives

Networked Sensors and Telecommunications Infrastructure

Today, operational effectiveness is based on rapid and accurate situational awareness obtained by properly collecting and processing real-time information. This is where the idea of a networked battlespace comes from, assuming fully extended surveillance, command and control capabilities, based on a common systems architecture. For this purpose, network-centric technology interconnects all available sensors, based on land or aboard vehicles, manned or unmanned, together with weapons and decision makers. From a telecommunications viewpoint, this interconnection treats every command and control, sensor and/or weapons platform, as a potentially interactive network node. In military terms, this approach is equivalent to a C4ISR system, which constitutes the connective material that unifies sensors and telecommunications into a common infrastructure, providing the operational commander with unprecedented capabilities of operational effectiveness. However, the integration of so many sensors, fixed or on land, airborne and seaborne vehicles, creates a huge telecommunications traffic flow, which demands extensive bandwidth resources. Moreover, the integration of so many platforms and users increases

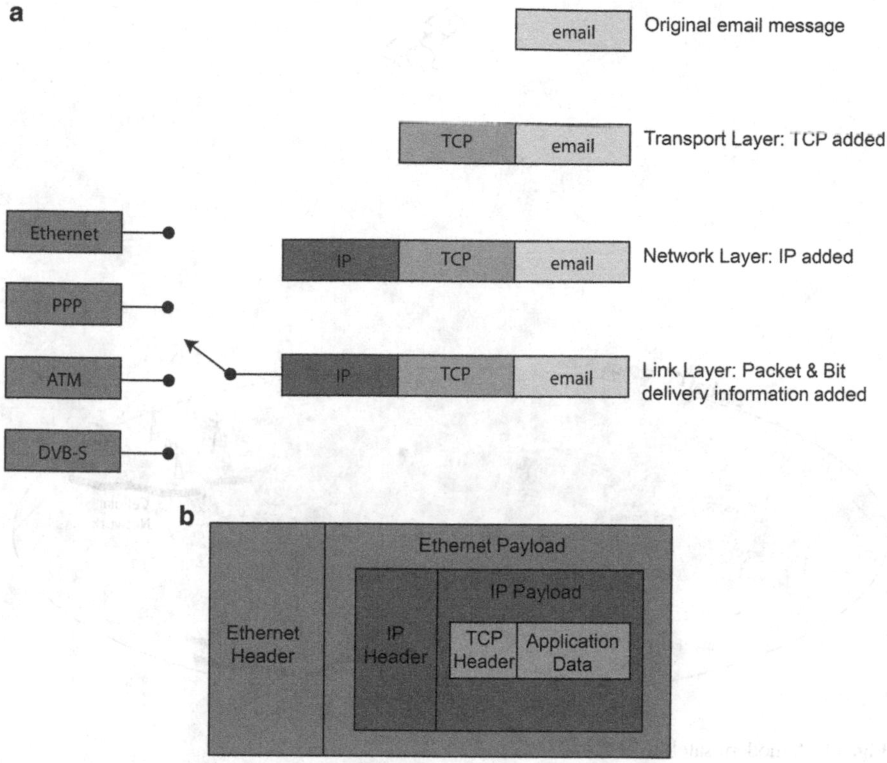

Fig. 10 Part (**a**) explains how each OSI layer adds its own overhead to the original data. Part (**b**) explains the encapsulation technique

the telecommunications security and reliability requirements. To handle this information volume, the telecommunications network architecture must combine all available high bandwidth telecommunications means, such as fiber optics terrestrial network, cellular networks and of course satellites [7]. Very efficient network protocols are also necessary to interconnect nodes belonging to heterogeneous networks.

A Modern Satellite Telecommunications Network

A modern design paradigm that implements the idea of the last paragraph would refer to a satellite equipped with multiple spot beams [1] that split the earth's surface coverage into several sectors. In order to effectively manage the limited available frequency spectrum, the satellite on-board signal processor supports frequency-band reuse among the beams and also flexibility in bandwidth and transmission power allocated to each user. Moreover, different areas are covered by different

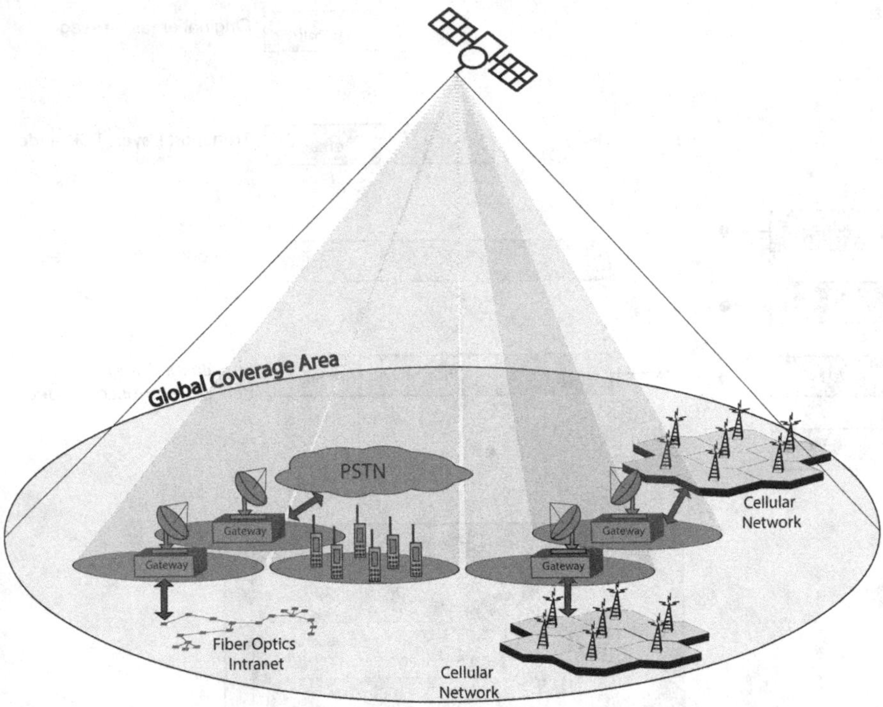

Fig. 11 A modern satellite network

service types that require different data rates and bandwidths. On-board switching permits to avoid duplication of transmission to different sectors by efficiently routing information among users. Data integrity is protected against channel losses or jammer action by the use of error correction coding techniques. Modern modulation techniques such as spread spectrum modulation provide both LPD and LPI capabilities to the transmitted signals, while also rejecting jammer signals on-board the satellite by the intervention of the on-board digital signal processor. The satellite is used to implement an IP-based network, which can provide direct connections among user terminals, connections for terminals to access terrestrial networks, and connections between terrestrial networks, fiber-optic or cellular. The proper fusion of heterogeneous networks into a common infrastructure combines the advantages of each particular network, while hiding their drawbacks (Fig. 11).

Conclusion

The telecommunications satellite is the only platform to achieve fixed and mobile users global connectivity, by overcoming distance and geographical barriers. Modern satellite technology is highly capable of implementing network-based defense telecommunication architecture, to link decision-making, information systems and weapon systems, into a single operational infrastructure. While the bandwidth and mobility demands of the military commander are constantly being increased, the satellite is the logical option to provide greater bandwidth with global coverage beyond the reach of civil terrestrial networks and far away to the open seas. However, the perfect exploitation of satellite communication benefits is achieved through a combined infrastructure that suitably integrates and internetworks all available means of broadband telecommunications and respective protocols, terrestrial and wireless, such as fiber optics networks and cellular networks. The idea is to strategically identify and construct satellite and/or terrestrial flexible telecommunications and sensor network relay-nodes to internetwork the entire extent of the battlefield.

References

1. Don Wilcoxson, *Advanced Commercial Satellite Systems Technology for Protected Communications*, The 2011 Military Telecommunications Conference.
2. L. Wood, W. Ivancic, D. Hodgson, E. Miller, B. Conner, S. Lynch, C. Jackson, A. da Silva Curiel, D. Cooke, D. Shell, J. Walke, D. Stewart, *Using Internet Nodes and Routers Onboard Satellites*, International Journal of Satellite Communications and Networking, 2007.
3. Zhili Sun, *Satellite Networking, Principles and Protocols*, John Wiley & Sons, Chichester, West Sussex, 2005.
4. Dennis Roddy, *Satellite Communications*, McGraw Hill, New York, 2001.
5. Louis J. Ippolito, Jr., *Satellite Communications Systems Engineering*, John Wiley & Sons, Chichester, West Sussex, 2008.
6. Marvin K. Simon, Jim K. Omura, Robert A. Scholtz, Barry K. Levitt Jr., *Spread Spectrum Communications Handbook*, McGraw Hill, New York, 1994.
7. *Cellular Backhaul Over Satellite*, Asia-Pacific Satellite Communication Council, Korea, 2012.

Stabilisation and Tracking for Swarm-Based UAV Missions Subject to Time-Delay

Georgios P. Kladis

Abstract It is well known that time-delay is often inherent in dynamic systems, which can be an important source of instability and degradation in the control performance. In particular, when safety is concerned for Unmanned Aerial Vehicle (UAV) applications, neglecting the presence of time-delay in the measurable states may jeopardise or result in catastrophic failures for operations. In this letter sufficient conditions for the existence of fuzzy state feedback gain are proposed for the stabilisation/tracking problem of swarm-based UAV missions subject to time-delays. The nonlinear model of the dynamics are represented by Takagi-Sugeno (TS) fuzzy models which offer a systematic analysis for stabilisation/tracking problems. Through a special property motivated by the Razumikhin theorem it allows the design of the distributed control law to be performed using tools from Lyapunov theory. The control law is composed of both node and network-level information. The design follows a two-step procedure. Firstly feedback gains are synthesised for the isolated UAVs ignoring interconnections among UAVs. The resulting common Lyapunov matrix is utilised at network level, to incorporate into the control law the relative differences in the states of the agents, to induce cooperative behaviour. Eventually stability is guaranteed for the entire swarm. The corresponding design criteria, proposed, are posed as Linear Matrix Inequalities (LMIs) where performance for the entire swarm is also stressed. The benefits of this analysis is that the design of the controller is decoupled from the size and topology of the network, and it allows a convenient choice of feedback gains for the term that is based on the relative state information. An illustrative example based on a UAV tracking scenario is included to outline the potential of the analysis.

G.P. Kladis (✉)
Department of Electrical and Computer Engineering, KIOS Research Center for Intelligent Systems and Networks, University of Cyprus, Nicosia, 1678, Cyprus.
e-mail: klamouf@gmail.com; kladis.georgios@ucy.ac.cy

N.J. Daras (ed.), *Applications of Mathematics and Informatics in Science and Engineering*, Springer Optimization and Its Applications 91, DOI 10.1007/978-3-319-04720-1_17,

Keywords Graph theory • Multi-agent systems • Distributed control • Time-delay • Parallel distributed compensation • Takagi-Sugeno fuzzy model • Linear matrix inequalities

Introduction

Large-scale multi-agent systems can be represented accurately by nonlinear models in a large domain of operation. However, this coupled with the dimensionality of the network means the task of designing a control law may be a far from trivial task. Most of the existing work in cooperative control has focused on the interconnection of systems with linear dynamics. For example, consensus was examined for multi-agent systems with general linear dynamics in [16, 20]. In [18, 23] consensus for agents with single/double or higher integrator dynamics were studied. In reference [6] the authors focused on the stabilisation of a network of identical agents with linear dynamics. Unlike the previous methodologies which consider first or higher order linear models for the vehicles' motion, in this work, a nonlinear representation of the dynamics of a group of UAV systems with constraints on angular and linear velocity is investigated.

In particular, motivated by work in [17] where the global stabilisation of a complex network of agents is considered by applying local decentralised output feedback control law, reference [13] developed a distributed control law for nonlinear systems based on a two-step procedure. This allowed a decoupled design procedure at both node and network level and offered a systematic analysis for stabilisation/tracking problems in a reasonably large class of networks of nonlinear systems represented in the Takagi-Sugeno (TS) framework [24]. The work in [13] was extended for a more general case of nonlinear systems with a focus on tracking for a swarm of UAVs, in [14]. However, if a delay perturbation is applied to the closed loop system used in article [14] it can be easily shown that the entire swarm leads to instability. Delay, which is often inherent in engineering processes, may compromise stability or lead to poor performance for the entire process. In particular, for the UAV application, neglecting its effect whilst designing a control law for tracking purposes may jeopardise swarm-based missions or even lead to catastrophic failures. In the literature the stabilisation problem of delayed systems has been dealt with a number of different ways. For example, in reference [22] stability is guaranteed for any value of the time-delays while authors in [7] suggest a stabilisation procedure with a maximum bound on time-delay. For retarded TS fuzzy models, conditions to guarantee stability for the entire system were investigated by authors in works [3, 25–27, 30], and references therein.

Thus both steps of the analysis illustrated in [14] need to be modified in order to accommodate the delay perturbation. In this letter such a modification is suggested to deal with retarded systems. This is possible, through a special property motivated by the Razumikhin theorem which allows the design of the distributed control law to be performed using tools from Lyapunov theory. Due to the structure of the TS model, which is a fuzzy *blending* of linear local models, this allows a systematic analysis for proving stability, in a Lyapunov sense, of a general class of nonlinear systems. Interesting work that addresses the design aspects for Takagi-Sugeno controllers exists in the literature: see for example, [25].

In this work, the model under investigation is the error dynamics of the UAV as developed in references [15] and [12]. Following the design procedure in [13] and [14], at the first step, the delayed error dynamics of the UAV system are isolated, and a node level control law is designed ignoring interconnections. The node level control law utilises a Parallel Distributed Compensation (PDC) structure as suggested in [29] and the feedback gains are synthesised, subject to certain design criteria posed as Linear Matrix Inequalities (LMIs). Subsequently in the second step, now including dependencies among the delayed UAVs, a distributed control law is introduced and it is shown that stability is guaranteed for the entire swarm under delay perturbation.

The novelty of this work is that it proposes a methodology for the analysis of a delayed network of nonlinear systems. An intermittent step (the creation of an equivalent TS representation form) allows a decoupled structure of the network into node level dynamics to be exploited. This structure facilitates a systematic analysis using Lyapunov theory for stabilisation/tracking. Eventually it is shown that the resulting common Lyapunov matrix, arising from node level analysis, can be used to create a Lyapunov function for the network level.

The benefit of the proposed approach is that the analysis and design is performed at node level, thus the problem of stabilisation/tracking is decoupled from the network's scale, topology, and complexity. Also the methodology can be applied to a reasonably large class of nonlinear systems.

The remainder of the paper is structured as follows: in section "Preliminaries" the graph theory tools which are used, and their relevance to a network of systems is presented. In section "UAV Model and the Takagi-Sugeno Model Representation" the Takagi-Sugeno model is described for a general network of nonlinear multi-agent systems subject to delay perturbation. Thereafter in section "Swarm Tracking and Control Law Description Subject to Time-Delay" the architecture of the controller and the LMI conditions to stabilise the system at node and network level are described. A swarm-based UAV tracking example is included in section "Simulation example" demonstrating the proposed analysis. In section "Conclusions" concluding remarks are stated.

Preliminaries

Graph Theory

In this section the graph theory preliminaries for the multi-agent systems application are stated. Adopting the notation in [19], a graph G is an ordered pair (V, E), where V is the set of nodes ($V = \{1, \ldots N\}$) and E is the set of edges, ($E = \{c_1, \ldots, c_l\}$), which represent every feasible connection among a pair of nodes. In this work a node coincides with a UAV within the swarm, and the set E denotes the communication links between UAVs i and j. A network topology G can be represented in the form of the adjacency matrix $\mathbf{A}(G) = [\alpha_{ij}] \in \mathbb{R}^{N \times N}$ and satisfies:

$$\alpha_{ij} = \begin{cases} 1, \forall (i,j) \in E \ and \ i \neq j \\ 0, \ otherwise \end{cases} \tag{1}$$

The degree $\mathbf{D}(G) = [d_{ij}] \in \mathbb{R}^{N \times N}$ of a graph is a diagonal matrix such that $d_{ii} = \sum_{i=1}^{N} \alpha_{ij}$ and $d_{ij} = 0, \ \forall i \neq j$. The Laplacian of a graph $\mathbf{L}(G) = [\ell_{ij}] \in \mathbb{R}^{N \times N}$ is defined by:

$$\mathbf{L}(G) = \mathbf{D}(G) - \mathbf{A}(G) = [\ell_{ij}] = \begin{cases} \sum_{j=1}^{N} \alpha_{ij}, \ i = j \\ -\alpha_{ij}, \quad i \neq j \end{cases} \tag{2}$$

According to [19], for undirected graphs (i.e. $\alpha_{ij} = \alpha_{ji}$) the Laplacian matrix is symmetric, positive semi-definite and satisfies $\sum_{j=1}^{N} \ell_{ij} = 0, \ \forall i \in V$. The Laplacian matrix is important for the swarm-based application, since it can reveal whether or not stabilisation/tracking or consensus can be reached by the swarm of UAVs. This depends on their communication topology, which is assumed bidirectional static in this article. For instance, consensus can be guaranteed provided that all eigenvalues of (2) are positive. This is possible if by construction the graph G is "connected" (refer to [4] for the definition). In this work by design the Laplacian matrix is positive semi-definite.

UAV Model and the Takagi-Sugeno Model Representation

Consider a group of systems $i = 1, \ldots, N$ described by:

$$\dot{e}^i(t) = f_i(e^i(t)) + g_i(e^i(t))u^i(t) \tag{3}$$

where $e^i(t) \in \mathbb{R}^n$, and $u^i(t) \in \mathbb{R}^m$ is the state, and input vector, respectively. Assume $f_i(e^i(t))$ and $g_i(e^i(t))$ are functions that are dependent on the state. The nonlinear model in (3) can be represented in a compact region of the state-space $\mathcal{X} \subseteq \mathbb{R}^n$ by a TS fuzzy model.

Adopting the notation in reference [25], for agent i, the TS fuzzy model is formed by κ local linear subsystems. The TS is represented by implications of **IF–THEN** form or Input–Output form. The general layout for the κth model rule is:
Model Rule κ [25]:

IF $z_1^i(t)$ is $M_{\kappa 1}$ **AND**... **AND** $z_q^i(t)$ is $M_{\kappa q}$ **THEN**

$$\dot{e}^i(t) = \mathbf{A}_\kappa e^i(t) + \mathbf{B}_\kappa u^i(t) \tag{4}$$

where $e^i(t) = col([e_1^i(t), \dots, e_n^i(t)]) \in \mathbb{R}^n$, and $\mathbf{A}_\kappa \in \mathbb{R}^{n\times n}$, $\mathbf{B}_\kappa \in \mathbb{R}^{n\times m}$ are constant matrices. The vector $z^i(t) = col([z_1^i(t), \dots, z_q^i(t)])$ is a known premise variable which may depend on the state vector. Every premise variable is a-priori bounded on a compact space (i.e. $z^i(t) \in [z_{\min}^i, z_{\max}^i]$) since the state is assumed to belong to $\mathcal{X}$. The symbol $M_{\kappa\mu}(z_\mu^i(t)) \in [0, 1]$ denotes the fuzzy sets and $r = 2^{|z|}$ the number of rules. The notation $|z|$ coincides with the length of the vector. The fuzzy sets $M_{\kappa\mu}(z_\mu^i(t))$ are generated utilising the sector nonlinearity approach [10].

In Input–Output form, the defuzzification process of system (4) can be represented by the following polytopic form:

$$\dot{e}^i(t) = \sum_{\kappa=1}^{r} \lambda_\kappa(z^i(t))[\mathbf{A}_\kappa e^i(t) + \mathbf{B}_\kappa u^i(t)] \tag{5}$$

where the $\lambda_\kappa(z^i(t))$ are normalised weighting functions defined by:

$$\begin{aligned} \lambda_\kappa(z^i(t)) &= w_\kappa(z^i(t)) / \sum_{\kappa=1}^{r} w_\kappa(z^i(t)) \\ w_\kappa(z^i(t)) &= \prod_{\mu=1}^{q} M_{\kappa\mu}(z_\mu^i(t)) \end{aligned} \tag{6}$$

The weighting terms $\lambda_\kappa(z^i(t))$ satisfy the convex sum property for all t. Provided that bounds on the state space are a-priori known, the TS model (5) is an exact representation of the nonlinear model (3) inside $\mathcal{X}$. Motivated by work in reference [13] it will be shown in the sequel that such a structure can be utilised in the UAV context.

It is well known that time-delay is often inherent in dynamic systems, which can be an important source of instability and degradation in the control performance. Due to this time-delay is included in the design.

Considering time-delay the TS is represented by implications **IF–THEN** with general layout for the κth model rule:
Model Rule κ [25]:
IF $z_1^i(t)$ is $M_{\kappa 1}$ **AND**... **AND** $z_q^i(t)$ is $M_{\kappa q}$ **THEN**

$$\dot{e}^i(t) = \mathbf{A}_\kappa^1 e^i(t) + \mathbf{A}_\kappa^2 e^i(t - \tau_\kappa^i(t)) + \mathbf{B}_\kappa u^i(t) \tag{7}$$

where $\mathbf{A}_\kappa^1 \in \mathbb{R}^{n\times n}$, $\mathbf{A}_\kappa^2 \in \mathbb{R}^{n\times n}$, and $\tau_\kappa^i(t) \leq \tau$ is the delay for all $i = 1, \dots, N$ and $\kappa = 1, \dots, r$, with $\tau > 0$. In this letter the assumption is that the premise variables do not depend on the input variables $u(t)$. In Input–Output form, the defuzzification process of delayed system (7) can be represented in the polytopic form:

$$\dot{e}^i(t) = \sum_{\kappa=1}^{r} \lambda_\kappa(z^i(t))[\mathbf{A}_\kappa^1 e^i(t) + \mathbf{A}_\kappa^2 e^i(t - \tau_\kappa^i(t)) + \mathbf{B}_\kappa u^i(t)] \tag{8}$$

where the $\lambda_\kappa(z^i(t))$ are defined previously in (6). In this letter, dynamical system (8) is used instead of (5) for the design of the control law.

Unmanned Aerial Vehicle Modelling: Error Posture Model

According to [28], the motion of the ith point-mass UAV, under assumptions that an electrically powered UAV is considered flying at constant altitude and ground speed, the thrust and velocity vector are collinear, and there is no slip in lateral direction, can be described by:

$$\begin{aligned} \dot{x}_c^i(t) &= v_{\text{er}}^i(t)\cos\theta_c^i(t) \\ \dot{y}_c^i(t) &= v_{\text{er}}^i(t)\sin\theta_c^i(t) \\ \dot{\theta}_c^i(t) &= w_{\text{er}}^i(t) \end{aligned} \tag{9}$$

In (9) x_c^i, y_c^i, are the position coordinates, θ_c^i is the heading angle, and v_{er}^i, w_{er}^i the linear and angular velocity.

In this analysis the error posture is utilised for the tracking problem for every agent in the network, as in references [15] and [12]. In particular, utilising the kinematics in (9) the tracking error is governed by:

$$e^i(t) = \begin{bmatrix} \cos(\theta_c^i(t)) & \sin(\theta_c^i(t)) & 0 \\ -\sin(\theta_c^i(t)) & \cos(\theta_c^i(t)) & 0 \\ 0 & 0 & 1 \end{bmatrix} (P_{\text{ref}}(t) - P_c^i(t)) \tag{10}$$

where $P_{\text{ref}}(x_{\text{ref}}, y_{\text{ref}}, \theta_{\text{ref}})$ and $P_c^i(x_c^i, y_c^i, \theta_c^i)$ the reference and current posture for the vehicle, $e^i(t) = [x_e^i(t), y_e^i(t), \theta_e^i(t)]^T$ is the tracking error in the state for the ith UAV in the $x - y$ plane and direction, respectively. The tracking error for a UAV in lateral motion is depicted in Fig. 1. Following the description in Sect. 3.1 of reference [12], taking the time derivative of (10), the error dynamics are generated. Hence, assuming that there is no side-slip (i.e. $\dot{x}_{\text{ref}}\sin(\theta_{\text{ref}}) = \dot{y}_{\text{ref}}\cos(\theta_{\text{ref}})$), and applying a control action vector $u_{\text{er}}^i(t) = u_F^i(t) + u^i(t)$ (proposed in [15]), where $u_F^i(t) = [v_{\text{ref}}(t)\cos(\theta_e^i(t)), w_{\text{ref}}(t)]^T$ the feedforward control action vector and $u^i(t) = [v^i(t), w^i(t)]^T$ the feedback elements, then the error dynamics satisfy:

$$\begin{bmatrix} \dot{x}_e^i(t) \\ \dot{y}_e^i(t) \\ \dot{\theta}_e^i(t) \end{bmatrix} = \begin{bmatrix} 0 & w_{\text{ref}}(t) & 0 \\ -w_{\text{ref}}(t) & 0 & v_{\text{ref}}(t)\text{sinc}(\theta_e^i(t)) \\ 0 & 0 & 0 \end{bmatrix} \times \begin{bmatrix} x_e^i(t) \\ y_e^i(t) \\ \theta_e^i(t) \end{bmatrix} + \begin{bmatrix} -1 & y_e^i(t) \\ 0 & -x_e^i(t) \\ 0 & -1 \end{bmatrix} u^i(t) \tag{11}$$

where $v_{\text{ref}}(t)$, $v^i(t)$ are the reference and current linear velocities, $w_{\text{ref}}(t)$, $w^i(t)$ the reference and current angular velocities.

The structure of the error posture dynamics in (11) allow its representation as a Takagi-Sugeno fuzzy model as shown by authors in [12]. Motivated by the work in [13] and [14], the two-step procedure is adopted for a network of delayed nonlinear error posture models in (11) which are structured into the TS form in (8). For the system, the control action vector $u^i(t)$ is designed based on TS concepts, and is shown in the sequel. The control law has the form referred to in the literature as PDC [29].

Swarm Tracking and Control Law Description Subject to Time-Delay

In this section the design of the control law for the stabilisation of the error dynamics in (8) subject to time-delay is described. The task is for the error state $e^i(t)$ for $i = 1, \ldots, N$ to converge to zero asymptotically at a local level. In this work the assumption is that individual systems have common $\mathbf{A}_\kappa^1, \mathbf{A}_\kappa^2, \mathbf{B}_\kappa, \forall\, \kappa = 1, \ldots, r$, and the communication topology is bidirectional static. Additionally it is assumed that for the delay term $0 \leq \tau_\kappa^i(t) \leq \tau$. As in [13], and [14] the control design for the stabilisation problem is treated in two steps.

Step 1: Node Level Tracking

The controller $u_\tau^i(e^i(t))$, used to stabilise the error dynamics, subject to delay, for the ith UAV system at node level, is designed from the rules of the TS fuzzy model and maintains the same structure as the model rules. Furthermore in this work it is assumed that the control law is not delay-dependent. The κth control rule at node level has the following structure:

Control Rule κ:
IF $z_1^i(t)$ is $M_{\kappa 1}$ **AND** ... **AND** $z_q^i(t)$ is $M_{\kappa q}$ **THEN** $u_\tau^i(e^i(t)) = -\mathbf{F}_\kappa e^i(t), \forall i, j = 1, \ldots, N$
for $\kappa = 1, \ldots, r$ and where $q = |z|$. In polytopic form the node level state feedback control law is equal to:

$$u_\tau^i(e^i(t)) = -\sum_{\kappa=1}^{r} \lambda_\kappa(z^i(t))\mathbf{F}_\kappa e^i(t) \tag{12}$$

where $\mathbf{F}_\kappa \in \mathbb{R}^{m\times n}$ are the feedback gains. By substitution of (12) into the dynamics in (8), the node level closed-loop error dynamics are equal to:

$$\dot{e}^i(t) = \sum_{\kappa=1}^{r}\sum_{\mu=1}^{r}\lambda_\kappa(z^i(t))\lambda_\mu(z^i(t))\left[\mathbb{A}_{\kappa\mu}e^i(t) + \mathbf{A}_\kappa^2 e^i(t-\tau_\kappa^i(t))\right] \tag{13}$$

where $\mathbb{A}_{\kappa\mu} = \mathbf{A}_\kappa^1 - \mathbf{B}_\kappa\mathbf{F}_\mu$. According to reference [25] the dynamics in (13) are expanded into (14) in order to use more relaxed conditions [25].

$$\begin{aligned}\dot{e}^i(t) &= \sum_{\kappa=1}^{r}\lambda_\kappa(z^i(t))^2\left[\mathbb{A}_{\kappa\kappa}e^i(t) + \mathbf{A}_\kappa^2 e^i(t-\tau_\kappa^i(t))\right]\\ &\ldots + 2\sum_{\kappa=1}^{r}\sum_{\kappa<\mu}^{r}\lambda_\kappa(z^i(t))\lambda_\mu(z^i(t))\\ &\times\left(\frac{\mathbb{A}_{\kappa\mu}e^i(t) + \mathbf{A}_\kappa^2 e^i(t-\tau_\kappa^i(t)) + \mathbb{A}_{\mu\kappa}e^i(t) + \mathbf{A}_\mu^2 e^i(t-\tau_\mu^i(t))}{2}\right)\end{aligned} \tag{14}$$

For the stabilisation of the node level error dynamics Lyapunov theory is utilised. The task is to determine the feedback gains $\mathbf{F}_\mu$, and a symmetric positive definite matrix $\mathbf{P} \in \mathbb{R}^{n\times n}$, such that a local performance criteria for stability is satisfied. It will be shown in the sequel that through the use of the Razumikhin theorem [8] a Lyapunov analysis is possible, feedback gains and a positive definite matrix can be calculated to satisfy stability conditions posed as Linear Matrix Inequalities (LMIs).

According to [2], the stabilisation of the error dynamics for the system (14) subject to time-delay is guaranteed via the PDC control law in (12) if there exists a symmetric positive matrix $\mathbf{P} > 0$ ($\mathbf{P} \in \mathbb{R}^{n\times n}$), $\mathbf{S}_\kappa > 0$ ($\mathbf{S} \in \mathbb{R}^{n\times n}$) and matrices $\mathbf{F}_\mu \in \mathbb{R}^{m\times n}$ for $\kappa, \mu = 1, \ldots, r$ such that the following conditions hold:

$$\begin{cases} \mathbf{P} > 0 \\ \mathbf{P} > \mathbf{S}_\kappa^{-1} \\ \mathbb{A}_{\kappa\kappa}^T\mathbf{P} + \mathbf{P}\mathbb{A}_{\kappa\kappa} + \zeta\mathbf{P} + \mathbf{P}\mathbf{A}_\kappa^2\mathbf{S}_\kappa\mathbf{A}_\kappa^{2\ T}\mathbf{P} < 0,\ \forall\kappa = 1,\ldots,r \\ \mathbb{A}_{\kappa\mu}^T\mathbf{P} + \mathbf{P}\mathbb{A}_{\kappa\mu} + \mathbb{A}_{\mu\kappa}^T\mathbf{P} + \mathbf{P}\mathbb{A}_{\mu\kappa} + 2\zeta\mathbf{P} \\ \quad\ldots + \mathbf{P}\mathbf{A}_\kappa^2\mathbf{S}_\kappa\mathbf{A}_\kappa^{2\ T}\mathbf{P} + \mathbf{P}\mathbf{A}_\mu^2\mathbf{S}_\mu\mathbf{A}_\mu^{2\ T}\mathbf{P} < 0 \end{cases} \tag{15}$$

with $\kappa < \mu\ s.t\ \lambda_\kappa(z^i(t)) \cap \lambda_\mu(z^i(t)) \neq \varnothing$,[1] for the third condition. Additionally, $\kappa, \mu = 1, \ldots, r$, and $\zeta > 1$. The proof of (15) follows directly from the Lyapunov analysis of the dynamics in (14).

[1]The notation $\lambda_\kappa(z^i(t)) \cap \lambda_\mu(z^i(t)) \neq \varnothing$ implies that the conditions hold for $\kappa < \mu$ except if $\lambda_\kappa(z^i(t)) \times \lambda_\mu(z^i(t)) = 0$ for all $z(t)$. The conditions are valid provided that two rules are active simultaneously.

Taking the time derivative of the positive definite function $v^i(t) = e^i(t)^T \mathbf{P} e^i(t),\ \forall i = 1, 2, \ldots, N$ the task is to show that this is negative definite for all $e^i(t) \neq 0$. Utilising the closed loop dynamics in (14) the time derivative of the Lyapunov function $v^i(t)$ is equal to:

$$
\begin{aligned}
\dot{v}^i(t) &= \sum_{\kappa=1}^{r} \lambda_\kappa^2(z^i(t)) \Big[e^i(t)^T (\mathbb{A}_{\kappa\kappa}^T \mathbf{P} + \mathbf{P}\mathbb{A}_{\kappa\kappa}) e^i(t) + 2e^i(t)^T \mathbf{P}\mathbf{A}_\kappa^2 e^i(t - \tau_\kappa^i(t)) \Big] \ldots \\
&+ \sum_{\kappa=1}^{r} \sum_{\kappa<\mu} \lambda_\kappa(z^i(t)) \lambda_\mu(z^i(t)) \Big[e^i(t)^T (\mathbb{A}_{\kappa\mu}^T \mathbf{P} + \mathbf{P}\mathbb{A}_{\kappa\mu} + \mathbb{A}_{\mu\kappa}^T \mathbf{P} + \mathbf{P}\mathbb{A}_{\mu\kappa}) e^i(t) \ldots \\
&+ 2e^i(t)^T \mathbf{P}\mathbf{A}_\kappa^2 e^i(t - \tau_\kappa^i(t)) + 2e^i(t)^T \mathbf{P}\mathbf{A}_\mu^2 e^i(t - \tau_\mu^i(t)) \Big] \\
&\leq \sum_{\kappa=1}^{r} \lambda_\kappa^2(z^i(t)) \Big[e^i(t)^T (\mathbb{A}_{\kappa\kappa}^T \mathbf{P} + \mathbf{P}\mathbb{A}_{\kappa\kappa} + \mathbf{P}\mathbf{A}_\kappa^2 \mathbf{S}_\kappa \mathbf{A}_\kappa^{2\ T} \mathbf{P}) e^i(t) \\
&\qquad + e^i(t - \tau_\kappa^i(t))^T \mathbf{S}_\kappa^{-1} e^i(t - \tau_\kappa^i(t)) \Big] \ldots \\
&+ \sum_{\kappa=1}^{r} \sum_{\kappa<\mu} \lambda_\kappa(z^i(t)) \lambda_\mu(z^i(t)) \Big[e^i(t)^T (\mathbb{A}_{\kappa\mu}^T \mathbf{P} + \mathbf{P}\mathbb{A}_{\kappa\mu} + \mathbb{A}_{\mu\kappa}^T \mathbf{P} + \mathbf{P}\mathbb{A}_{\mu\kappa} \ldots \\
&+ \mathbf{P}\mathbf{A}_\kappa^2 \mathbf{S}_\kappa \mathbf{A}_\kappa^{2\ T} \mathbf{P} + \mathbf{P}\mathbf{A}_\mu^2 \mathbf{S}_\mu \mathbf{A}_\mu^{2\ T} \mathbf{P}) e^i(t) \ldots \\
&+ e^i(t - \tau_\kappa^i(t))^T \mathbf{S}_\kappa^{-1} e^i(t - \tau_\kappa^i(t)) + e^i(t - \tau_\mu^i(t))^T \mathbf{S}_\mu^{-1} e^i(t - \tau_\mu^i(t)) \Big] \\
&\leq \sum_{\kappa=1}^{r} \lambda_\kappa^2(z^i(t)) \Big[e^i(t)^T (\mathbb{A}_{\kappa\kappa}^T \mathbf{P} + \mathbf{P}\mathbb{A}_{\kappa\kappa} + \mathbf{P}\mathbf{A}_\kappa^2 \mathbf{S}_\kappa \mathbf{A}_\kappa^{2\ T} \mathbf{P}) e^i(t) + v^i(t - \tau_\kappa^i(t)) \Big] \ldots \\
&+ \sum_{\kappa=1}^{r} \sum_{\kappa<\mu} \lambda_\kappa(z^i(t)) \lambda_\mu(z^i(t)) \Big[e^i(t)^T (\mathbb{A}_{\kappa\mu}^T \mathbf{P} + \mathbf{P}\mathbb{A}_{\kappa\mu} + \mathbb{A}_{\mu\kappa}^T \mathbf{P} + \mathbf{P}\mathbb{A}_{\mu\kappa} \ldots \\
&+ \mathbf{P}\mathbf{A}_\kappa^2 \mathbf{S}_\kappa \mathbf{A}_\kappa^{2\ T} \mathbf{P} + \mathbf{P}\mathbf{A}_\mu^2 \mathbf{S}_\mu \mathbf{A}_\mu^{2\ T} \mathbf{P}) e^i(t) + v^i(t - \tau_\kappa^i(t)) + v^i(t - \tau_\mu^i(t)) \Big]
\end{aligned}
\tag{16}
$$

Based on the Razumikhin theorem [8] for the positive definite function $v^i(t)$ there exist σ_1 and σ_2 such that:

$$
\sigma_1 \|e^i(t)\|^2 \leq v^i(t) \leq \sigma_2 \|e^i(t)\|^2 \tag{17}
$$

where σ_1 and σ_2 are the minimum and maximum eigenvalues of $\mathbf{P}$. By using (17), it is assumed that there exists a real $\zeta > 1$ such that:

$$
v^i(t - \theta) < \zeta v^i(t) \tag{18}
$$

for $\theta \in [0, \tau]$, then from (18), (16) reduces to:

$$\dot{v}^i(t) \leq \sum_{\kappa=1}^{r} \lambda_\kappa^2(z^i(t)) \Big[e^i(t)^T \big(\mathbb{A}_{\kappa\kappa}^T \mathbf{P} + \mathbf{P}\mathbb{A}_{\kappa\kappa} + \mathbf{P}\mathbf{A}_\kappa^2 \mathbf{S}_\kappa \mathbf{A}_\kappa^{2\ T} \mathbf{P} + \zeta \mathbf{P} \big) e^i(t) \Big]$$

$$\ldots + \sum_{\kappa=1}^{r} \sum_{\kappa<\mu} \lambda_\kappa(z^i(t)) \lambda_\mu(z^i(t)) \Big[e^i(t)^T \big(\mathbb{A}_{\kappa\mu}^T \mathbf{P} + \mathbf{P}\mathbb{A}_{\kappa\mu} + \mathbb{A}_{\mu\kappa}^T \mathbf{P} + \mathbf{P}\mathbb{A}_{\mu\kappa}$$

$$\ldots + \mathbf{P}\mathbf{A}_\kappa^2 \mathbf{S}_\kappa \mathbf{A}_\kappa^{2\ T} \mathbf{P} + \mathbf{P}\mathbf{A}_\mu^2 \mathbf{S}_\mu \mathbf{A}_\mu^{2\ T} \mathbf{P} + 2\zeta \mathbf{P} \big) e^i(t) \Big] \tag{19}$$

Thus for (19) to be negative definite it is only sufficient to find $\mathbf{P} > 0$, $\mathbf{S}_\kappa > 0$ and matrices $\mathbf{F}_\mu \in \mathbb{R}^{m \times n}$ for $\kappa, \mu = 1, \ldots, r$ such that the conditions (15) hold, and the proof is completed. □

In order to solve the previous problem, the bilinear matrix inequalities (BMIs) need to be recast into LMIs. This can be performed subject to a congruence transformation of $\mathbf{X}$, where $\mathbf{X} = \mathbf{P}^{-1}$, and the definition of $\Xi_\mu = \mathbf{F}_\mu \mathbf{X}$. Thus conditions (15) are transformed to LMIs:

$$\begin{cases} \mathbf{X} > 0 \\ \mathbf{S}_\kappa > \mathbf{X} \\ \mathbf{\Pi}_{\kappa\kappa}^1 < 0, \ \forall \kappa = 1, \ldots, r \\ \mathbf{\Pi}_{\kappa\mu}^2 < 0, \ \kappa < \mu \ s.t \ \lambda_\kappa(z^i(t)) \cap \lambda_\mu(z^i(t)) \neq \varnothing \end{cases} \tag{20}$$

where $\mathbf{\Pi}_{\kappa\kappa}^1$ and $\mathbf{\Pi}_{\kappa\mu}^2$ are equal to:

$$\mathbf{\Pi}_{\kappa\kappa}^1 = \mathbf{X}\mathbf{A}_\kappa^{1\ T} + \mathbf{A}_\kappa^1 \mathbf{X} - \Xi_\kappa^T \mathbf{B}_\kappa^T - \mathbf{B}_\kappa \Xi_\kappa + \zeta \mathbf{X} + \mathbf{A}_\kappa^2 \mathbf{S}_\kappa \mathbf{A}_\kappa^{2\ T} \tag{21}$$

and

$$\mathbf{\Pi}_{\kappa\mu}^2 = \mathbf{X}\mathbf{A}_\kappa^{1\ T} + \mathbf{A}_\kappa^1 \mathbf{X} - \Xi_\mu^T \mathbf{B}_\kappa^T - \mathbf{B}_\kappa \Xi_\mu + \mathbf{X}\mathbf{A}_\mu^{1\ T} + \mathbf{A}_\mu^1 \mathbf{X} - \Xi_\kappa^T \mathbf{B}_\mu^T - \mathbf{B}_\mu \Xi_\kappa$$

$$\ldots + 2\zeta \mathbf{X} + \mathbf{A}_\kappa^2 \mathbf{S}_\kappa \mathbf{A}_\kappa^{2\ T} + \mathbf{A}_\mu^2 \mathbf{S}_\mu \mathbf{A}_\mu^{2\ T} \tag{22}$$

Provided that the LMIs in (20) are feasible, then a solution can be recovered from:

$$\mathbf{F}_\mu = \Xi_\mu \mathbf{X}^{-1} \tag{23}$$

Remark 1. A faster response for the closed loop system (14) can be considered. This can be performed if a decay rate η is included in conditions (20). This is equivalent to ensuring $\dot{v}^i(t) + 2\eta v^i(t) < 0$ and replacing the previous constraints $\mathbf{\Pi}_{\kappa\kappa}^1$ and $\mathbf{\Pi}_{\kappa\mu}^2$ with $\overline{\mathbf{\Pi}}_{\kappa\kappa}^1 = \mathbf{\Pi}_{\kappa\kappa}^1 + 2\eta \mathbf{X}$ and $\overline{\mathbf{\Pi}}_{\kappa\mu}^2 = \mathbf{\Pi}_{\kappa\mu}^2 + 4\eta \mathbf{X}$,respectively.

Remark 2. Moreover a generalised eigenvalue problem subject to (20), with modified $\mathbf{\Pi}^1_{\kappa\kappa}$, $\mathbf{\Pi}^2_{\kappa\mu}$ and $\eta > 0$, can be used as suggested in [1]. Provided that the initial conditions are a-priori known, the control effort can be constrained to satisfy $\|\xi_\tau(e(t))\|_2 \leq \nu$. This can be enforced by means of the optimisation problem

$$\min_{\mathbf{X},\Xi_1,\ldots,\Xi_r} \nu \tag{24}$$

subject to the modified LMIs of (20) and

$$\begin{cases} \begin{bmatrix} 1 & e(0)^T \\ e(0) & \mathbf{X} \end{bmatrix} \geq 0 \\ \begin{bmatrix} \mathbf{X} & \Xi_\kappa^T \\ \Xi_\kappa & \nu^2\mathbf{I} \end{bmatrix} \geq 0 \end{cases} \tag{25}$$

for $\kappa = 1, \ldots, r$.

Remark 3. Note that conditions (25) are dependent to $e(0)$ which is a limitation. This can be overcome by assuming an upper bound ϕ for $\|e(0)\|$. Then as suggested in [25] conditions (25) can be replaced by

$$\phi^2\mathbf{I} \leq \mathbf{X} \tag{26}$$

The modified conditions in (20), and (26) or (20), (24) and (25) lead to a good compromise between complexity and conservatism. It should be noted that the performance constraints are chosen according to functional and physical limitations of the aircraft involved. Hence provided the feedback gains $\mathbf{F}_\kappa$ are chosen for a common Lyapunov matrix $\mathbf{P}$ satisfying conditions (20) (with the modified $\mathbf{\Pi}^1_{\kappa\kappa}$, $\mathbf{\Pi}^2_{\kappa\mu}$), (24) and (25), $\dot{v}^i(t) + 2\eta v^i(t) < 0$. Thus stability can be guaranteed for any set of initial conditions $e^i(0) \in \mathcal{X}$ for the delayed system. Thereafter based on the node level stabilisation, a second step is undertaken illustrated in the next section.

Step 2: Tracking at Network Level

At a network level an additional term which represents the relative state information among neighbouring UAVs and the reference trajectory is introduced in the control law so that:

$$u(e^i(t)) = -\sum_{\kappa=1}^{r} \lambda_\kappa(z^i(t))\mathbf{F}_\mu e^i(t) + \gamma\overline{\mathbf{F}} \sum_{j,\, i\neq j}^{N} \ell_{ij} e^j(t) \tag{27}$$

where $\overline{\mathbf{F}} \in \mathbb{R}^{m \times n}$ and γ a positive scalar. Using the control law in (27), at a network level the error-dynamics, subject to delay, are equal to:

$$\dot{e}^i(t) = \sum_{\kappa=1}^{r}\sum_{\mu=1}^{r}\lambda_\kappa(z^i(t))\lambda_\mu(z^i(t))\left(\mathbb{A}_{\kappa\mu}e^i(t)+\mathbf{A}_\kappa^2 e^i(t-\tau_\kappa^i(t))+\gamma\mathbf{B}_\kappa\overline{\mathbf{F}}\sum_{j=1}^{N}\ell_{ij}e^j(t)\right) \tag{28}$$

In a compact form (28) can be conveniently written using the Kronecker product notation [9] (refer to properties in Appendix A), as:

$$\dot{\mathrm{e}}(t) = [\mathcal{A}(z(t)) + \gamma\mathcal{B}(z(t))(\mathbf{L} \otimes I_n)]\mathrm{e}(t) + \mathcal{A}_2(z(t))\mathrm{e}(t - \tau(t)) \tag{29}$$

where

$$\mathcal{A}(z(t)) = \mathrm{diag}\left\{\sum_{\kappa=1}^{r}\sum_{\mu=1}^{r}\lambda_\kappa^1\lambda_\mu^1\mathbb{A}_{\kappa\mu},\ldots,\sum_{\kappa=1}^{r}\sum_{\mu=1}^{r}\lambda_\kappa^N\lambda_\mu^N\mathbb{A}_{\kappa\mu}\right\} \tag{30}$$

$$\mathcal{A}_2(z(t)) = \mathrm{diag}\left\{\sum_{\kappa=1}^{r}\lambda_\kappa^1\mathbf{A}_\kappa^2,\ldots,\sum_{\kappa=1}^{r}\lambda_\kappa^N\mathbf{A}_\kappa^2\right\} \tag{31}$$

and

$$\mathcal{B}(z(t)) = \mathrm{diag}\left\{\sum_{\kappa=1}^{r}\lambda_\kappa^1\mathbf{B}_\kappa\overline{\mathbf{F}},\ldots,\sum_{\kappa=1}^{r}\lambda_\kappa^N\mathbf{B}_\kappa\overline{\mathbf{F}}\right\} \tag{32}$$

and $\mathrm{e}(t)$, $\mathrm{e}(t-\tau(t))$ is the concatenation of the state vectors $e^i(t)$, $e^i(t-\tau_\kappa^i(t))$ so that $\mathrm{e}(t) = col([e^1(t),\ldots,e^N(t)])$, and $\mathrm{e}(t-\tau(t)) = col([e^1(t-\tau_\kappa^1(t)),\ldots,e^N(t-\tau_\kappa^N(t))])$, respectively. Using Lyapunov theory, a candidate Lyapunov function for the swarm is defined as

$$V(t) = \sum_{i=1}^{N} e^i(t)^T\mathbf{P}e^i(t) \tag{33}$$

where the symmetric positive definite matrix $\mathbf{P}$ is from the earlier node level synthesis in section "Step 1: Node Level Tracking". Taking the time derivative of (33), and substituting for the control law (27) in (28), it results in

$$\dot{V}(t) = V_1 + V_2 \tag{34}$$

where

$$V_1 = \sum_{i=1}^{N}\sum_{\kappa=1}^{r}\sum_{\mu=1}^{r}\lambda_\kappa(z^i(t))\lambda_\mu(z^i(t))\left[e^i(t)^T(\mathbb{A}_{\kappa\mu}^T\mathbf{P}+\mathbf{P}\mathbb{A}_{\kappa\mu})e^i(t)+2e^i(t)^T\mathbf{P}\mathbf{A}_\kappa^2e^i(t-\tau_\kappa^i(t))\right] \tag{35}$$

and

$$V_2 = 2\gamma e^T(t)(I_N \otimes \mathbf{P})\mathcal{B}(z(t))(\mathbf{L} \otimes I_n)e(t) \tag{36}$$

For the swarm of UAVs, subject to delay, to track the virtual leader system, which is moving according to a prescribed reference trajectory, it is sufficient to show that $\dot{V}(t) < 0$. Utilising the stabilisation procedure from the first step of the design process in section "Step 1: Node Level Tracking", for the choice of a common Lyapunov matrix $\mathbf{P}$ and feedback gains $\mathbf{F}_\mu$, $V_1 < 0$. Hence all that needs to be shown is that V_2 is negative semi-definite for all $e(t) \neq 0$. It is evident from the TS model that the input matrix $\mathbf{B}_\kappa$ is time-varying because of (11); however, the first column is constant: i.e. $\mathbf{B}_\kappa = [\mathbf{B}_1, \mathbf{B}_{2\kappa}]$. Here by choice:

$$\overline{\mathbf{F}} = -[\mathbf{B}_1, 0]^T \mathbf{P} \tag{37}$$

which means that $\mathcal{B}(z(t)) = I \otimes \mathbf{B}_1\mathbf{B}_1^T\mathbf{P}$. As a result of this choice in (36):

$$V_2 = -2\gamma e^T(t)(\mathbf{L} \otimes \mathbf{P}\mathbf{B}_1\mathbf{B}_1^T\mathbf{P})e(t) \tag{38}$$

The Laplacian $\mathbf{L}$ is positive semi-definite, by definition in (2), and by construction $\mathbf{P}\mathbf{B}_1\mathbf{B}_1^T\mathbf{P} \geq 0$, it follows that $-(\mathbf{L}\otimes\mathbf{P}\mathbf{B}_1\mathbf{B}_1^T\mathbf{P}) \leq 0$ by Corollary 4.2.13 [9]. Thus (34) is negative definite for all $e(t) \neq 0$ and the error dynamics of the swarm, subject to delay, is stable.

Simulation Example

In this section a tracking scenario is considered where a swarm of UAVs is deployed to collectively follow the prescribed trajectory of a virtual leader from any initial conditions satisfying bounds on the state space. The path is assumed to be a-priori known. The reference track considered for the virtual leader in this example is referred to in the literature as the Dubins path [5].

The Dubins path comprises of line segments and circular arcs of type CLC or CCC (C = Circular arc, L = Line segment) or another combination of the previous two. By the former the two segments of the circumference of circles are joined by their common tangent. By the latter, CCC is formed by three consecutive tangential circular arcs. The physical interpretation of the particular path is a combination of the shortest line for rectilinear motion and the shortest circular arc for turning. In this work the path is constructed with the use of principles of Euclidean Geometry. The design procedure can be found in thesis [21] for the interested reader. Example trajectories for CLC paths with common external and internal tangents, respectively, are depicted in Fig. 2. The two waypoints depicted have poses of $P_{\text{start}}(100, 100, 80^o)$ and $P_{\text{final}}(150, 150, 45^o)$. It should be noted that for the previous poses there exist four different Dubins paths, the left to left turn

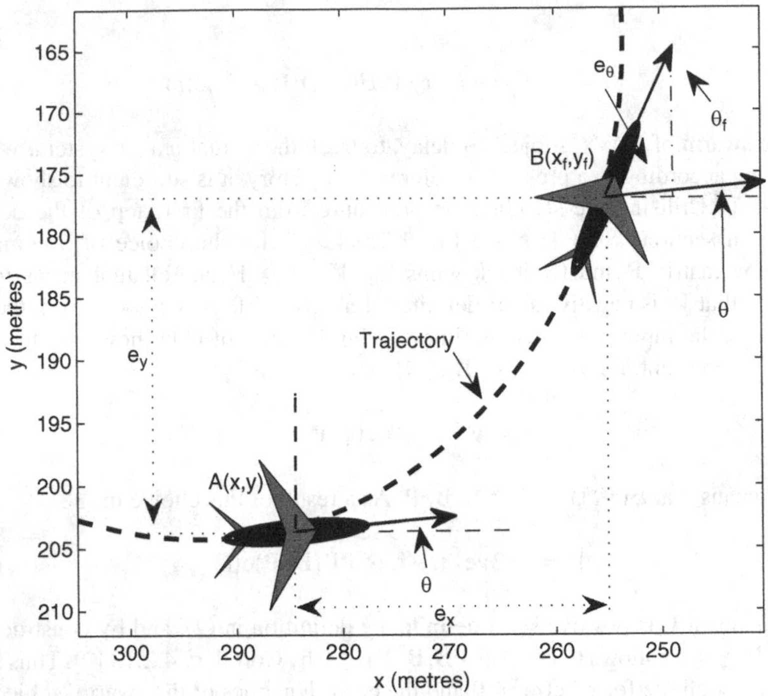

Fig. 1 Posture error $P(x_e, y_e, \theta_e)$ for an aerial vehicle in lateral motion

(LSL), the left to right turn (LSR), the right to right turn (RSR) and the right to left turn (RSL). From the previous four trajectories for the node to node path the one that yields the minimum energy requirements motivated by the work in [11] is chosen as the virtual leader trajectory.

Description of the Takagi Sugeno UAV Model

For the purpose of illustration consider a swarm of identical UAV models. The TS fuzzy model has been derived as described in section "UAV Model and the Takagi-Sugeno Model Representation". For the model illustrated in (11) $z_1^i(t) = w_{\text{ref}}(t)$, $z_2^i(t) = v_{\text{ref}}(t)\text{sinc}(\theta_e^i(t))$, $z_3^i(t) = y_e^i(t)$ and $z_4^i(t) = x_e^i(t)$ are chosen as the premise variables with $z_1^i(t) \in [-0.513, 0.513]$, $z_2^i(t) \in [18.0048, 20]$, $z_3^i(t) \in [-10, 10]$ and $z_4^i(t) \in [-10, 10]$. In addition $\theta_e^i(t) \in [-\pi/4, \pi/4]$. Hence the number of rules of the fuzzy system is equal to $r = 16$ and the length of the premise vector is equal to $q = 4$. It should be noted that the latter bounds are not chosen in an arbitrary manner, and are selected in order not to lose controllability of the system. Utilising the sector nonlinearity approach in [10] the membership functions

$M_{\kappa\mu}(z^i(t))$ are determined and the weighting terms are calculated according to (6). All sixteen rules are developed as prescribed in section "UAV Model and the Takagi-Sugeno Model Representation" Eq. (4), where $\mathbf{A}_\kappa$ and $\mathbf{B}_\kappa$ are shown in the sequel. Finally, the defuzzification is carried out with respect to Eq. (5). Hence the equivalent TS fuzzy model (5) for the full nonlinear is derived. For model (5):

$$\mathbf{A}_\kappa = \begin{bmatrix} 0 & -\varepsilon_\kappa^1 w_{r,\max} & 0 \\ \varepsilon_\kappa^1 w_{r,\max} & 0 & \overline{\mu}_\kappa^i \\ 0 & 0 & 0 \end{bmatrix} \quad \mathbf{B}_\kappa = \begin{bmatrix} -1 & \varepsilon_\kappa^3 e_{\max}^i \\ 0 & \varepsilon_\kappa^4 e_{\max}^i \\ 0 & -1 \end{bmatrix} \tag{39}$$

where $w_{r,\max} = 0.513$ (rad/s), $e_{\max}^i = 10$ (m) and

$$\varepsilon_\kappa^1 = \begin{cases} -1, & \text{for } 1 \le \kappa \le 8 \\ +1, & \text{otherwise} \end{cases} \quad \varepsilon_\kappa^4 = (-1)^{\kappa+1}$$

$$\varepsilon_\kappa^3 = \begin{cases} +1, & \text{otherwise} \\ -1, & \text{for } \kappa \in \{1, 2, 5, 6, 9, 10, 13, 14\} \end{cases}$$

$$\overline{\mu}_\kappa^i = \begin{cases} 18.0048, & \text{for } 1 \le \kappa \le 4 \text{ and } 9 \le \kappa \le 12 \\ 20, & \text{otherwise} \end{cases}$$

The membership function are depicted in Fig. 3. To illustrate the proposed results on the time-delay systems it is assumed the ith UAV error posture model is perturbed by time-delay and the delayed system is given as:

$$\begin{bmatrix} \dot{x}_e^i(t) \\ \dot{y}_e^i(t) \\ \dot{\theta}_e^i(t) \end{bmatrix} = \begin{bmatrix} 0 & a w_{\text{ref}}(t) & 0 \\ -a w_{\text{ref}}(t) & 0 & a v_{\text{ref}}(t)\text{sinc}(\theta_e^i(t)) \\ 0 & 0 & 0 \end{bmatrix} \begin{bmatrix} x_e^i(t) \\ y_e^i(t) \\ \theta_e^i(t) \end{bmatrix}$$
$$+ \begin{bmatrix} 0 & (1-a) w_{\text{ref}}(t) & 0 \\ -(1-a) w_{\text{ref}}(t) & 0 & (1-a) v_{\text{ref}}(t)\text{sinc}(\theta_e^i(t-\tau(t))) \\ 0 & 0 & 0 \end{bmatrix} \begin{bmatrix} x_e^i(t-\tau(t)) \\ y_e^i(t-\tau(t)) \\ \theta_e^i(t-\tau(t)) \end{bmatrix}$$
$$+ \begin{bmatrix} -1 & y_e^i(t) \\ 0 & -x_e^i(t) \\ 0 & -1 \end{bmatrix} u^i(t) \tag{40}$$

where retarded coefficient a satisfies $a \in [0, 1]$. It should be noted that the bounds on a correspond for a completely retarded system if a equals zero and otherwise non-delayed system. It is assumed that $a = 0.7$ and $\tau(t) = 10\sin(10t)$ in this example. In polytopic form the TS is equal to:

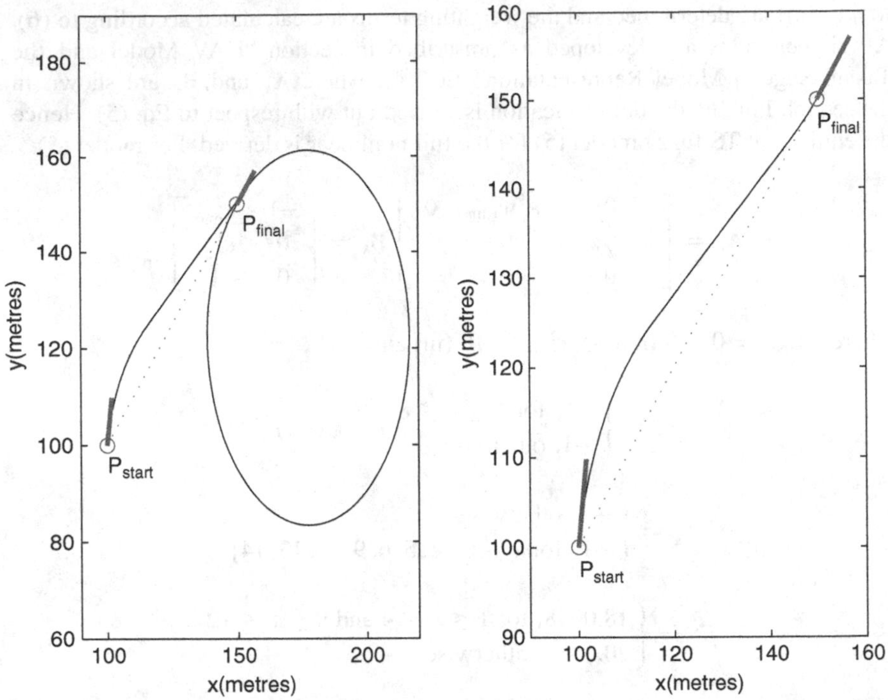

Fig. 2 Common external (*left*) and internal (*right*) tangent between two circular arcs. The *solid red lines* in source and destination waypoints depict the entry and exit heading angles at the particular poses (Colour figure online)

$$\dot{e}^i(t) = \sum_{\kappa=1}^{r} \lambda_\kappa(z^i(t))[\alpha \mathbf{A}_\kappa e^i(t) + (1-\alpha)\mathbf{A}_\kappa e^i(t-\tau(t)) + \mathbf{B}_\kappa u^i(t)] \tag{41}$$

where $\mathbf{A}_\kappa$ and $\mathbf{B}_\kappa$ are defined in (39), for the specified parameters. It can be easily shown that the closed system is not stable under delay perturbation and with the control law used in article [14]. Thus the analysis illustrated in section "Step 1: Node Level Tracking" is applied to synthesise the feedback gains and the positive definite matrix, at node level, and to be later utilised for the stabilisation of the entire swarm.

Tracking a Virtual Leader

In this example a swarm of UAVs ($N = 20$) is deployed to collectively follow a virtual leader system. It is assumed that the leader is moving according to a reference track (Dubins path) which is known a-priori from the preflight planning. The UAVs are interconnected through control law (27), and the contribution factor for the global information is chosen as $\gamma = 0.7$. The adjacency matrix is depicted in Fig. 4, and the Laplacian matrix has the form in (2). The graph considered here is $G(20, 184)$ and the task is for the error state $e^i(t) \to 0$ as $t \to \infty$.

Following the procedure introduced in section "Swarm Tracking and Control Law Description Subject to Time-Delay", firstly the LMIs are synthesised at node level for the closed loop error posture model (13). This leads to the choice of the feedback gains $\mathbf{F}_\mu$ and a common positive definite matrix $\mathbf{P}$ by minimising (24) subject to the LMI conditions in (25), and the transformed conditions from (20). The gains $\mathbf{F}_\kappa$ and $\mathbf{S}_\kappa$ for $\kappa = 1, \ldots, 16$ are shown in Appendix B and the positive definite matrix returned is:

$$\mathbf{P} = 10^{-5} \begin{bmatrix} 0.0029 & -0.0002 & -0.0028 \\ -0.0002 & 0.0009 & 0.0101 \\ -0.0028 & 0.0101 & 0.1344 \end{bmatrix}$$

From the minimisation problem of (24) subject to LMIs (25), and the transformed conditions from (20), $\eta = 0.1$, $\zeta = 1$, and $\nu = 9.6768e - 007$. Altering the elements in η, and ζ results in different responses of the system. This gives the designer the possibility of obtaining another control performance according to design specifications.

From Fig. 7 the bounds on the state space $e^i(t)$ are not violated and thus the TS model represents exactly the nonlinear retarded model of the error dynamics of the UAV. Utilising the stabilisation procedure at node level, $\overline{\mathbf{F}}$ is chosen as (37) at the second step according to section "Step 2: Tracking at Network Level". Hence the overall control law (27) is synthesised and is added to the feed-forward control action vector $u_F^i(t) = [v_{\text{ref}}(t)\cos(\theta_e^i(t)), w_{\text{ref}}(t)]^T$ to generate $u_{\text{er}}^i(t)$. The control input $u_{\text{er}}^i(t)$ consists of the angular $w_{\text{er}}^i(t)$ and the linear $v_{\text{er}}^i(t)$ velocities which are fed to the delayed ith UAV model (9). Thereafter the measured state of the vehicle is used to calculate the tracking error as in (10). The initial conditions for each UAV were chosen in a random manner (whilst satisfying the a-priori assumed bounds on the state space).

The swarm trajectories which are converging to the virtual leader reference track are depicted in Fig. 5. Figure 6 shows the heading angle of each UAV versus the virtual leader's. The states of the tracking error ($e^i(t) = [x_e^i(t), y_e^i(t), \theta_e^i(t)]^T$) are given in Fig. 7. The firing of the weighting functions $\lambda_\kappa(z^i(t))$ are depicted in Fig. 8. The control action vector $u_{\text{er}}^i(t)$ is given in Fig. 9.

The benefit of the proposed analysis is that the design of the controller is decoupled from the size of the network and its topology. This is due to the fact that there are only r LMIs that are utilised to stabilise each node locally. Additionally, due to the decoupled structure of the network it allows a convenient choice for gain $\overline{\mathbf{F}}$. The advantage is that through the special choice of feedback gains (37) for the relative state information, the methodology can be applied to a large class of nonlinear large-scale network of systems, subject to delay perturbation.

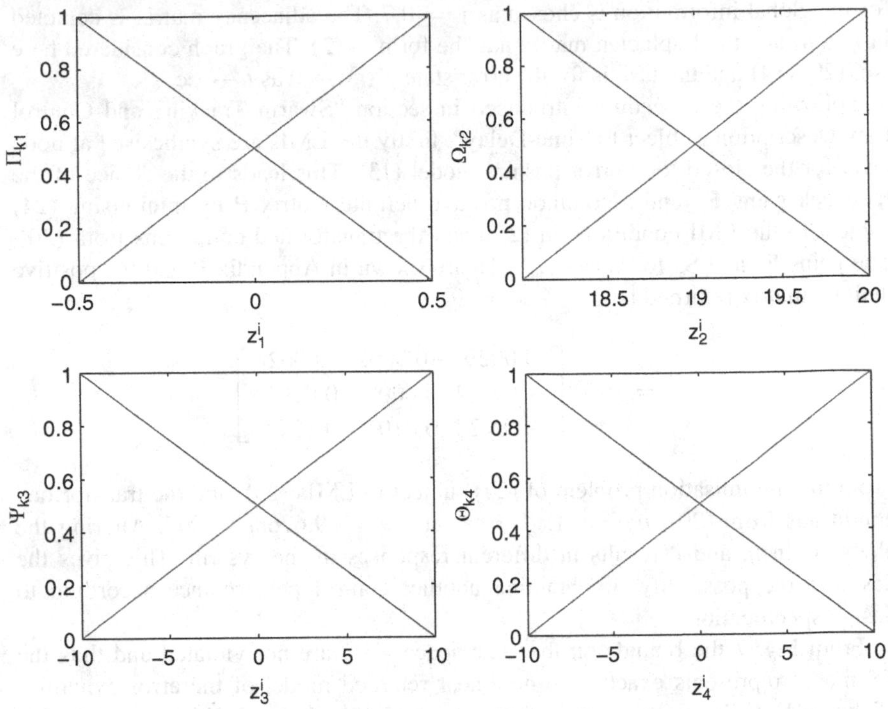

Fig. 3 Membership functions for $z^i(t)$ premise vector with saturation points not shown

Conclusions

This work proposes a systematic analysis for tracking problems in swarm-based UAV missions with linear and angular velocity constraints, subject to time-delay. The communication topology among the UAVs is represented using graph theory tools. The intermediate step of representing a network of nonlinear systems with TS models circumvents the difficulty in designing a control law when dependencies among the time-delayed UAVs are considered. A special choice of feedback gains for the relative state information allows the methodology to be applied to a reasonably large class of nonlinear systems. The distributed control law which is proposed, is composed of both node and network level information. The two-step design procedure is performed subject to criteria, posed as Linear Matrix Inequalities (LMIs). An illustrative example, where a swarm of UAVs, subject to delay, is deployed to collectively follow the track of a virtual leader, was included to demonstrate the potential of the analysis.

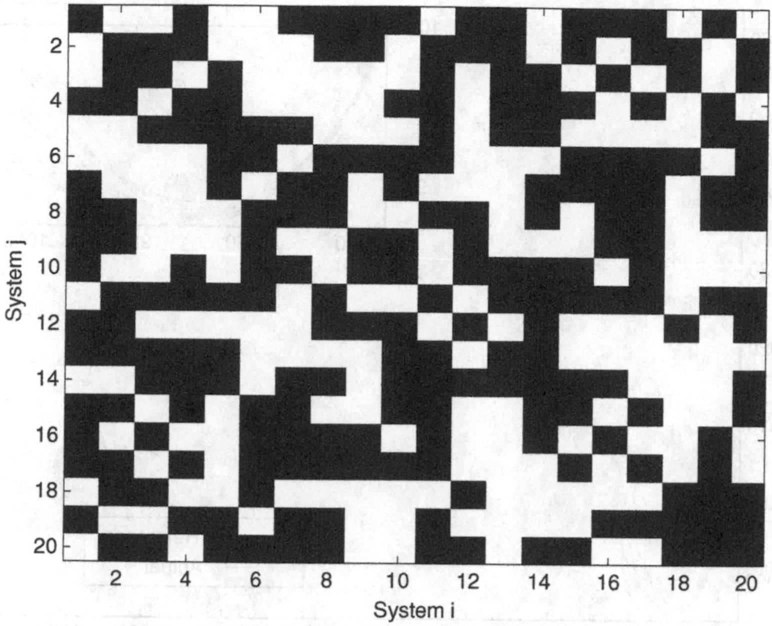

Fig. 4 Adjacency matrix **A** for network $G(V, E)$. Nodes that are interconnected ($a_{ij} = 1$) coincide with the white colour

Appendix

A-Kronecker Product Properties

The Kronecker product of **A** and **B** denoted as $\otimes$ is a block matrix **C** with entries $\mathbf{C} = \mathbf{A} \otimes \mathbf{B} = [c_{ij}] = [\mathbf{A}_{ij}\mathbf{B}]$. The Kronecker product is a special case of the tensor product and satisfies the following identities according to [9]:

1. $\mathbf{A} \otimes (\mathbf{B} + \mathbf{C}) = \mathbf{A} \otimes \mathbf{B} + \mathbf{A} \otimes \mathbf{C}$
2. $(k\mathbf{A}) \otimes \mathbf{B} = \mathbf{A} \otimes (k\mathbf{B}) = k(\mathbf{A} \otimes \mathbf{B})$
3. $(\mathbf{A} \otimes \mathbf{B}) \otimes \mathbf{C} = \mathbf{A} \otimes (\mathbf{B}) \otimes \mathbf{C}$
4. $(\mathbf{A} \otimes \mathbf{B})(\mathbf{C} \otimes \mathbf{D}) = \mathbf{AC} \otimes \mathbf{BD}$
5. $(\mathbf{A} \otimes \mathbf{B})^{-1} = \mathbf{A}^{-1} \otimes \mathbf{B}^{-1}$
6. $(\mathbf{A} \otimes \mathbf{B})^T = \mathbf{A}^T \otimes \mathbf{B}^T$

B-Results Calculated for Tracking Scenario

The gains calculated for the tracking scenario are :

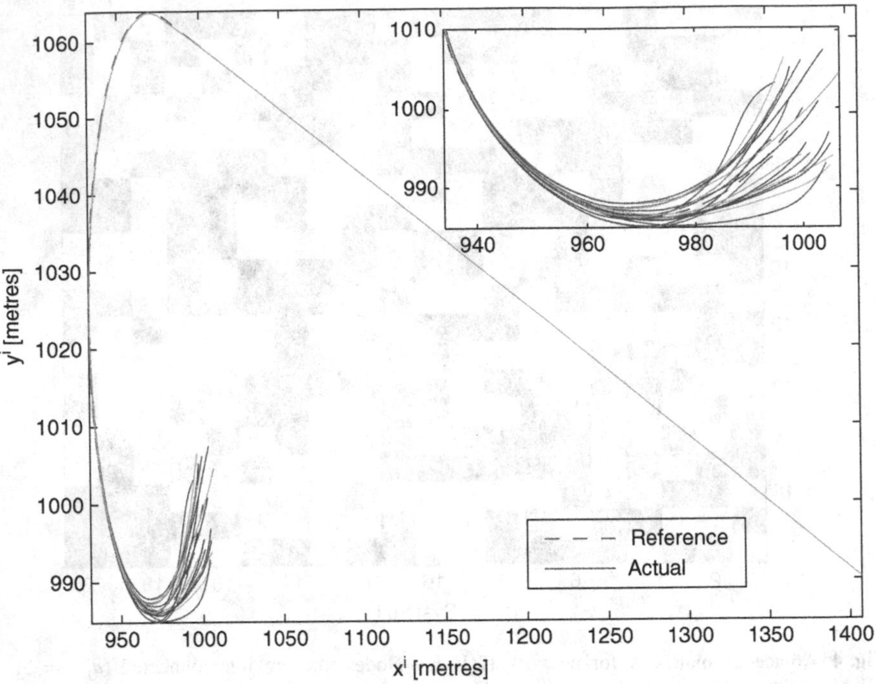

Fig. 5 Trajectories of the delayed swarm in $x - y$ plane (*solid lines*) versus the virtual leader's trajectory (*dashed line*)

$$
\begin{aligned}
&\mathbf{F}_1 = \begin{bmatrix} -5.7787 & 6.9570 & 91.5205 \\ 0.1910 & -0.6166 & -8.2395 \end{bmatrix} \mathbf{F}_2 = \begin{bmatrix} -4.4262 & 1.4297 & 18.2658 \\ 0.1705 & -0.6984 & -9.2015 \end{bmatrix} \mathbf{F}_3 = \begin{bmatrix} -2.6445 & -5.2920 & -70.7494 \\ 0.1836 & -0.6185 & -8.2680 \end{bmatrix} \\
&\mathbf{F}_4 = \begin{bmatrix} -1.0755 & -10.9130 & -145.4940 \\ 0.1656 & -0.6800 & -8.9622 \end{bmatrix} \mathbf{F}_5 = \begin{bmatrix} -5.5718 & 6.7533 & 88.9730 \\ 0.1894 & -0.6342 & -8.4770 \end{bmatrix} \mathbf{F}_6 = \begin{bmatrix} -4.3860 & 2.1222 & 27.6472 \\ 0.1776 & -0.7314 & -9.6451 \end{bmatrix} \\
&\mathbf{F}_7 = \begin{bmatrix} -2.3516 & -5.6580 & -75.4223 \\ 0.1934 & -0.6346 & -8.4904 \end{bmatrix} \mathbf{F}_8 = \begin{bmatrix} -1.0922 & -10.2054 & -135.8902 \\ 0.1897 & -0.7089 & -9.3582 \end{bmatrix} \mathbf{F}_9 = \begin{bmatrix} -5.6573 & 5.4646 & 73.8794 \\ 0.1550 & -0.6188 & -8.2624 \end{bmatrix} \\
&\mathbf{F}_{10} = \begin{bmatrix} -7.2677 & 11.4144 & 152.9849 \\ 0.2063 & -0.6779 & -8.9350 \end{bmatrix} \mathbf{F}_{11} = \begin{bmatrix} -2.1162 & -6.8384 & -89.1303 \\ 0.1484 & -0.6145 & -8.2023 \end{bmatrix} \mathbf{F}_{12} = \begin{bmatrix} -3.9435 & -0.8725 & -10.0146 \\ 0.2061 & -0.6930 & -9.1316 \end{bmatrix} \\
&\mathbf{F}_{13} = \begin{bmatrix} -5.5398 & 5.7460 & 77.3921 \\ 0.1552 & -0.6345 & -8.4794 \end{bmatrix} \mathbf{F}_{14} = \begin{bmatrix} -6.8592 & 10.6092 & 142.0409 \\ 0.1969 & -0.7049 & -9.3047 \end{bmatrix} \mathbf{F}_{15} = \begin{bmatrix} -1.9847 & -6.6967 & -87.4434 \\ 0.1602 & -0.6344 & -8.4698 \end{bmatrix} \\
&\mathbf{F}_{16} = \begin{bmatrix} -3.4437 & -1.7102 & -21.3904 \\ 0.2196 & -0.7292 & -9.6178 \end{bmatrix}
\end{aligned} \tag{42}
$$

From the solution of LMIs the positive definite matrices $1.0e + 007\mathbf{S}_\kappa$ for $\kappa = 1, \ldots, 16$ are equal to:

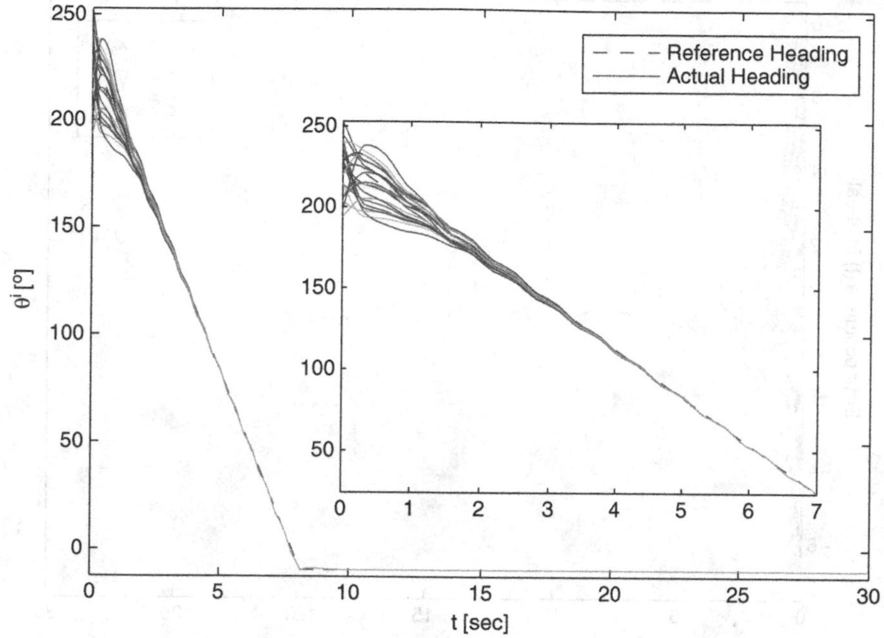

Fig. 6 Heading angle profile for each delayed UAV (*solid lines*) and heading angle of the virtual leader (*dashed line*)

$$\mathbf{S}_1 = \begin{bmatrix} 1.1368 & 0.3553 & -0.0904 \\ 0.3553 & 2.1092 & -0.0600 \\ -0.0904 & -0.0600 & 0.0145 \end{bmatrix} \mathbf{S}_2 = \begin{bmatrix} 1.2647 & 0.1497 & -0.0439 \\ 0.1497 & 2.2958 & -0.1288 \\ -0.0439 & -0.1288 & 0.0304 \end{bmatrix} \mathbf{S}_3 = \begin{bmatrix} 1.1351 & 0.3542 & -0.0907 \\ 0.3542 & 2.0749 & -0.0579 \\ -0.0907 & -0.0579 & 0.0143 \end{bmatrix}$$

$$\mathbf{S}_4 = \begin{bmatrix} 1.2596 & 0.1325 & -0.0440 \\ 0.1325 & 2.3397 & -0.1366 \\ -0.0440 & -0.1366 & 0.0310 \end{bmatrix} \mathbf{S}_5 = \begin{bmatrix} 1.2039 & 0.3137 & -0.0774 \\ 0.3137 & 2.1515 & -0.0664 \\ -0.0774 & -0.0664 & 0.0159 \end{bmatrix} \mathbf{S}_6 = \begin{bmatrix} 1.3674 & 0.0482 & -0.0140 \\ 0.0482 & 2.4149 & -0.1641 \\ -0.0140 & -0.1641 & 0.0392 \end{bmatrix}$$

$$\mathbf{S}_7 = \begin{bmatrix} 1.2043 & 0.3141 & -0.0774 \\ 0.3141 & 2.1193 & -0.0641 \\ -0.0774 & -0.0641 & 0.0157 \end{bmatrix} \mathbf{S}_8 = \begin{bmatrix} 1.3584 & 0.0321 & -0.0153 \\ 0.0321 & 2.4575 & -0.1719 \\ -0.0153 & -0.1719 & 0.0395 \end{bmatrix} \mathbf{S}_9 = \begin{bmatrix} 1.3488 & -0.2552 & 0.1091 \\ -0.2552 & 2.0853 & -0.0646 \\ 0.1091 & -0.0646 & 0.0193 \end{bmatrix}$$

$$\mathbf{S}_{10} = \begin{bmatrix} 1.3920 & -0.1348 & 0.0856 \\ -0.1348 & 2.3035 & -0.1258 \\ 0.0856 & -0.1258 & 0.0313 \end{bmatrix} \mathbf{S}_{11} = \begin{bmatrix} 1.3495 & -0.2546 & 0.1088 \\ -0.2546 & 2.1216 & -0.0674 \\ 0.1088 & -0.0674 & 0.0196 \end{bmatrix} \mathbf{S}_{12} = \begin{bmatrix} 1.3923 & -0.1484 & 0.0865 \\ -0.1484 & 2.2525 & -0.1171 \\ 0.0865 & -0.1171 & 0.0304 \end{bmatrix}$$

$$\mathbf{S}_{13} = \begin{bmatrix} 1.3952 & -0.2173 & 0.0974 \\ -0.2173 & 2.1246 & -0.0694 \\ 0.0974 & -0.0694 & 0.0201 \end{bmatrix} \mathbf{S}_{14} = \begin{bmatrix} 1.4554 & -0.0381 & 0.0604 \\ -0.0381 & 2.4659 & -0.1731 \\ 0.0604 & -0.1731 & 0.0415 \end{bmatrix} \mathbf{S}_{15} = \begin{bmatrix} 1.3950 & -0.2163 & 0.0975 \\ -0.2163 & 2.1614 & -0.0724 \\ 0.0975 & -0.0724 & 0.0203 \end{bmatrix}$$

$$\mathbf{S}_{16} = \begin{bmatrix} 1.4588 & -0.0482 & 0.0598 \\ -0.0482 & 2.4212 & -0.1654 \\ 0.0598 & -0.1654 & 0.0410 \end{bmatrix} \qquad (43)$$

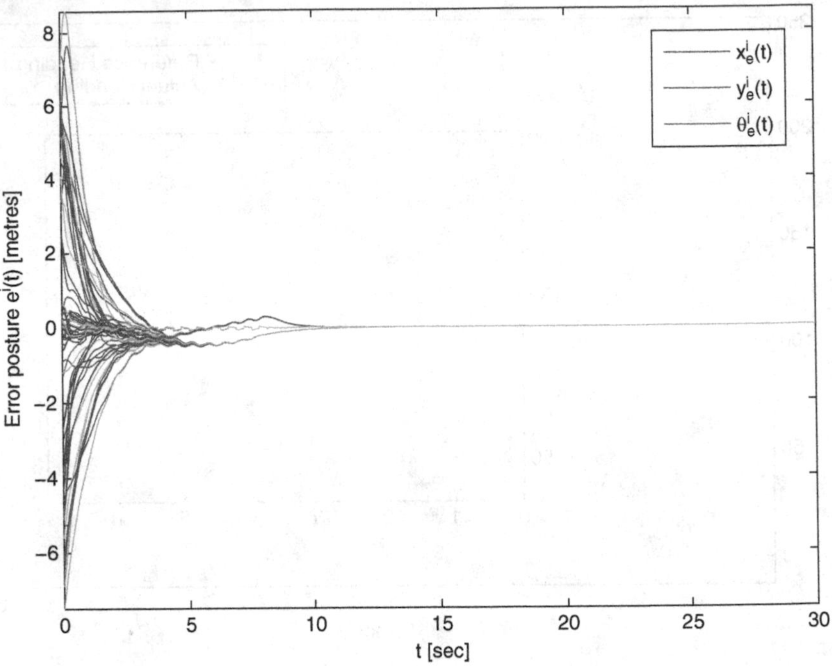

Fig. 7 Tracking error $e^i(t) = [x_e^i(t), y_e^i(t), \theta_e^i(t)]^T$ of every delayed UAV within the swarm

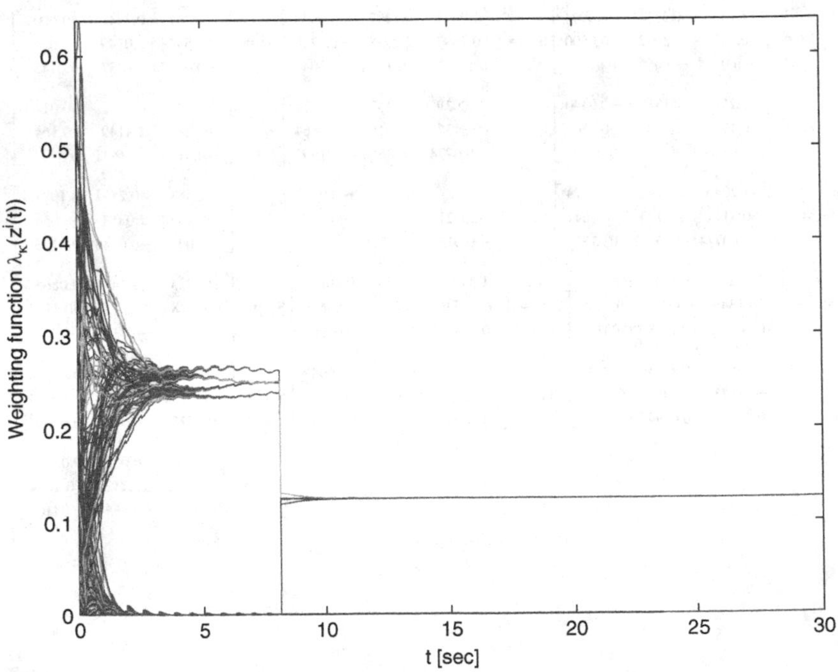

Fig. 8 Weighting functions $\lambda_\kappa(z_i(t))$

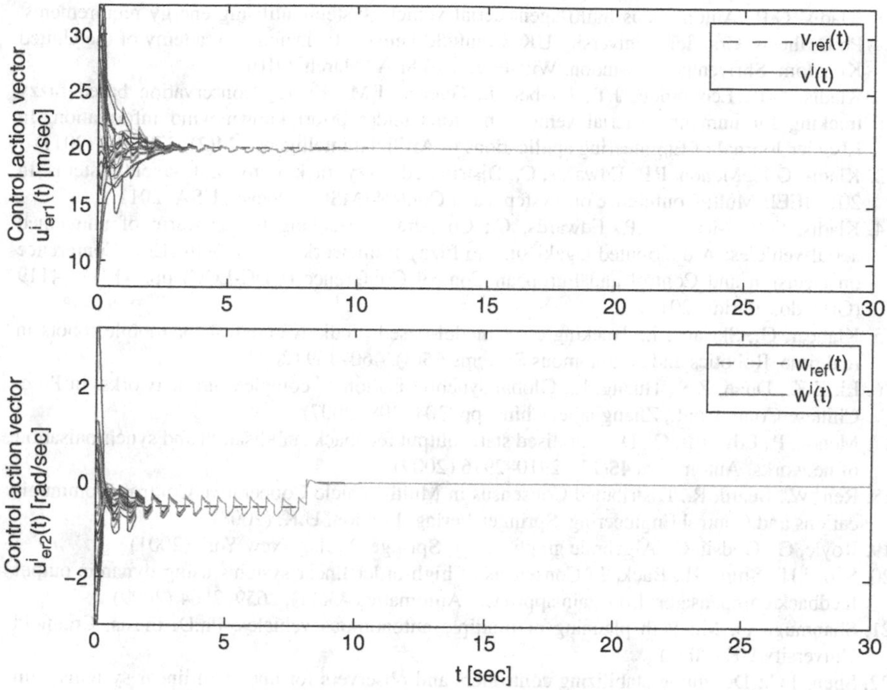

Fig. 9 Control action vector $u^i_{er}(t) = [v^i_{er}(t), w^i_{er}(t)]$

References

1. Boyd, S., El Ghaoui, L., Féron, E., Balakrishnan, V.: Linear matrix inequalities in system and control theory. Studies in Applied Mathematics, Philadelphia. (1994)
2. Cao, Y.Y., Frank, P.M.: Stability analysis and synthesis of nonlinear time-delay systems via takagi-sugeno fuzzy models. Fuzzy sets and Systems **124**(2), 213–229 (2001)
3. Chen, B., Liu, X., Lin, C., Liu, K.: Robust hinf control of takagi-sugeno fuzzy systems with state and input time delays. Fuzzy Sets and Systems **160**, 403–422 (2009)
4. Christofides, N.: Graph theory an algorithmic Approach. Acadamic Press INC. (London) LTD (1975)
5. Dubins, L.E.: On curves of minimal length with a constraint on average curvature and with prescribed initial and terminal positions and tangent. American Journal of Mathematics **79**, 497–516 (1957)
6. Fax, J.A., Murray, R.M.: Information flow and cooperative control of vehicle formations. IEEE Transactions on Automatic Control **49**, 1465–1476 (2004)
7. Gu, Y., Wang, S., Li, Q., Cheng, Z., Qian, J.: On delay-dependent stability and decay estimate for uncertain systems with time-varying delay. Automatica **34**(8), 1035–1039 (1998)
8. Hale, J.: Theory of Functional Differential Equations. Springer, New York (1977)
9. Horn, R.A., Johnson, C.R.: Topics in Matrix Analysis. Cambridge U.P., New York (1985)
10. Kawamoto, S., Tada, K., Ishigame, A., Taniguchi, T.: An approach to stability analysis of second order fuzzy systems. Proceedings of First IEEE International Conference on Fuzzy Systems **1**, 1427–1434 (1992)

11. Kladis, G.P.: Autonomous multi-agent aerial vehicle systems utilising energy requirements. Ph.D. thesis, Cranfield University UK, Cranfield University, Defence Academy of the United Kingdom, Shrivenham, Swindon, Wiltshire, SN6 8LA (March 2010)
12. Kladis, G.P., Economou, J.T., Lauber, J., Guerra, T.M.: Energy conservation based fuzzy tracking for unmanned aerial vehicle missions under priori known wind information. In Elsevier Journal of Engineering applications in Artificial intelligence **24(2)**, 278–294 (2011)
13. Kladis, G.P., Menon, P.P., Edwards, C.: Distributed fuzzy tracking for multi-agent systems. In 2011 IEEE Multi-Conference on Systems and Control (MSC) (Denver, USA, 2011.)
14. Kladis, G.P., Menon, P.P., Edwards, C.: Cooperative tracking for a swarm of unmanned aerial vehicles: A distributed tagaki-sugeno fuzzy framework design. 50th IEEE Conference on Decision and Control and European Control Conference (CDC-ECC) pp. 4114 – 4119 (Orlando, Florida, 2011)
15. Klančar, G., Skrjanc, I.: Tracking-error model-based predictive control for mobile robots in real time. Robotics and Autonomous Systems **55**(3), 460–469 (2007)
16. Li, Z.K., Duan, Z.S., Huang, L.: Global synchronization of complex lure networks. In Proc. Chinese Contr. Conf., Zhangjiajie, China, pp. 304–308 (2007)
17. Menon, P., Edwards, C.: Decentralised static output feedback stabilisation and synchronisation of networks. Automatica **45**(12), 2910–2916 (2009)
18. Ren, W., Beard, R.: Distributed Consensus in Multi-Vehicle Cooperative Control. Communications and Control Engineering. Springer-Verlag, London, U.K. (2008)
19. Royle, G., Godsil, C.: Algebraic graph theory. Springer Verlag, New York (2001)
20. Seo, J.H., Shim, H., Back, J.: Consensus of high-order linear systems using dynamic output feedback compensator: Low gain approach. Automatica **45**(11), 2659–2664 (2009)
21. Shanmugavel, M.: Path planning of multiple autonomous vehicles. Ph.D. thesis, Cranfield University UK (2007)
22. Shen, J.C.: Designing stabilizing controllers and observers for uncertain linear systems with time-varying delay. Automatica **33**(4), 331–333 (1997)
23. Sun, Y.G., Long, W.: Consensus problems in networks of agents with double-integrator dynamics and time-varying delays. Int. Journal Control **82**(10), 1937–1945 (2009)
24. Takagi, T., Sugeno, M.: Fuzzy identification of systems and its application to modeling and control. IEEE Trans. On System Man and Cybernetics **15**(1), 116–132 (1985)
25. Tanaka, K., Wang, H.O.: Fuzzy control systems design and analysis: A Linear Matrix Inequality Approach. 2001 John Wiley and Sons, Inc, New York (2001)
26. Tian, E., Yue, D., Zhang, Y.: Delay-dependent robust hinf control for t-s fuzzy system with interval time-varying delay. Fuzzy Sets and Systems **160**(12), 1708–1719 (2009)
27. Ting, C.H.: An observer-based approach to controlling time-delay chaotic systems via takagi-sugeno fuzzy model. Information Sciences **177**(20), 4314–4328 (2007)
28. Vinh, N.X.: Flight mechanics of high-performance aircraft, 5, vol. Cambridge Aerospace Series. Cambridge, United Kingdom: Cambridge University Press (1995)
29. Wang, H.O., Tanaka, K., Griffin, M.F.: Parallel distributed compensation of nonlinear systems by Takagi-Sugeno fuzzy model. In Proc. FUZZ-IEEE/IFES'95 pp. 531–538 (1995)
30. Yoneyama, J.: New delay-dependent approach to robust stability and stabilization for takagi-sugeno fuzzy time-delay systems. Fuzzy Sets and Systems **158**(20), 2225–2237 (2007)

About Model Complexity of 2-D Polynomial Discrete Systems: An Algebraic Approach

Stelios Kotsios and Dionyssios Lappas

Abstract By means of special operators and operations, the so-called D-operators and the star-product, a special algebraic description for Nonlinear Polynomial Discrete Systems in two dimensions is developed. By using this description we can check if these nonlinear systems are "similar" or "equivalent" with linear systems, in the sense that the evolution of both systems, under the same initial conditions, are related to each other. Different kinds of solutions to the problem seem to determine different degrees of complexity for the original nonlinear systems.

Introduction

Difference equations or Discrete systems of equations which are in use for the creation of Models in a variety of domains are, in principle, non-linear. On the other hand, most of the existing results refer upon linear systems and various linearization processes are in practice, not always successfully. The reason is that the initial systems possess complexities and due to this fact, basic characteristics do not inherit into linearization. It is therefore a necessity to rethink about linearization processes, their tools and degrees of acceptance for the obtained results. Mathematical Control Theory provides a unifying framework for posing and studying such problems [1, 4]. In this respect, we treat equations or systems known as non-linear discrete systems of polynomial type and deal with non-linearities by using mainly algebraic

S. Kotsios (✉)
Department of Economics, National and Kapodistrian University of Athens,
8 Pesmazoglou Street, 10559 Athens, Greece
e-mail: skotsios@econ.uoa.gr

D. Lappas
Department of Mathematics, National and Kapodistrian University of Athens,
15784, Athens, Greece
e-mail: dlappas@math.uoa.gr

N.J. Daras (ed.), *Applications of Mathematics and Informatics in Science and Engineering*, Springer Optimization and Its Applications 91, DOI 10.1007/978-3-319-04720-1_18,

tools based upon the so-called star-product (cf. section "Preliminaries"). The star-product corresponds to the composition of polynomial functions, in other words to the substitution of one polynomial into another. This star-product allows to describe the evolution of the system along naturally defined operations, the D-operator (cf. section "2D Polynomial Discrete Systems"). This operation is compatible with the cascade connection of one system with another. In a series of papers, problems of evolution and stability of those systems have been studied [3, 5, 8]. In the present note, inspired by similar problems in Control Theory [1, 7], we set down the problem of equivalence of two such systems, in the framework of D-operators, and we look for conditions in order to transform one system into a (sometimes given) equivalent one, with the same future evolution. We deal with the equation $\mathbf{F} * \mathbf{T} = \mathbf{T} * \mathbf{G}$ of D-operators, and we are looking for solutions $\mathbf{T}$, when $\mathbf{F}$ and $\mathbf{G}$ are given (cf. section "T-Similarity"). For a specific system $\mathbf{F}$ and when the given system $\mathbf{G}$ is a linear one, the problem of Model Complexity arises. It turns out that in this case a notion of complexity could be introduced, which realizes the intrinsic non-linearity of the system. The solution $\mathbf{T}$ may be a polynomial operator, a series of operators, a series of series, to be invertible or not and to converge or not. Each one of these situations determines a type of non-linearity complexity for the underlying Model (cf. section "Levels of Model Complexity"). Here are the contents of this work. In the beginning we give the preliminary notion of a D-operator and develop the algebraic tools which allow the transformation of the given equation in an algebraic-like object. After that, we deal with the main object of study, the 2D-Nonlinear Discrete-Polynomial Systems. Initially we define an equivalence relation among D-operations, which turns out to be the appropriate one to characterize the evolution of the underlying systems (Theorem 2). This relation is used to define the notion the $\mathbf{T}$-similarity (Definition 2), between two pairs of sequences and reduce this algebraically to corresponding D-operators (Theorem 2). The determination of the operator $\mathbf{T}$ in the equation $\mathbf{F} * \mathbf{T} = \mathbf{T} * \mathbf{G}$, requires a lot of machinery in order to solve the occurring linear-like systems. This is achieved in an algorithmic manner and in each stage of this process, a set of initial conditions should be chosen. Theorem 3.4 ensures that under middle restriction, for a given nonlinear discrete polynomial system the linear T-similarity problem accepts a series-solution. Along the same considerations, a table for the levels of Model Complexity is established. All the above situations are illustrated through some indicative arithmetic examples, which conclude this presentation. Exact proofs as well as applications to specific problems would be given in a forthcoming work [7].

Preliminaries

In this section we shall work with algebraic tools, they will be used later in order nonlinear polynomial discrete systems of dimension two to be described. The cornerstone of our approach is the so-called D-operator. It has been introduced in [5], and transforms a pair of sequences to a pair of sequences. In order to present these ideas in a comprehensive way we shall follow a constructive method, starting from simpler operators and proceeding gradually.

Consider a sequence $x(t), t \in Z$ with $x(t) = 0$, for $t < 0$. Let us further consider the $\boldsymbol{\delta}_{\mathbf{i}}$ operator, $\mathbf{i} = (i_1, i_2, \ldots, i_n)$ be a given vector of integers, named multi-index. This operator defines a new sequence as follows:

$$\boldsymbol{\delta}_{\mathbf{i}} x(t) = x(t - i_1)x(t - i_2)\cdots x(t - i_n)$$

If $\mathbf{i} = i$ is just a positive integer then $\boldsymbol{\delta}_i x(t) = x(t - i)$, which means that $\boldsymbol{\delta}_i$ coincides with the well-known shift operator. A special case is the operator $\boldsymbol{\delta}_0$, which leaves a sequence unchanged, i.e. $\boldsymbol{\delta}_0 x(t) = x(t)$. It is called the identity operator. For the sake of completeness we define by convention that $\boldsymbol{\delta}_e x(t) = 1$. Using this action of the $\boldsymbol{\delta}$-operators upon sequences, we can define an external operation among $\boldsymbol{\delta}$-operators, named addition, as follows: $(\boldsymbol{\delta}_{\mathbf{i}} + \boldsymbol{\delta}_{\mathbf{j}})x(t) = \boldsymbol{\delta}_{\mathbf{i}} x(t) + \boldsymbol{\delta}_{\mathbf{j}} x(t)$. An internal operation, named star-product, is defined as the composition of two sequences. Indeed, if $w(t) = \boldsymbol{\delta}_{\mathbf{i}} x(t)$, then $\boldsymbol{\delta}_{\mathbf{j}} * \boldsymbol{\delta}_{\mathbf{i}} x(t) = \boldsymbol{\delta}_{\mathbf{j}} w(t) = \boldsymbol{\delta}_{\mathbf{j}}(\boldsymbol{\delta}_{\mathbf{i}} x(t))$. It can be proved [6] that $\boldsymbol{\delta}_{\mathbf{k}} * (\boldsymbol{\delta}_{\mathbf{i}} + \boldsymbol{\delta}_{\mathbf{j}}) \neq \boldsymbol{\delta}_{\mathbf{k}} * \boldsymbol{\delta}_{\mathbf{i}} + \boldsymbol{\delta}_{\mathbf{k}} * \boldsymbol{\delta}_{\mathbf{i}}$. The latter relation indicates that the set $(\Delta, +, *)$ of the $\boldsymbol{\delta}$-operators, equipped with the operations of addition and the star-product, is not a ring. Expressions of the form $A = \sum_{n=0}^{w} \sum_{\mathbf{i} \in \mathbf{I_n}} a_{\mathbf{i}} \boldsymbol{\delta}_{\mathbf{i}}$ are called $\boldsymbol{\delta}$-polynomials, where by $\mathbf{I_n}$ we denote the set of multi-indexes with n elements. By convention $\mathbf{I_0} = \{\boldsymbol{\delta}_e\}$. The $\boldsymbol{\delta}$-polynomials also work as functions transforming sequences to sequences as follows: Let A be a $\boldsymbol{\delta}$-polynomial and $x(t)$ a sequence, then

$$Ax(t) = \sum_{n=0}^{w} \sum_{\mathbf{i} = (i_1, \ldots, i_n) \in \mathbf{I}_n} a_{\mathbf{i}} x(t - i_1)x(t - i_2)\cdots x(t - i_n)$$

The star-product between $\boldsymbol{\delta}$-polynomials corresponds, as before, to the composition, in other words to the substitution of one polynomial into another. Indeed, if A, B are two $\boldsymbol{\delta}$-polynomials, then $A * By(t) = A \circ By(t) = A(B(y(t)))$. An addition of $\boldsymbol{\delta}$-polynomials is defined as $(A + B)x(t) = Ax(t) + Bx(t)$ and the following property holds: $C * [A + B] \neq C * A + C * B$. All the above are applied straightforward in the case of $\boldsymbol{\delta}$-series, too, which is nothing but polynomials with an infinite number of terms. We can also extend the whole methodology so that to act not to a single sequence but to a pair of sequences. We can achieve that by means of the $\boldsymbol{\delta\epsilon}$-operator. Indeed, let $\boldsymbol{\delta}_{\mathbf{i}} \boldsymbol{\epsilon}_{\mathbf{j}}$ be a $\boldsymbol{\delta\epsilon}$-operator, $\mathbf{i} = (i_1, i_2, \ldots, i_n)$, $\mathbf{j} = (j_1, j_2, \ldots, j_m)$ two multi-indexes. This operator works as follows:

$$\boldsymbol{\delta}_{\mathbf{i}} \boldsymbol{\epsilon}_{\mathbf{j}} [x(t), y(t)] = x(t - i_1)\cdots x(t - i_n) y(t - j_1)\cdots y(t - j_m)$$

Therefore, the $\boldsymbol{\delta}$-part of the $\boldsymbol{\delta\epsilon}$-operator acts exclusive on the first sequence and the $\boldsymbol{\epsilon}$-part on the second. If either $\mathbf{j} = \{e\}$ or $\mathbf{i} = \{e\}$ then $\boldsymbol{\delta}_{\mathbf{i}} \boldsymbol{\epsilon}_e [x(t), y(t)] = \boldsymbol{\delta}_{\mathbf{i}} x(t)$, $\boldsymbol{\delta}_e \boldsymbol{\epsilon}_{\mathbf{j}} [x(t), y(t)] = \boldsymbol{\epsilon}_{\mathbf{j}} x(t)$. We can define the addition as follows:

$$(\boldsymbol{\delta}_{\mathbf{i}} \boldsymbol{\epsilon}_{\mathbf{j}} + \boldsymbol{\delta}_{\mathbf{i'}} \boldsymbol{\epsilon}_{\mathbf{j'}})[x(t), y(t)] = \boldsymbol{\delta}_{\mathbf{i}} \boldsymbol{\epsilon}_{\mathbf{j}} [x(t), y(t)] + \boldsymbol{\delta}_{\mathbf{i'}} \boldsymbol{\epsilon}_{\mathbf{j'}} [x(t), y(t)]$$

Let $A = \sum_{n=0}^{\nu} \sum_{m=0}^{\mu} \sum_{(\mathbf{i},\mathbf{j})\in \mathbf{I}_n \times \mathbf{J}_m} c_{\mathbf{ij}} \delta_{\mathbf{i}} \epsilon_{\mathbf{j}}$ be a $\delta\epsilon$-polynomial. This polynomial acts on a pair of sequences as follows:

$$A[(x(t), y(t)] = \sum_{n=0}^{\nu} \sum_{m=0}^{\mu} \sum_{(\mathbf{i},\mathbf{j})\in \mathbf{I}_n \times \mathbf{J}_m} c_{\mathbf{ij}} x(t-i_1)\cdots x(t-i_n) y(t-j_1)\cdots y(t-j_m)$$

If A is a $\delta\epsilon$-series, then $A[x(t), y(t)]$ is a Volterra series, containing products among delays of $x(t)$ and $y(t)$. In the case of linear polynomials (or linear series), $A[x(t), y(t)]$ is a linear polynomial (or a linear series) of delays of $x(t)$ and $y(t)$. The star-product among $\delta\epsilon$-operators (or $\delta\epsilon$-polynomials or $\delta\epsilon$-series) corresponds to the composition among maps. Indeed, let B, C, A be $\delta\epsilon$-polynomials, if we substitute the polynomial B into the δ-part of A and C into the ϵ-part of A, we get a $\delta\epsilon$-polynomial which corresponds to the composition, $A \circ [B, C]$ and is called the star-product of the polynomials A, B, C and is denoted by $A * [B, C]$.

We present now the D-operators. They are nothing but a pair of $\delta\epsilon$-polynomials, in other words:

$$D = \begin{bmatrix} A \\ B \end{bmatrix} = \begin{bmatrix} \sum_{(\mathbf{i},\mathbf{j})\in \mathbf{I_a}\times \mathbf{J_a}} a_{\mathbf{ij}} \delta_{\mathbf{i}} \epsilon_{\mathbf{j}} \\ \sum_{(\mathbf{i},\mathbf{j})\in \mathbf{I_b}\times \mathbf{J_b}} b_{\mathbf{ij}} \delta_{\mathbf{i}} \epsilon_{\mathbf{j}} \end{bmatrix}$$

If the above $\delta\epsilon$-polynomials are linear, then we speak about a linear D-operator. If instead of the $\delta\epsilon$-polynomials A and B we have the $\delta\epsilon$-series A and B, then the D-operator is a called a D-series.

Definition 1. Let G and F be two D-operators:

$$G = \begin{bmatrix} \sum_{(\mathbf{i},\mathbf{j})\in \mathbf{I}_{\mathbf{g},1}\times J_{g,1}} g^{(1)}_{\mathbf{ij}} \delta_{\mathbf{i}} \epsilon_{\mathbf{j}} \\ \sum_{(\mathbf{i},\mathbf{j})\in \mathbf{I}_{\mathbf{g},2}\times J_{g,2}} g^{(2)}_{\mathbf{ij}} \delta_{\mathbf{i}} \epsilon_{\mathbf{j}} \end{bmatrix}, \quad F = \begin{bmatrix} \sum_{(\mathbf{i},\mathbf{j})\in \mathbf{I}_{\mathbf{f},1}\times \mathbf{J}_{\mathbf{f},1}} f^{(1)}_{\mathbf{ij}} \delta_{\mathbf{i}} \epsilon_{\mathbf{j}} \\ \sum_{(\mathbf{i},\mathbf{j})\in \mathbf{I}_{\mathbf{f},2}\times \mathbf{J}_{\mathbf{f},2}} f^{(2)}_{\mathbf{ij}} \delta_{\mathbf{i}} \epsilon_{\mathbf{j}} \end{bmatrix}$$

We say that $G = F$ if and only if $I_{g,k} = I_{f,k}$, $J_{g,k} = J_{f,k}$, $g^{(k)}_{ij} = f^{(k)}_{ij}$, $k = 1, 2$. In other words they have the same sets of multi-indexes and the same coefficients.

The next operations is a generalization of the foregoing definitions.

Definition 2. Let us have two D-operators:

$$D_1 = \begin{bmatrix} A_1 \\ B_1 \end{bmatrix}, \quad D_2 = \begin{bmatrix} A_2 \\ B_2 \end{bmatrix}$$

their dot-product and star-product are defined as:

$$D_1 \cdot D_2 = \begin{bmatrix} A_1 \cdot A_2 \\ B_1 \cdot B_2 \end{bmatrix}, \quad D_1 * D_2 = \begin{bmatrix} A_1 * [A_2, B_2] \\ B_1 * [A_2, B_2] \end{bmatrix}$$

We can extend all the above to the case of $\delta\epsilon$ series, in a similar way.

2D Polynomial Discrete Systems

In this section we present how we can use the D-operators in order to describe non-linear polynomial discrete systems. Let us start with polynomial discrete systems involving only one sequence. They have the form:

$$x(t) = \sum_{k=1}^{\theta} \sum_{\substack{\mathbf{i}\in\mathbf{I}_k \\ \mathbf{i}=(i_1,i_2,\ldots,i_k)}} c_{\mathbf{i}} x(t-1-i_1)x(t-1-i_2)\cdots x(t-1-i_k) \tag{1}$$

with $c_{\mathbf{i}} \in \mathbf{R}$ and $\mathbf{I}_k$, a finite set of multi-indexes of dimension k. We say that we assign to this system a set of initial conditions $I = \{\gamma_0, \gamma_1, \ldots, \gamma_{s-1}\}$ if and only if $x(0) = \gamma_0$, $x(1) = \gamma_1$, ..., $x(s-1) = \gamma_{s-1}$, where s is the maximum delay appeared in (1). Starting from these initial conditions and using (1), we can calculate all the future evolution of the system, that is the quantities $x(s), x(s+1), x(s+2), \ldots$ Now, by using the $\boldsymbol{\delta}$-polynomial $A = \sum_{k=1}^{\theta} \sum_{\substack{\mathbf{i}\in\mathbf{I}_k \\ \mathbf{i}=(i_1,i_2,\ldots,i_k)}} c_{\mathbf{i}}\boldsymbol{\delta}_{\mathbf{i}}$, we can re-write the above system, shortly as $x(n) = Ax(n-1)$. By means of this notation the evolution of the system is described through the star-product. Indeed, it can be proved [3] that:

Theorem 1. *The evolution of the system 1 can be calculated by the formula:* $x(t) = \underbrace{A * A * \cdots * A}_{n-times} x(t-n) = A^n x(t-n)$, $t = s, s+1, s+2, \ldots$, *under the assumption that the same set of initial conditions* I*, has been used.*

Let us come now to 2D Polynomial Discrete Systems, that is systems transforming a pair of sequences to a pair of sequences in a nonlinear polynomial way. Let us have the sequences $x_1(t), x_2(t)$ and the system:

$$x_1(t) = \sum_{\alpha=1}^{\alpha'} \sum_{\beta=1}^{\beta'} \sum_{\substack{(\mathbf{i},\mathbf{j})\in\mathbf{I}_\alpha\times\mathbf{J}_\beta \\ \mathbf{i}=(i_1,\ldots,i_\tau) \\ \mathbf{j}=(j_1,\ldots,j_\xi)}} c^{(1)}_{\mathbf{ij}} x_1(t-i_1)\cdots x_1(t-i_\tau)x_2(t-j_1)\cdots x_2(t-j_\xi)$$

$$x_2(t) = \sum_{\alpha=1}^{\alpha''} \sum_{\beta=1}^{\beta''} \sum_{\substack{(\mathbf{i}',\mathbf{j}')\in\mathbf{I}'_\alpha\times\mathbf{J}'_\beta \\ \mathbf{i}'=(i'_1,\ldots,i'_{\tau'}) \\ \mathbf{j}'=(j'_1,\ldots,j'_{\xi'})}} c^{(2)}_{\mathbf{i}'\mathbf{j}'} x_1(t-i'_1)\cdots x_1(t-i'_{\tau'})x_2(t-j'_1)\cdots x_2(t-j'_{\xi'}) \tag{2}$$

where $\mathbf{I}_a, \mathbf{I}_\beta, \mathbf{J}'_a, \mathbf{J}'_\beta$ sets of multi-indexes with α and β elements respectively. We say that we assign to this system, the following sets of initial values:

$$I_1 = \{a_0, a_1, \ldots, a_{\rho-1}\} \quad , \quad I_2 = \{b_0, b_1, \ldots, b_{\sigma-1}\}$$

if $x_1(0) = a_0, x_1(1) = a_1, \ldots, x_1(\rho - 1) = a_{\rho-1}$ and $x_2(0) = b_0, x_2(1) = b_1, \ldots, x_2(\sigma - 1) = b_{\sigma-1}$, where ρ and σ are the maximum delays of the $x_1(t)$ and $x_2(t)$ sequences correspondingly.

By means of the D-operators we can rewrite (2) as follows:

$$\mathbf{x}(t) = \mathbf{G}\mathbf{x}(t-1), \quad \mathbf{x}(t) = \begin{bmatrix} x_1(t) \\ x_2(t) \end{bmatrix}, \quad \mathbf{G} = \begin{bmatrix} G_1 \\ G_2 \end{bmatrix}$$

where G_1, G_2 are proper $\delta\epsilon$-polynomials and $\mathbf{G}$ the corresponding D-operator.

The next definition ensures that two systems have the same dynamic behaviour.

Definition 3. We say that two systems $\mathbf{x}(t) = \mathbf{G}\mathbf{x}(t-1)$ and $\mathbf{z}(t) = \mathbf{F}\mathbf{z}(t-1)$, $\mathbf{F}, \mathbf{G}$, D-operators, are equivalent, if $\mathbf{x}(t) = \mathbf{z}(t)$, $t = 1, 2, \ldots$, whenever they operate under identical initial conditions.

It is trivial to be seen that this notion is an equivalence relation. The next theorem combines equivalence of dynamical systems with equality of D-operators.

Theorem 2. *[7] Let us have the systems* $\mathbf{x}(t) = \mathbf{G}\mathbf{x}(t-1)$ *and* $\mathbf{y}(t) = \mathbf{F}\mathbf{y}(t-1)$. *These systems are equivalent if and only if the D-operators* $\mathbf{G}$ *and* $\mathbf{F}$ *are equal.*

Finally, we can obtain a result similar to that of theorem 1, in the case of 2D Polynomial Discrete Systems. Indeed, the time evolution of the system (2) can be given by the formula:

$$\begin{bmatrix} x_1(t) \\ x_2(t) \end{bmatrix} = \underbrace{D * D * \cdots * D}_{n-times}\,\mathbf{x}(t-n) = D^n\mathbf{x}(t-n), \quad t = s, s+1, s+2, \ldots$$

T-Similarity

In this section we establish conditions which guarantee that the output (evolution) of a system is identically equal with the output of another system, under the same initial conditions through a proper change of coordinations procedure, obtained by means of the star-product and D-series [7, 8]. This will help us later to classify the nonlinear systems with respect to this property.

Let us present now the relevant definitions.

Definition 1. A D-series $\mathbf{T}$ is called invertible if we can find another D-series $\mathbf{T}'$, such that $\mathbf{T}' * \mathbf{T} = \begin{bmatrix} \delta_0 \\ \epsilon_0 \end{bmatrix}$

Definition 2. Two pairs of sequences $\mathbf{x}(t) = \begin{bmatrix} x_1(t) \\ x_2(t) \end{bmatrix}$ and $\mathbf{y}(t) = \begin{bmatrix} y_1(t) \\ y_2(t) \end{bmatrix}$, are called T-Similar, if there exists a nonsingular (invertible) series $\mathbf{T} = \begin{bmatrix} T_1 \\ T_2 \end{bmatrix}$, such that $\mathbf{y}(t) = \mathbf{T}\mathbf{x}(t)$.

The meaning of the above definition is that by means of T we can go from $\mathbf{x}(t)$ to $\mathbf{y}(t)$ and vice-versa. Let us now see how we can extend this notion in order for D-operators to be involved.

Definition 3. Let $\mathbf{G} = \begin{bmatrix} G_1 \\ G_2 \end{bmatrix}$, $\mathbf{F} = \begin{bmatrix} F_1 \\ F_2 \end{bmatrix}$ be two D-operators. They are called T-similar, if we can find a series $\mathbf{T} = \begin{bmatrix} T_1 \\ T_2 \end{bmatrix}$, such that: $F_1 * [T_1, T_2] = T_1 * [G_1, G_2]$, $F_2 * [T_1, T_2] = T_2 * [G_1, G_2]$ or shortly $\mathbf{F} * \mathbf{T} = \mathbf{T} * \mathbf{G}$.

Theorem 1. **T**-*similarity is an equivalence relation among D-operators.*

If **F** and **G** are **T**-similar we write $\mathbf{F} \overset{\mathbf{T}}{\sim} \mathbf{G}$. Equivalent classes are denoted by [**F**].

Theorem 2. *[7] Let* $\mathbf{x}(t) = \mathbf{G}\mathbf{x}(t-1)$, $\mathbf{y}(t) = \mathbf{F}\mathbf{y}(t-1)$ *be two 2D Polynomial Discrete Systems. The sequences* $\mathbf{x}(t), \mathbf{y}(t)$ *are* **T**-*similar, if and only if the D-operators* $\mathbf{G}, \mathbf{F}$ *are* **T**-*similar.*

The most interesting situation is when the D-operator **F**, is a linear one. In this case we speak for the linear T-similarity. In other words: Let us suppose that we have the given nonlinear D-operator **G** and the linear one **L**. We want to find a D-series **T**, such that $\mathbf{L} * \mathbf{T} = \mathbf{T} * \mathbf{G}$. Now two fundamental questions arise: First, what is the construction of T? Is it a simple series (and hence its convergence can be checked by classical techniques) or series of series (and thus its convergence cannot be easily checked). Second, how can we obtain the T-series? We shall establish two theorems dealing with the first question, that of the T-series construction.

- Before we proceed with the calculations we need some terminology.

$$L_\theta = \sum_{a=0}^{1} L_\theta^{(a,1-a)} \quad , \quad L_\theta^{(a,1-a)} = \sum_{i=0}^{\nu} l_{\theta,i}^{(a,1-a)} \delta_i^a \epsilon_i^{1-a} \quad , \quad \theta = 1,2$$

$$T_\theta = \sum_{a=0}^{\infty} \sum_{b=0}^{\infty} T_\theta^{(a,b)} \quad , \quad T_\theta^{(a,b)} = \sum_{(\mathbf{i},\mathbf{j}) \in \mathbf{I} \times \mathbf{J}} t_{\theta,(\mathbf{i},\mathbf{j})}^{(a,b)} \boldsymbol{\delta}_{\mathbf{i}} \boldsymbol{\epsilon}_{\mathbf{j}} \quad , \quad \theta = 1,2$$

$$G_\theta = \sum_{a=0}^{a'} \sum_{b=0}^{b'} G_\theta^{(a,b)} \quad , \quad G_\theta^{(a,b)} = \sum_{(\mathbf{i},\mathbf{j}) \in \mathbf{I} \times \mathbf{J}} g_{\theta,(\mathbf{i},\mathbf{j})}^{(a,b)} \boldsymbol{\delta}_{\mathbf{i}} \boldsymbol{\epsilon}_{\mathbf{j}} \quad , \quad \theta = 1,2$$

- By Q_0 we denote the matrix:

$$Q_0 = \begin{pmatrix} L_1^{(1,0)} - G_1^{(1,0)} & L_1^{(0,1)} & -G_2^{(1,0)} & 0 \\ L_2^{(1,0)} & L_2^{(0,1)} - G_1^{(1,0)} & 0 & -G_2^{(1,0)} \\ -G_1^{(0,1)} & 0 & L_1^{(1,0)} - G_2^{(0,1)} & L_1^{(0,1)} \\ 0 & -G_1^{(0,1)} & L_2^{(1,0)} & L_2^{(0,1} - G_2^{(0,1)} \end{pmatrix}$$

- By $\mathcal{A}$ we denote the matrix:

$$\mathcal{A} = \begin{bmatrix} l_{1,0}^{(1,0)} - g_{1,0}^{(1,0)} & l_{1,0}^{(0,1)} & -g_{2,0}^{(1,0)} & 0 \\ l_{1,0}^{(2,0)} & l_{2,0}^{(0,1)} - g_{1,0}^{(1,0)} & 0 & -g_{2,0}^{(1,0)} \\ -g_{1,0}^{(0,1)} & 0 & l_{1,0}^{(1,0)} - g_{2,0}^{(0,1)} & l_{1,0}^{(0,1)} \\ 0 & -g_{1,0}^{(0,1)} & l_{2,0}^{(1,0)} & l_{2,0}^{(0,1)} - g_{2,0}^{(0,1)} \\ l_{1,1}^{(1,0)} - g_{1,1}^{(1,0)} & l_{1,1}^{(0,1)} & -g_{2,1}^{(1,0)} & 0 \\ l_{1,1}^{(2,0)} & l_{2,1}^{(0,1)} - g_{1,1}^{(1,0)} & 0 & -g_{2,1}^{(1,0)} \\ -g_{1,1}^{(0,1)} & 0 & l_{1,1}^{(1,0)} - g_{2,1}^{(0,1)} & l_{1,1}^{(0,1)} \\ 0 & -g_{1,1}^{(0,1)} & l_{2,1}^{(1,0)} & l_{2,1}^{(0,1)} - g_{2,1}^{(0,1)} \\ & \vdots & & \vdots \\ l_{1,\nu}^{(1,0)} - g_{1,\nu}^{(1,0)} & l_{1,\nu}^{(0,1)} & -g_{2,\nu}^{(1,0)} & 0 \\ l_{1,\nu}^{(2,0)} & l_{2,\nu}^{(0,1)} - g_{1,\nu}^{(1,0)} & 0 & -g_{2,\nu}^{(1,0)} \\ -g_{1,\nu}^{(0,1)} & 0 & l_{1,\nu}^{(1,0)} - g_{2,\nu}^{(0,1)} & l_{1,\nu}^{(0,1)} \\ 0 & -g_{1,\nu}^{(0,1)} & l_{2,\nu}^{(1,0)} & l_{2,\nu}^{(0,1)} - g_{2,\nu}^{(0,1)} \end{bmatrix}$$

- The pair of equations:

$$L_i^{(1,0)} * T_1^{(n,m)} + L_i^{(0,1)} * T_2^{(n,m)} - T_i^{(n,m)} * [G_1^{(1,0)}, G_2^{(0,1)}] - T_i^{(m,n)} * [G_1^{(0,1)}, G_2^{(1,0)}] -$$

$$- \sum_{\substack{a+b=k, a\neq n, b\neq m \\ a(x_1+y_1)+b(x_2+y_2)=k \\ a,b,x_1,y_1,x_2,y_2 \in \mathbf{N}}} T_i^{(a,b)} * [G_1^{(x_1,y_1)}, G_2^{(x_2,y_2)}] =$$

$$= \sum_{\substack{a+b<k \\ a(x_1+y_1)+b(x_2+y_2)=k \\ a,b,x_1,y_1,x_2,y_2 \in \mathbf{N}}} T_i^{(a,b)} * [G_1^{(x_1,y_1)}, G_2^{(x_2,y_2)}], \quad i = 1,2$$

with the coefficients of the $T_i^{(n,m)}$, $n + m = k$ as unknowns, is called the basic nonlinear k-degree system.

- The matrix of the coefficients of the term $\delta_{\mathbf{i}}\epsilon_{\mathbf{j}}$, which arises from the left hand part of the above equation, is denoted by $C_{\mathbf{i},\mathbf{j}}$.

- The matrix of coefficients of the above system is denoted by Q_k. The corresponding augmented matrix is denoted by $Q_k^*(T)$, where we use this notation to indicate the dependence from the polynomials $T_i^{(a,b)}, a+b<k$.
- The set of the solutions of the 1-degree system is denoted by Γ.
- The set $\mathcal{S}$ is defined as:

$$\mathcal{S} = \{T \in \Gamma : rank(Q_k) = rank(Q_k^*(T)), k = 1, 2, 3, \ldots\}$$

We present now the main theorem:

Theorem 3. *[7] Let L be a given linear two dimension discrete system and* **G** *a polynomial one. Let* **T** *be the series that solves the T-similarity problem, i.e.* $\mathbf{L} * \mathbf{T} = \mathbf{T} * \mathbf{G}$. *Then,*

(i) If $|Q_0| = 0$ *and* $\mathcal{S} \neq \emptyset$ *then the* **T***-series is a simple series.*
(ii) If $rank(\mathcal{A}) < 4$ *and* $det(C_{\mathbf{i},\mathbf{j}}) \neq 0$ *for every* $\mathbf{i}, \mathbf{j}$, *then the* **T***-series is a series of series.*

Let us pass now to the second problem that of calculating the different parts of the series **T**. To achieve that we use the next procedure:

- By solving the system:

$$L_1^{(1,0)} * T_1^{(1,0)} + L_1^{(0,1)} * T_2^{(1,0)} = T_1^{(1,0)} * G_1^{(1,0)} + T_1^{(0,1)} * G_2^{(1,0)}$$
$$L_1^{(1,0)} * T_1^{(0,1)} + L_1^{(0,1)} * T_2^{(0,1)} = T_1^{(1,0)} * G_1^{(0,1)} + T_1^{(0,1)} * G_2^{(0,1)} \quad (3)$$

$$L_2^{(1,0)} * T_1^{(1,0)} + L_2^{(0,1)} * T_2^{(1,0)} = T_2^{(1,0)} * G_1^{(1,0)} + T_2^{(0,1)} * G_2^{(1,0)}$$
$$L_2^{(1,0)} * T_1^{(0,1)} + L_2^{(0,1)} * T_2^{(0,1)} = T_2^{(1,0)} * G_1^{(0,1)} + T_2^{(0,1)} * G_2^{(0,1)} \quad (4)$$

  we get the linear parts of the requested series. Since we have to do with a homogeneous system the relation $|Q_0| = 0$ guarantees that we get an infinite number of polynomial solutions ($T_i^{(1,0)}, T_j^{(0,1)}$ are polynomials). Otherwise a series solution is obtained ($T_i^{(1,0)}, T_j^{(0,1)}$ are series).
- Now, we go to the quadratic part. It consists from the next equations:

$$L_1^{(1,0)} * T_1^{(2,0)} + L_1^{(0,1)} * T_2^{(2,0)} = T_1^{(1,0)} * G_1^{(2,0)} + T_1^{(0,1)} * G_2^{(2,0)} +$$
$$+T_1^{(2,0)} * G_1^{(1,0)} + T_1^{(0,2)} * G_2^{(1,0)} + T_1^{(1,1)} * [G_1^{(1,0)}, G_2^{(1,0)}] \quad (5)$$

$$L_2^{(1,0)} * T_1^{(2,0)} + L_2^{(0,1)} * T_2^{(2,0)} = T_2^{(1,0)} * G_1^{(2,0)} + T_2^{(0,1)} * G_2^{(2,0)} +$$
$$+T_2^{(2,0)} * G_1^{(1,0)} + T_2^{(0,2)} * G_2^{(1,0)} + T_2^{(1,1)} * [G_1^{(1,0)}, G_2^{(1,0)}] \quad (6)$$

$$L_1^{(1,0)} * T_1^{(0,2)} + L_1^{(0,1)} * T_2^{(0,2)} = T_1^{(1,0)} * G_1^{(0,2)} + T_1^{(0,1)} * G_2^{(0,2)} + \\ + T_1^{(2,0)} * G_1^{(0,1)} + T_1^{(0,2)} * G_2^{(0,1)} + T_1^{(1,1)} * [G_1^{(0,1)}, G_2^{(0,1)}] \quad (7)$$

$$L_2^{(1,0)} * T_1^{(0,2)} + L_2^{(0,1)} * T_2^{(0,2)} = T_2^{(1,0)} * G_1^{(0,2)} + T_2^{(0,1)} * G_2^{(0,2)} + \\ + T_2^{(2,0)} * G_1^{(0,1)} + T_2^{(0,2)} * G_2^{(0,1)} + T_2^{(1,1)} * [G_1^{(0,1)}, G_2^{(0,1)}] \quad (8)$$

$$L_1^{(0,1)} * T_1^{(1,1)} + L_1^{(0,1)} * T_2^{(1,1)} = T_1^{(0,1)} * G_1^{(1,1)} + T_1^{(0,1)} * G_2^{(1,1)} + \\ + T_1^{(1,1)} * [G_1^{(0,1)}, G_2^{(1,0)}] + T_1^{(1,1)} * [G_1^{(1,0)}, G_2^{(0,1)}] \quad (9)$$

$$L_2^{(0,1)} * T_1^{(1,1)} + L_2^{(0,1)} * T_2^{(1,1)} = T_2^{(0,1)} * G_1^{(1,1)} + T_2^{(0,1)} * G_2^{(1,1)} + \\ + T_2^{(1,1)} * [G_1^{(0,1)}, G_2^{(1,0)}] + T_2^{(1,1)} * [G_1^{(1,0)}, G_2^{(0,1)}] \quad (10)$$

Relations (5), (6) arise by comparing the $\boldsymbol{\delta}_i\boldsymbol{\delta}_j$ terms, (7), (8) by comparing the $\boldsymbol{\epsilon}_i\boldsymbol{\epsilon}_j$ terms and (9), (10) the $\boldsymbol{\delta}_i\boldsymbol{\epsilon}_j$ terms. Substituting the solutions we have already found from the linear part we get the quadratic quantities $T_i^{(2,0)}, T_j^{(0,2)}, T_k^{(1,1)}$. We repeat the procedure for the cubic terms and so on. This method will finally endow us with the desired series **T**.

An interesting result, connected with the above iteration, is the next corollary:

Corollary 1. *[7] If the linear equations, i.e. (3),(4) accept a series as solution, then the* **T** *is a series of series.*

Levels of Model Complexity

Complex systems appears in many fields of contemporary science, and different communities have different aspects about complexity and how they ranked it [1,2]. In this section we shall try to approach this issue for 2D Polynomial Discrete Systems, using the mathematical tools developed previously. Specifically, we have described a procedure for checking the equivalence of a nonlinear discrete system with a linear one. This was achieved via a D-series, named **T**. The construction of **T** determines the kind of the model complexity or how "hard" the nonlinearity is. If, for instance, **T** converges, then we speak for a "light" complexity, otherwise for a strong one. If **T** is a simple series or consists from an infinite sum of series (series of series), this will influence the kind of complexity since checking convergence in the latter case is a very difficult task. The nature of **L** also plays an important role. If, for instance, it is stable, then the level of complexity is less than the level of complexity which corresponds to an unstable **L**. We summarize the different cases of complexity degrees in the next table:

T-Series	Complexity degree	
	L Stable	L Unstable
A polynomial	0	0+
An invertible, convergence, simple series	1	1+
A convergence simple series	1.5	1.5+
A simple series	2	2+
An invertible, convergence, series of series	3	3+
A convergence series of series	3.5	3.5+
A series of series	4	4+

Examples

Example 1. Let us have the linear system:

$$x(t) = x(t-1) + 2x(t-2) + \frac{1}{2}y(t-1)$$

$$y(t) = \frac{7}{2}x(t-1) - 2y(t-1) + 2y(t-2)$$

We want to see how this will be equivalent to another linear one. This is just to understand the procedures and to see how our approach fits with well-known cases. The linear system "target", will be:

$$u(t) = -\frac{3}{2}u(t-1) + 2u(t-2) - 3v(t-1)$$

$$v(t) = -u(t-1) + \frac{1}{2}v(t-1) + 2v(t-2)$$

Using the D-operators, we get the next descriptions:

$$\begin{bmatrix} x(t) \\ y(t) \end{bmatrix} = \begin{bmatrix} \delta_0 + 2\delta_1 + \frac{1}{2}\epsilon_0 \\ \frac{7}{2}\delta_0 - 2\epsilon_0 + 2\epsilon_1 \end{bmatrix} \begin{bmatrix} x(t-1) \\ y(t-1) \end{bmatrix}$$

$$\begin{bmatrix} u(t) \\ v(t) \end{bmatrix} = \begin{bmatrix} -\frac{3}{2}\delta_0 + 2\delta_1 - 3\epsilon_0 \\ -\delta_0 + \frac{1}{2}\epsilon_0 + 2\epsilon_1 \end{bmatrix} \begin{bmatrix} u(t-1) \\ v(t-1) \end{bmatrix}$$

and shortly $\mathbf{x}(t) = \mathbf{G}\mathbf{x}(t-1)$, $\hat{\mathbf{x}}(t) = \mathbf{L}\hat{\mathbf{x}}(t-1)$. We want to find series T_1, T_2, such that the following equations hold:

$$\mathbf{L} * \mathbf{T} = \mathbf{T} * \mathbf{G} \Rightarrow \begin{cases} L_1 * [T_1, T_2] = T_1 * [G_1, G_2] \\ L_2 * [T_1, T_2] = T_2 * [G_1, G_2] \end{cases}$$

For the sake of the computation we arbitrarily set: $T_1 = w_{1,0}\delta_0 + w_{1,1}\delta_1 + h_{1,0}\epsilon_0 + h_{1,1}\epsilon_1$, $T_2 = w_{2,0}\delta_0 + w_{2,1}\delta_1 + h_{2,0}\epsilon_0 + h_{2,1}\epsilon_1$; we could of course, take any other number of terms for the series T_1, T_2. By equating the coefficients and solving the corresponding system of equations we get:

$$w_{1,0} = h_{1,0} - 6h_{2,0} \quad , \quad w_{1,1} = h_{1,1} - 6h_{2,1}$$

$$w_{2,0} = -2h_{1,0} + 5h_{2,0} \quad , \quad w_{2,1} = -2h_{1,1} + 5h_{2,1}$$

and thus a transformation which solves the problem is:

$$T_1 = (h_{1,0} - 6h_{2,0})\delta_0 + (h_{1,1} - 6h_{2,1})\delta_1 + h_{1,0}\epsilon_0 + h_{1,1}\epsilon_1$$

$$T_2 = (-2h_{1,0} + 5h_{2,0})\delta_0 + (-2h_{1,1} + 5h_{2,1})\delta_1 + h_{2,0}\epsilon_0 + h_{2,1}\epsilon_1$$

with $h_{ij} \in \mathbf{R}$. Since the **T**-similarity problem in this case accepts a polynomial solution, and **L** is unstable, we say that the original linear system has a complexity degree equal to 0+. If we could solve the problem with a stable **L**, then complexity degree would be equal to 0.

Example 2. Let us consider now the nonlinear system:

$$x(t+1) = x(t) + y(t) - x^2(t)$$

$$y(t+1) = x(t)$$

we want to examine if it can be equivalent with the next linear system (the "target") and thus to find its complexity degree.

$$z(n+1) = z(n) - z(n-1) + w(n) + \frac{1}{2}(1 - \sqrt{5})w(n-1)$$

$$w(n+1) = z(n) + \frac{1}{2}(1 + \sqrt{5})z(n-1) + w(n-1)$$

Using the D-operators, we get the next descriptions: $\mathbf{x}(t+1) = \mathbf{G}\mathbf{x}(t)$, $\hat{\mathbf{x}}(t+1) = \mathbf{L}\hat{\mathbf{x}}(t)$ where:

$$\mathbf{G} = \begin{bmatrix} \delta_0 + \epsilon_0 - \delta_0^2 \\ \delta_0 \end{bmatrix} \quad , \quad \mathbf{L} = \begin{bmatrix} \delta_0 - \delta_1 + \epsilon_0 + \frac{1}{2}(1 - \sqrt{5})\epsilon_1 \\ \delta_0 + \frac{1}{2}(1 + \sqrt{5})\delta_1 + \epsilon_1 \end{bmatrix}$$

First of all we see that $|Q_0| \neq 0$ and thus the problem accepts a simple series as a solution. This means that it will be of complexity degree either 1 or 3. To calculate the series T_1, T_2, such that the following equations hold: $\mathbf{L} * \mathbf{T} = \mathbf{T} * \mathbf{G}$ we follow the procedure of the previous section and we take:

$$\mathbf{T}_1 = (\Gamma + A)\delta_0 + \left(-\frac{1}{2}(1+\sqrt{5})\Gamma + \Delta - \frac{1}{2}(1+\sqrt{5})A + B\right)\delta_1 + A\epsilon_0$$
$$+\left(\Gamma - \frac{1}{2}(1+\sqrt{5})A + B\right)\epsilon_1 + \frac{1}{2}A\delta_0^2 + \frac{1}{2}\Gamma\epsilon_0^2 + (A+\Gamma)\delta_0\epsilon_0$$
$$-\frac{1}{6}A\delta_0^3 - \frac{1}{6}\Gamma\epsilon_0^3 + \frac{1}{2}(A+\Gamma)\delta_0\epsilon_0^2 + \frac{1}{2}(3A+\Gamma)\delta_0^2\epsilon_0 + \cdots\cdots$$
$$\mathbf{T}_2 = A\delta_0 + B\delta_1 + \Gamma\epsilon_0 + \Delta\epsilon_1 + \frac{1}{2}\Gamma_0^2 + \frac{1}{2}(A-\Gamma)\epsilon_0^2 + A\delta_0\epsilon_0 - \frac{1}{6}\Gamma\delta_0^3 + \frac{1}{6}(\Gamma - A)\epsilon_0^3$$
$$+\frac{1}{2}A\delta_0\epsilon_0^2 + \left(\frac{1}{2}A + \Gamma\right)\delta_0^2\epsilon_0 + \cdots\cdots$$

where A, B, Γ, Δ arbitrary parameters take real values. If we are able to find values for these parameters which can guarantee the convergence of the series, the complexity degree will be equal to 1+.

References

1. N. Karcanias, "Structure evolving systems and control in integrated design" Annual Reviews in Control 32 (2008) 161–182
2. S. Wiggins. Introduction to applied nonlinear dynamical systems and chaos. Springer-Verlag, New York, 1990.
3. Kotsios, S. and Lappas, D.1996. A description of 2-Dimensional discrete polynomial dynamics. *IMA Journal of Mathematical Control and Information, Vol. 13, 409–428.*
4. S.T.Glad (1989), "Differential algebraic modeling of nonlinear Systems", In proceedings of MTNS-89, Amsterdam, 97–105.
5. S. Kotsios-D. Lappas "A Stability Result for "Separable" Nonlinear Discrete Systems." IMA Journal of Mathematical Control and Information, 18, 325–339, (2001).
6. S. Kotsios "An Application on Ritt's Remainder Algorithm to Discrete Polynomial Control Systems." Journal of Mathematical Control and Information, 18, 19–29, (2001).
7. Stelios Kotsios - Dionyssios Lappas "The Nonlinearity of 2-D Discrete Systems. An Algebraic Approach." Work in progress, (2013).
8. Stelios Kotsios, Dionyssios Lappas, "Linear Similarity of Nonlinear Polynomial Discrete Systems. An Algebraic Approach." Conference of Modern Mathematical Methods in Science and Technology (M3ST), (2012), Kalamata, Greece.

A Control Scheme Towards Accurate Firing While Moving for a Mobile Robotic Weapon System with Delayed Resonators

Fotis N. Koumboulis and Nikolaos D. Kouvakas

Abstract In the present paper a mobile robotic weapon system is considered. The system comprises of a ground vehicle equipped with a robotic manipulator carrying a gun. The goal is to perform accurate firing while moving and despite the vibrations due to uneven ground and reaction weapon forces. The vehicle is considered to be equipped with passive and active suspension systems. The active suspension involves delayed resonators feeding back the resonators' acceleration. Using the Euler-Lagrange approach, the model of the system is derived in the form of a nonlinear neutral time delay mathematical description. From the system design point of view, the goal is formulated as a command following problem with simultaneous disturbance attenuation, under appropriate constraints. To achieve these goals, an algebraic control scheme based on the linear approximant of the system's model is proposed. The controller is of the measurable output feedback dynamic type. Despite the complexity of the system's model, the derived controller is realizable in the sense that no predictors are required and is simple enough to be implemented to low level computer platforms. Thus the proposed controller offers itself to upgrade traditional armed ground vehicles. This upgrade appears to be of low cost. The good performance of the proposed controller is demonstrated through computational experiments upon the nonlinear model of the system.

Introduction

Vibration absorption is of great importance for several applications ranging from ride comfort improvement and passenger safety in conventional vehicles (see [1–6] and the references therein) and industrial applications (see [7–10] and the references

---

F.N. Koumboulis (✉) • N.D. Kouvakas
Department of Automation, Halkis Institute of Technology, 34400, Psahna Evias, Greece
e-mail: koumboulis@teihal.gr

N.J. Daras (ed.), *Applications of Mathematics and Informatics in Science and Engineering*, Springer Optimization and Its Applications 91, DOI 10.1007/978-3-319-04720-1_19,

therein) to military applications (see [11, 12] and the references therein). Besides the difficulty that arises from performance constraints, the designer has to take into account the actuator influence [13–17]. Of special interest, as far as military applications are concerned, are anti-mine robotic manipulators (see [18–20]). Also, such robotic systems are used for patrols around a military base or other government installation or for convoy support operations. Here we focus on such robotic systems where accurate firing is required while moving. Clearly, vibration minimization and compensation of the gun reaction force together with target following are required to be satisfied via an appropriate control scheme.

In the present paper, the mathematical description of a simplified robotic weapon system is presented. The system comprised of a ground vehicle equipped with a robotic manipulator carrying a gun. A two-stage control scheme towards performing accurate firing while moving and despite vibrations due to uneven ground and reaction weapon forces is designed. The control scheme is developed on the basis of the linear approximant of the system. The scheme is analyzed using an inner and an outer control loop. The inner loop controller achieves diagonalization for the transfer matrix mapping the external commands to the performance outputs as well as it satisfies a mixed disturbance rejection/disturbance attenuation criterion in order to reduce the influence of reaction weapon forces and road unevenness. The outer loop controller produces an appropriate command for the inner controller in order to follow fast and accurately a moving target. The performance of the proposed control scheme is demonstrated through simulations. The control scheme appears to contribute to force multiplication, expansion of the battle-space, extension of the warfighter's reach, and last but not least casualty reduction. Finally, it is noted that the present paper is a generalization of the results presented in [21].

Mathematical Description of a Mobile Robotic Weapon System

Consider the mobile robotic weapon presented in Fig. 1. Assuming that the vehicle moves on a straight path, the whole motion can be faced as a 2-D problem. The mobile robot consists of a platform carrying a two-joint robot which includes a prismatic and a revolute joint. In Fig. 1 the gun is represented by an abstractive arrow shape. The vehicle is assumed to move on a horizontal level due to a force that is always parallel to the road. Ground unevenness is modeled as a force disturbance. The vibrations of the vehicle are absorbed by four absorbers, two of which act as active absorbers and the rest two as passive absorbers. The passive absorbers (vehicle suspension) are two identical conventional spring-damper structures that connect the platform of the vehicle with the front and rear wheel, respectively. They are considered to be locked, i.e., no independent motion is allowed. Consequently, the platform cannot rotate. The active vibration absorbers are two identical mass-spring-damper trios that utilize acceleration feedback with controlled delay and are placed on top of the vehicle's platform at the position of the front and the rear

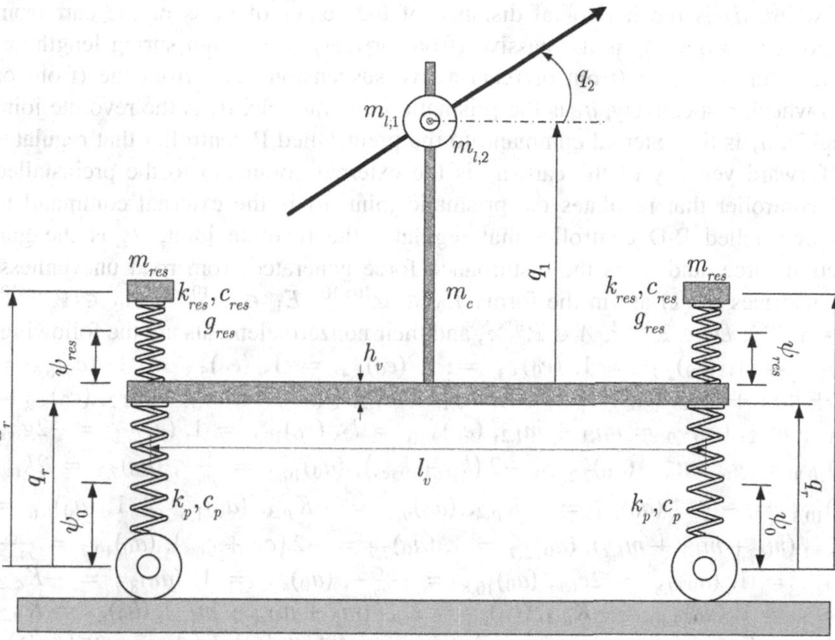

Fig. 1 A mobile robotic weapon system

wheel, respectively (see [7–9]). At the center of the platform, a robotic manipulator that carries a gun is considered. The "gripper" holds the gun from its center of mass. Three independent preinstalled controllers are considered to be used to regulate the forward velocity of the vehicle and the robot joints. These controllers are of the P/PD type with gravity compensation. The first controller feeds back the error between the respective commands and the forward velocity by appropriate gain $K_{p,1}$. The second and the third controller feed back the error between the respective commands and the joint variables as well as the derivatives of the joint variables, both multiplied by appropriate gains, $K_{p,i}$ and $K_{d,i}$ $(i = 2, 3)$, respectively.

The mathematical model of the robotic system is in the following general neutral multi-delay form

$$E_0\dot{x}(t) + E_1\dot{x}(t-\tau) = A_0x(t) + B_0u(t) + D_0(x)\xi(t) + \Lambda_0 \tag{1}$$

where

$$\begin{aligned} x &= \left[x_1\ x_2\ x_3\ x_4\ x_5\ x_6\ x_7\ x_8\ x_9\ x_{10}\right]^T \\ &= \left[d_f\ q_r\ x_r\ q_1\ q_2\ \ \dot{d}_f\ \dot{q}_r\ \dot{x}_r\ \dot{q}_1\ \dot{q}_2\right]^T \end{aligned} \tag{2}$$

$$u = \left[u_1\ u_2\ u_3\right]^T \tag{3}$$

$$\xi = \left[\xi_1\ \xi_2\right]^T = \left[f_g\ f_r\right]^T \tag{4}$$

and where d_8 is the horizontal distance of the center of mass of the cart from the point of origin, q_r is the passive (front or rear) suspension spring length, x_r is the distance of the (front or rear) active suspension mass from the (front or rear) wheel, respectively, q_1 is the prismatic joint variable, q_2 is the revolute joint variable, u_1 is the external command to the preinstalled P controller that regulates the forward velocity of the cart, u_2 is the external command to the preinstalled P-D controller that regulates the prismatic joint, u_3 is the external command to the preinstalled P-D controller that regulates the revolute joint, f_g is the gun reaction force, and f_r is the disturbance force generated from road unevenness. The matrices in (1) are in the forms $E_0 \in \mathbb{R}^{10\times10}$, $E_1 \in \mathbb{R}^{10\times10}$, $A_0 \in \mathbb{R}^{10\times10}$, $B_0 \in \mathbb{R}^{10\times3}$, $D_0 \in \mathbb{R}^{10\times2}$, $\Lambda \in \mathbb{R}^{10\times2}$, and their nonzero elements are the following: $(e_0)_{1,1} = 1$, $(e_0)_{2,2} = 1$, $(e_0)_{3,3} = 1$, $(e_0)_{4,4} = 1$, $(e_0)_{5,5} = 1$, $(e_0)_{6,6} = m_c + m_{l,1} + m_{l,2}$, $(e_0)_{7,7} = m_c + m_{l,1} + m_{l,2}$, $(e_0)_{8,7} = m_{l,1} + m_{l,2}$, $(e_0)_{8,9} = m_{l,1} + m_{l,2}$, $(e_0)_{7,9} = m_{l,1} + m_{l,2}$, $(e_0)_{9,10} = I_2$, $(e_0)_{10,8} = 1$, $(e_1)_{7,8} = -2g_{\text{res}}$, $(e_1)_{10,8} = g_{\text{res}}/m_{\text{res}}$, $(a_0)_{7,2} = -2\left(k_p + k_{\text{res}}\right)$, $(a_0)_{10,2} = \frac{k_{\text{res}}}{m_{\text{res}}}$, $(a_0)_{7,3} = 2k_{\text{res}}$, $(a_0)_{10,3} = -\frac{k_{\text{res}}}{m_{\text{res}}}$, $(a_0)_{8,4} = -K_{p,2}$, $(a_0)_{9,5} = -K_{p,3}$, $(a_0)_{1,6} = 1$, $(a_0)_{6,6} = -K_{p,1}\left(m_c + m_{l,1} + m_{l,2}\right)$, $(a_0)_{2,7} = 1$, $(a_0)_{7,7} = -2\left(c_p + c_{\text{res}}\right)$, $(a_0)_{10,7} = \frac{c_{\text{res}}}{m_{\text{res}}}$, $(a_0)_{3,8} = 1$, $(a_0)_{7,8} = 2c_{\text{res}}$, $(a_0)_{10,8} = -\frac{c_{\text{res}}}{m_{\text{res}}}$, $(a_0)_{4,9} = 1$, $(a_0)_{8,9} = -K_{d,2}$, $(a_0)_{5,10} = 1$, $(a_0)_{9,10} = -K_{d,3}$, $(b_0)_{6,1} = K_{p,1}\left(m_c + m_{l,1} + m_{l,2}\right)$, $(b_0)_{8,2} = K_{p,2}$, $(b_0)_{9,3} = K_{p,3}$, $(d_0)_{6,1} = \cos\left(x_5(t)\right)$, $(d_0)_{7,1} = \sin\left(x_5(t)\right)$, $(d_0)_{8,1} = \sin\left(x_5(t)\right)$, $(d_0)_{6,2} = 1$, $(\lambda_0)_{7,1} = 2\psi_0 k_p - 2\left(h_v + \psi_{\text{res}}\right)k_{\text{res}} - g\left(m_c + m_{l,1} + m_{l,2}\right)$, $(\lambda_0)_{10,1} = -g + \frac{(h_v + \psi_{\text{res}})k_{\text{res}}}{m_{\text{res}}}$.

Note that m_c is the cart mass, $m_{l,1}$ is the first link mass, $m_{l,2}$ is the second link mass, m_{res} is the resonator mass, I_2 is the second link moment of inertia, h_v is the vehicle height, ψ_{res} is the resonator spring's free length, ψ_0 is the passive absorber's free length, k_{res} is the resonator spring constant, k_p is the suspension spring constant, c_{res} is the resonator dumping factor, c_p is the suspension dumping factor, g_{res} is the resonator feedback gain, τ is the resonator feedback delay, and g is the gravity acceleration. The performance outputs of the system description (1) are the forward velocity of the cart $y_1(t) = x_6(t)$, the distance of the center of mass (c.m.) of the gun from the ground $y_2(t) = h_v + r_0 + x_2(t) + x_4(t)$, and the inclination of the gun: $y_3(t) = x_5(t)$. Note that r_0 denotes the radius of the wheels. The performance output vector is given by the relation

$$y(t) = \begin{bmatrix} y_1(t) \\ y_2(t) \\ y_3(t) \end{bmatrix} = Cx(t) + \begin{bmatrix} 0 \\ h_v + r_0 \\ 0 \end{bmatrix}$$

where

$$C = \begin{bmatrix} 0\,0\,0\,0\,0\,1\,0\,0\,0\,0 \\ 0\,1\,0\,1\,0\,0\,0\,0\,0\,0 \\ 0\,0\,0\,0\,1\,0\,0\,0\,0\,0 \end{bmatrix} \tag{5}$$

The measurable outputs of the system are denoted by

$$\psi(t) = \begin{bmatrix} \psi_1(t) \\ \psi_2(t) \\ \psi_3(t) \\ \psi_4(t) \end{bmatrix} = Lx(t)$$

where

$$L = \begin{bmatrix} 0\,1\,0\,0\,0\,0\,0\,0\,0\,0 \\ 0\,0\,0\,1\,0\,0\,0\,0\,0\,0 \\ 0\,0\,0\,0\,1\,0\,0\,0\,0\,0 \\ 0\,0\,0\,0\,0\,1\,0\,0\,0\,0 \end{bmatrix} \tag{6}$$

To derive the linear approximant of the robotic system, first consider the following trim conditions

$$\bar{u}_1 = \bar{w}_1, \bar{u}_2 = \bar{w}_2, \bar{u}_3 = \bar{w}_3, \bar{\xi}_1 = 0, \bar{\xi}_2 = 0$$

$$\bar{x}_1 = v_n t, \bar{x}_2 = \psi_0 - \frac{g\,(m_c + m_{l,1} + m_{l,2} + 2m_{\text{res}})}{2k_p}$$

$$\bar{x}_3 = \psi_0 + \psi_{\text{res}} + h_v - \frac{g}{2k_p}(m_c + m_{l,1} + m_{l,2}) - \frac{g}{k_p k_{\text{res}}}(k_p + k_{\text{res}})\, m_{\text{res}}$$

$$\bar{x}_4 = \bar{u}_2, \bar{x}_5 = \bar{u}_3, \bar{x}_6 = v_n, \bar{x}_7 = 0, \bar{x}_8 = 0, \bar{x}_9 = 0, \bar{x}_{10} = 0$$

Second, consider the perturbations of the state and input variables

$$\delta y\,(t) = y\,(t) - \bar{y}, \delta x\,(t) = x\,(t) - \bar{x}, \delta u\,(t) = u\,(t) - \bar{u}$$

$$\delta\xi\,(t) = \xi\,(t) - \bar{\xi}, \delta\psi\,(t) = \psi\,(t) - \bar{\psi}$$

where $\bar{y} = C\bar{x}$ and $\bar{\psi} = L\bar{x}$. Thus, the linear approximant of the system is of the following generalized neutral form

$$E_0\delta\dot{x}\,(t) + E_1\delta\dot{x}\,(t - \tau) = A_0\delta x\,(t) + B_0\delta u\,(t) + D_0\,(\bar{x})\,\delta\xi\,(t) \tag{7}$$

$$\delta y\,(t) = C\delta x\,(t) \tag{8}$$

$$\delta\psi\,(t) = L\delta x\,(t) \tag{9}$$

Controller Design

Here, the accurate performance of the system is analyzed in two design goals. The first is to achieve a desired closed loop transfer matrix while simultaneously achieving a mixed disturbance rejection/disturbance attenuation scheme and the second is accurate following of a target. The control scheme will be analyzed using an inner and an outer control loop. The inner control loop will satisfy the first design goal while the outer loop will satisfy the second design goal.

Inner Control Loop

The inner loop controller is considered to be of the dynamic multi-delay measurement output type:

$$\delta U(s) = K(s,z)\,\delta\Psi(s) + G(s,z)\,\delta R(s) \tag{10}$$

where $\delta U(s)$, $\delta\Psi(s)$, and $\delta R(s)$ are the Laplace transforms of the vector signals $\delta u(t)$, $\delta\psi(t)$, and $\delta r(t)$, respectively, while $\delta r(t) = \left[\delta r_1(t)\ \delta r_2(t)\ \delta r_3(t)\right]^T$ is the 3×1 vector of external inputs. Clearly, it holds that $\delta R(s) = \left[\delta R_1(s)\ \delta R_2(s)\ \delta R_3(s)\right]^T$. The Laplace transform of the perturbation of the measurement output vector is expressed by the relation $\delta\Psi(s) = L\delta X(s)$, where $\delta X(s)$ is the Laplace transform of the perturbation of the state vector $\delta x(t)$. The elements of the controller matrices are rational functions of s with coefficients being rational functions of $z = e^{-s\tau}$, i.e., $k_{i,j}(s,z)$, $g_{i,j}(s,z) \in \mathbb{R}(s,z)$, where $\mathbb{R}(s,z)$ is the field of rational functions of s with coefficients being rational functions of z. It is important to mention that for the implementation of the controller (10) the elements of the controller matrices should be realizable (see [22–24]). After substituting the controller (10) to system described by (7) to (9) the forced response of the closed loop system is given by the relation

$$\begin{aligned}\delta X(s) = {} & [I_{10} - H_u(s,z)K(s,z)L]^{-1}H_u(s,z)G(s,z)\,\delta R(s)\\ & + [I_{10} - H_u(s,z)K(s,z)L]^{-1}H_\xi(s,z)\,\delta\Xi(s)\end{aligned} \tag{11}$$

where

$$H_u(s,z) = [s(E_0 + zE_1) - A_0]^{-1}B \tag{12}$$

$$H_d(s,z) = [s(E_0 + zE_1) - A_0]^{-1}D_0(\bar{x}) \tag{13}$$

The desired closed loop matrix should have the following three characteristics: First, the closed loop transfer matrix relating the external inputs to the performance outputs is diagonal and invertible. This is the I/O decoupling design requirement. This design requirement is formally expressed as follows:

$$C\left[I_{10}-H_u(s,z)\,K(s,z)\,L\right]^{-1}H_u(s,z)\,G(s,z) = \begin{bmatrix} h_{m,1}(s,z) & 0 & 0 \\ 0 & h_{m,1}(s,z) & 0 \\ 0 & 0 & h_{m,1}(s,z) \end{bmatrix} \tag{14}$$

where the rational functions $h_{m,1}(s,z)$, $h_{m,2}(s,z)$, and $h_{m,3}(s,z)$ are different than zero and they belong to $\mathbb{R}(s,z)$.

Second, the second row–first column element and the third row elements of the closed loop transfer matrix relating the disturbances to the performance outputs are equal to zero. This is a partial disturbance rejection design requirement. This design requirement is formally expressed as follows:

$$C\left[I_{10}-H_u(s,z)\,K(s,z)\,L\right]^{-1}H_d(s,z) = \begin{bmatrix} (h_d)_{1,1}(s,z) & (h_d)_{1,2}(s,z) \\ 0 & (h_d)_{2,2}(s,z) \\ 0 & 0 \end{bmatrix} \tag{15}$$

According to the above design requirement, the gun reaction force being the main disturbance does not influence the distance of the center of mass (c.m.) of the gun from the ground and the inclination of the gun. Also, the inclination of the gun, having the greater influence to the accuracy of the shot, is not influenced by the disturbance force generated from road unevenness.

Third, the norm of the nonzero elements of the closed loop transfer matrix relating the disturbances to the performance outputs is enough small, i.e.,

$$\max\left\{\left\|(h_d)_{1,1}(s,z)\right\|_\infty, \left\|(h_d)_{1,2}(s,z)\right\|_\infty, \left\|(h_d)_{2,2}(s,z)\right\|_\infty\right\} < \varepsilon \tag{16}$$

where ε is an enough small positive real guaranteeing attenuation of the influence of the disturbances to the performance outputs.

In what follows the elements of the controller matrices in (10) will be selected to satisfy all three of the above design requirements. The solution of the precompensator to satisfy the I/O decoupling design requirement is

$$G(s,z) = \left\{C\left[I_{10}-H_u(s,z)\,K(s,z)\,L\right]^{-1}H_u(s,z)\right\}^{-1} \times \begin{bmatrix} h_{m,1}(s,z) & 0 & 0 \\ 0 & h_{m,1}(s,z) & 0 \\ 0 & 0 & h_{m,1}(s,z) \end{bmatrix} \tag{17}$$

The solution of the feedback matrix to satisfy the partial disturbance rejection problem in (15) is the following

$$\begin{aligned}
k_{2,4}(s,z) = &\Big(\big(s(H_u)_{1,1}(s,z)\,k_{1,4}(s,z)-1\big)\\
&\times\big\{(H_d)_{2,1}(s,z)\big[1+(H_u)_{4,2}(s,z)\,(k_{2,1}(s,z)\\
&-k_{2,2}(s,z))\big]+(H_d)_{4,1}(s,z)\big[1+(H_u)_{2,2}(s,z)\,(k_{2,2}(s,z)-k_{2,1}(s,z))\big]\big\}\Big)\,/\\
&\big\{s\big[(H_d)_{1,1}(s,z)\big((H_u)_{2,2}(s,z)+(H_u)_{4,2}(s,z)\big)\\
&-(H_u)_{1,1}(s,z)\big((H_d)_{4,1}(s,z)\,(H)_{2,2}(s,z)\\
&-(H_d)_{2,1}(s,z)\,(H_u)_{4,2}(s,z)\big)\,(k_{1,1}(s,z)-k_{1,2}(s,z))\big]\big\}
\end{aligned} \tag{18}$$

$$\begin{aligned}
k_{3,2}(s,z) = &\big[(H_d)_{4,1}(s,z)\,(H_u)_{2,2}(s,z)\,k_{2,2}(s,z)\\
&+(H_d)_{2,1}(s,z)\big(1-(H_u)_{4,2}(s,z)\,k_{2,2}(s,z)\big)\big]\\
&\times k_{3,1}(s,z)\,/\,\big[(H_d)_{4,1}(s,z)\big((H_u)_{2,2}(s,z)\,k_{2,1}(s,z)-1\big)\\
&-(H_d)_{2,1}(s,z)\,(H_u)_{4,2}(s,z)\,k_{2,1}(s,z)\big]
\end{aligned} \tag{19}$$

$$\begin{aligned}
k_{3,4}(s,z) = &\Big(\big(\big((H_d)_{2,1}(s,z)\,(H_u)_{4,2}(s,z)\\
&-(H_d)_{4,1}(s,z)\,(H_u)_{2,2}(s,z)\big)\\
&\times\big(s(H_u)_{1,1}(s,z)\,k_{1,4}(s,z)-1\big)\big\{(H_d)_{2,1}(s,z)\\
&\times\big[1+(H_u)_{4,2}(s,z)\,(k_{2,1}(s,z)-k_{2,2}(s,z))\big]\\
&+(H_d)_{4,1}(s,z)\big[1+(H_u)_{2,2}(s,z)\,(k_{2,2}(s,z)-k_{2,1}(s,z))\big]\big\}\,k_{3,1}(s,z)\Big)\,/\\
&\big\{s\big[(H_d)_{1,1}(s,z)\big((H_u)_{2,2}(s,z)+(H_u)_{4,2}(s,z)\big)\\
&-(H_u)_{1,1}(s,z)\big((H_d)_{4,1}(s,z)\,(H_u)_{2,2}(s,z)\\
&(H_d)_{2,1}(s,z)\,(H_u)_{4,2}(s,z)\big)\,(k_{1,1}(s,z)-k_{1,2}(s,z))\big]\\
&\big[(H_d)_{2,1}(s,z)\,(H_u)_{4,2}(s,z)\,k_{2,1}(s,z)\\
&+(H_d)_{4,1}(s,z)\big(1-(H_u)_{2,2}(s,z)\,k_{2,1}(s,z)\big)\big]\big\}
\end{aligned} \tag{20}$$

where $(H_u)_{i,j}(s,z)$, denotes the (i,j) element of $H_u(s,z)$ and $(H_d)_{i,j}(s,z)$ denotes the (i,j) element of $H_d(s,z)$.

To satisfy the requirement of disturbance attenuation, the rest of the elements of the feedback matrices is selected to be real, i.e.,

$$k_{1,1}(s,z)=k_{1,1}\in\mathbb{R},\, k_{1,2}(s,z)=k_{1,2}\in\mathbb{R},\, k_{1,3}(s,z)=k_{1,3}\in\mathbb{R}$$

$$k_{1,4}(s,z)=k_{1,4}\in\mathbb{R},\, k_{2,1}(s,z)=k_{2,1}\in\mathbb{R},\, k_{2,2}(s,z)=k_{2,2}\in\mathbb{R}$$

$$k_{2,3}(s,z) = k_{2,3} \in \mathbb{R}, k_{3,1}(s,z) = k_{3,1} \in \mathbb{R}, k_{3,3}(s,z) = k_{3,3} \in \mathbb{R}$$

For the determination of the above real parameters, a heuristic algorithm is proposed (see [25–27]).

It is important to mention that a controller selection as in (18) to (20) with the remaining elements being static can guarantee the solvability of the closed loop system, i.e.,

$$\det\left[s\left(E_0 + zE_1\right) - A_0 - B_0 K(s,z) L\right] \not\equiv 0 \tag{21}$$

It can be verified that the same selection guarantees the realizability of the feedback matrix. Furthermore, it can be observed that the rational matrix $C\left[I_{10} - H_u(s,z) K(s,z) L\right]^{-1} H_u(s,z)$ resulting after the above selection of the elements of the feedback matrix is birealizable. Hence, for the precompensator in (17) to be invertible and realizable, it suffices to choose the rational functions $h_{m,1}(s,z)$, $h_{m,2}(s,z)$, and $h_{m,3}(s,z)$ to be delayless and different than zero, i.e.,

$$h_{m,1}(s,z) = h_{m,1}(s)$$

$$h_{m,2}(s,z) = h_{m,2}(s)$$

$$h_{m,3}(s,z) = h_{m,3}(s)$$

Finally, note that for the precompensator to be at least strictly proper, a possible selection for the models' transfer functions is

$$h_{m,1}(s) = \frac{p_{1,0}}{s + p_{1,0}}$$

$$h_{m,2}(s) = \frac{p_{2,0}}{s^2 + p_{2,1}s + p_{2,0}}$$

$$h_{m,3}(s) = \frac{p_{3,0}}{s^2 + p_{3,1}s + p_{3,0}}$$

where $p_{i,j} > 0$. It is mentioned that the above selection of the decoupled closed loop transfer functions guarantees asymptotic command following.

Outer Control Loop

As already mentioned the design goal of the outer loop is to follow accurately a target. Let $\alpha(t)$ be the inclination of the gun with respect to the vertical axis, i.e., it holds that $\alpha(t) = x_5(t) - \frac{3\pi}{2}$ (see Fig. 2). Let $l(t)$ be the distance between the

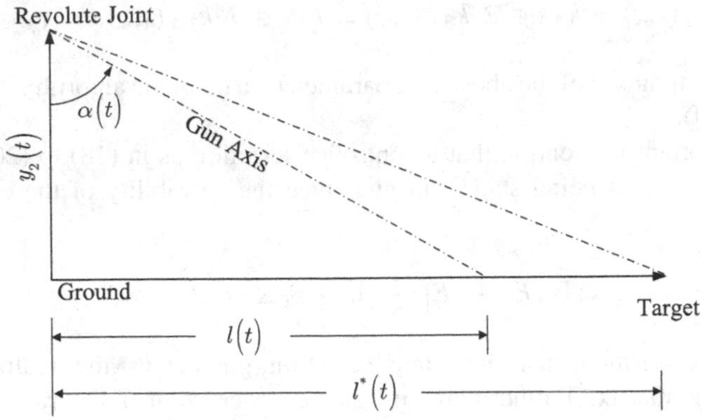

Fig. 2 A geometric interpretation of the distance between the aiming point and the target

projection of the revolute joint to the ground and the intersection between the gun axis and the ground level. Clearly it holds that $l\,(t) = y_2\,(t)\tan\left(\alpha\,(t)\right)$. Finally, let $l^*(t)$ be the distance between the projection of the revolute joint to the ground and the target.

Let $\delta l(t) = l(t) - \bar{l}$ be the perturbation of $l(t)$, where

$$\bar{l} = \bar{y}_2 \tan\left(\bar{\alpha}\right) = \left(\bar{x}_2 + \bar{x}_4 + h_v + r_0\right) \tan\left(\bar{x}_5 - \frac{3\pi}{2}\right) \tag{22}$$

Thus, it holds that

$$\begin{aligned} \delta L\,(s) = &-\cot\left(\bar{u}_3\right) h_{m,2}\,(s,z)\,\delta R_2\,(s) + \\ &\left(h_v + r_0 + \bar{x}_2 + \bar{x}_4\right) \csc\left(\bar{u}_3\right)^2 h_{m,3}\,(s,z)\,\delta R_3\,(s) \end{aligned} \tag{23}$$

where $\delta L(s)$ denotes the Laplace transform of $\delta l(t)$. It is important to mention that after substituting the trim condition of the state variables presented in section "Mathematical Description of a Mobile Robotic Weapon System" to the relation (22) we get

$$\begin{aligned} \bar{l} = -\Big[h_v - 0.5gk_p^{-1}\left(m_c + m_{l,1} + m_{l,2} + 2m_{\text{res}}\right) \\ +\psi_0 + r_0 + \bar{u}_2\Big] \cot\left(\bar{u}_3\right) \end{aligned} \tag{24}$$

Target following will be achieved by adjusting the distance of the revolute joint from the ground while preserving constant inclination of the gun, i.e., $\delta R_3\,(s) = 0$. Using $\delta R_3\,(s) = 0$ as well as (14) and (15) we observe (with respect to the linear model) that the inclination of the gun remains constant. Also, from (23) we observe (with respect to the linear model) that the influence of the disturbances to $\delta l(t)$ has been eliminated. With respect to the nonlinear model, it is significant to mention that through the disturbance rejection controller the influence of the gun reaction to the distance of the center of mass of the gun from the ground and the inclination of the gun have been eliminated while the disturbance due to road unevenness has significantly been reduced.

Let $\delta l^*(t) = l^*(t) - \bar{l}$ be the perturbation of $l^*(t)$. Let $\delta L^*(s)$ be the Laplace transform of $\delta l^*(t)$. The outer loop controller is proposed to be of the following PID form

$$\delta R_2(s) = \left(f_p + \frac{1}{s} f_i + s f_d\right)\left(\delta L^*(s) - \delta L(s)\right) \tag{25}$$

Let $l(0-) = \bar{l}$. Hence using $\delta R_3(s) = 0$, (22) to (24) as well as the definition of the decoupled closed loop transfer function presented in section "Inner Control Loop," the perturbation of the aiming point position is related to the perturbation of the target positions by the relation

$$\delta L(s) = H_l(s)\delta L^*(s) \tag{26}$$

where

$$\begin{aligned} H_l(s) = &-\left(f_d p_{2,0} \cot(\bar{u}_3) s^2 + f_p p_{2,0} \cot(\bar{u}_3) s\right. \\ &\left.+ f_i p_{2,0} \cot(\bar{u}_3)\right) / \left[s^3 + \left(p_{2,1} - f_d p_{2,0} \cot(\bar{u}_3)\right) s^2\right. \\ &\left.+ p_{2,0}\left(1 - f_p \cot(\bar{u}_3)\right) s - f_i p_{2,0} \cot(\bar{u}_3)\right] \end{aligned} \tag{27}$$

The transfer function (27) is stable if and only if the PID controller parameters satisfy the following inequalities

$$f_p > \tan(\bar{w}_3) \tag{28}$$

$$f_i > 0 \tag{29}$$

$$f_d > \frac{f_i - f_p p_{2,1} + p_{2,1} \tan(\bar{u}_3)}{p_{2,0} - f_p p_{2,0} \cot(\bar{u}_3)} \tag{30}$$

To satisfy (28) to (30) the PID controller parameters are selected to be

$$f_p = \frac{\left[p_{2,0} - \lambda\left(1 + \lambda + \lambda^2\right)\rho^2\right] \tan(\bar{u}_3)}{p_{2,0}} \tag{31}$$

$$f_i = \frac{\lambda^3 \rho^3 \tan(\bar{u}_3)}{p_{2,0}} \tag{32}$$

$$f_d = \frac{\left[p_{2,1} - \left(1 + \lambda + \lambda^2\right)\rho\right] \tan(\bar{u}_3)}{p_{2,0}} \tag{33}$$

where $\lambda > 0$, $\rho > 0$ are free parameters. Thus, (27) takes on the form

$$H_l(s) = \{[(1+\lambda+\lambda^2)\rho - p_{2,1}]s^2 + [\lambda(1+\lambda+\lambda^2)\rho^2 - p_{2,0}]s + \lambda^3\rho^3\} / [(s+\rho)(s+\lambda\rho)(s+\lambda^2\rho)] \quad (34)$$

To complete the determination of the outer control loop, it suffices to find appropriate $\lambda > 0$ and $\rho > 0$ such that the gun follows quickly and accurately any change to the distance of the target from the vehicle while simultaneously satisfying actuator constraints.

Step-Wise Supervisor

The above design scheme is based mainly upon adjustments of the distance of the revolute joint from the ground. To handle also the cases where the prismatic joint reaches its upper or lower bound, the following step-wise supervisor is proposed:

Case 1: If the distance of the target from the vehicle increases and the prismatic joint reaches its upper bound then
Step 1.1: lower the prismatic joint,
Step 1.2: increase the angle of the gun,
Step 1.3: reevaluate the linearized model and controller parameters and apply the design scheme presented in sections "Inner Control Loop" and "Outer Control Loop."
Case 2: If the distance of the target from the vehicle decreases and the prismatic joint reaches its lower bound then
Step 2.1: lift the prismatic join,
Step 2.2: decrease the angle of the gun,
Step 2.3: reevaluate the linear model and controller parameters and apply the design scheme presented in sections "Inner Control Loop" and "Outer Control Loop."

Simulation Results

The model parameters are considered to be:

$$m_c = 15.2(\text{kg}), m_{l,1} = 0.81(\text{kg}), m_{l,2} = 0.628(\text{kg})$$

$$m_{\text{res}} = 0.177(\text{kg}), k_p = 62{,}000(\text{N/m}), k_{\text{res}} = 3{,}490{,}000(\text{N/m})$$

$$c_p = 2{,}500(\text{kg/s}), c_{\text{res}} = 81.8(\text{kg/s}), g_{\text{res}} = 0.01833(\text{kg})$$

$$I_2 = 0.00081(\text{kg}\times\text{m}^2), \psi_0 = 0.3\,(\text{m}), r_0 = 0.02\,(\text{m})$$

$$h_v = 0.01(\mathrm{m}), g = 9.81(\mathrm{m/sc}^2), \tau = 0.000387(\mathrm{s})$$

$$K_{p,2} = 100, K_{p,3} = 100, K_{d,2} = 10, K_{d,3} = 10$$

$$K_{p,1} = 2, v_n = 1(\mathrm{m/s})$$

The nonzero trim conditions are:

$$\bar{u}_1 = 1(\mathrm{m/s}), \bar{u}_2 = 0.25(\mathrm{m}), \bar{u}_3 = 6.152(\mathrm{rad})$$

$$\bar{x}_1 = t(\mathrm{m}), \bar{x}_2 = 0.2987(\mathrm{m}), \bar{x}_3 = 0.4087(\mathrm{m})$$

$$\bar{x}_4 = 0.25(\mathrm{m}), \bar{x}_5 = 6.152(\mathrm{rad}), \bar{x}_6 = 1(\mathrm{m/s})$$

The inner closed loop I/O transfer matrix parameters are selected to be:

$$p_{1,0} = 10, p_{2,0} = 50, p_{2,1} = 15, p_{3,0} = 50, p_{3,1} = 15$$

With respect to the feedback matrix, we select $\varepsilon = 5e-5$. Then a set of controller parameters satisfying the inequality in (16) are:

$$k_{1,1} = 0, k_{1,2} = -466.015, k_{1,3} = -2.1622, k_{1,4} = -586.9273$$

$$k_{2,1} = -27.3172, k_{2,2} = -565.5233, k_{2,3} = -30.3138$$

$$k_{3,1} = -4.2357, k_{3,3} = -555.3803$$

With respect to the outer closed loop, it is observed that for the transfer function in (34) to have a maximum rise time of 0.575 s, it suffices to choose $\rho = 3$ and $\lambda = 2$ thus yielding $f_p = 0.407283$, $f_i = 1.15754$, and $f_d = 0.0321539$.

With respect to the gun reaction force (see Fig. 3) it will be assumed that it simulates the reaction force generated from a rifle shot (e.g., [28]). We will further assume that the force pattern is repeated every 2 s. With respect to the road disturbance, it is assumed that it is a random uniform noise in the form presented in Fig. 4. The target distance is considered to be $l^*(t) = (h_v + r_0 + \bar{x}_2 + \bar{x}_4)\tan\left(\bar{x}_5 - \frac{3\pi}{2}\right) + \sin(4t)$. The external commands are selected to be $\delta r_1(t) = 0$ and $\delta r_3(t) = 0$. The distances of the aiming point from the vehicle and the target point from the vehicle are presented in Fig. 5. Note that both responses are visually identical. The forward velocity is presented in Fig. 6. It is important to mention that the inclination of the gun is not affected neither by the disturbances nor variations of the other variables. The distance of the center of mass of the gun from the ground is presented in Fig. 7. The rest of the state variables of the system remains within acceptable limits.

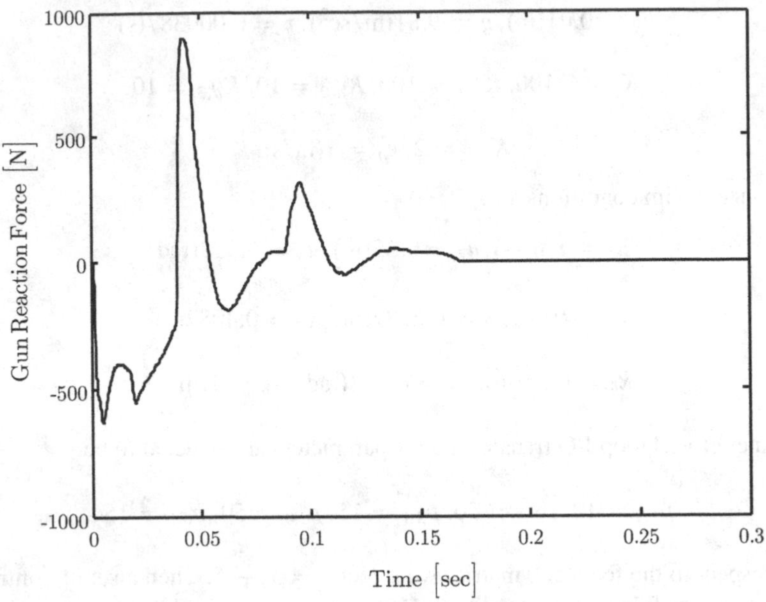

Fig. 3 Gun reaction force

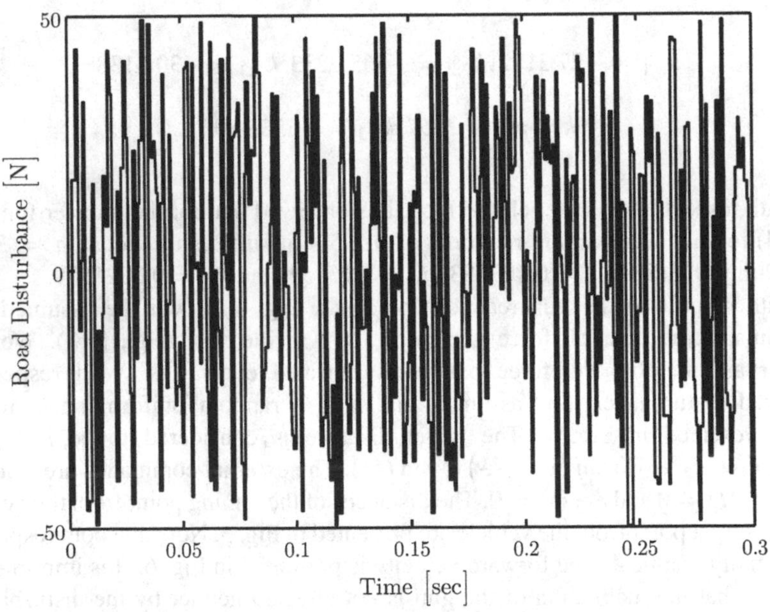

Fig. 4 Road disturbance force

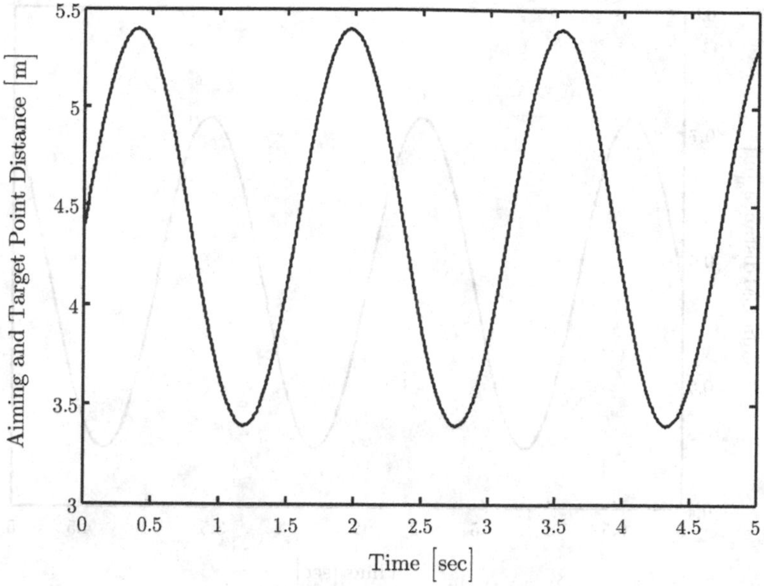

Fig. 5 Distance of the aiming and target points (cont. aiming, dotted target, visually identical)

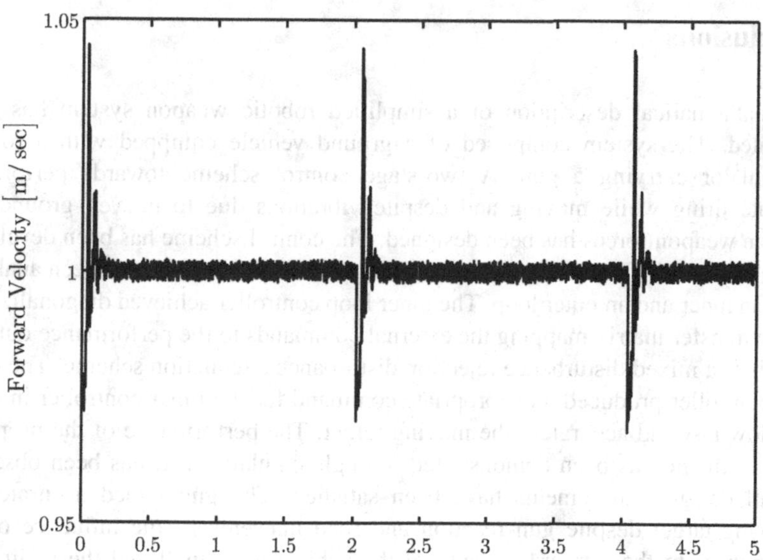

Fig. 6 Forward velocity of the cart

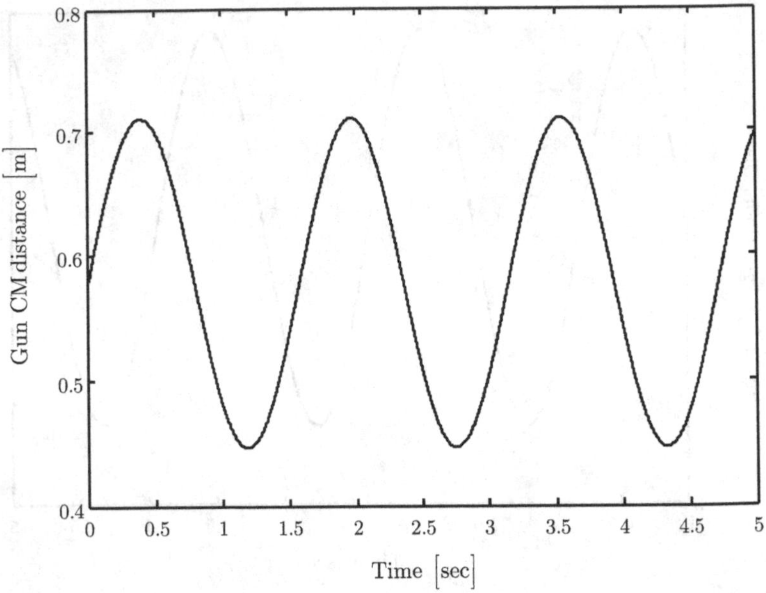

Fig. 7 Distance of the center of mass of the gun from the ground

Conclusions

The mathematical description of a simplified robotic weapon system has been presented. The system comprised of a ground vehicle equipped with a robotic manipulator carrying a gun. A two-stage control scheme towards performing accurate firing while moving and despite vibrations due to uneven ground and reaction weapon forces has been designed. The control scheme has been developed on the basis of the linear approximant of the system. The scheme has been analyzed using an inner and an outer loop. The inner loop controller achieved diagonalization for the transfer matrix mapping the external commands to the performance outputs, as well as a mixed disturbance rejection/disturbance attenuation scheme. The outer loop controller produced an appropriate command for the inner controller in order to follow fast and accurately the moving target. The performance of the proposed control scheme has been demonstrated through simulations. It has been observed that all design requirements have been satisfied. The gun aimed accurately at a moving target despite gun reaction and road unevenness, the influence of the disturbances to the forward velocity of the vehicle was small and the inclination of the gun remained unaffected.

References

1. C. Kim and P. Ro, 'A sliding mode controller for vehicle active suspension systems with non-linearities', *Proceedings of the Institution of Mechanical Engineers, Part D: Journal of Automobile Engineering*, vol. 212, pp. 79–92, 1998.
2. D. Karnopp, 'Active damping in road vehicle suspension systems', *Vehicle System Dynamics*, vol. 12, pp. 291–311, 1983.
3. I. Fialho and G. J. Balas, 'Road adaptive active suspension design using linear parameter-varying gain-scheduling', *IEEE Transactions on Control Systems Technology*, vol. 10, pp. 43–54, 2002.
4. M. Rao and V. Prahlad, 'A tunable fuzzy logic controller for vehicle-active suspension systems', *Fuzzy sets and systems*, vol. 85, pp. 11–21, 1997.
5. T. Yoshimura, A. Kume, M. Kurimoto, and J. Hino, 'Construction of an active suspension system of a quarter car model using the concept of sliding mode control', *Journal of Sound and Vibration*, vol. 239, pp. 187–199, 2001.
6. F. N. Koumboulis and N. D. Kouvakas, 'A three term controller for ride comfort improvement', *Proceedings of the 19th Mediterranean Conference on Control & Automation (MED 2011)*, June 20–23, 2011, Corfu, Greece.
7. M. P. Tzamtzi, F. N. Koumboulis, N. D. Kouvakas and M. G. Skarpetis, 'Robustness in Liquid Transfer Vehicles with Delayed Resonators', *WSEAS Transactions on Systems*, vol. 7, no. 11, pp. 1359–1370, 2008
8. M. P. Tzamtzi, F. N. Koumboulis, N. D. Kouvakas and G. E. Panagiotakis, 'A Simulated Annealing Controller for Sloshing Suppression in Liquid Transfer with Delayed Resonators', *Proceedings of the 14th IEEE Mediterranean Conference on Control and Automation (MED 2006)*, June 28–30, 2006, Ancona, Italy.
9. M. P. Tzamtzi, F. N. Koumboulis and N. D. Kouvakas, 'A Two Stage Robot Control for Liquid Transfer', *Proceedings of the IEEE Conference on Emerging Technology and Factory Automation (ETFA 2007)*, September 26–28, 2007, Patras, Greece.
10. S. Futami, N. Kyura, and S. Hara, 'Vibration absorption control of industrial robots by acceleration feedback', *IEEE Transactions on Industrial Electronics*, vol. 30, no. 3, pp. 299–305, 1983.
11. S. R. Walker and F. J. Andrews, 'Shock and Vibration Isolators for Military Applications', *Journal of the American Defence Preparedness Association*, March 1986, pp. 2–5.
12. H. Wang, R. Li, H. Xue, H. Pan, 'The vibration identification of military vehicle seat based on wavelet-scale energy coefficients', *Proceedings of the 10th IEEE International Conference on Signal Processing (ICSP 2010)*, October 24–28, 2010, Beijing, China.
13. D. Hrovat, 'Optimal active suspension structures for quarter-car vehicle models', *Automatica*, vol. 26, no. 5, pp. 845–860, 1990
14. A. Hac and I. Youn, 'Optimal semi-active suspension with preview based on a quarter car model', *Proceedings of the American Control Conference*, pp. 433–438, Boston, USA, 26–28 June 1991
15. M. Ahmadian and C.A. Pare, 'A quarter-car experimental analysis of alternative semiactive control methods', *Journal of Intelligent Material Systems and Structures*, vol. 11, no. 8, pp. 604–612, 2000
16. G. Verros, S. Natsiavas and C. Papadimitriou, 'Design optimization of quarter-car models with passive and semi-active suspensions under random road excitation', *Journal of Vibration and Control*, vol. 11, no. 5, pp. 581–606, 2005
17. Y.M. Sam, J.H.S. Osman and M. Ghani, 'A class of proportional-integral sliding mode control with application to active suspension system', *System and Control Letters*, vol. 51, pp. 219–223, 2004
18. J. Trevelyan, 'Robots and landmines', *Industrial Robot: An International Journal*, vol. 24, pp. 114–125, 1997.

19. N. Andersson, C. P. da Sousa, and S. Paredes, 'Social cost of land mines in four countries: Afghanistan, Bosnia, Cambodia, and Mozambique', *BMJ: British Medical Journal*, vol. 311, p. 718, 1995.
20. M. A. Cameron, R. J. Lawson, and B. W. Tomlin, *To walk without fear: the global movement to ban landmines*, Cambridge Univ Press, 1998.
21. F.N. Koumboulis and N.D. Kouvakas, 'A Control Scheme Towards Accurate Firing While Moving for a Mobile Robotic Weapon System with Delayed Resonators', *2nd International Conference in Applications of Mathematics and Informatics in Military Science*, April 11–12, 2013, Vari, Athens, Greece.
22. F. N. Koumboulis, G. E. Panagiotakis and P. N. Paraskevopoulos, 'Exact Model Matching of Left Invertible Neutral Time Delay Systems', *International Journal of Modelling, Identification and Control*, Vol. 3, no. 4, pp. 376–384, 2008 / *Proceedings of the 13th Mediterranean Conf. on Control and Automation (2005 ISIC–MED)*, Limassol, Cyprus, pp. 1548–1555, 2005.
23. F.N. Koumboulis, G.E. Panagiotakis, 'Transfer Matrix Approach to the Triangular Decoupling of General Neutral Multi-Delay Systems', *Proceedings of the 17th IFAC World Congress*, Seoul, Corea, pp. 1279–1286, 2008.
24. F. N. Koumboulis, N. D. Kouvakas, and P. N. and Paraskevopoulos, 'Dynamic Disturbance Rejection Controllers for Neutral Time Delay Systems with Application to a Central Heating System', *Proc. International Conference on Modeling, Identification and Control*, Shanghai, China, June 29 – July 2, 2008 / *Science in China Series F: Information Sciences*, vol. 52, no. 7, pp. 1084–1094, 2009.
25. F.N. Koumboulis and M.P. Tzamtzi, 'A metaheuristic approach for controller design of multivariable processes', *12th IEEE International Conference on Emerging Technology on Factory Automation*, Rio-Patras, Greece, 2007, pp. 1429–143.
26. F. N. Koumboulis, N. D. Kouvakas, M. P. Tzamtzi and A. Stathaki, Metaheuristic control of substrate concentration for an activated sludge process, *International Journal of Modelling, Identification and Control*, vol. 10, no. 1, pp. 117–125, 2010.
27. F.N. Koumboulis and M.P. Tzamtzi, 'A generalised control metaheuristic framework for industrial processes', *International Journal of Automation and Control*, vol. 4, no. 2, pp. 218–233, 2010.
28. M. Pedzisz and M. Wojtyra, 'Dynamical Analysis of Semi-Automatic Sniper Rifle', *16th European ADAMS User Conference*, November 14–15, 2001, Berchtesgaden, Germany.

Reliability Analysis of Coherent Systems with Exchangeable Components

M.V. Koutras and I.S. Triantafyllou

Abstract In this paper we study reliability properties of coherent systems consisting of n exchangeable components. We focus on the aging behavior of a reliability structure and several results are reached clarifying whether a system displays the *IFR/DFR* property or not. More specifically, a necessary and sufficient condition is deduced for a system's lifetime to be *IFR*, while additional signature-based conditions aiming at the same direction are also delivered. For illustration purposes, special cases of well-known reliability systems and specific lifetimes' distributions are considered and studied in detail.

Keywords Coherent systems • Increasing failure rate • Samaniego's signature • Minimal and maximal signatures

Introduction

Over the past three decades, much effort has been devoted to the study of reliability characteristics of coherent systems and the special class of consecutive-type structures. This can be attributed to the fact that such systems have been used to model and establish optimal designs of telecommunication networks, oil pipeline systems, vacuum systems in accelerators, spacecraft relay stations, etc. A survey of consecutive-type systems and their generalizations may be found in [1, 3, 5].

Beyond the traditional ways of studying the operation of a reliability system, i.e., through its reliability function or the mean time to failure, a useful tool to characterize reliability systems is the concept of signature which was introduced in [10] (for a review on signatures and their applications see the excellent monograph [9]).

M.V. Koutras (✉) • I.S. Triantafyllou
Department of Statistics and Insurance Science, University of Piraeus, 80 Karaoli and Dimitriou Street, 18534, Piraeus, Greece
e-mail: mkoutras@unipi.gr; itrantal@unipi.gr

N.J. Daras (ed.), *Applications of Mathematics and Informatics in Science and Engineering*, Springer Optimization and Its Applications 91, DOI 10.1007/978-3-319-04720-1_20,

If the component lifetimes of a coherent system are independent and identically distributed with common absolutely continuous reliability function F, then, the system's reliability function can be expressed as a mixture of the reliability functions of the order statistics corresponding to F, with the mixture coefficients being the coefficients in the signature vector [10]. This representation was extended by Navarro and Rychlik [6] to the case of coherent systems with exchangeable component lifetimes and will play an important role in the development of our results. Furthermore, interesting results concerning the aging properties of a structure and stochastic relationships between systems' lifetimes can be established by exploiting properties of the signature vector of the systems.

Most of the published works in the area deal with reliability systems comprising independent components. In the present article we shall present results concerning systems with exchangeable components. We recall that a random vector $(X_1, X_2, \ldots, X_n)$ is called exchangeable if the following equality holds true

$$P(X_1 \leq t_1, X_2 \leq t_2, \ldots, X_n \leq t_n) = P(X_{\pi(1)} \leq t_1, X_{\pi(2)} \leq t_2, \ldots, X_{\pi(n)} \leq t_n),$$

for any permutation $\pi = (\pi(1), \pi(2), \ldots, \pi(n))$ of $\{1, 2, \ldots, n\}$, i.e., the joint distribution (or survival function) of $X_1, X_2, \ldots, X_n$ is symmetric in $t_1, t_2, \ldots, t_n$. The exchangeability means that the components have identical distributions, but they are not necessarily independent, that is they may affect one another within the system.

It is worth stressing that the signature of a system with identical components does not depend on the individual distribution of $X_1, X_2, \ldots, X_n$ since the equality $P(X_1 < X_2 < \ldots < X_n) = P(X_{\pi(1)} < X_{\pi(2)} < \ldots < X_{\pi(n)})$ holds true for any permutation $\pi = (\pi(1), \pi(2), \ldots, \pi(n))$. Therefore, a system with exchangeable components has the same signature vector as the respective system with independent and identical components.

Generally speaking, the results presented in this article have the potential to be applied in military systems, such as telecommunication networks or artillery target coverage. In section "Aging Properties of a Reliability System," we discuss several results pertaining to aging properties of a reliability system. More precisely, we establish conditions, under which the lifetime T of a structure displays an increasing failure rate, namely $T \in IFR$. In section "Signature-Based Sufficient Conditions for *IFR*/*DFR* Property," some easier to apply sufficient conditions, based on system's signatures, are derived (and proved in detail) referring to the aging behavior of a reliability structure with n exchangeable components. For illustration purposes, several examples and special cases of the main results are also presented.

Aging Properties of a Reliability System

Let $X_1, X_2, \ldots, X_n$ denote the component lifetimes of a reliability structure with n components and $X_{1:n} \leq X_{2:n} \leq \ldots \leq X_{n:n}$ the order statistics of $X_1, X_2, \ldots, X_n$.

If T denotes the system's lifetime, the signature vector of the system is defined as the probability vector $(s_1(n), s_2(n), \ldots, s_n(n))$ with entries

$$s_i(n) = P(T = X_{i:n}), \qquad i = 1, 2, \ldots, n.$$

In words, $s_i(n)$ is the proportion of permutations, among the $n!$ possible permutations of $X_1, X_2, \ldots, X_n$, that result in a minimal cut set failure when exactly i components break down (i.e., at time $X_{i:n}$ which represents the lifetime of a i-out-of-n: F system). If $X_1, X_2, \ldots, X_n$ are exchangeable, the signature vector of the system will depend only on its structure and not on the underlying distribution of X_i, $i = 1, 2, \ldots, n$ (which in fact are identically distributed but not necessarily independent).

Samaniego [10] proved that any coherent system with independent and identical components having absolutely continuous distribution functions can be written as a mixture of i-out-of-n: F structures, while Navarro and Rychlik [6] showed that this representation holds also true whenever the components are exchangeable. In addition, Navarro *et al.* [8] proved that any coherent system (in the exchangeable case) can be expressed as a generalized mixture of series or parallel systems. Recapitulating the aforementioned well-known results, we may state that the reliability function $\bar{F}$ of a coherent structure with n exchangeable components can be expressed as follows

$$\bar{F}(t) = P(T > t) = \sum_{i=1}^{n} c_i(n) g_i(t), \text{ for all } t > 0, \tag{1}$$

where

$$c_i(n) = s_i(n) \text{ and } g_i(t) = \bar{F}_{i:n}(t) \tag{2}$$

or

$$c_i(n) = a_i(n) \text{ and } g_i(t) = \bar{F}_{1:i}(t) \tag{3}$$

or

$$c_i(n) = \beta_i(n) \text{ and } g_i(t) = \bar{F}_{i:i}(t) \tag{4}$$

($a_i(n)$, $\beta_i(n)$ are the so-called minimal component signatures and maximal component signatures, respectively). The above representations will be proved useful in the sequel for the study of the failure rate of a reliability system. Generally speaking, the failure rate of a continuous distribution is one of its crucial characteristics and has attracted a lot of research interest because of the wealth of its applications. This notion is of great importance in reliability theory, biostatistics, actuarial science, etc., and due to the wide spectrum of its applications in a diverse number of research areas, many different names such as hazard rate or force of mortality have been used for it in the literature.

The typical definition of the failure rate for an absolutely continuous random variable T with cumulative density function F is given by the formula

$$r(t) = \frac{f(t)}{\bar{F}(t)},$$

where $\bar{F}(t) = 1 - F(t)$ is the survival (or reliability) function, while $f(t) = \bar{F}'(t)$ denotes the corresponding probability density function.

The next proposition offers a necessary and sufficient condition for a reliability system T to display an increasing failure rate function ($T \in IFR$).

Proposition 1. *Assume that T is the lifetime of a system with respective reliability function*

$$\bar{F}(t) = \sum_{i=1}^{n} c_i(n) g_i(t) \tag{5}$$

where $c_i(n), g_i(t)$ are as described in (1)–(4). Then, $T \in IFR$ if and only if the following condition holds true

$$\sum\nolimits_{1 \le i < j \le n} c_i(n) c_j(n) g_{ij}(t) \le 0, \textit{ for all } t \ge 0$$

where

$$g_{ij}(t) = g_i''(t) g_j(t) + g_j''(t) g_i(t) - 2 g_i'(t) g_j'(t). \tag{6}$$

Proof. It is obvious that $T \in IFR$ if and only if $r(t) = -\frac{\bar{F}'(t)}{\bar{F}(t)}$ is increasing in t, for all $t \ge 0$. This means that $r'(t) \ge 0$ for all $t \ge 0$, and making use of (5) we conclude that $T \in IFR$ if and only if the following inequality holds true

$$\left(\sum\nolimits_{i=1}^{n} c_i(n) g_i''(t)\right)\left(\sum\nolimits_{i=1}^{n} c_i(n) g_i(t)\right) \le \left(\sum\nolimits_{i=1}^{n} c_i(n) g_i'(t)\right)^2, \text{ for all } t \ge 0$$

or equivalently

$$\sum\nolimits_{i,j} c_i(n) g_i''(t) c_j(n) g_j(t) - \sum\nolimits_{i,j} c_i(n) g_i'(t) c_j(n) g_j'(t) \le 0, \text{ for all } t \ge 0.$$

We next observe that the last inequality can be rewritten as

$$\sum_{i<j} c_i(n)c_j(n)(g_i''(t)g_j(t)-g_i'(t)g_j'(t))+\sum_{i>j} c_i(n)c_j(n)(g_i''(t)g_j(t)-g_i'(t)g_j'(t))\leq 0$$

or

$$\sum_{i<j} c_i(n)c_j(n)(g_i''(t)g_j(t)-g_i'(t)g_j'(t))+\sum_{i<j} c_j(n)c_i(n)(g_j''(t)g_i(t)-g_j'(t)g_i'(t))\leq 0.$$

Finally, using a single sum for the terms of the LHS of the above expression we get

$$\sum_{i<j} c_i(n)c_j(n)(g_i''(t)g_j(t) - g_i'(t)g_j'(t) + g_j''(t)g_i(t) - g_j'(t)g_i'(t)) \leq 0$$

and the proof is complete. □

The next corollary deals with the special case where the random vector $X_1, X_2, \ldots, X_n$ follows a multivariate *Pareto* distribution.

Corollary 1. *Let us consider a reliability structure whose component lifetimes' random vector* $(X_1, X_2, \ldots, X_n)$ *follows a multivariate Pareto distribution with joint survival distribution function*

$$\bar{F}_a(x_1, x_2, \ldots, x_n) = P(X_1 > x_1, X_2 > x_2, \ldots, X_n > x_n)$$
$$= \left(\sum_{i=1}^{n} x_i - n + 1\right)^{-a}, \quad x_1, x_2, \ldots, x_n > 0,$$

(a is a positive parameter) and denote by $a_i(n), i = 1, 2, \ldots, n$ *the minimal signatures of the n components. Then* $T \in IFR$ *if and only if*

$$\frac{a+1}{a}\sum_{i=1}^{n}\frac{i^2 a_i(n)}{(1+i(t-1))^{a+2}}\sum_{i=1}^{n}\frac{a_i(n)}{(1+i(t-1))^{a}} \leq \left(\sum_{i=1}^{n}\frac{i a_i(n)}{(1+i(t-1))^{a+1}}\right)^2, \; \text{for } t>1.$$

Proof. We shall apply Proposition 1 for the case where coefficients $c_i(n)$ are the minimal signatures of the reliability structure, i.e., $c_i(n) = a_i(n), i = 1, 2, \ldots, n$. Then, under the multivariate *Pareto* distribution, the functions $g_i(t), i = 1, 2, \ldots, n$ (and its first and second derivative) take on the following form

$$g_i(t) = (i(t-1)+1)^{-a}, g_i'(t) = -ia(i(t-1)+1)^{-a-1},$$
$$g_i''(t) = i^2 a(a+1)(i(t-1)+1)^{-a-2}.$$

The result is readily deduced by replacing all these quantities in the necessary and sufficient condition of Proposition 1 and carrying out some trivial algebraic manipulations. □

Another interesting case emerges if we consider a reliability structure consisting of n exchangeable components that follow the multivariate *Farlie-Gumbel-Morgenstern* distribution. In this case we have the following result.

Corollary 2. *Let us consider a reliability structure whose component lifetimes' random vector $(X_1, X_2, \dots, X_n)$ follows a multivariate Farlie-Gumbel-Morgenstern distribution with joint survival distribution function*

$$\bar{F}_a(x_1, x_2, \dots, x_n) = P(X_1 > x_1, X_2 > x_2, \dots, X_n > x_n)$$

$$= \left(1 + a\prod_{i=1}^{n}(1 - e^{-x_i})\right)\exp\left(-\sum_{i=1}^{n} x_i\right), \quad |a| \le 1, \ , \ x_1, x_2, \dots, x_n > 0.$$

and denote by $a_i(n), i = 1, 2, \dots, n$ the minimal signatures of the n components. Then $T \in IFR$ if and only if

$$\sum_{i=1}^{n} a_i(n) i e^{it} \{(i + a(1 - e^t)^{i-2}(i + e^t(4i(e^t - 1) - 1))\}$$

$$\times \sum_{i=1}^{n} a_i(n) e^{it}(1 + a(1 - e^t)^i)$$

$$\le \left(\sum_{i=1}^{n} a_i(n) i e^{it}(1 - a(1 - e^t)^{i-1}(2e^t - 1))\right)^2.$$

Proof. If we choose the minimal signature to play the role of the coefficients $c_i(n), i = 1, 2, \dots, n$ in Proposition 1, then under the multivariate Farlie-Gumbel-Morgenstern distribution we have

$$g_i(t) = (1 + a(1 - e^t)^i)e^{it}, \ g_i'(t) = i e^{it}(1 - a(1 - e^t)^{i-1}(2e^t - 1)),$$

$$g_i''(t) = i e^{it}(e^t - 1)^{-2}((e^t - 1)^2 i + a(1 - e^t)^i(i + e^t(4i(e^t - 1) - 1)).$$

The desired sufficient and necessary condition for the above system to display the IFR property is readily derived by employing analogous arguments as in Corollary 1. □

In the next proposition, we provide a more convenient sufficient condition for a system to have an *IFR* lifetime. The charm of this condition is that in its LHS the part related to the structure of the system (i.e., to the signatures) is separated from the part related to the component lifetimes.

Proposition 2. *Assume that T is the lifetime of a system with respective reliability function*

$$\bar{F}(t) = \sum_{i=1}^{n} c_i(n) g_i(t)$$

where $c_i(n), g_i(t)$ are as described in (1)–(4). If the following holds true

$$\left(\sum_{i=1}^{n} c_i^2(n)\right)\left(\sum_{i=1}^{n} g_i(t) g_i''(t) + \left(\sum_{i=1}^{n} g_i^2(t)\right)^{\frac{1}{2}} \left(\sum_{i=1}^{n} (g_i''(t))^2\right)^{\frac{1}{2}}\right)$$

$$\leq 2\left(\sum_{i=1}^{n} c_i(n) g_i'(t)\right)^2$$

for all $t \geq 0$, then $T \in IFR$.

Proof. Making use of the following bound of the product of two linear forms (see, [11], p. 59)

$$\left(\sum_{i=1}^{n} u_i x_i\right)\left(\sum_{i=1}^{n} v_i x_i\right) \leq \frac{1}{2}\left(\sum_{i=1}^{n} x_i^2\right)\left(\sum_{i=1}^{n} u_i x_i + (\sum_{i=1}^{n} u_i^2)^{1/2} (\sum_{i=1}^{n} v_i^2)^{1/2}\right)$$

we may write

$$\left(\sum_{i=1}^{n} c_i(n) g_i''(t)\right)\left(\sum_{i=1}^{n} c_i(n) g_i(t)\right)$$

$$\leq \frac{1}{2}\left(\sum_{i=1}^{n} c_i^2(n)\right)\left(\sum_{i=1}^{n} c_i(n) g_i''(t) + (\sum_{i=1}^{n} g_i^2(t))^{1/2} (\sum_{i=1}^{n} (g_i''(t))^2)^{1/2}\right).$$

The result follows immediately by combining the last inequality with the necessary and sufficient condition stated in Proposition 1. □

As an illustration, let us consider a series system consisting of two exchangeable components with lifetimes X_1, X_2 that follow the bivariate Log-logistic distribution with parameters β, θ. The joint survival function of the lifetimes vector (X_1, X_2) can be expressed as (see, e.g., [4])

$$\bar{F}(t,t) = \frac{1}{1 + 2(\theta t)^{\beta}},$$

where $\beta > 1, \theta > 0$. The maximal signatures of the aforementioned structure are given by (see, e.g., [7])

$$\beta_1(2) = 2, \beta_2(2) = -1$$

and the corresponding functions $g_i(t)$ will be (see (4))

$$g_i(t) = \frac{1}{1 + i(\theta t)^{\beta}}, i = 1, 2.$$

Since $c_i(n)$ were selected as $c_i(n) = \beta_i(n),\ i = 1, 2, \ldots, n$, in the special case $\beta = \theta = 2$, inequality (6) takes on the form

$$\frac{128t^2(2304t^8 + 96t^6 - 256t^4 - 54t^2 - 3)}{(32t^4 + 12t^2 + 1)^4} \leq 0.$$

It is not difficult to verify that the last formula holds true for all real values of t, such that $0 < t < 0.6335821$, therefore we deduce, by applying Proposition 1, that a series system consisting of two exchangeable log-logistic components belongs to the *IFR* class for the aforementioned range of t. It is worth mentioning that the above result agrees with the one established for the same reliability structure by Eryilmaz [4] (see Proposition 3.3 therein).

Signature-Based Sufficient Conditions for *IFR/DFR* Property

In this section, we provide sufficient conditions for a reliability system to own the *IFR* (*DFR*) property. More specifically, we use the notion of Samaniego's signature of a reliability system, in order to reach several results that refer to its aging properties.

Proposition 3. *Assume that T is the lifetime of a system with respective reliability function $\bar{F}(t)$. If the following holds true*

$$\min(1, n\bar{F}(t)) \sum_{i=1}^{n} s_i(n)\bar{F}''_{i:n}(t)$$

$$\leq \left(\sum_{i=1}^{n} s_i(n)\bar{F}'_{i:n}(t)\right)^2$$

for all $t \geq 0$, then $T \in IFR$.

Proof. Since $0 \le s_i(n), \bar{F}_{i:n}(t) \le 1$, for all $i = 1, 2, \ldots, n$, we have

$$\sum_{i=1}^{n} s_i(n)\bar{F}_{i:n}(t) \le \begin{cases} \sum_{i=1}^{n} s_i(n) = 1 \\ \sum_{i=1}^{n} \bar{F}_{i:n}(t) = n\bar{F}(t) \end{cases} \tag{7}$$

We next apply the main result of Proposition 1 for $c_i(n) = s_i(n)$ and $g_i(t) = \bar{F}_{i:n}(t)$, and the result we are chasing is immediately reached by using the above inequalities. □

In the next proposition, we provide a sufficient condition for the *IFR* property of a reliability system, based on the lower bound of its signatures.

Proposition 4. *Assume that T is the lifetime of a system with respective reliability function $\bar{F}(t)$ and denote by b_n a lower bound of the sequence of non-zero entries of its signature vector, i.e.,*

$$s_i(n) \ge b_n, \text{ for all } i \text{ such that } s_i(n) \ne 0.$$

If the following holds true

$$\left(\sum_{i=1}^{n} s_i(n)\bar{F}''_{i:n}(t)\right)\left(\sum_{i=1}^{n} s_i(n)\bar{F}_{i:n}(t)\right)$$
$$\le b_n^2\left(n + \sum_{i=1}^{n}\sum_{r=n-i+1}^{n} (-1)^{r-n+i-1}\binom{r-1}{n-i}\binom{n}{r}\bar{F}'_{1:r}(t)\right)^2$$

for all $t \ge 0$, then $T \in IFR$.

Proof. Since

$$s_i(n) \ge b_n, \text{ for } i = 1, 2, \ldots, n,$$

we conclude that

$$\sum_{i=1}^{n} s_i(n)\bar{F}'_{i:n}(t) \ge b_n \sum_{i=1}^{n} \bar{F}'_{i:n}(t).$$

Recalling the following well-known formula (see, e.g., [2]. p. 46)

$$\bar{F}'_{i:n}(t) = \sum_{r=n-i+1}^{n} (-1)^{r-n+i-1}\binom{r-1}{n-i}\binom{n}{r}\bar{F}'_{1:r}(t) \tag{8}$$

the result is a straightforward application of Proposition 1. □

A simpler sufficient condition, compared to the one given in Proposition 4, is offered by the following corollary.

Corollary 3. *Assume that T is the lifetime of a system with respective reliability function $\bar{F}(t)$ and b_n a lower bound of the sequence of non-zero entries of its signature vector. If the following condition holds true*

$$\min(1, n\bar{F}(t)) \sum_{i=1}^{n} s_i(n)\bar{F}''_{i:n}(t)$$

$$\leq b_n^2 \left(n + \sum_{i=1}^{n} \sum_{r=n-i+1}^{n} (-1)^{r-n+i-1} \binom{r-1}{n-i} \binom{n}{r} \bar{F}'_{1:r}(t) \right)^2$$

for all $t \geq 0$, then $T \in IFR$.

Proof. Recalling inequality (7), it is not difficult to verify that

$$\left(\sum_{i=1}^{n} s_i(n)\bar{F}''_{i:n}(t) \right) \left(\sum_{i=1}^{n} s_i(n)\bar{F}_{i:n}(t) \right) \leq (\min(1, \bar{F}_{i:n}(t)) \left(\sum_{i=1}^{n} s_i(n)\bar{F}''_i(t) \right).$$

The desired result is now immediately reached by combining the last formula with (8) and Proposition 1. □

In the next proposition, we provide a sufficient condition for the *IFR* / *DFR* property of a reliability system, based on an upper bound of its signatures.

Proposition 5. *Assume that T is the lifetime of a system with respective reliability function $\bar{F}(t)$ and let B_n denote an upper bound of the system's signatures, i.e.,*

$$s_i(n) \leq B_n, \text{ for } i = 1, 2, \ldots, n.$$

(i) If the following condition holds true

$$(nB_n\bar{F}(t)) \left(\sum_{i=1}^{n} s_i(n)\bar{F}''_{i:n}(t) \right) \leq \left(\sum_{i=1}^{n} s_i(n)\bar{F}'_{i:n}(t) \right)^2$$

for all $t \geq 0$, then $T \in IFR$.

(ii) If the following condition holds true

$$\left(\sum_{i=1}^{n} s_i(n)\bar{F}''_{i:n}(t) \right) \left(\sum_{i=1}^{n} s_i(n)\bar{F}_{i:n}(t) \right)$$

$$\geq B_n^2 \left(\sum_{i=1}^{n} \sum_{r=n-i+1}^{n} (-1)^{r-n+i-1} \binom{r-1}{n-i} \binom{n}{r} \bar{F}''_{1:r}(t) \right)^2$$

for all $t \geq 0$, then $T \in DFR$.

Proof. Since

$$s_i(n) \leq B_n, \text{ for } i = 1, 2, \ldots, n,$$

it is obvious that

$$\sum_{i=1}^{n} s_i(n)\bar{F}'_{i:n}(t) \le B_n \sum_{i=1}^{n} \bar{F}'_{i:n}(t)$$

and

$$\sum_{i=1}^{n} s_i(n)\bar{F}_{i:n}(t) \le B_n \sum_{i=1}^{n} \bar{F}_{i:n}(t) = nB_n\bar{F}(t).$$

In view of expression (8) we may state the following

(i) If

$$nB_n\bar{F}(t)\sum_{i=1}^{n} s_i(n)\bar{F}''_{i:n}(t) \le \left(\sum_{i=1}^{n} s_i(n)\bar{F}'_{i:n}(t)\right)^2$$

then, $(nB_n\bar{F}(t))\left(\sum_{i=1}^{n} s_i(n)\bar{F}''_{i:n}(t)\right) \le \left(\sum_{i=1}^{n} s_i(n)\bar{F}'_{i:n}(t)\right)^2$ and Proposition 3 guarantees that $T \in IFR$.

(ii) If

$$\left(\sum_{i=1}^{n} s_i(n)\bar{F}''_{i:n}(t)\right)\left(\sum_{i=1}^{n} s_i(n)\bar{F}_{i:n}(t)\right) \ge B_n^2\left(\sum_{i=1}^{n} \bar{F}'_{i:n}(t)\right)^2$$

then,

$$\left(\sum_{i=1}^{n} s_i(n)\bar{F}''_{i:n}(t)\right)\left(\sum_{i=1}^{n} s_i(n)\bar{F}_{i:n}(t)\right)$$

$$\ge B_n^2\left(\sum_{i=1}^{n}\sum_{r=n-i+1}^{n} (-1)^{r-n+i-1}\binom{r-1}{n-i}\binom{n}{r}\bar{F}''_{1:r}(t)\right)^2$$

and Proposition 3 guarantees that $T \in DFR$.

□

References

1. Chao, M. T., Fu, J. C. and Koutras, M. V.: Survey of reliability studies of consecutive−k−out−of−n: F & Related Systems, *IEEE Transactions on Reliability,* **44**, 120–127 (1995).
2. David, H. A. and Nagaraja, H. N.: *Order Statistics*, 3rd Edition, John Wiley & Sons, N. J (2003).

3. Eryilmaz, S.: Review of recent advances in reliability of consecutive-k-out-of-n: F and related systems, *Proceedings of the Institution of Mechanical Engineering-Part O- Journal of Risk and Reliability,* **224**, 225–237 (2010).
4. Eryilmaz, S.: Reliability properties of systems with two exchangeable log-logistic components, *Communication in Statistics-Theory and Methods*, **41**, 3416–3427 (2012).
5. Kuo, W. and Zuo, M. J.: *Optimal Reliability Modeling: Principles and Applications*, John Wiley & Sons, N.J (2003).
6. Navarro, J. and Rychlik, T.: Reliability and expectation bounds for coherent systems with exchangeable components, *Journal of Multivariate Analysis*, **98**, 102–113 (2007).
7. Navarro, J. and Eryilmaz, S.: Mean residual lifetimes of consecutive-k-out-of-n: F systems, *Journal of Applied Probability*, **44**, 82–98 (2007).
8. Navarro, J., Ruiz, J. M. and Sandoval, C. J.: Properties of coherent systems with dependent components, *Communication in Statistics- Theory and Methods*, **36**, 175–191 (2007).
9. Samaniego, F. J.: *System Signatures and their Applications in Engineering Reliability*, Vol.110, Springer, N. Y (2007).
10. Samaniego, F. J.: On closure of the *IFR* class under formation of coherent systems, *IEEE Transactions on Reliability*, **34**, 69–72 (1985).
11. Steele, J. M.: *The Cauchy-Schwarz Master Class*, Cambridge University Press, N. Y (2004).

Abductive Reasoning in 2D Geospatial Problems

Achilleas Koutsioumpas

Abstract Spatial Analysis has been using so far Spatial Reasoning, but it mainly confines itself to spatial statistical analysis of the observed phenomena searching for pattern analysis, geostatistical indices etc. Researchers Shakarian P., Subrahmanian V. S., and Sapino M. L. are the first who examined the possibility of extending Spatial Analysis in finite, discretized, 2D space with the incorporation of Abductive Reasoning, which originates from the cognitive field of Artificial Intelligence, and is related to the analysis of causation of the phenomena under consideration. The new class of Geospatial Problems was named point-based Geospatial Abduction Problems (or point-based GAPs). They primarily focused on a version of GAPs named Improvised Explosive Devices Cache Detection Problem, or IED Cache Detection Problem, and they carried out experiments with real-world data from Baghdad. In this paper a technique which reduces the total computational cost in any version of point-based GAPs will be introduced, and an exact algorithm for the natural optimization problem of point-based GAPs will be presented along with its computational complexity results.

Keywords Abductive reasoning • Geospatial abduction problems • Spatial problems • IED cache detection problem

[1]This research has been conducted in the context of the writer's MSc Thesis, which was supervised by Professor Marinos Kavouras, in the field of Geoinformatics at the National Technological University of Athens (NTUA).

A. Koutsioumpas (✉)
Department of Mathematics and Engineering Sciences, Hellenic Military Academy, 166 73, Vari, Greece
e-mail: axilleas_kouts@yahoo.com

N.J. Daras (ed.), *Applications of Mathematics and Informatics in Science and Engineering*, Springer Optimization and Its Applications 91, DOI 10.1007/978-3-319-04720-1_21,

Introduction

Spatial Analysis has been using so far Spatial Reasoning, but it mainly confines itself to spatial statistical analysis of the observed phenomena searching for pattern analysis, geostatistical indices etc. A trivial example is the spatial analysis of the crime phenomenon in a given region, which will result in finding subregions that are statistically characterized by high rates of crime. The most common strategy in response is the increase of policing in those subregions. However, that kind of strategy does not affect the cause of the phenomenon under consideration, i.e. the criminals, but only the observations of the phenomenon. Thus, the crime will either shift or spread out to the remaining space, as the increase of policing in a set of subregions implies the decrease in the rest of the remaining space, which in turn implies that criminals are offered now new opportunities to carry out their illegal activities in those exact under-policing subregions. On the contrary, the best strategy for law enforcement would be the use of a spatial relationship, if such exists, between the locations where the crimes were committed and the locations of the hideouts of the criminals which obviously would pin-point the latter, given the fact that the locations of crime incidents are known to the police, and the criminologists can distinguish the crime signatures, and thus allocate the crimes accordingly. That kind of reasoning can be applied in a wide variety of real-world problems named Geospatial Abduction Problems or GAPs such as the detection of terrorist caches given the location of their attacks, the detection of dwellings of a wild animal given the location of its attacks, the detection of environmental pollutants, etc. [13]. Informally, a GAP is a problem in geospatial space of finding a set of partner locations that best "explain" a given set of locations of observations through a given spatial relationship. The spatial relationship for any GAP, even for instances of the same problem, is unique as it expresses the correlation between partner and observations in the context of a specific domain, in a specific geographical space, and for a specific period of time.

The complexity of human cognition and the difficulty of its simulation by mechanic means are proven by the fact that since the Aristotelian logic was formulated till nowadays various inference rules and kinds of reasoning have been proposed. Informally, reasoning defines a rough theoretical context in which valid inference rules can be applied, whereas inference rules are formally defined mechanisms, algorithms, which produce new valid knowledge [16], or in other words, logical transformations used to produce conclusions from a premise set. Some of the most common inference rules are:

- The Modus Ponens rule which is used in Mathematical Logic, where if one formula implies another one and the first formula is TRUE, then the second one is also TRUE. Formally: $A \rightarrow B, A \vdash B$.
- The Modus Tollens rule, which is also used in Mathematical Logic, where if one formula implies another one and the latter formula is FALSE, then the first formula is also FALSE. Formally: $A \rightarrow B, \neg B \vdash \neg A$.
- The generalized Modus Ponens and Modus Tollens inference rules in the domain of Fuzzy Logic.

- The Generalization rule, where if an attribute is TRUE for some instances of a class, then the attribute is TRUE for any instance of the class.
- The Specification rule, where if an attribute is TRUE for a class, then the attribute is TRUE for any instance of the class.

On the other hand, there are three fundamental kinds of reasoning:

- Deductive Reasoning
- Inductive Reasoning
- Abductive Reasoning

More specifically, Deductive Reasoning infers conclusions of what is assumed, thus given the truth of the assumptions, the conclusion is guaranteed to be TRUE. Assumptions are given in the form of a rule and a fact. For example:

<Rule> All my children are boys.

<u><Fact> This child is mine.</u>

<Conclusion> This child is a boy.

Inductive Reasoning infers a rule from a set of facts. Even if the set of facts are TRUE and the derived rule seems to be reasonable, this does not ensure the soundness of the rule. It must be noted that this kind of reasoning has nothing in common with the mathematical induction since the latter is a form of Deductive Reasoning. For example:

<Fact 1> This boy is my child.

<u><Fact 2> This boy is also my child.</u>

<Rule> All my children are boys.

Abductive Reasoning infers a hypothesis given a set of rules, which consists a knowledge basis, and a set of facts, that consists a set of observations. For example:

<Rule> All my children are boys.

<u><Observation> This child is a boy.</u>

<Hypothesis> This child is mine.

Obviously, there is no such thing as "best" reasoning, since the above kinds of reasoning are not competing against each other. On the contrary, they are complementary since each one's start point differs, thus unavoidably each one infers a different kind of output.

Despite the fact that the term Abductive Reasoning was proposed by the great American philosopher Peirce in 1890 [6], in order to describe the kind of reasoning used for evaluating explanatory hypotheses for a given set of observations, philosophers and scientists were familiar with the context of the term from the Renaissance at least [1]. Some thinkers have been skeptical that a hypothesis should be accepted merely, but others have argued that this kind of reasoning is a legitimate part of scientific reasoning [15]. Clearly any feasible explanatory hypothesis, or simply explanation, even if it is supposed to be the inference to the best explanation, is fraught with uncertainty [15] due to the fact that abduction is equivalent to the logical fallacy affirming the consequent, or else in Latin "post hoc ergo propter hoc." Nevertheless, Abductive Reasoning plays a key role in human cognition. For example when: scientists produce new theories to explain their data, criminologists compose the evidence from criminal activities, mechanics try to find out what problem is responsible for a mechanical breakdown, doctors try to infer which disease explains a patient's symptom, even when people generate hypotheses to explain the behavior of others [15], they all use abduction. What is left is the formal semantic definition of the term explanation. In philosophy and cognitive science, there have been at least six approaches; explanations have been viewed as deductive arguments, statistical relations, schema applications, analogical comparisons, causal relations and finally, linguistic acts [14]. Geospatial Abduction uses the approach of causal relation in the context of a specific domain. In other words, Geospatial Abduction is used to infer as the best explanation in a specific domain the relation between the locations of the observed phenomena and the partner locations which cause, facilitate, support or are somehow correlated with the observations [13].

Formalism-Technical Preliminaries

To address the Geospatial Abduction for the new class of problems named point-based GAPs, researchers Shakarian, Subrahmanian and Sapino proposed an appropriate formalism [8, 12, 13], which is adopted in this paper, as follows:

Definition 1 (Space). Geospatial universe or universe or simply space S is a finite two-dimensional grid of size $M \times N \in \mathbb{N}^2$, where $M, N \geq 1$ and $\mathbb{N}$ is the set of natural numbers.

Definition 2 (Point). A point p of a finite 2D space S is a representation of a unit square on the grid in the form (x, y) where $x, y \in \mathbb{N}$. Therefore, $S = \{x \mid x \in [0, M)\} \times \{y \mid y \in [0, N)\}$.

In addition, conventionally the coordinates (x, y) in each such unit cell are not assigned to its center, but to its lower left corner, and the point with coordinates (0,0) is the lower left point of the grid. Thus, the space is used to represent a specific region of interest on the ground and the size of the point in ground distance units defines the space resolution. The suitable extents of the space are normally easily defined by the GAP application itself whereas the resolution of the space is defined "arbitrarily" by the application developer or the analyst.

Definition 3 (Observation—Set of Observations). An observation o is a point of a finite 2D space S where a phenomenon under consideration appeared. A set of observations $O = \{o_1, o_2 \ldots o_k\}$ is any finite subset of space S where each element o_i is an observation of the phenomenon under consideration.

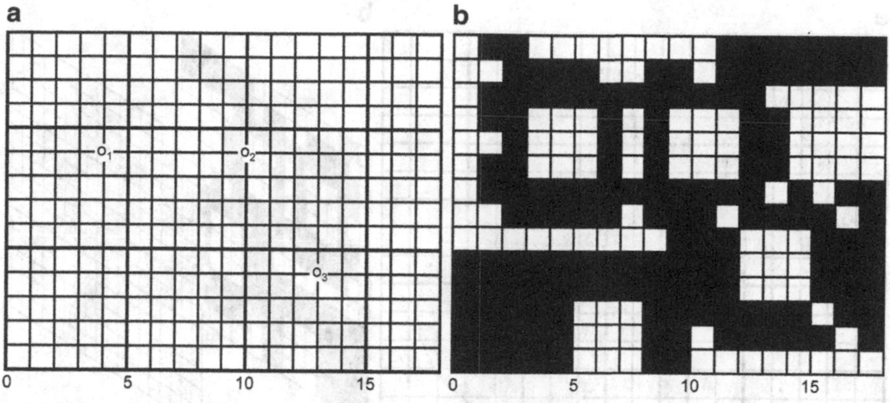

Fig. 1 Examples of a set of observations and a feasibility predicate in space. (**a**) A set of observations $O = \{o_1, o_2, o_3\}$ in space S. (**b**) A feasibility predicate in space S

Figure 1a illustrates the above concepts in a space 18×14. Furthermore, any space S has an associated distance function d that satisfies the following definition.

Definition 4 (Distance Function). A distance function d in metric space (S, d) is a mapping $S \times S \to \mathbb{R}$, where $\mathbb{R}$ is the set of real numbers, such that for any $x, y \in S$ the following axioms are satisfied:

- $d(x, x) = 0$, coincidence axiom.
- $d(x, y) = d(y, x)$, symmetry axiom.
- $d(x, y) + d(y, z) \geq d(x, z)$, triangle inequality axiom.

There are numerous distance functions that satisfy the above three axioms. The more familiar are the Euclidean distance, where $d_e = [(x_1 - x_2)^2 + (y_1 - y_2)^2]^{1/2}$; the Manhattan distance, where $d_m = |x_1 - x_2| + |y_1 - y_2|$, which are special cases of the Minkowski or L_p-metrics [7]; and the road distance $d_r((x_1, y_1), (x_2, y_2))$ that is defined as the shortest path between two points in a road network.

Definition 5 (Feasibility Predicate). A feasibility predicate $feas$ in a finite 2D space S is a function $feas : S \to [TRUE, FALSE]$ such that $feas(p) = TRUE$ for any point $p \in S$ if and only if (iff) that point can be included in a feasible solution of the GAP or $feas(p) = FALSE$ otherwise.

It is noted that the feasibility predicate *feas* is an arbitrary, yet a fixed predicate. It is an arbitrary predicate in the sense that both the semantics for feasibility and its spatial assignment in space are assessed by the analyst. This means that for any space S there are offered 2^n distinct alternative process choices for an analyst, given that $|S| = n$. Thus, two analysts, A and B, even for the same geospatial problem in the space S, will most likely come up to $feas_A \neq feas_B$. At the same time it is a fixed predicate in the mathematical sense, since for any point $p \in S$ is either $feas(p) = TRUE$ or $feas(p) = FALSE$. In terms of computational complexity,

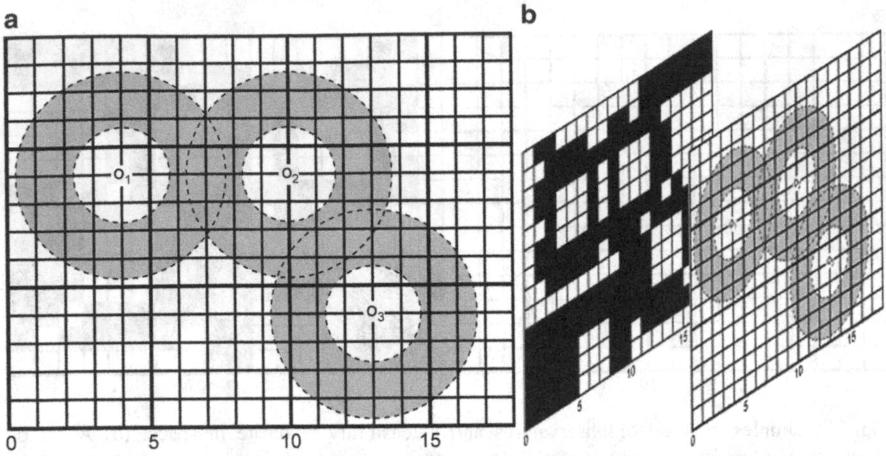

Fig. 2 Examples of geometric restrictions and their logical composition with feasibility predicate. (**a**) Geometric restrictions form $|O|$ concentric rings in space S. (**b**) The (α, β)-explanation is a logical composition in space S

since it is user-defined, it is assumed that it has an O(1) complexity cost. Figure 1b above illustrates in color a feasibility predicate in space S, white indicates that $feas(p) = TRUE$ and black indicates that $feas(p) = FALSE$.

At this point we can define the concept of an explanation in the framework of geospatial abduction. Intuitively, given a set of observations an explanation is a set of points in space such that every point is feasible and such that for every observation there is a point in explanation that satisfies certain geometric, or distance, restrictions. The geometric restrictions, which are set by the analyst or the domain expert, define the minimum and the maximum allowed metric distances between the location of an observation and its partner location. Formally:

Definition 6 ((α, β)-Explanation). Let O be a set of observations in a 2D finite space S, E be a finite set of points where $E \subset S$ and $\alpha, \beta \in \mathbb{R}$ such that $0 \leq \alpha < \beta$, then E is an (α, β)-explanation of O iff:

$$[\forall p \in E | feas(p) = TRUE] \wedge [\forall o \in O \exists p \in E | \alpha \leq d(p, o) \leq \beta]$$

In Fig. 2 are illustrated, for the running example, the geometric restrictions in space (Fig. 2a) and the logical composition of the geometric restrictions and the feasibility predicate (Fig. 2b), given that the size of the grid cell is $500m$, $\alpha = 850m$ and $\beta = 1850m$. The composition produces the set of partner locations for the given set of observations. Any subset of this set that explains all the observations is a possible solution.

The simplest form of point-based GAPs is the decision problem named the Simple (α, β)-Explanation Problem (SEP), where given a finite 2D space S, a finite set of observations O, a feasibility predicate $feas$ and $\alpha, \beta \in \mathbb{R}$, such that

$0 \leq \alpha < \beta$, the question is if there is an (α, β)-explanation for O. Another variant of GAPs is the k-SEP problem, which defines an upper bound in the cardinality of any possible solution. More specifically, it is required that $|E| \leq k$ for a number $k \in \mathbb{N}$, given that $k < |O|$.

Generally, the GAPs are highly non-trivial to solve for four reasons [13]. First of all, the cardinality of the solution is unknown. Secondly, the feasibility predicate defines highly irregular in shape areas for partner locations, thus simple geometric reasoning cannot be applied. Thirdly, in case that the phenomenon under consideration involves a rational adversary who will take any action to evade detection (in terms of Game Theory this is an adaptive adversary), then a simple algorithmic solution would not be sufficient and more sophisticated algorithms must be applied. Finally, the most GAPs are proven to be NP-complete, in terms of computational complexity, or in other words intractable to compute in practice.

Since there are many possible explanations for a given set of observations the truly intriguing variants of point-based GAPs are the ones which try to find an "optimal" explanation according to some cost function χ, where a cost function is a mapping from the set of explanations to non-negative reals.

Definition 7 (Optimal (α, β)-Explanation). Suppose S is a finite 2D space, $O \subset S$ is a finite set of observations, $E \subset S$ is a finite set of points, $\alpha, \beta \in \mathbb{R}$ such that $0 \leq \alpha < \beta$ and χ is a cost function. Then E is said to be an optimal (α, β)-explanation iff E is an (α, β)-explanation for O, and there is no other (α, β)-explanation E' for O such that $\chi(E') < \chi(E)$.

One can easily prove that standard classification problems like k-means [4, 5] are a special case of GAPs by making three simple assumptions: a) $\alpha = 0$, b) $\beta > \max(M, N)$ and c) $\nexists p \in S | feas(p) = FALSE$ [13].

Normally in philosophy and in science additional requirements are enforced to an explanation, the parsimony requirements. In SEP no such requirement is enforced, since any explanation will suffice. On the other hand, k-SEP and cost-based SEP enforce parsimony requirements. Another parsimony requirement, which expresses the famous principle known as the Occam's razor: "Pluritas non est ponenda sine necessitate," is irredundancy. In Geospatial Abduction this means that between two explanations the simplest one must be chosen. The formal definition follows:

Definition 8 (Irredundant (α, β)-Explanation). An (α, β)-explanation E is irredundant iff no strict subset of E is an (α, β)-explanation.

One may think that by enforcing the irredundancy requirement in GAPs the uniqueness of the solution is guaranteed, but this is far from being true. A closer look easily shows that there still exist an exponential number of such solutions, thus an algorithm that simply returns irredundant solutions could produce non-deterministic results [13]. The best one can do is to attempt to find explanations of minimal cardinality. This requirement poses the Minimal Simple (α, β)-explanation Problem (MINSEP) which is the natural optimization problem associated with k-SEP [8]. Formally:

Definition 9 (MINSEP).

INPUT: a finite 2D space S, a finite set of observations O, a feasibility predicate *feas* and $\alpha, \beta \in \mathbb{R}$ such that $0 \leq \alpha < \beta$.

OUTPUT: An (α, β)-explanation E such that $|E| = minimal$.

Algorithms

In this chapter a technique which reduces the total computational cost in any version of GAPs will be proposed, and an exact algorithm for the natural optimization problem of GAPs, the MINSEP, will be presented.

Even though mathematically for point-based GAPs, it is sufficient to require for the relation between α, β and space S that $\alpha, \beta < inf\{M, N\}$, this obviously does not maximize the utility of the set of observations. In order to accomplish that the requirement for the relation between α, β and space S has to be rewritten as follows:

$$\forall p(x, y) \in O \text{ then } M, N | [\lfloor x+\beta \rfloor \leq M] \wedge [\lfloor y+\beta \rfloor \leq N] \wedge [\lceil x-\beta \rceil \geq 0] \wedge [\lceil y-\beta \rceil \geq 0]$$

But in reality, in most applications, the analyst faces the exact opposite problem; the size of the given space is bigger than the necessary. Thus, in order to succeed effective computation, space must somehow be reduced to the exact size needed. A technique that reduces the given space S to the absolutely necessary space S' is: given that $x_{\min}$, $x_{\max}$, $y_{\min}$ and $y_{\max}$ are the minimum and the maximum values of coordinates for the observation set, then it is easy to show that any explanation, with respect to the geometric restrictions, is strictly in the space S' which equals to the Minimum Bounding Rectangular box (MBR) with coordinates in space S:

$$(X_{\min}, Y_{\min}) = (\lceil x_{\min} - \beta \rceil, \lceil y_{\min} - \beta \rceil)$$

$$(X_{\max}, Y_{\max}) = (\lfloor x_{\max} + \beta \rfloor, \lfloor y_{\max} + \beta \rfloor)$$

The reduction is completed with the parallel shift of the coordination system to the point with coordinates $(\lceil x_{\min} - \beta \rceil, \lceil y_{\min} - \beta \rceil)$, in order for space S' to have coordinates (0,0) in its lower left point of the grid. One can easily show that the size of space $S' = M' \times N'$ is:

$$M' = (\lfloor x_{\max} + \beta \rfloor - \lceil x_{\min} - \beta \rceil), \text{ where } M' \in [1, M]$$

$$N' = (\lfloor y_{\max} + \beta \rfloor - \lceil y_{\min} - \beta \rceil), \text{ where } N' \in [1, N]$$

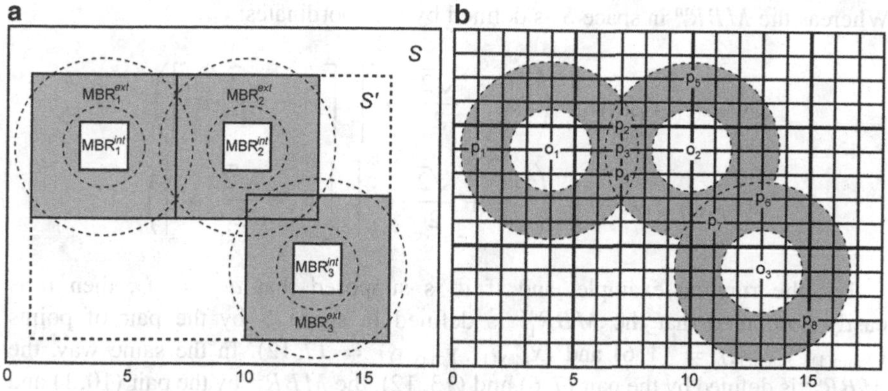

Fig. 3 Examples of the reduction for the given space, the MBRs for the set of observations and the set of partner locations. (**a**) The reduction of space S to space S' and the MBRs of the set of observations. (**b**) The set of partner locations for the (1.7, 3.7)-explanation

Practically, for the running example since $\beta = 1850/500 = 3.7$ metric units in space S, then the coordinates of space S' are calculated as follows and the result is illustrated in Fig. 3a.

$$(X_{\min}, Y_{\min}) = (\lceil 4 - 3.7 \rceil, \lceil 4 - 3.7 \rceil) = (\lceil 0.3 \rceil, \lceil 0.3 \rceil) = (1, 1)$$

$$(X_{\max}, Y_{\max}) = (\lfloor 13 + 3.7 \rfloor, \lfloor 9 + 3.7 \rfloor) = (\lfloor 16.7 \rfloor, \lfloor 12.7 \rfloor) = (16, 12)$$

In order to achieve more effective computation the use of MBRs can be applied in the set of observations as well. Instead of searching for partner locations generally in space S' for any observation o_i, the search can be focused in an appropriate subspace $S_i \subseteq S'$. Suppose that MBR_i^{ex} is an MBR which contains any point that is no more than β metric units away from the observation o_i and that MBR_i^{in} is an MBR which contains any point that is at least α' metric units away from the observation o_i, where $\alpha' \leq \alpha$ such that $\alpha' = mind_e(o_i, p_j) | [d(o_i, p_j) = \alpha]$. Obviously, if the distance function used in space is the Euclidean distance d_e, then $\alpha' = \alpha$. Now, it is easy to see that space S_i is defined as $S_i = MBR_i^{\text{ex}} - MBR_i^{\text{in}}$. Trivially it can be proven that both MBR_i^{ex} and MBR_i^{in} are squares, where geometrically the first one equals to a square in which a circle of radius β is inscribed, while the latter equals to a square inscribed in a circle of radius α'. The MBR_i^{ex} in space S is defined by the coordinates:

$$(x_{\min(i)}, y_{\min(i)}) = (\lceil x_i - \beta \rceil, \lceil y_i - \beta \rceil)$$

$$(x_{\max(i)}, y_{\max(i)}) = (\lfloor x_i + \beta \rfloor, \lfloor y_i + \beta \rfloor)$$

Whereas the MBR_i^{in} in space S is defined by the coordinates:

$$(x_{\min(i)}, y_{\min(i)}) = \left(\left\lceil x_i - \frac{\sqrt{2}}{2}\cdot\alpha'\right\rceil, \left\lceil y_i - \frac{\sqrt{2}}{2}\cdot\alpha'\right\rceil\right)$$

$$(x_{\max(i)}, y_{\max(i)}) = \left(\left\lfloor x_i + \frac{\sqrt{2}}{2}\cdot\alpha'\right\rfloor, \left\lfloor y_i + \frac{\sqrt{2}}{2}\cdot\alpha'\right\rfloor\right)$$

For the running example, and if it is supposed that $d = d_e$, then it is easily computed that the MBR_1^{ex} is defined in space S by the pair of points $(x_{\min(1)}, y_{\min(1)}) = (1, 6)$ and $(x_{\max(1)}, y_{\max(1)}) = (7, 12)$. In the same way, the MBR_2^{ex} is defined by the pair $(7, 6)$ and $(13, 12)$, the MBR_3^{ex} by the pair $(10, 1)$ and $(16, 7)$, whereas MBR_1^{in} by the pair $(3, 8)$ and $(5, 10)$, the MBR_2^{in} by the pair $(9, 8)$ and $(11, 10)$, and finally the MBR_3^{in} by the pair of points $(12, 3)$ and $(14, 5)$. These results are illustrated in Fig. 3a. Additionally in Fig. 3b is illustrated the $(1.7, 3.7)$-explanation of maximum cardinality, or in other words the set of partner locations, in space S for the running example.

Despite the fact that the SEP problem has a proven computational complexity in PTIME [8, 12, 13], its outputs are not reliable because any (α, β)-explanation E, independently of its cardinality, suffices. Thus, an algorithm for SEP would produce, with high probability, an explanation that is redundant. On the other hand, real-world data from Baghdad for the IED Cache Detection Problem, which were experimentally tested in order to show that the Geospatial Abduction framework is viable, proved that the cardinality of the solution plays a key role in the accuracy [8, 12, 13]. But shifting from a SEP problem to a k-SEP problem (or in other words the GAP variant which imposes cardinality constraints to the acceptable explanations) is not cost-free, since it automatically implies NP-completeness. So, since an exact solution to k-SEP takes exponential time (in k), it follows that any efficient algorithm would be nothing more than a fundamental trade-off between accuracy and computational cost. In the end, among the three approximation algorithms that were tested, the algorithm which was chosen to be implemented in SCARE (Social-Cultural Abductive Reasoning Engine) was not the fastest one, but the algorithm that consistently returned the solution of the smallest cardinality [8, 12, 13].

The algorithm NAIVE-MINSEP-EXACT below includes the technique that was presented above, which reduces the total computational cost, and provides an exact solution for MINSEP by leveraging the NAIVE-KSEP-EXACT algorithm [8].

It must be noted that the NAIVE-MINSEP-EXACT algorithm is flexible enough to include the domain's expert assessment, if such exists, of the expected cardinality of the solution, k. If this kind of information is unavailable or its use bears high risk, then a simple modification will be needed in the body of the algorithm. Just a simple replacement in the 5th and 6th step of the number k with the cardinality of the set of observations $|O|$ suffices. Furthermore, the search for the size of the minimal size explanation is performed by applying the technique "divide and conquer," which guarantees that no more than $\log_2 k$ (respectively $\log_2 |O|$) trials will be needed at the most [2].

Algorithm 1 NAIVE-MINSEP-EXACT

INPUT: Space S, set of observations O, feasibility predicate $feas$, distance function $d \in \mathbb{R}$, $\alpha, \beta \in \mathbb{R}$ such that $0 \leq \alpha < \beta$ and $k \in \mathbb{N}$.
OUTPUT: Set $E \subseteq S$ such that $|E| \leq k$ and $E = minimal$.

1. Define distance α' and space S'.
2. Let M be a matrix array of pointers to binary string $\{0, 1\}^{|O|}$. M is of the same dimensions as S'.
   Each element in M is initialized to NULL. For a given $p \in S'$, $M[p]$ is the place in the array.
3. Let L be a list of pointers to binary strings. L is initialized as NULL.
4. For each $o_i(x_i, y_i) \in O$ do the following:
   a. Determine all points $p_j(x_j, y_j) \in S'$ such that
   $(x_j, y_j) \in [\lceil x_i - \beta \rceil, \lfloor x_i + \beta \rfloor] \wedge [\lceil y_i - \beta \rceil, \lfloor y_i + \beta \rfloor]$ and
   $(x_j, y_j) \notin \left[\left\lceil x_i - \frac{\sqrt{2}}{2} \cdot \alpha' \right\rceil, \left\lceil y_i - \frac{\sqrt{2}}{2} \cdot \alpha' \right\rceil\right] \wedge \left[\left\lfloor x_i + \frac{\sqrt{2}}{2} \cdot \alpha' \right\rfloor, \left\lfloor y_i + \frac{\sqrt{2}}{2} \cdot \alpha' \right\rfloor\right]$ and
   $feas(p_j) = TRUE$.
   b. For each of these points $p_j(x_j, y_j)$:
   If $M[p] = NULL$, then
   (1) Initialize a new array where only bit i is set to 1.
   (2) Add a pointer to $M[p]$ in L.
   Else, set bit i of the existing array to 1.
5. Let $n \in \mathbb{N}^*$ such that $n \leq k$. For any n elements of L (actually the n elements pointed to by elements of L), we shall designate $l_1, l_2 \ldots l_j \ldots l_n$ as the elements. The i-th bit of element $l_j(i)$ will be referred as $l_j(i)$.
6. Exhaustively generate all possible combinations of n elements, for every $n \leq k$, starting from $n = \frac{k}{2}$ and by applying the technique "divide and conquer" until one such combination of minimal n is found where $\forall i \in [1, |O|]| \sum_{j=1}^{n}(l_j(i)) > 0$.
7. If no such combination is found, then
   return NO
   Else, return the first combination that was found.

Complexity

In this section the computational complexity of the NAIVE-MINSEP-EXACT algorithm is studied by leveraging the complexity results of the NAIVE-KSEP-EXACT algorithm [8, 12, 13]. It is noted that k-SEP has a proven NP-completeness [13] via a reduction from the well known NP-complete problem Geometric Covering by Discs (GCD) [3].

Proposition. *The complexity of the NAIVE-MINSEP-EXACT algorithm is:*

$$O\left(\frac{\log_2 k}{(k-1)!} \cdot (\lceil \pi \cdot (\beta^2 - \alpha^2) \rceil \cdot |O|)^{(k+1)}\right)$$

Proof. Obviously, the computations $S', \alpha' \in$ PTIME. In addition the cost to set up the matrix array of pointers M is O(1) and not the size of the matrix, since all the pointers in M are initially null, thus there is no need to iterate every element in M, and only lists in M can be initialized as needed. □

Since each observation $o_i \in O$ has at most $\lceil \pi \cdot (\beta^2 - \alpha^2) \rceil$ partner points, which are located in space $S_i = MBR_i^{ex} - MBR_i^{in}$, then list L has at most $\lceil \pi \cdot (\beta^2 - \alpha^2) \rceil \cdot |O|$ elements. Thus, there are $\binom{\lceil \pi \cdot (\beta^2 - \alpha^2) \rceil \cdot |O|}{n}$ iterations taking place at most in the sixth step, which are as many as the n-tuples in the list L. Each iteration has a complexity cost of $n \cdot |O|$, as there must be compared $|O|$ bits of each n-bit string. Hence, the combined cost of computations equals to: $\binom{\lceil \pi \cdot (\beta^2 - \alpha^2) \rceil \cdot |O|}{n} \cdot n \cdot |O|$.

Additionally, since we are looking for the minimal $n \in \mathbb{N}^*$, where $n \leq k$, and we use the "divide and conquer" technique we will need at the most $\log_2 k$ trials , thus $\log_2 k$ times the combined cost of computations. It is noted that if we have not adopted the "divide and conquer" technique, then the computational cost would be $\sum_{n=1}^{k} \binom{\lceil \pi \cdot (\beta^2 - \alpha^2) \rceil \cdot |O|}{n} \cdot n \cdot |O|$. Next, we are searching for an upper bound for the computational cost of the algorithm's 6th step.

Lemma. *$\forall n \in \mathbb{N}^*$, where $n < k$, can be proven that:*

$$\binom{\lceil \pi \cdot (\beta^2 - \alpha^2) \rceil \cdot |O|}{n} \cdot n \cdot |O| < \binom{\lceil \pi \cdot (\beta^2 - \alpha^2) \rceil \cdot |O|}{k} \cdot k \cdot |O|$$

Proof. Let $A = \lceil \pi \cdot (\beta^2 - \alpha^2) \rceil \cdot |O|$, then it must be proven that for any $n < k$:

$$\binom{A}{n} \cdot n \cdot |O| < \binom{A}{k} \cdot k \cdot |O| \quad \Leftrightarrow$$

$$\frac{A!}{n! \cdot (A-n)!} \cdot n \cdot |O| < \frac{A!}{k! \cdot (A-k)!} \cdot k \cdot |O| \quad \Leftrightarrow$$

$$\frac{1}{(n-1)! \cdot (A-n)!} < \frac{1}{(k-1)! \cdot (A-k)!} \quad \Leftrightarrow$$

$$(k-1)! \cdot (A-k)! < (n-1)! \cdot (A-n)! \tag{1}$$

Since $n < k$, where $n \in \mathbb{N}^*$, then there exists $t \in \mathbb{N}^*$ such that:

$$n = k - t \tag{2}$$

From (1), (2) $\Rightarrow$

$$(k-1)! \cdot (A-k)! < (k-(t+1))! \cdot (A-(k-t))! \quad \Leftrightarrow$$

$$(k-(t+1))! \cdot (k-(t+1)+1) \cdot (k-(t+1)+2) \ldots \cdot (k-2) \cdot (k-1) \cdot (A-k)!$$

$$< (k - (t + 1))! \cdot (A - k)! \cdot (A - k + 1) \cdot (A - k + 2) \ldots \cdot (A - (k - t)) \quad \Leftrightarrow$$

$$(k-t)\cdot(k-t+1)\ldots\cdot(k-2)\cdot(k-1) < (A-k+1)\cdot(A-k+2)\ldots\cdot(A-(k-t)) \quad (3)$$

In the products of the inequality $k - 1$ terms, which are naturals, appear at the most. Obviously, the inequality holds if each term of the right product is larger from the corresponding term of the left product, or that each one of the following inequalities holds:

$$k - t < A - k + 1 \quad (4.1)$$

$$k - t + 1 < A - k + 2 \quad \Leftrightarrow \quad k - t < A - k + 1 \quad (4.2)$$

....

$$k - 2 < A - k + t - 1 \quad \Leftrightarrow \quad k - t < A - k + 1 \quad (4.k\text{-}2)$$

$$k - 1 < A - k + t \quad \Leftrightarrow \quad k - t < A - k + 1 \quad (4.k\text{-}1)$$

It is observed that each one of the above inequalities (4.1)...(4.k-1) collapses to the same expression:

$$2 \cdot k < A + t + 1 \quad (5)$$

Hence, since $(t + 1) \in \mathbb{N}^*$ and $(t + 1) \geq 2$ (directly from the definition of t) it suffices, from (5), to prove that for any $n < k$, where $n \in \mathbb{N}^*$:

$$2 \cdot k < A \quad \Leftrightarrow \quad 2 \cdot k < \lceil \pi \cdot (\beta^2 - \alpha^2) \rceil \cdot |O| \quad \Leftrightarrow$$

$$k < \frac{\lceil \pi \cdot (\beta^2 - \alpha^2) \rceil}{2} \cdot |O| \quad (6)$$

But

$$\frac{\lceil \pi \cdot (\beta^2 - \alpha^2) \rceil}{2} > \frac{\pi}{2} > 1 \quad (7)$$

And

$$k < |O| \quad (8)$$

From (7),(8) the inequality in (6) holds. Thus, it is proven that for every $n \in \mathbb{N}^*$, where $n < k$, the following inequality holds:

$$\binom{\lceil \pi \cdot (\beta^2 - \alpha^2) \rceil \cdot |O|}{n} \cdot n \cdot |O| < \binom{\lceil \pi \cdot (\beta^2 - \alpha^2) \rceil \cdot |O|}{k} \cdot k \cdot |O|$$

□

Now, the proof of the proposition can be continued. Since the above lemma is proven, then trivially it can be proven that $\forall n \in \mathbb{N}^*$, where $n \le k$, the following inequality stands:

$$\binom{\lceil \pi \cdot (\beta^2 - \alpha^2) \rceil \cdot |O|}{n} \cdot n \cdot |O| \le \binom{\lceil \pi \cdot (\beta^2 - \alpha^2) \rceil \cdot |O|}{k} \cdot k \cdot |O|$$

Thus, an upper bound for the computational cost of the algorithm's 6th step is:

$$\log_2 k \cdot \binom{\lceil \pi \cdot (\beta^2 - \alpha^2) \rceil \cdot |O|}{k} \cdot k \cdot |O| = \log_2 k \cdot \frac{(\lceil \pi \cdot (\beta^2 - \alpha^2) \rceil \cdot |O|)!}{k! \cdot (\lceil \pi \cdot (\beta^2 - \alpha^2) \rceil \cdot |O| - k)!} \cdot k \cdot |O|$$

$$= \log_2 k \cdot \frac{(\lceil \pi \cdot (\beta^2 - \alpha^2) \rceil \cdot |O| - k)! \cdot (\lceil \pi \cdot (\beta^2 - \alpha^2) \rceil \cdot |O| - (k-1)) \ldots \cdot (\lceil \pi \cdot (\beta^2 - \alpha^2) \rceil \cdot |O|)}{(k-1)! \cdot (\lceil \pi \cdot (\beta^2 - \alpha^2) \rceil \cdot |O| - k)!} \cdot |O|$$

$$< \mathrm{O}\left(\frac{\log_2 k}{(k-1)!} \cdot (\lceil \pi \cdot (\beta^2 - \alpha^2) \rceil \cdot |O|)^{(k+1)} \right)$$

It must be noted that, if number k is unavailable, then it suffices to replace k with $|O|$ in the above expression. As this term dominates the complexity of the other steps, this will be the complexity of the NAIVE-MINSEP-EXACT algorithm. ■

Conclusions

In this paper a technique which reduces the total computational cost in any version of point-based Geospatial Abduction Problems (point-based GAPs for short) was introduced, and an exact algorithm for the natural optimization problem of point-based GAPs, named MINSEP, was presented along with its computational complexity results.

Even though the use of GAPs has already been extended from point-based explanations to region-based explanations [9, 10, 13] and Geospatial Abduction has even been examined as a game between adaptive adversaries (in terms of game theory) [11], the field is very promising for future work. First of all, the extension of Geospatial Abduction must be explored in dimensions greater than two [8]. In particular, priority should be given to the 3D surface of earth, which is produced by geospatial digital data such as Digital Elevation Model (DEM), Digital Terrain Model (DTM) etc. Secondly, the possibility of replacing the deterministic feasibility predicate with a probability distribution function (PDF) must be explored, or analogously the replacement of the deterministic α, β distances with a PDF based on distance [8]. Finally, Geospatial Abduction should be able to capture topological relations between each observation and the set of partner locations, for example, [8] an observation cannot have a partner across a body of water.

References

1. Blake, R. M., Ducasse, C. J., Madden, E. H., 1960. Theories of scientific method: The renaissance through the nineteenth century, Reprint 1989, ISBN 2881243517, Gordon and Breach Science Publishers S.A.
2. Cormen, H. T., Leiserson, E. C., Rivest, L. R., Stein, C., 2002. Introduction to Algorithms, Second Edition, ISBN 0262032937, MIT Press.
3. Johnson, D. S., 1982. The np-completeness column: An ongoing guide, Journal of Algorithms, Volume 3(2), pp. 182 – 195.
4. Lloyd, S., 1982. Least Squares Quantization in PCM, IEEE Transactions on Information Theory, Volume IT-28(2), pp. 129 – 137.
5. MacQueen, B. J., 1967. Some Methods for classification and Analysis of Multivariate Observations, Proceedings of 5-th Berkeley Symposium on Mathematical Statistics and Probability, Berkeley, University of California Press, Volume 1, pp. 281 – 297.
6. Peirce, C. S., 1931–1958. Collected Papers, Cambridge, MA: Harvard University Press.
7. Preparata, P. F., Shamos, I. M., 1985. Computational Geometry: An Introduction, ISBN 03879611313, Springer-Verlag, New York.
8. Shakarian, P., Subrahmanian, V. S., Sapino, M. L., 2009. SCARE: A Case Study with Baghdad, Proc.2009 Intl. Conf. on Computational Cultural Dynamics, Dec. 2009, Nau D. and Mannes A., AAAI Press.
9. Shakarian, P., Subrahmanian, V. S., 2011. Region-based Geospatial Abduction with Counter-IED Applications, accepted for publication in Wiil, U.K. (ed.) Counterterrorism and Open Source Intelligence, Springer Verlag Lecture Notes on Social Networks, to appear.
10. Shakarian, P., Nagel, M., Schuetzle, B., Subrahmanian, V. S., 2011. Abductive Inference for Combat: Using SCARE-S2 to Find High-Value Targets in Afghanistan, in Proc. 2011 Intl. Conf. on Innovative Applications of Artificial Intelligence, Aug. 2011, AAAI Press.
11. Shakarian, P., Dickerson, P. J., Subrahmanian, S. V., 2012a. Adversarial Geospatial Abduction Problems ACM Transactions on Intelligent Systems and Technology (TIST) Volume 3 (2), Article No. 34.
12. Shakarian, P., Subrahmanian, V. S., Sapino, M. L., 2012b. GAPS: Geospatial Abduction Problems, ACM Transactions on Intelligent Systems and Technology (TIST), Volume 3(1), to appear.
13. Shakarian, P., Subrahmanian, V. S., 2012. GeoSpatial Abduction: Principles and Practice, ISBN 9781461417934, Springer.
14. Thagard, P., 1992. Conceptual revolutions, ISBN 9780691024905, Princeton, Princeton University Press.
15. Thagard, P., 2007. Abductive Inference: From Philosophical Analysis to Neural Mechanisms, in Inductive Reasoning: Experimental, Developmental and Computational Approaches by Freeny A. and Heit E., ISBN-9780521856485, Cambridge University Press, pp. 226 – 247.
16. Vlahavas, I., Bassiliades, N., Kefalas, P., Kokkoras, F., Sakellariou, I., 2011. Artificial Intelligence, 3rd Edition, ISBN 978-960-8396-64-7, University of Macedonia Press, Greece.

Computational Number Theory and Cryptography

Preda Mihăilescu and Michael Th. Rassias

Abstract This is a succinct survey of the development of cryptography with accent on the public key age. The paper is written for a general, technically interested reader. We also review some fundamental mathematical ideas of computational number theory that play an important role in present time cryptography.

Keywords Computational number theory • Cryptography • Elliptic curves over finite fields • Diffie-Hellman algorithm

2000 Mathematics Subject Classification: 11Y11, 11G05, 11Y16, 11Y40, 68Q17, 68Q25

Introduction

Cryptography is the collection of methods and approaches for concealing information in communications from the access by unwished or unauthorized parties. A logical art for dealing with this problem is known from early Antiquity and

---

This paper substantially improves and extends the former article of the authors, that appeared in [MR].

P. Mihăilescu
Mathematisches Institut der Universität Göttingen, Germany
e-mail: preda@uni-math.gwdg.de

M.Th. Rassias (✉)
Department of Mathematics, ETH-Zürich, Rämistrasse 101, 8092 Zürich, Switzerland
e-mail: michail.rassias@math.ethz.ch

N.J. Daras (ed.), *Applications of Mathematics and Informatics in Science and Engineering*, Springer Optimization and Its Applications 91, DOI 10.1007/978-3-319-04720-1_22,

it developed along the centuries, mostly in the frame in which two parties, say nobleman and general, or concealed lovers, communicated in writing by sending each other messages, which could only be understood when knowing some additional data—*secret keys*—and the details for the procedure of encrypting and decrypting the messages—*algorithm*. Algorithms were often assembled from a collection of useful basic ideas, known by tradition.

Traditional Secret Key Cryptography

Transposing the alphabet of a spoken language into a sequence of numeric codes is always useful for discussing cryptographic ideas. Suppose thus that the Latin alphabet $a, b, \ldots, z$ is encoded in ascending order by the numbers $0, 1, \ldots, 24$. The idea of permuting the letters cyclically by a constant σ was purportedly used by Caesar in the Gallic wars—hence the name of *Caesar* code. For instance, for $\sigma = 3$, the word ATHENS becomes DWKHQV. For decryption, use $\sigma = 25 - 3 = 22$. One can improve the security of this code, by using context specific keys, key sequences, and other well-defined combination—such variations were investigated in the 16th century by the French diplomat Blaise de Vigenère. The purpose was to counter the obvious weakness of the Caesar code with respect to *frequency* attacks: provided a sufficiently large cipher code, and knowing that letters like e, a, m occur much more frequently than z, h, q, one can easily determine the value of σ, thus compromising the whole encryption. Since these ideas can in addition be combined with some commonly known text modifications, the bag of tricks for artisanal cryptography offered *sufficient* variety for satisfying the needs until the advent of the 20th century. In parallel with the development of new, particular algorithms of encryption, the analysis of methods for discovering both keys and the particularities of an encryption procedure—like for instance, the frequency analysis mentioned for the Caesar cipher—developed itself into the science of *cryptanalysis*. Today, cryptanalysis and cryptography are regarded as the two complementary aspects of the science of *cryptology*. While the creation of private codes and keys could be considered to some extent as a playful, even enjoyable undertaking, which requires some rigor though, for preventing countermeasures of the cryptanalyst, the classical encryption has one more important limitation: the peers need to be in anticipated agreement regarding both of the encryption algorithm and the keys. This leads to several consequences: the first is that one would wish the algorithm to be so strong, that it suffices to exchange the keys while keeping the same algorithm over longer periods. The second is that one needs well-trained and faithful couriers for the keys.

In order to illustrate the methods and challenges of classical ciphers, we propose to the reader to try and decrypt the following small text, which was encrypted by a scheme developed by the 10-year-old daughter of the first author starting from a children's game encryption, found in a book, and which they use for discussing within a gang of good friends. The cipher text is:

TGGMCITGWKKNKVCTZCOKNUKECFGOZCCFCWK

It is obtained by a combination of the ideas discussed above.

The Advent of Computing Machines . . .

In the 20th century, military confrontations became more devastating, and the power of data processing with the help of machines increased without precedent. However, until World War II and even later, the basic scheme for secure communication remained the same: the secret communicators shared some common algorithm, which could eventually be performed by a machine, and they were using some *shared secret key*, for the diffusion of which many lives were put in danger. One of the most documented episodes of warfare use of cryptography was the German development of their encryption machine Enigma II on the basis of a simpler, earlier version Enigma I, which had been a commercial product before the war.[1] Little did they know that this unfortunate combination of economic and military application had led to the fact that a team of young Polish mathematicians from Poznan were in possession of a means for breaking Enigma I. When hired by the British authorities, the work for breaking the enhanced version was, against the German expectation, an achievable one, and the breaking of Enigma had its important consequences for the outcome of the war [10, 19].

. . . and of Personal Computers and Networked Communications

The advent of computers brought on the one hand the massive improvement of computational capacities, then, in the early 1970s, and on the other as the US-army built the ARPA-net, the advent of networked communications, an ancestor of the Internet. In light of this progress, cryptography was led into simplifying the definition of its object and tasks. Some very useful principles have been established, which stay to hold. First, it was understood that there is little security in the use of proprietary, secret algorithms—the choice of cryptographers going in the direction of simple, publicly known and well-understood and cryptanalized algorithms. As long as the bag of tricks is known, it can even happen more easily that a flaw escapes in the design of a proprietary algorithm. As a consequence the assumed gain of security obtained from keeping a secret of the cryptographic procedure is counterbalanced by the insecurity stemming from the lack of reliable cryptanalysis. In simple words, the modern attitude to security is resumed in the

[1] The development of Enigma I during the early days of mechanical office machine, shows that there has always existed an important requirement for cryptography also in business.

paradigm publically known and cryptanalized algorithm and secret keys. As a consequence, the protection of keys becomes the center of the security concerns and is offered the due attention: the system is as secure as the keys are. In addition, the new approach to cryptography promises that, due to the collective scrutiny of the cryptographic community, in time the most efficient and reliable algorithms are naturally selected, while weaknesses and possible attacks eventually show up in the processes. An algorithm is more reliable when it has longly resisted public scrutiny by the community, and *not* when it is based on sophisticated *secret tricks*.

We mentioned that in early times of cryptography, secret keys were transported by couriers, which brought their life in danger for this purpose, while even later, in times of telegraphic transmission of keys, the problem of building secure channels for key transmission was a crucial one. In the seventies, the first networked system of computers became conceivable. It physically realized by the American army, in the form of the ARPA net, which first connected between 1972–1974 a number of Universities on the East and West Coast, for research and experimental purposes. The notion of remote computer-communication became tangible for the users of the net.

Under these conditions, it became obvious that the old systems for secure key distribution could no longer satisfy the needs of security for this technological advance and some new ideas were called for, in order to solve the problem in a simple, time-efficient, and reliable way.

The idea was provided by the concept of public-key cryptography, which was born in Stanford from the joint work of W. Diffie and R. Hellman who studied public key infrastructures, and R. Merkle who studied secret key distribution. Here is the way Diffie and Hellman presented the problem in [8], which mentions the joint work with Merkle: *In turn, such applications (fast computers) create a need for new types of cryptographic systems, which minimize the necessity of secure key distribution channels and supply the equivalent of a written signature.*

Public Key Cryptography Arises

The idea was remarkably simple and elegant. Its natural properties were strikingly reflected 30 years later, when it became publicly known and verified, that J. Ellison, an engineer and cryptographer working for MI5s General Communication Head-Quarters GCHQ, had developed exactly the same concepts and schemes as Diffie, Hellman and Merkle, yet seven years earlier. The research was only declassified after the year 2000; it was a matter of academic debate, if a person working for secret services, outside the academic community should be granted credit for scientific developments. Beyond these it is in any way remarkable that the same ideas could be developed twice in a totally independent way.

Traditionally, a protected communication was established by using secret key cryptography. In a wide area communication network, in which numerous peers could communicate over large distances, the chances for establishing a common secret key prior to communication are low, so there was demand for a procedure which would allow a pair of peers A and B—Alice and Bob, as cryptographers often used to name them—to dispose of a *shared secret key*, without any prior communication, either direct or by means of a parallel, secured channel. Only some public known data base Δ, and algorithm could be accepted as premise for achieving the purpose.

The concept of *public key cryptography*, introduced by the three authors mentioned above, is simply described by the following: If X is a peer who wants to engage in secure network communication, he should start by generating a set of data, which is bundled into his own *secret key* S_X. A subset of this data, bundled in the *public key* P_X will be made public to all peers he might be wishing to communicate with—it will be, for instance, part of the data base Δ, or it may be transmitted over any unsecured channel. The two keys should enjoy the following two properties:

1. Both keys can be used for encrypting texts according to some algorithm yet to be defined.
2. Messages encrypted by S_X can be decrypted by P_X and vice versa. Moreover, the keys should be sufficiently random: the chances for two peers generating accidentally the same secret key should be close to zero.
3. It should be computationally unfeasible to derive S_X from P_X.

For ascertaining the third condition, one usually derives the public key P_X from the secret one, by using some kind of *trap-door* function f. Under this term, one understands an invertible function, such that the value of f is very easy to compute, but the inverse is computable in theory, but infeasible in practice, provided the data is sufficiently large. A typical such example is the map $f : \mathbb{N} \times \mathbb{N} \to \mathbb{N}$ which associates two primes p, q to their product $n = p \cdot q$. This can be computed very efficiently even for quite large primes. However, the inverse problem, of factoring n is assumed. There is no proof for the fact that there cannot exist some fast—e.g. *polynomial* algorithm, thus one whose run-time is a polynomial in the number $m = \log_2(n)$ of bits of the input number n—for factoring integers. However, the problem is one of the most intensively researched ones in algorithmic number theory; after decades of collective work, the most efficient algorithm for factoring, the Number Field Sieve (NFS) requires the order of

$$e^{cm^{1/3}}$$

binary operations, to be *very hard.*

On the basis of the premises 1.–3., if Alice and Bob want to communicate, then Alice sends to Bob messages encrypted by P_B, which she may retrieve from the public key repository. However, only Bob can decrypt the message, so the communication is secured. Based on this idea, a further useful application emerged: it is often useful to be able to certify the ownership of some message, to *sign* the message in a unique and non-repudiable way. In this case, secrecy is less of a concern than ownership is. The solution consists of associating a short cryptographic *hash-value* H to the message, which is encrypted by the secret key S_A. Any receiver will then be able to regenerate the hash value on his own, decrypt the encrypted hash with P_A, and then compare the two results. If they match, Bob has a proof that it was Alice who sent the message.

Within the next 20 years the public key cryptography and the academic paradigm of cryptology spread out and reached probably even most of the banks and diplomatic transmissions, which traditionally used to consider the use of private algorithm as a particularly welcome increase of security.[2]

Classical Public Key Cryptosystems

In the next two years after the abstract definition of public key cryptography, two major algorithms that implement this idea and are still in use today, were invented.[3]

The first one was using the *discrete logarithm* problem in the multiplicative group of finite fields as a trap door. If p is some large prime and $g \in \mathbb{F}_p^\times$ generates the multiplicative group modulo p and $a = g^c \in \mathbb{F}_p^\times$, then it is easy to compute

$$f(x) = b = a^x, \text{ for arbitrary } x.$$

However, to recover x from b, the Discrete Logarithm Problem in finite fields, is a computationally hard problem—thus adequate for a trap door function. For the factoring problem, there is no proof that no faster algorithms can be found—however, the best one discovered until today has a comparable asymptotic complexity to the number field sieve for factoring integers, mentioned above.

Diffie and Hellman proposed an algorithm for exchange of a shared secret over an insecure channel, and is widely known as the *Diffie-Hellman key exchange algorithm*. It functions as follows: If p and $g \in \mathbb{F}_p^\times$ are like before—these being public data—then Alice and Bob start by choosing some random one-time keys A_R, B_R, which are elements of $\mathbb{Z}/((q-1)\cdot\mathbb{Z})$. Then Alice sends to Bob $M_A = g^{A_R}$ and receives from Bob $M_B = g^{B_R}$. The reader can verify that by using the private data and the data received, both Alice and Bob may retrieve $S = g^{A_R \cdot B_R}$, which is the data from which the common secret key is extracted. However an eavesdropper, who is always called Eve in cryptography, would only know g^{A_R} and g^{B_R}, but not A_R or B_R. The system can be broken by breaking the Discrete Logarithm. But, does the converse also hold? This is not known. The particular, more special

[2]This fact was reflected again in the fact that the producers of cryptographic machinery were involved in customer tailoring algorithms for this purpose. In the late nineties, manufacturers of cryptographic hardware still had only a precious few customers insisting on the "privilege" of purchasing machines which run according to some unique and "secret" algorithm.

[3]It is also noteworthy that, after J. Ellis had defined the abstract notion of public key cryptosystems, in a similar way to Diffie, Hellman and Merkle, the same algorithms were discovered in MI5 too, by C. Crook and M. Williamson, only in the reverse order.

problem in which one should retrieve $g^{xy} \in \mathbb{F}_p^\times$ from g^x, g^y has received the name DH-Problem, for obvious reasons. More recently, variants of the Diffie–Hellman key exchange have been proposed, which can be *proved* to be equivalent to the DH problem: i.e., they can be broken if and only if the DH problem is broken. The key exchange algorithm does not offer the possibility to generate signatures; however, J. L. Massey and J. K. Omura proposed in 1983, a variant based also on the discrete logarithm trap door function, which allows also public key encryption, and thus signatures.

We should mention that in general public key algorithms are much slower than secret key encryption. Therefore it is most likely that one would use them for establishing a shared secret key, after which a communication session can be encrypted with a common agreed secret key algorithm, using the established key. For this purpose the original Diffie–Hellman algorithm is sufficient. Incidentally this two-step approach to encryption is the core idea in the SSL/TLS protocol, developed between 1992–2002, and which is currently used in all confidential https communications on the Internet—for instance, when you book an electronic flight ticket, or buy a book from Amazon.

The first proper public key *encryption* algorithm was provided one year later, in 1977, by R. Rivest, A. Shamir, and L. Adleman at MIT. Their algorithm, widely known as RSA after the initials of their names, uses the problem of factoring integers as a trap door. A secret key consists of $S_A = \{p, q, d\}$, where p, q are two large primes satisfying some additional randomness conditions and

$$0 < d < (p-1)(q-1), \quad \text{with} \quad (d, pq(p-1)(q-1)) = 1$$

is a random number; if $e \in \mathbb{N}$ is such that

$$ed \equiv 1 \bmod (p-1)(q-1),$$

the public key consists only of $P_A = (n, e)$, with $n = p \cdot q$. In some instances, e is a fixed number for the whole system, so d will be determined by the holder of the secret key using the same defining congruence. With these prerequisites, if M is a short message it will be identified with a number in $\mathbb{Z}/(n \cdot \mathbb{Z})$ and its public key encryption $M_e \equiv M^e \bmod n$ can be computed in the open, but can only be decrypted by Alice, the holder of d, since

$$M \equiv M_e^d = M^{ed} \bmod n.$$

Conversely, if Alice encrypts M with d, then anyone can recover M and upon doing so will have a proof Alice having produced the encryption: indeed, only the owner of the secret key could produce this encryption, which can thus act as a private signature of Alice.

Despite initial attempts of the NSA to inhibit the publicizing of the ideas of public key encryption and RSA, these were brought to the public already in 1977 by Martin Gardner in his widely read column "Mathematical Games" in the Scientific

American magazine and were eventually published in the communications of the ACM [31]: the way to public key cryptography was open!

In the same year 1978, R. McEliece proposed a somewhat different cryptosystem, which was inspired from coding theory. The trap door function is drawn in this case from general linear codes, a context in which the parameters of a linear code are specially adapted to the purpose of public key cryptography. The resulting algorithm has an advantage compared to the number theory-based algorithm mentioned above and some further, based on elliptic curves, that we shall discuss below, since it is faster. However, the keys may be as large as $1MB$, which compares poorly to the $128B$ required by RSA for a comparable level of security.[4]

Cryptanalysis

In 1978, Hellman and Merkle invented a public key cryptosystem that did not rely on number theory, but rather on the NP-complete *knapsack* problem.

The first major success of public key cryptography was that the expectation became true, and the domain of cryptanalysis—concerned with the analysis of possible attacks against cryptographic schemes—became a flourishing academic domain of investigation. One of the most spectacular successes was due to the development of the *lattice reduction* algorithm by A. Lenstra, H. Lenstra Jr., and L. Lòvasz, the LLL-algorithm. Given a lattice $\mathscr{L} \subset \mathbb{Z}^n$, there exists a base consisting of the shortest vectors. Classical algorithms for finding such a base are known from the work of Charles Hermite. Only, in the case when the base is presented by an initial generating system of very large vectors, the process is exponential. The algorithm was developed from techniques used by Lòvasz in integer programming; the idea was to use an approximate Gram-Schmidt-orthogonalization which provides some *close* to minimal vectors in $\mathscr{L}$. The advantage is that the algorithm runs in polynomial time and has therefore a wide variety of applications both in cryptography and in number theory itself. One of the first applications of LLL was in showing that the keys of the knapsack cryptosystem could be cracked in polynomial time: in order to do so, one had only to solve a particularly simple *subfamily* of problems belonging to the knapsack family. This result showed the advantage of public academic scrutiny of cryptographic schemes, since it had only taken five years to reveal the weaknesses of one of them. But it also blocked the way for applications of the knapsack. Some improved versions have been presented, that could never be attacked—but they never made it to public applications.

The most important effect of cryptanalysis was less visible. The community quickly developed its own language and defined a variety of subtle *attack scenarios*,

[4]One compares the security of two fundamentally different algorithms, by estimating the parameter sets required, such that breaking the given algorithms by means of the best state-of-the-art algorithm would require comparably large amounts of time.

in which the eavesdropper *Eve* was offered increasing levels of advantages: thus Eve can simply tap a wire communication, but she might also collect large amounts of data signed by Alice, or even induce her into signing a chosen suite of messages. Thus, possible attacks could be investigated for these various levels of disclosure. The procedure is very fruitful, since the algorithms to which no attack is found, even under the most generous premises for Eve, is for good reasons assumed to offer reliable security.

Later, the encryption hardware began being regarded as a point of attack, as it was observed that physical measurements on a chip while it is computing an RSA encryption, for instance, may reveal some bits of the secret key. Additional measures were then developed to protect from these *side channel attacks*. This way, well-defined attack scenarios are used for checking the security of various cryptosystems and protocols.

The development of cryptography is triggered by the two opposite demands, for efficiency and for security. It occurred more than once, that the wish for efficiency led to the use of some extreme key configurations. These provided particularly efficient arithmetic, thus effective computation of the cryptographic scheme. However, as in the case of the knapsack problem, the question could have been asked, if by restricting to particular families of the key space, one did not move into a particular instance of the general, hard problem, to which the trap door function was associated. The question was first answered by the observation that no algorithms are currently known that could take advantage of the particular family of keys used. But eventually, an attack was discovered, which discarded the use of certain keys, or even whole cryptographic schemes. As an example, it is for instance, useful to have a universal, short public exponent e for the RSA scheme. This had been used in practice in the late 1980s. But M. Hagstad and then D. Coppersmith showed that if e is too small, it is easy to gather sufficiently many messages signed by the same key S_A, and then use simple arithmetic in order to crack that key. Therefore, the smallest fixed key currently allowed by standards is $e = 2^{16} + 1$, and this may change with the growth of computing and storage capacities.

We have already discussed the fact that for the number theoretical public key systems introduced so far, an efficient attack of the underlying number theoretic problem (factoring or discrete logarithm) breaks the schemes. Conversely however, it is not known if general attacks can be found that break the scheme *without* offering an efficient general solution for the inversion of the trap door function. Such questions about *provable* security became actual in the late nineties. We have already mentioned that by modelling the DH-problem, which is a particular form of the discrete logarithm, the best results available in this direction were obtained by U. Maurer [23] and V. Shoup, and various coauthors, e.g. in [6].

Dickman's Theorem and the Trap Door Functions

In the thirties of the last century, J. Dickson considered the question of estimating the largest prime factors of some random integer n. Using heuristic estimates on the repartition of primes, he found for instance that if $p|n$ is the largest prime dividing n, then $p = O(n^{\ln 2})$. More generally, an integer $n > 1$ is defined to be y-smooth if none of its prime factors exceeds y. The function

$$\psi(x, y) = \sharp\{\ 1 \leq n \leq x\ :\ n \text{ is } y\text{-smooth}\ \}$$

counts the smooth numbers less than x. With these definitions, Dickman also proved that for all $u > 0$ there is a real number $\rho(u)$ such that

$$\psi(x, x^{1/u}) \sim \rho(u)x.$$

The function $\rho(u)$ was described in terms of a differential equation, in which u was fixed for $x \to \infty$.

Half a century later, the gap was filled by Canfield, Erdős and Pomerance [5], who proved that

Theorem 1 (Canfield, Erdős and Pomerance). *For all real sequences with $u \to \infty$ under the constraint $u < (1 - \epsilon) \ln x / \ln \ln x$, one has*

$$\psi(x, x^{1/u}) = xu^{-u+o(u)} \tag{1}$$

As a consequence one concludes that with probability $P > 1/2$ one out of

$$L_n[1/2, 1] := e^{\sqrt{\log(n) \log\log(n)}}$$

random integers belonging to the interval $(0, n)$ will be y-smooth, for $y = O(L[1/2, 1])$. Bounds of the type

$$L_n[c, d] = e^{d \log(n)^c \log\log(n)^{1-c}},\ 0 < c < 1$$

are called *subexponential* for obvious reasons: they grow much faster than any polynomial in $m = \log(n)$ but substantially slower than e^m. All the state-of-the-art, subexponential algorithms for solving either a variant of the discrete logarithm problem, or for factoring integers, take advantage in some way of this consequence or variants thereof.

We exemplify here the ideas on the instance of the *quadratic sieve method*, which is a classical fast algorithm for factoring integers. It has its origin in the following simple observation of Fermat: if m is a composite integer, then the congruence

$$x^2 \equiv c \bmod m$$

will have at least four solutions, and there are x, y such that $x \not\equiv \pm y \bmod m$, but $x^2 \equiv y^2 \bmod m$. Then $(x+y, m)$ is a non-trivial factor of m. Theorem 1 helps find such pairs x, y, as follows: for numbers $x(i)$ in some interval $\lceil \sqrt{n} \rceil + i, 0 \leq i \leq B$, one computes the remainder[5]

$$r(i) = x(i)^2 \text{ rem } m$$

and retains only those values of x, for which r is a B-smooth number. After gathering sufficiently many such relations, one may hope that the product of some $r(i)$ is a square: namely, that there is an index subset $J \subset [0, B]$ such that

$$\prod_{i \in J} r(i) = R^2, \ R \in \mathbb{Z}.$$

Letting then

$$X = \prod_{i \in J} x(i),$$

we obtain the congruence

$$X^2 \equiv R^2 \bmod m.$$

If in addition, $X \not\equiv \pm R \bmod m$, which should happen with probability $\geq 1/2$, then $(X \pm R, m)$ is a non-trivial factor of m. The method relies on some empirical assumptions on the repartition of factors of $r(i)$: namely, that the distribution of these residues is such that one may apply the relation (1) for estimating the probability that one of these numbers is B-smooth. These allow to establish an *optimal* bound

$$B \sim \exp(\sqrt{\log(m) \log\log(m)}) = L_n[1/2, 1].$$

In our case $B = L(n; 1/2)$ and the quadratic sieve runs in time polynomial in B-experience having so far confirmed the underlying heuristical assumptions.

The following nice example is taken from the book of R. Crandall and C. Pomerance [7]: let $m = 1649$, with $41 = \lceil \sqrt{m} \rceil$. We find

$$41^2 \equiv 32 \bmod m; \quad 42^2 \equiv 115 \bmod m; \quad 43^2 \equiv 200 \bmod m.$$

Since $32 \cdot 200 = 2^{5+3} \cdot 5^2 = 80^2$, we let $R = 80$ and

$$X = 41 \cdot 43 = 42^2 - 1 \equiv 114 \bmod m,$$

finding that $114^2 \equiv 80^2 \bmod m$ and eventually $17 = (114 - 80, 1649)$, which is a non-trivial factor.

---

[5]In computational algebra, the notation x rem y stands for the unique representative of the equivalence class of $x \bmod y$ which lays in the interval $[0, y)$.

For the discrete logarithm problem in $\mathbb{F}_p^\times$, which consists of determining x such that $g^x \equiv b \bmod p$, one uses smooth numbers as follows: Fix a smoothness bound y and let $q_1, \ldots, q_r < y$ be all the primes up to y. For random values of m, one computes $u = g^m$ rem p and keeps only those values of u which are y-smooth. After collecting sufficiently many relations, one will then be able to compute the discrete logarithms l_i such that $q_i \equiv g^{l_i} \bmod p$. Next, one tries random values of k searching for ones that make $v = bg^{-k}$ rem p be a y-smooth number. The precomputed values l_i will then help determine $x = k + \log_p(v)$ from the prime decomposition of v. This algorithm also relies on heuristic assumptions, on the basis of which the running time is $L_p[1/2, \sqrt{2}]$.

At the end of the 1980s, John Pollard found a way for applying the idea of the quadratic sieve to integers in number fields rather than $\mathbb{Q}$. The method was first applied to the factorization of the Fermat number $F_9 = 2^{2^9} + 1$. In the following years, it was generalized and improved by a series of mathematicians, starting with A. Lenstra and M. Manasse. The resulting *number field sieve* is currently the asymptotically fastest factoring method and it runs in time $O(L_n[1/3, c])$, for some constant $c < 2$.

Similar methods are known for the discrete logarithm method: they use number fields in case of larger characteristics, and function fields for small characteristics. Like in the case of factoring, their running time is also $O(L_n[1/3, c])$. Current records reach as high as 7–800 binary digits for factoring composite of general form and $\sim$ 5–600 for the discrete logarithm in prime fields. During more than one decade, the discrete logarithm was hardest in finite fields $\mathbb{F}_{p^\ell}$ for which $\ell \sim \log(p)$: these orders of magnitude could not be attacked by either number or function field sieves.

Recently A. Joux from INRIA Nancy developed a series of new ideas for improving discrete logarithms in finite non-prime extensions. There are several versions and applications of these ideas. First, they succeed in filling in the gap that existed between the function field and the number field sieve, by providing algorithms in the order of $L_p[1/3, c]$, and also for the case of extension fields with $\ell \sim \log(p)$. They allow to solve the discrete logarithm problem in *quasi-polynomial* for field $\mathbb{F}_{p^\ell}$ when $\ell \sim p$; the result has been presented at Eurocrypt 2013 and is published in [1]. The ideas find another application in discrete logarithm in the fields of characteristic two extension degree $\mathbb{F}_{2^{q \cdot k}}$ with q a prime and k an integer related to q. Joux also announced a variant of his method to yield an algorithm for discrete logarithms in general fields of characteristic two, running in $L_q[1/4, c]$, where q is the size of the field [18]. This would be the first known algorithm of this efficiency. The developments in this field are still quite fluid, but certainly within the following months to a few years, some important and efficient versions of discrete logarithm algorithms in a variety of fields will be well described and understood.

Elliptic Curves

The cryptographic schemes discussed so far use multiplicative groups $(\mathbb{Z}/n \cdot \mathbb{Z})^*$ or $\mathbb{F}_q^\times$ and related trap door function. Having (computational) access to a larger family of well-understood abelian groups would certainly enlarge the possibilities for cryptographic and algorithmic applications.

In 1984 René Schoof made the way by opening the discovery of a polynomial time algorithm for counting the number of points on an elliptic curve over a finite field. This brought the groups of algebraic geometry in the realm of applications and algorithms. Within one year, H. W. Lenstra Jr. proposed an important variant of Pollard's rho-method for factoring, based on elliptic curves: the *elliptic curve method* or ECM. Also, V. Miller and N. Koblitz proposed independently the use of elliptic curves for cryptography. The ECM method has a run-time comparable to the quadratic sieve, but it behaves particularly well for numbers m which have some small prime factors, i.e. sensibly smaller than $\sqrt{m}$: the run time is namely estimated to be $L_p[1/2, \sqrt{2}]$, where p is the smallest prime dividing m.

We recall that an ordinary elliptic curve over a finite field $\mathbb{F}_q = \mathbb{F}_{p^\ell}$ of characteristic $p > 3$ is the set of solutions

$$E_q(a,b) := \{P = (X,Y) : Y^2 = X^3 + aX + b, X, Y \in \mathbb{F}_q\} \subset \mathbb{F}_q^2 .$$

There is an abelian addition $\oplus$ defined on this curve, which has the *point at infinity* $\mathscr{O}$ as neutral element. The neutral element can be understood as arising when the addition law, which is based on rational functions, leads to a division by zero. The formally correct definition is obtained by embedding the curve in a projective space. The curve is ordinary, if it is not singular and not supersingular, two conditions that can be verified in terms of $q, b, \mathbb{F}_q$. Thus

$$\mathscr{E}_q(a,b) = (E_q(a,b), \oplus)$$

becomes an abelian group. The classical theorem of Hasse gives the following bounds for the size of this finite group:

$$|\mathscr{E}_q(a,b) - q + 1)| < 2\sqrt{q}. \tag{2}$$

An elliptic curve can be defined in a similar way over the algebraic closure $\overline{\mathbb{F}}_q$. Its N-torsion is

$$\mathscr{E}[N] = \{P \in \mathscr{E} : [N]P = \mathscr{O}\},$$

where $[N]P$ denotes the N-fold addition of P to itself. The torsion subgroup is—with one exception—a two-dimensional free $\mathbb{Z}/(N \cdot \mathbb{Z})$-module, and a vector space, for prime N. If $\zeta \in \overline{\mathbb{F}}_q$ is a primitive N−th root of unity, there is a non-degenerate bilinear, skew symmetric pairing:

$$\langle \cdot, \cdot \rangle : \mathscr{E}[N] \times \mathscr{E}[N] \to \langle \zeta \rangle, \tag{3}$$

the Weil pairing. In particular, if P, Q are linear independent torsion points, then

$$\langle P, [x]Q\rangle = (\langle P, Q\rangle)^x, \tag{4}$$

an identity in the multiplicative group $(\mathbb{F}_q[\zeta])^{\times}$.

The idea of the ECM factoring method of Lenstra adapts an older algorithm of Pollard, which was designed to work in multiplicative groups, to the larger family of elliptic curves. It can be described briefly as follows: if n is a number to be factored, one draws random numbers a, b such that a point $P = (X, Y)$ is known with

$$Y^2 = X^3 + aX + b, 0 \leq X, Y < n.$$

Assume now that n has a prime divisor p such that $m := |E_p(a, b)|$ is a B-smooth integer for some fixed, not too large integer B. If $K = B\,!$, then in the process of computing the multiple $[K]P$ by additions and doublings on the curve modulo n, one will *most probably* encounter a factorization of n: some denominator will be divisible by p (point at infinity!), but not by all the primes dividing n. Lenstra proved that for uniform randomly distributed a, b, the numbers m are close to being uniformly distributed in the Hasse interval (2). Theorem 1 then implies that by choosing $B = L_p[1/2, 1]$ random curves, one will find a curve for which m is B-smooth with probability $> 1/2$. This explains the main steps of the algorithm and of its proof. The interested reader may use Silverman's [34] and Washington's [35] textbooks for a detailed rigorous introduction to elliptic curves and their applications to cryptography.

Counting Points

The idea of Schoof is both elegant and important, beyond even the immediate algorithmic and cryptographic applications: it opened a new area of research for practical algorithms for counting points on finite abelian varieties. This research area is still growing, while the main domain of application goes beyond the limits of cryptography, since at least a decade. The algorithms are more and more used for larger computations related to mathematical questions such as the Birch Swinnerton-Dyer conjecture, and other properties of L-series. See also [30] for an elementary theoretical application of point counting.

Initially, Schoof [32] started from the following simple remark: if

$$E_p(a, b) \; : \; Y^2 = X^3 + aX + b$$

is an elliptic curve defined over the finite field $\mathbb{F}_p$, of which one assumes that it is ordinary, then Riemann's conjecture for elliptic curves implies that, in the endomorphism ring of the curve $\text{End}(E_p, \overline{\mathbb{F}}_p)$ defined over the algebraic closure of $\mathbb{F}_p$, the Frobenius verifies the quadratic equation

$$\Phi^2 - t\Phi + p = 0. \tag{5}$$

Since E_p is fixed by Φ, we have

$$|E_p(a,b)| = p - t + 1$$

for the number of points fixed by the Frobenius. Counting the points is thus equivalent to determining the value of the *trace of the Frobenius* t; since the Hasse inequality (2) states that

$$t < 2\sqrt{p},$$

it suffices to determine the remainder $t \operatorname{rem} \ell$ for a set of small primes with:

$$L = \prod \ell > 2\sqrt{p}.$$

Therefore, the core step of the algorithm consists in modeling the ℓ-torsion $E_p[\ell]$ into an algebra

$$\mathbb{B} = \mathbb{F}_p[X,Y]/\left(\psi_\ell(X),\ Y^2 - (X^3 + aX + b)\right),$$
$$P = \left(X + (\psi_\ell(X)),\ Y + (Y^2 - (X^3 + aX + b))\right) \in \mathbb{B}.$$

in which $\psi_\ell(X)$ is the ℓ-division polynomial which has as roots all the x-coordinates of ℓ-division points. Therefore, any such point enjoys the properties which define the *generic* ℓ-torsion point $P \in \mathbb{B}$. It is then a straightforward computation, to determine $t \operatorname{rem} \ell$ from the identity

$$\Phi^2 P + pP = t\Phi P.$$

The seminal idea of Schoof, to determine the parameters of the Riemann ζ-function from projections in torsion spaces, and thus counting points on varieties over finite fields was both improved for simple varieties, such as elliptic curves, and extended to more general abelian varieties. In the first case, the primary thing to do was to reduce the size of the algebra $\mathbb{B}$—which can be done by finding smaller factors of $\psi_\ell(X)$ mod p.

The breakthrough in this direction was indicated by Noam Elkies (cf. [9, 33]), who brought modular forms in the game, thus showing how to find in half of the cases some factors $f(X)|\psi_\ell(X)$ of linear degree, compared to the quadratic degree in ℓ of the division polynomial. The ℓ-torsion $E_p[\ell] \cong \mathbb{F}_\ell^2$ as a vector space; fixing two linear independent points $P, Q \in E_p[\ell]$, we see that $G := \operatorname{Gal}(\mathbb{B}/\mathbb{F}_p)$ acts on the vector space $E_p[\ell]$ by acting on the base P, Q. We obtain herewith a representation $\rho : G \to \mathrm{GL}_2(\mathbb{F}_\ell)$, with respect to which $\rho(\Phi)$ verifies the same quadratic equation. Let δ be the discriminant of the quadratic polynomial in (5), which is the same as the characteristic polynomial of the image of $\rho(\Phi) \in \mathrm{GL}_2(\mathbb{F}_\ell)$. Then, according to the value of the Legendre symbol $\left(\frac{\delta}{\ell}\right) \in \{1, 0, -1\}$, the matrix $\rho(\Phi)$ is diagonalizable, has normal upper triangular form or has eigenvalues in $\mathbb{F}_{\ell^2}$.

In the first case, there are two *eigenpoints* P, Q of the Frobenius and the orbit of their x coordinates under multiplication on the curve is galois invariant. We obtain herewith the *eigenpolynomials*

$$f_P(X) = \prod_{k=1}^{(\ell-1)/2} (X - ([k]P)_x) \mid \psi_\ell(X), \quad \text{where}$$

$$\deg(f_P) = (\ell - 1)/2, \quad \text{and} \quad \deg(\psi_\ell) = (\ell^2 - 1)/2,$$

together with a new algebra $\mathbb{B}'$, obtained by replacing ψ_ℓ with f_P. For the computation of F_P, Elkies considered the function field $\mathbb{C}[[j(q)]]$. Some classical arguments on Eisenstein series and $\Gamma_0(\ell)$-modular forms, imply that for each j-invariant j_m of an ℓ-isogenous curve to E_p—or also, for each zero of the modular equation $\Phi_\ell(X, j(q))$—there is a polynomial $f_j(X) \in C[[j(q)]][X]$ which has the x-coordinates of the kernel of the respective isogeny as zeroes. The polynomials can be constructed in the function field by manipulations of q-expansions and they have the useful property that all the coefficients are algebraic integers. The insight of Elkies was to show that one can substitute for j_m the value of some zero $\Phi_\ell(X, j(E_p))$ mod p and reduce the coefficients of $f_j(X)$ modulo p, thus obtaining some eigenpolynomial corresponding to the value of j_m. Indeed, if $\mathbf{E}$ is any curve over $\overline{\mathbb{Q}}$ which reduces to E_p at some prime ideal above p, then its j-invariant reduces to the one of E_p and so do the invariants of its ℓ-isogenies. Therefore, if the modular equation has linear factors j_m over $\mathbb{F}_p$, by inserting these in the expression for $f_j(X)$, upon reduction at the same prime, the coefficients of the polynomial f_j map to the ones of some eigenpolynomial. Using improved algorithms for manipulation of series [4], one can compute the eigenpolynomials in time $O(\log^3(p))$, the running time being dominated by the computation of zeroes of $\Phi_\ell(X, j(E_p))$ mod p. Further improvements can be achieved by using the galois structure of the resulting algebras [25]. The galois theory of finite, commutative algebras has wider applications in algorithmic context and was generalized in [27].

For curves defined over finite fields of small characteristic p, it is possible to project (5) in the p^N-torsion group. Using different flavors of cohomology combined with Newton iterations, various authors starting with T. Satoh, K. Kedlaya and A. Lauder developed in this way, the most efficient point counting algorithms for elliptic curves. Some of them are generalized to super elliptic curves, elliptic surfaces, etc. However, this approach works best only for very small characteristics.

Cryptography

The elliptic curve-based cryptographic schemes which have survived scrutiny and became part of current standards on public key cryptography are essentially variants of the Diffie-Hellman key exchange scheme and are based on the difficulty of solving the discrete logarithm problem: find x such that

$$[x]P = Q, \quad \text{for} \quad P, Q \in E_p(a, b)$$

being points on an elliptic curve, such that Q is known to generate a cyclic group of high order. Unlike in finite fields, the discrete logarithm problem on elliptic curves is not known to allow any sub-exponential time solutions. The best known methods have run time $O(\sqrt{p})$, where p is the characteristic of the (prime) field over which E_p is defined. As a consequence, one can work in much smaller groups than in the case of the multiplicative groups of finite fields, still achieving the same estimated security of a scheme, with respect to state-of-the-art attacks. This advantage led to a new wave of interest for elliptic curve cryptography in connection with the security of mobile phones.

The Weil pairing requires certain caution though. One may in principle use the identity (4) in order to reduce the discrete logarithm problem on the elliptic curve to one in the multiplicative group of the field $\mathbb{F}_r := \mathbb{F}_q[\zeta]$. Since discrete logarithms in multiplicative groups allow for subexponential algorithms, being thus much more efficient, the size of this extension $\mathbb{F}_r$ plays an important role and the reduction might cause problems when $\mathbb{F}_r$ is not too large. The use of the Weil pairing for the discrete logarithm on *supersingular* elliptic curves was pointed out for the first time by Gerhard Frey. The problem came to light when Frey was asked to estimate a software using these curves—on which a particularly efficient implementation of the group laws is possible—for its security. He showed that for these specific curves, the Weil pairing reduced the elliptic curve logarithm problem to one in finite fields of critically small size—$\mathbb{F}_r = \mathbb{F}_{q^k}$ for $k \in \{2, 3, 6\}$, thus leading to serious security problems. The idea was taken over by A. J. Menezes, P. C. van Oorschot and S. A. Vanstone and is currently known in the literature under the name of *MOV attack*. The attack is in general inefficient, but discarded the use of supersingular curves for cryptographic purposes, for the reasons mentioned above. Interestingly, more than a decade later, due to the increasing demand for efficient cryptography using short bandwidth, in application to securing cell phone communications, the supersingular curves found a revival. Recently, some research is invested in finding good combinations of finite fields and supersingular curves, such that on the one hand time savings can be made in the arithmetic, and on the other hand the field $\mathbb{F}_r = \mathbb{F}_{q^6}$ is intractable for the number field sieve discrete logarithm. This example shows that there still is a certain volatility about development of practical cryptographic system, which however overlaps the reliable overall results of cryptanalysis.

A further example where efficiency is sought at the critical border line of the MOV attacks are the so-called Koblitz curves, defined over fields $\mathbb{K} = F_{p^\ell}$ of small characteristic and having $a, b \in \mathbb{F}_p$. Since p is small, it is of course likely that the field $\mathbb{F}_r \supset \mathbb{K}$ required for a MOV attack is a not very large extension of $\mathbb{K}$, even when the curves are not supersingular.

In the last years, D. Boneh and A. Joux developed the idea of *identity-based* cryptography. In order to cope with increasing demand of various cryptographic keys, the idea is to provide the possibility in some limited networks for the user to have access to his secret key, essentially by means of his own identity. The most spread implementation of this idea also uses Weil pairing, and is thus called *pairing-based cryptography*. The recent developments in discrete logarithms for fields of small characteristics described above have thus an important impact, requiring significant increases in the size of the keys used.

Despite standardization, which made cryptographic developments obsolete on the Internet, there are thus reasons why research in this particular area is still very fertile. We recommend the detailed and lively survey of Heß et. al. [16]. The interested reader is referred to [20–22, 24, 26, 28] for further reading.

Key Management and Biometry

Since the security of a cryptosystem relies on its keys, it is an important task to manage these keys in a secure and efficient way. In a public-key environment, one discerns the following essentially distinct aspects:

A. *Managing secret keys*. Since these are data without meaning for humans, they should necessarily be stored on some electronic media, thus leading to the security concern that only the authorized key possessor should have access to the use of these keys.
B. *Trusting public keys*. We have seen that in the public key setting, Alice needs to use some public key of Bob. This can either be provided by Bob during the communication, or read from a common, public data base. But in both cases, since the key is obtained over the network, Alice wishes to be certain that the public key received really belongs to Bob. Otherwise, Eve might for instance provide an own key, while convincing Alice that she obtained the public key of Bob. In this way Eve would be in the position of decrypting messages that Alice had encrypted in the assumption they should only be accessible to Bob, the rightful owner of the secret key belonging to the public one that she received.

There are various solutions for solving both of the above problems. For the first, keys can be stored on some card device, that needs to be activated by some password. Alternatively, the same principle can be replicated on any variety of secure storage media, including an encrypted hard disk. Alternatively, the user may have access to secure *applications* that manage keys locally on his behalf. In this case, the activation password will be application-dependent.

For the second problem, the key idea is called *certification*. Some *trusted authority*, which has verified the physical identity of Bob matching to his pubic key, will add a signature on this public key, made with the secret key of the authority. The signature put by the trusted authority upon Bob's key is also called a *certificate*. The trusted authority's public key will be accessible in a non-forgeable way, so Alice can verify the signature, thus gaining trust for the fact that Bob's key is genuine. In practice, in order to generate a chain of trust reaching from Bob to Alice it may sometimes be necessary to build up a chain of certificates: trusted authority T_1 certifies Bob's key, T_2 certifies the one of T_1, reaching to T_k which is the last authority the key of which is unconditionally trusted by Alice.

Public Key Infrastructure

The principle is very useful and works well in local networks belonging to an environment which has an own hierarchy of trust which can be naturally mapped to the certificate hierarchy. Such are, for instance, large enterprises, administrations and government institutions. Since auxiliary problems of secure key generation, certificate production and verification, secure storage, etc. follow from these key management problems, producing professional solutions to the key management problem of large intranets became a market and the typical software solutions are called Public Key Infrastructures (PKI), being systems that allow to implement all the above-mentioned functionalities within the intranet of some institution.

Note that in this case the fact of having a common institutional frame is a major help, since it allows to distribute the trust according to well-defined rules that belong to the institution and are very likely to exist independently of the cryptographic setting. It is, however, not always the case that secured communication needs to be established within a closed intranet. In that case, although numerous major companies offer the facility of key generation and distribution, thus offering themselves like some kind of trusted authority for the customer, the level of trust that can be offered to such commercial solutions is rather low and would not suffice for offering reliable confidentiality.

The Open System Approach

An alternative idea was invented by Paul Zimmerman, who has developed a public domain software for secure mail exchange, called Pretty Good Cryptography, and which is meanwhile available also as professional software. Zimmerman's idea of trust in an open network is strikingly simple: it is likely to assume that the communicating peers—Alice and Bob—can agree upon some commonly trusted instance, say Tim. In that case Bob can either already hold a certificate signed by Tim, or one signed by a person that holds a certificate signed by Tim, and so on. If this chain of verifications breaks up, then Bob will be able to provide Alice with a set of certificates that convince her that Tim *indirectly* trusts Bob. Otherwise, Bob will have to ask Tim for a certificate which shall be provided a posteriori. In this way the ring of certificates of each peer grows dynamically, by request and need. While the trust system here is perfectly non-hierarchical and symmetric—now peer has an unconditional level of trust, some other problems must be taken into consideration. For instance, the fact that the trust chain can be quite unreliable, especially when growing too long. Instance A may trust B within a certain frame, and B may trust C, but at the end A might not have a sufficient level of trust in C at all, and would not have signed a certificate if directly asked for one.

These elementary concerns have not been mentioned here with the aim of an exhaustive discussion, but rather in order to raise the awareness about the multiple facets of the problem of secure key management, while indicating the most important approaches for a solution, with their known advantages and disadvantages.

Passwords and Biometry

We have mentioned that in the case of problem A above, Alice may end up having a multitude of secret keys distributed through various applications she may work with on a permanent basis. And the access to her secret keys will be granted by some password, that should sufficiently identify her. This and other contexts in which access is granted based on passwords leads to new issues. First, in order to grant the password with sufficient security, there should exist both a minimal dynamics—requiring periodical password changes—as a sufficient randomness in the passwords themselves, which is seldom granted when using passwords that can be memorized by humans. Add to this the expectation that the password of the same peer, for different applications or environments should differ—so that the compromising of one password does not put in danger the whole range of domains accessed by Alice. We see that the access control by means of passwords poses problems itself.

The identification of persons by means of their physical body or dynamics—called *biometric recognition*—is a specialty that grew from forensic needs developing itself in the computer era into a self-contained branch of computer science at the intersection of image processing, pattern recognition and security. Whether the biometry of concern is provided by fingerprints, iris or face traits, voice or writing patterns, biometric recognition has always the following specific characteristics:

a. Identification is a *stochastic process* and not a deterministic one, as for instance in the case of a password verification by means of some one-way function. Since the biometrics of a person are sampled at two distinct places in time and space, they will not be identical. Due to this and a series of additional factors of incertitude introduced by the physical and computer-processing, identification will always be subject to error. The standard way to measure these average errors is by overlapping the two possible error sources: *false accept*, when another person is falsely accepted for Alice and *false reject* when Alice's identity is not accepted on the basis of her biometry and she is rejected. The equal error rate (EER) is the optimal performance of a system in which the two error rates are identical.

b. Unless the data caption system has a reliable method for distinguishing live, natural biometrics from artificial counterfacts, impersonation attacks are possible.
c. Biometrics are unique, so a biometric trait oncc compromised for a certain type of application, is irreplaceable and ulterior use of that biometrics has lost its security.

Despite these quite restrictive conditions of use, biometric identification has the important advantage of commodity: it can make the necessity of multiple, dynamic passwords obsolete. As a consequence, biometry is already in use for access control applications of low security sensitiveness: access to lounges, clubs, hotel rooms or as replacement for visitor's cards. It can also replace the login password for personal computers. When it comes to security applications, neither the potential uses nor the attacks are so well delimited and classified as is the case in cryptography. Consequently the security claims one encounter in the vast literature of the field does not offer the reliability expected from the context of cryptography. One should therefore recall as a rule of thumb the fact that the probability of a successful attack against a biometric system is quite well approximated by the EER of the system. Since EER of one in a million are seldom—being claimed for some systems using iris recognition, it can be seen that biometric identification is practical and comfortable, but yet not acceptable in conjunction with cryptographic applications. The use of multiple biometrics—including multifinger recognition—is therefore an area of active research, in which one of the subtle issues to consider is the fact that it should not be possible to uncouple the individual biometrics.

Quantum Technology and Other Cryptosystems

The main intensively used public key cryptography methods rely on the number theoretic problems described above. There have been numerous interesting attempts to use the large list of NP complete problems in order to derive some trap door function—the knapsack problem is only one of the most famous ones. We can hardly go into the detail necessary in order to pay justice both to the interest of the attempts and the reasons for their failure or restricted use.

Before discussing below several alternative cryptosystems which survived the scrutiny of cryptananlyst and are still discussed as possible alternatives, we turn here our attention to the contribution of physics.

The Advent of Quantum Theory

Since the early 1980s the Canadian mathematicians G. Brassard and C. Crépeau suggested the use of quantum effects for security applications: the simple idea was that Eve could not tap a quantum communication wire, without destroying the information content transmitted, so security would be provided by a *self-destruction mechanism* introduced by quantum mechanics in the confidential information transmitted. The physical and cryptographical aspects of the idea have been in active research ever since. Unlike the mathematical systems already described, or also others that follow, which can be conceived and analyzed on paper, after which their practical realization reduces to quite a simple task of programming, the difficulties encountered in this case were and remain of physical nature. In the first decade of this century, several practical implementations of quantum[6] cryptography have been announced, reaching over distances of up to 100 km. It is thus the distance and the stability of quantum transmission via fiber-optics which is the bottleneck for this system.

In the nineties of the last century, various ingenious experiments and ideas for alternative computing infrastructures were imagined or even tested. One may mention along these lines, L. Adleman's—one of the inventors of RSA—experiments for computing with bacteria.[7] Perhaps the most persisting future projection in this context is the concept of *quantum computing*; in this case there is a physical idea behind, which is stable enough in order to lead to formal mathematical models of computations that might be performed on quantum computers; one can use for a start the short introduction given in [10]. Using existing models of quantum computers, mathematicians since more than a decade have been developing algorithms that run according to the given model. It is, for instance, known that quantum computers *can* invert all the trap door functions used in the cryptographic schemes described above, in polynomial time. Developing models for quantum computing is an ongoing area of intensive research activity in which some of the most eminent theoretical mathematicians and physicists find appealing questions. For instance, the Fields medalist Michael H. Freedman leads the Q-Section of Microsoft where he applies topological methods to quantum computation (cf. [3, 11–15]).

The quantum computers information unit is a *qubit*; unlike a bit, a qubit can, simply speaking, carry any superposition of the states 0 and 1. The calculation on a quantum computer with n qubits ends with measurement of all the states, collapsing each qubit into one of the two pure states. It is the fact that computations happen in a state of superposition of all quantum states which leads to the distinct superior capacities of quantum computers. Somehow similar to the case of quantum cryptography, there is a major physical problem in the realization of quantum computers, and that is realizing *stable* qubits, stability being with respect to the influence of the environment and in particular other qubits. There are persistent announcements of small progress in the technology of quantum computing, keeping

[6]The reader should not confuse *quantum cryptography* with *quantum computing*, where quantum effects are wished to help computations, not only secure information transmission: the physical challenges are even larger in the latter case.

[7]The idea showed to be in principle feasible, but never reached more than the representation of the decimal digits on such "computers."

the hope alive that one might live the day when first experimental quantum computers carrying more than 3–4 qubits will be routinely available. For instance, in order to factor an RSA key of the currently standard length of 1024 bits, a quantum computer should have in the order of magnitude of 1024 qubits. With this prerequisite however, the number would be factored within milliseconds.

Alternative Cryptosystems

Public key cryptography is sensibly slower than secret key encryption, by a factor of roughly 1,000, as a thumb rule. This led to the wish to design some fast asymmetric schemes that do not use the kind of arithmetics that are the bottlenecks for the DH and RSA systems.

A successful solution in this respect was invented by three number theorists: J. Hoffstein, J. Pipher and J.H. Silverman [17]; they designed the cryptosystem NTRU (Number Theorists are Us), which uses arithmetic in a ring of truncated polynomials, such that decryption—the slower operation—can be done in $O(n \log(n))$ rather than $O(n^2 \log(n))$ or more operations, as is the case for RSA. Here, the constant n is roughly the key size. In the case of NTRU, this is slightly larger for comparable security; for instance, a comparable security to the one provided by RSA keys of 1024 bits may require NTRU bits of 4000 bits. This key increase is affordable, for the performance advantage gained. The security of the system is based on the problem of finding shortest vectors in large lattices. While the best methods for solving this problem continuously improve, this fact can be easily compensated for, by accordingly small increases of the key sizes. The system NTRU has been developed a lot during the last 15 years and was accepted five years ago also as an IEEE standard.

Recently, Dan Bernstein gave a new revival to McElieces algorithms, by developing a variant which is technically improved for efficiency and uses, among others, some algorithms for polynomial simultaneous evaluation and interpolation, which developed in part after the original invention of the cryptosystem. Bernstein refers to his variant as Mcbits [2] and uses the argument that unlike the number theoretical cryptosystems, this scheme is resistant to the state-of-the-art models of quantum computing. One may of course argue that the day when quantum computers become routinely available, it should be expectable that quantum encryption is available too, thus making mathematical cryptography somehow obsolete. The practical bottleneck of Mcbits in present days is the size of the keys, with ranges to several megabytes. It is otherwise an efficient algorithm which can be taken into consideration in environments in which communicating large keys is less of a bottleneck than the computation time for encryption/decryption.

A further family of interesting public key schemes uses non-commutative groups—such as for instance *braid groups*, (e.g. cf. [29]). Their developers also make a point out of the fact that the scheme is resistant to quantum computing.

Conclusion

Cryptography was born in the early ages as a skill of mental combinations put at the service of privacy and military protection. It developed over time into a highly mathematized discipline, which unites the science of concealing with the analysis of attacks into one single unit, *cryptorology*. While the last decades of research and the development of computers have offered satisfactorily wide methods for solving the elementary needs of security, it seems that the prognoses for the future are more captivated by the advent of physical solutions offered by quantum mechanics, both to the cryptanalysis of the most widely spread public key schemes but also, constructively, for the implementation of new, purely physical cryptosystems.

Acknowledgements We would like to express our thanks to Professor Joseph Silverman for his useful remarks on the manuscript.

References

1. R. Bărbulescu, P. Gaudry, A. Joux, É. Thomé, *A quasi-polynomial algorithm for discrete logarithm in finite fields of small characteristic*, http://eprint.iacr.org/2013/400
2. D. J. Bernstein, Tung Chou, Peter Schwabe, *McBits: fast constant-time code-based cryptography,* CHES 2013, to appear.
3. M. Bordewich, M. H. Freedman, L. Lovász and D. Welsh, *Approximate counting and quantum computation*, Combinatorics, Probability and Computing, **14**(2005), 737–754.
4. A. Bostan, F. Morain, B. Salvy and É Schost: *Fast algorithms for computing isogenies between elliptici curves*, Math. Comp. **77** (2008), 1755–1778.
5. E. R. Canfield, P. Erdős, C. Pomerance, *On a problem of Oppenheim concerning Factorisatio Numerorum*, J. Number Theory 17 (1983) 1–28.
6. R. Cramer and V. Shoup, *Signature Schemes based on strong RSA assumptions*, Extended abstract in Proc. ACM CCS 1999.
7. R. Crandall and C. Pomerance, *Prime Numbers – A Computational Perspective*, Springer, 2004.
8. Whitfield Diffie and Martin Hellman, *New Directions in Cryptography*, IEEE Transactions on Information Theory; Nov. 1976.
9. N. D. Elkies, *Elliptic and modular curves over finite fields and related computational issues*, Computational Perspectives on Number Theory: Proc. Conf. in honor of A. O. L. Atkin (D. A. Buell and J. T. Teitelbaum, eds.), AMS/International Press, 1998, 21–76.
10. *Cryptanalysis of ENIGMA in Wikipedia* : http://en.wikipedia.org/wiki/Cryptanalysis_of_the_Enigma
11. M. H. Freedman, *Complexity classes as mathematical axioms*, Annals of Math., **170**(2009), 995–1002.
12. M. H. Freedman, A. Kitaev and Z. Wang, *Simulation of topological field theories by quantum computers*, Commun. Math. Phys., **227**(2002), 587–603.
13. M. H. Freedman, A. Kitaev, M. J. Larsen and Z. Wang, *Topological quantum computation*, Bull. Amer. Math. Soc., **40**(2003), 31–38.
14. M. H. Freedman, M. J. Larsen and Z. Wang, *Density representations of braid groups and distribution of values of Jones invariants*, Commun. Math. Phys. **228**(2002), 177–199.
15. M. H. Freedman, M. J. Larsen and Z. Wang, *A modular functor which is universal for quantum computation*, Commun. Math. Phys., **227**(2002), 605–622.

16. F. Heß, A. Stein, S. Stein and M. Lochter, *The Magic of Elliptic Curves and Public Key Cryptography*, Jahresbericht Deutsch Math.-Ver. **114** (2012), 59–88.
17. J. Hoffstein, J. Pipher and J.H. Silverman: *An Introduction to Mathematical Cryptography*, Springer (2008)
18. A. Joux: *A new index calculus algorithm with complexity $L(1/4 + o(1))$ in very small characteristic*, http://eprint.iacr.org/2013/095
19. D. Kahn: *The Codebreakers: The Comprehensive History of Secret Communication from Ancient Times to the Internet*, Scribner (1997).
20. N. Koblitz, *Elliptic curve cryptosystems*, Math. Comp., **48**(1987), 203–209.
21. N. Koblitz, *Course in Number Theory and Cryptography*, Springer-Verlag, New York, 1994.
22. H. W. Lenstra, *Factoring integers with elliptic curves*, Annals Math., **126**(3)(1987), 649–673.
23. U. Maurer: *Towards the equivalence of breaking the Diffie-Hellman protocol and computing discrete logarithms.* Advances in Cryptology - Crypto '94, Springer-Verlag, (1994), 271–281.
24. R. J. McEliece (January and February 1978), *A Public-Key Cryptosystem Based On Algebraic Coding Theory*, DSN Progress Report. 42–44: 114. Bibcode:1978DSNPR..44..114M.
25. Mihăilescu, P., Morain, F., and Schost, E.: *Computing the eigenvalue in the Schoof-Elkies-Atkin algorithm using Abelian lifts.* In ISSAC '07: Proceedings of the 2007 international symposium on Symbolic and algebraic computation (New York, NY, USA, 2007), ACM Press, pp. 285–292.
26. P. Mihăilescu and M. Th. Rassias, *Public key cryptography, number theory and applications*, Newsletter of the European Mathematical Society, **86**(2012), 25–30.
27. P. Mihăilescu and V. Vuletescu, *Elliptic Gauss sums and applications to point counting.* J. Symb. Comput. 45, **8**(2010), 825–836.
28. V. Miller, *Uses of elliptic curves in cryptography*, Advances in Cryptology: Proc. of Crypto '85, Lecture Notes in Computer Science, **218**(1986), Springer-Verlag, New York, pp. 417–426.
29. A. Myasnikov, V. Shpilrain and A. Ushakov, *Group-based Cryptography*, Advanced Courses in Math. CRM Barcelona, Birkhäuser Verlag (2008)
30. M. Th. Rassias, *On the representation of the number of integral points of an elliptic curve modulo a prime number*, http://arxiv.org/abs/1210.1439
31. R. Rivest, A. Shamir and L. Adleman, *A method for obtaining signatures and public key cryptography*, Communications of the ACM, **21** (1978), 121–126.
32. R. Schoof, *Elliptic Curves over Finite Fields and Computation of Square Roots mod p*, Math. Comp. **43**(1985), 483–494.
33. R. Schoof, *Counting Point on Elliptic Curves over Finite Fields*, Journal de Th. des Nombres Bordeaux,
34. J. Silverman, *The Arithmetic of Elliptic Curves*, Graduate Texts in Mathematics **106**, Springer-Verlag, New York, 1986.
35. L. C. Washington, *Elliptic Curves-Number Theory and Cryptography*, CRC Press, London, New York, 2008.

Improvement of Order Performance of a Supplier

Christodoulos Nikou and Socrates J. Moschuris

Abstract Defence procurement is a multi-criteria decision problem and a large part of a Department of Defence (DoD) annual budget. For example, in the fiscal year 2009, contract obligations for the US DoD included $370 billion for defence-related supplies and services [3], and in Great Britain the respective amount for 2011–2012, was 20.1 billion pounds (UK NAO-National Audit Office (2013), *Improving Government Procurement*, London, United Kingdom). This paper develops a procedure for the evaluation/improvement of the ordering process in Military Critical Items to help procurement personnel report quickly and accurately to the Hierarchy of an agency/company. In this attempt, Perfect Order (PO) concept, Principal Component Analysis and Multivariate EWMA control chart were combined onto real data collected from members of the Armed Forces with the required confidentiality.

Introduction

A key starting point in developing and implementing an effective results-oriented management framework is an agency's strategic planning effort [56], and in that planning, strategies are set forth by its Leaders, in order to serve appropriately the needs of its "customers" [15]. Strategic and tactical decisions have a horizon of 2–10 years, and of few months to 2 years respectively [5]. [57] included supplier agenda in military leadership. Usually leading/senior staff in an agency/company

C. Nikou (✉)
Department of Industrial Management and Technology, University of Piraeus, 80 Karaoli and Dimitriou Street, 18534 Piraeus, Greece
e-mail: chrisnikou@gmail.com; cnikou@unipi.gr

S.J. Moschuris
Associate Professor of Logistics and Supply Management, Department of Industrial Management and Technology, University of Piraeus, 80 Karaoli and Dimitriou Street, 18534 Piraeus, Greece
e-mail: smosx@unipi.gr

N.J. Daras (ed.), *Applications of Mathematics and Informatics in Science and Engineering*, Springer Optimization and Its Applications 91, DOI 10.1007/978-3-319-04720-1_23,

does have limited time for strategic and tactical decisions; consequently, suggesting something to them to decide, should be precise, schemed and presented in an understandable way. Decision Analysis assists in reducing complexity and provides a formal mechanism for integrating the results so that a course of action can be provisionally selected [17].

In this paper, by taking into account the above-mentioned role of Decision Analysis, we evaluated real data collected through confidential questionnaires of members of the Hellenic armed forces. Our attempt was to examine how data in Defence Procurement could be simplified and presented more clearly to the hierarchy through the use of descriptive statistics, Confidence Intervals and Multivariate Statistical Analysis. Additionally, we drew attention to the variability of special causes/noise (unusual occurrences that are not normally part of the process) that may indicate an out of control process by using multivariate control charts. All this could provide the decision makers a tool that would reduce the subjectivity of their decisions, and this could be used to report quickly and accurately to Hierarchy about the supplier performance monitoring process. It is formed after a survey in the fields of supplier performance evaluation (vendor rating) and supplier selection methods/models, the armed forces logistic principles and the use of statistical methods, in a professional area where to the best of our knowledge there is still work to be done with advanced statistical techniques. These fields are both selected firstly because suppliers should be evaluated not only at the selection stage, but through their cooperation with an agency [30]. Secondly, because performance evaluation procedures focus on the criteria of quality, delivery, total cost and service [49, 55] seen also in supplier selection literature as key success factors [9, 35] and as basic targets for an acquisition strategy [30].

The rest of the paper is organized as follows: In the next section we review parts of relevant literature and present our conceptual framework. Then, the phases that comprise the evaluation procedure are described, and conclusions, limitations and directions for future research are cited.

Literature Review

Generally speaking, dealing with suppliers is a task that needs to draw much of an agency's attention. Suppliers account for 50–80 % of a major item's value and much of the technical innovation incorporated into a new weapon, comes from the suppliers [54]. Several papers deal with the supplier selection issue [3, 6, 8, 11, 18, 31, 33, 44], and it is considered a crucial process that addresses how organizations select strategic suppliers to enhance their competitive advantage [22]. Supplier performance evaluation is also important in order for an agency to be in place to verify suppliers initially stated skills, and includes contract management, periodic vendor rating as well as programs/actions for their development [30]. It is also an issue that has been seen several times in the relevant academic literature [4, 7, 23, 36, 46, 49, 58]. Rating a supplier who co-operates with a company is a mechanism toward the development of the business relations between that agency and its suppliers, who can also pinpoint ways to improve their levels of co-operation [40]. Two of the four principles that Procter & Gamble has adopted in its effort to integrate/coordinate its supply chain, are the real-time communication with the suppliers that the company has established long-term relations with, and the use of commonly accepted metrics, focused on the delivery of items/services to the customer [40].

In this article we focus on a specific area of defence procurement that has to do with suppliers of Military Critical Items (MCI). MCI are a subpart of the list of defence-related products as these are mentioned in the [13]. MCI are items with special technical characteristics, requiring special treatment and maintenance procedures, whose failure could cause loss of life, permanent disability or major injury, loss of a system, or significant equipment damage [12, 25] . From a supply positioning model perspective, MCI may correspond to critical and bottleneck items [31]. In a short sentence, MCI are items that have a major impact to the safety and accomplishment of a mission. Material procured under Urgent Operational Requirement/Urgent Sustainment Requirement (UOR,USR) may be included in MCIs. UOR,USR procedures are used for the rapid purchase of new or additional equipment, or for an enhancement or essential modification to any existing equipment, in order to support a current or imminent military operation [52].

When reviewing [61], which is the result of transposing European Directive 81,2009 into Greek Law for the public defence procurement, we saw that in chapters dealing with selection and performance evaluation of potential suppliers, a set of indicative criteria is mentioned for the award of a contract, while the supplier who receives the contract is monitored by its contractual obligations. [22] stated that their review of supplier selection literature indicated a lack of consensus in providing definitive guidance to supply managers involved in strategic purchasing. [21], in their seminal work for supplier evaluation and selection problem indicate a variety of methods applied, showing that no unanimity exists for the best one. Furthermore, [50] argued for the need to enhance public sector with the tools that are currently at the disposal of private sector procurement offices. He also stated that it is vital for the public sector to be characterized by transparency. Military Logistics, part of which is the acquisition of military items, are important for the success of a mission, due to the fact that they act as a force multiplier by increasing the timeliness and endurance of an acting force [51]. Greek Land Forces Doctrine [20] urges for simplicity and agility in the SCM procedures, as time is one of the critical factors in military operations. [35] provided us with a list of common vendor rating methods, where no Multivariate Statistical Methods (MSM) are referred and with a matrix of supplier evaluation factors where the ordering process is seen as one of the most frequent ones. Additionally, in NSPA procurement regulations [41], no use of MSM is observed. [45] argued that statistics can assist in truth by converting knowledge into useful knowledge. Consequently, many criteria that are seen in the supplier selection field could be used in the MCI supplier performance evaluation under statistical methods, in an attempt to enhance methods applied in the public sector using ideas from the private sector, aiming at meritocracy, transparency, and efficient MCI logistics.

Conceptual Framework

It is supposed that we operate within the frame of a defence agency under public procurement law dealing with an MCI supplier who is under evaluation. The theoretical basis for the construct of our tool is provided by Principal Component Analysis (PCA), used for supplier selection issues [2, 44, 47] in order to reduce variables under study without losing valuable information from the initial data [24].

Real data were evaluated, collected through confidential questionnaires filled in by members of the Armed Forces under the Statistical confidentiality-Principle No 5 of the European Statistics Code of Practice transposed into Greek Law via [60], and the National Security Regulation. Then we applied some preliminary statistical tools to examine the shape and spread of sample data and confidence intervals to have an estimate on the population's answers. Afterwards, by using PCA, we made an effort to represent data as simply as possible without sacrificing valuable information. We also tried to monitor quality aspects with control charts so that occurrences of special causes can be identified.

Figure 1 depicts schematically our intention about what we call in this article, The Order Performance Improvement Procedure.

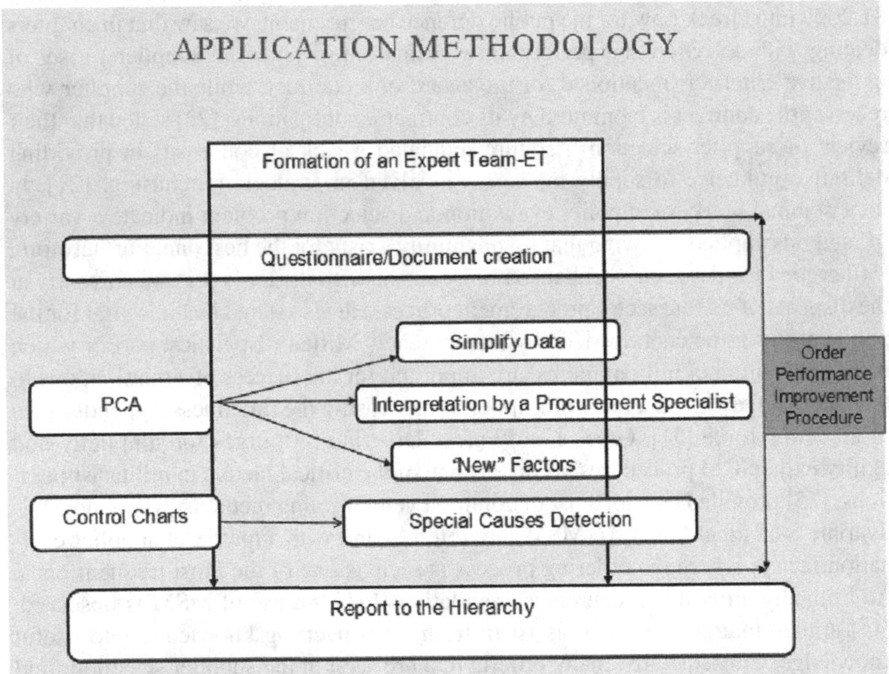

Fig. 1 The order performance improvement procedure

Synthesis Based on Perfect Order Concept in 5 point Likert-Scale

1.	Fault entries in e-ordering.	V1
2.	Lack of stock	V2
3.	Funding Problems	V3
4.	Inability of contractual/arranged lead time	V4
5.	Not sufficient Documentation at the collection	V5
6.	Shipment delays	V6
7.	Arrival delays	V7
8.	Defective material at the collection	V8
9.	Inaccurate Invoices	V9

Fig. 2 Factors of the perfect order concept

Question Synthesis

Data collected, are answers to one of the questions included in the questionnaire. We constructed the specific question, based on the Perfect Order (PO) concept [30, 38] and the most common mistakes in an ordering process [30]. The factors that the members of the armed forces were asked to rate by their importance and rate of appearance, on a five point Likert Scale are depicted in Fig. 2 and represent variables 1 to 9 (V1..V9). PO was chosen as the basis for this question, for the reasons mentioned below:

- It is considered as a complex index for evaluating the capability of a supply chain to work properly [30].
- Experience plays a vital role in a successful human capital function in acquisition planning [55]. Thus, we followed the practice of a big multinational company such as P&G, which, puts a lot of importance to the measurement of its POs as it considers them having much importance at their client's opinion. Clients in our case are considered to be the Operational Units supported by the military SCM.

The suitability of the factors in Fig. 2 can also be verified by the fact that many of them could be related to the factors mentioned by [14] as the most usual in case studies referring to supplier scoring and assessment (i.e. on-time performance/shipment-arrival delays, supply quality/defective material, delivery frequency/lack of stock, supplier viability/funding problems, information coordination capability/inability of contractual lead time).

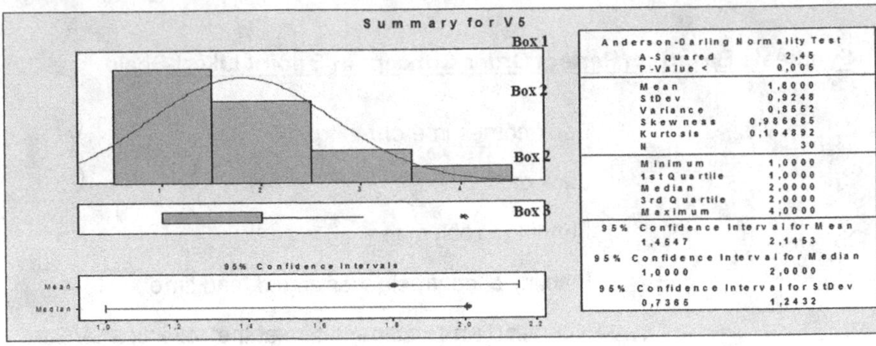

Fig. 3 An example of the descriptive statistics used in this paper

Evaluation Procedure

In [10] it is argued that in order to implement effective strategies for sourcing, multifunctional teams should be used (i.e. teams that include members from purchasing, manufacturing, engineering and planning). Intra-departmental collaboration is promoted as a key factor for the success of acquisition strategies [20, 55]. The idea of using evaluation multifunctional teams in the supplier selection procedure is supported by several authors [6, 11, 47]. We suggest the formation of an expert team-ET from members of the agency, based on their knowledge of the military procurement area and relevant post graduate studies. The evaluation of a supplier in cases of complex items should be done by a team having a legal advisor [30]. As MCIs are important for the success of a mission and usually complex items, the ET proposed in this article may include a representative of the agency's legal department and a member from the financial department due to the prerequisites set in Greek Law [59], where it is stated that for any action that has an economic burden, the first compulsory step for the initialization of a planning process is the reassurance of the budgetary financial capital. Its main tasks are the following:

- To apply the Order Performance Improvement Procedure under the idea of Perfect Order and of PCA, along with the provision of an interpretation to its (PCA) results.
- To examine the feasibility of applying the idea of the Order Performance Improvement Procedure or the PO concept itself, adjusted to military procurement particularities, in the national military logistics software.

Descriptive Statistics

Descriptive Statistics is used in order to derive conclusions about the characteristics of the population under study, excluding any other generalizations and predictions [1]. Figure 3 presents synoptically the selected descriptive statistics of a Variable (V5), retrieved from MINITAB Statistical Software. In the first box, it is examined whether or not the observations of the sample follow a normal distribution, by using the P-value and the Anderson Darling (A-D) test (a smaller A-D value indicates that the distribution fits the data better). In V5 it is clear that observations do not follow a normal distribution. Consequently, if normality was a prerequisite for applying other statistical methods to reduce variability which is an obstacle for the achievement of Quality [34], a BOX-COX transformation would be necessary. In this case, we could study V5 by using distributions for data that do not follow a normal distribution, such as the Polynomial Logistic Distribution [28]. In the second box, statistics that provide a general idea of the shape and spread of sample data are presented so that the hierarchy of the agency can get a quick idea of them. In the third box, Confidence Intervals (CIs) are presented. Mean CI, which is of our interest, can be used with no assumption for the population distribution, since in our case Central Limit Theorem applies [29]. In its practical application within the Armed Forces field, the hierarchy can get a fast and efficient estimation for the mean of the population under study, and with that to explore ways to improve the ordering process. For example, a CI with a low upper level may imply that the subject variable is not highly ranked as vital for the improvement of the ordering process. A broad range between low and upper level at a high level of Confidence (95 %) may mean that no consensus exists among the personnel of the military agency about the importance of the variable under study. Alternatively, it could also imply that this Variable plays a role whose importance depends on the position that the members serve their duty in the agency. Furthermore, CIs may appear very useful for alerting Hierarchy in cases that involve human factors such as moral or training factors, which can affect significantly a military procurement system [20, 48].

Principal Component Analysis

In military decision making, Analytical Hierarchical Procedure (AHP) is often seen as a method that integrates the judgment, experience, and intuition of decision makers [32]. In this article we preferred PCA, having in mind what [27] mentioned for inadequate measurement systems, considering them the main reason why changes are not completed successfully. PCA is one of the most widely used multivariate statistical techniques for dimensionality reduction, identifying a lower uncorrelated dimensional variable set that can explain most of the variability of the original variable set [32]. It is useful in cases of many variables with few observations and it benefits from AHP since PCA outputs are derived from real data and loadings are assigned to factors in proportion with the degree of variance/information they contain.

Algebraically, are particular linear combinations of p random variables that can explain most of the variability of the original variable set [24]. Let the random vector

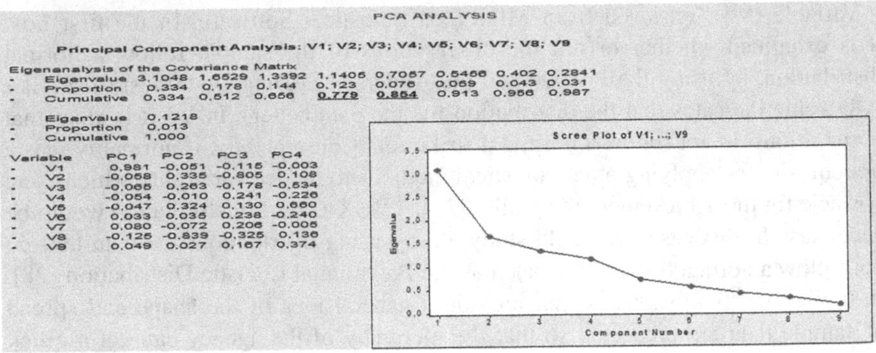

PCA ANALYSIS

Principal Component Analysis: V1; V2; V3; V4; V5; V6; V7; V8; V9

Eigenanalysis of the Covariance Matrix

- Eigenvalue	3,1048	1,6529	1,3392	1.1405	0,7057	0.5466	0.402	0,2841	0,1218
- Proportion	0,334	0.178	0.144	0,123	0,076	0,059	0.043	0,031	0,013
- Cumulative	0,334	0,512	0,656	0.779	0.854	0.913	0,956	0,987	1,000

Variable	PC1	PC2	PC3	PC4
- V1	0,981	-0,051	-0.115	-0.003
- V2	-0,058	0,335	-0,805	0,108
- V3	-0,065	0,263	-0.178	-0,534
- V4	-0,054	-0,010	0,241	-0,226
- V5	-0,047	0,324	0.130	0,660
- V6	0,033	0,035	0.238	-0.240
- V7	0,080	-0,072	0,206	-0,005
- V8	-0,125	-0,839	-0,325	0,136
- V9	0,049	0.027	0,167	0,374

Fig. 4 PCA results in variables 1–9

X that has a sample of observations (n) for a set of p variables (i.e. Xi = [X1,X2, . . ., Xp]) and its covariance (or correlation) matrix R. PC development does not require multivariate normal assumptions. The original variables X are transformed into linear combinations Z = aTXi..p that are uncorrelated, where Zi = [Z1, Z2, . . ., Zp]. By Eigen value analysis of R, these Zi (i = 1,. . . p) are ordered so that the Zi with the highest Eigen value of R corresponds to the first PC and describes the largest amount of the variability in the original data and so on. A visual aid to determine an appropriate number of PCs is the scree plot [24] and PCs that may remain in the model are those with an Eigen value greater than one criterion [2]. Alternatively, a desired limit for the percentage of total variance explained in the model could be set out, and by that to choose the number of PCs to retain in the model. In this paper, MINITAB statistical software is used to perform PCA to 9 variables (V1–V9) that represent factors of the PO concept. Figure 4 provides the results of the PCA at a sample of 30 questionnaires. The conclusions derived from the above figure are the following:

- Scree Plot and relevant academic literature [2, 26] allow us to reduce the number of variables up to 4 or 5 (5 if a very high percentage of variance to explain is required). Four components could be an acceptable solution since the residual components have Eigen Values less than one.
- The first 4 principal components account for 77.9 % of the total variance and respectively the first 5 for the 85.4 %. The percentage of variance explained by each component represents its relative importance.
- The coefficients listed under each PC show how to calculate the PC scores.

Correlation Matrix provided in Fig. 5 tests whether the variables reveal a sufficient level of correlation.

The interpretation of the principal components is subjective [24], thus it should be performed by a person or a team (i.e. the ET) relevant with the specific professional area. This subjectivity enhanced us to select an austere level of

Correlation Matrix, Significance Level (0.01)

Corrélations: V1, V2, V3, V4, V5, V6, V7, V8, V9, SCPC1, SCPC2, SCPC3, SCPC4

	V1	V2	V3	V4	V5	V6	V7	V8	V9
SCPC1	0.995	-0.096	-0.132	-0.118	-0.089	0.079	0.211	-0.184	0.128
	0.000	0.615	0.486	0.536	0.638	0.679	0.264	0.331	0.500
SCPC2	-0.038	0.401	0.390	-0.017	0.451	0.061	-0.139	-0.897	0.051
	0.842	0.028	0.033	0.930	0.012	0.748	0.464	0.000	0.790
SCPC3	-0.077	-0.868	-0.237	0.347	0.163	0.378	0.357	-0.313	0.286
	0.688	0.000	0.207	0.061	0.389	0.040	0.053	0.092	0.125
SCPC4	-0.002	0.108	-0.657	-0.301	0.762	-0.353	-0.008	0.121	0.592
	0.993	0.572	0.000	0.106	0.000	0.056	0.967	0.524	0.001

Fig. 5 Correlations matrix

significance-a=0.01 [43], in order to determine the variables that are most highly correlated with the PCs. These variables may constitute the basis for PC's interpretation to something objectively close/relevant to them. Figures 4 and 5 lead to the below mentioned conclusions which were extracted with the assistance of two senior officers of the military procurement area who had at their disposal a variety of supplier selection criteria in the form of quality, delivery and cost-related attributes [21]. The attempt aimed to describe the information of initial variables as well as possible [42]:

- PC1 could represent "The level of training in e-ordering system and its simplicity for use", as it is strongly and directly related to V1.
- PC2 could correspond to "Material Perfect Rate" as it is strongly and conversely related to V8.
- PC3 could imply a factor called as the "Percentage of Orders Delivered by the Due Date" or "Military Unit (MU) Waiting Time" as it is strongly and conversely related to V2.
- PC4 could be strongly and directly related to V5 and V9, but strongly and conversely related to V3. It could be thought of as a contrasting level of V5, V9 with V3 to some extent, thus a combination of factors like "Cost Reduction Efforts" and "MU Rejections".

Additionally, [35] summarized the ten Cs of the effective supplier evaluation. Among them, Consistency, Commitment, Cost, and Control systems are those that may be corresponded to PC1–PC4, since all of them refer to matters of quality, service/availability, delivery and cost as PCs do.

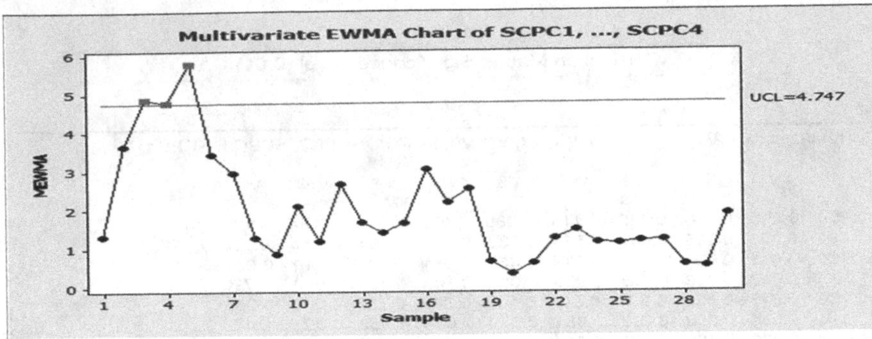

Fig. 6 MEWMA of PC1–PC4

Monitoring Quality with Multivariate Exponentially Weighted Moving Average Control Chart

Multivariate control charts are used to monitor quality with Principal Components [24]. A multivariate exponentially weighted moving average (MEWMA) chart is used to simultaneously monitor two or more related process characteristics in an exponentially weighted control chart. It shows how several variables jointly influence a process by a single control limit which determines whether the process is in control. In this article, EWMA was preferred because it is less affected by the skewness of a Distribution than Shewhart-style charts, relatively robust in the face of non-normally distributed characteristics [16], and better in detecting small and moderate sustained shifts [37]. However, it should be noticed that the scale on multivariate charts is unrelated to the scale of any of the variables. Thus, out-of-control signals in MEWMA do not reveal which initial variable/s caused the signal, consequently MEWMA is considered to be more difficult to interpret than classic Shewhart control charts. In Fig. 6 MEWMA of PC1 to PC4 is provided. Value of r (the weight of the multivariate exponentially weighted moving average) must be a number between 0 and 1 and the influence/importance of the past observations in the shape of MEWMA is decreased exponentially as r values rise [16]. The value used in this paper was 0.1, in order to keep past and present observations at the same level of importance.

MEWMA in our case serves as a tool that quickly detects large or small shifts in a process so that they can be presented to the Hierarchy of the agency aiming to draw the attention to the appearance of special causes. It indicated that the Test failed at points 3, 4 and 5; those points fall above the upper control limit indicating that differences in weight and length over time are not due to common causes. Special causes seem to appear in questionnaires 3, 4 and 5, which may need additional examination, i.e. the members that filled in these questionnaires could be interviewed to see why their replies show unusual differentiation from the rest.

Managerial Implications and Conclusions

In an era where defence spending may be used as a component of a quality indicator of a country's governance [19], this paper combines real data with conceptual framework based on principal component analysis, aiming at assisting the hierarchy of an agency to take quick, but less subjective decisions by reducing the volume of data that need to be evaluated. It pointed out that aspects of Quality, Delivery and Cost are considered to be very important for the improvement of a supplier's order procedure, regardless of him being in the public or private sector. Consequently, in the public sector, this improvement (order performance) could be achieved by the renegotiation of monitoring procedures in existing contracts and the introduction of stricter terms and motives for success in terms of quality and delivery, such as the factors described with PCs 1 to 4. Alternatively, a quality award could be set up for the "best" supplier as far as the contractual obligations are concerned. The paper also suggested the creation of an expert team promoting co-operation by applying intradepartmental co-operation, a necessity when dealing with suppliers of critical items [30]. Additionally, it demonstrated the idea of Perfect Order and its utility as a future evaluation tool of a military supply chain, using MSA techniques to retrieve data from existing software.

While this study has provided the frame for the achievement of low cost with insignificant loss of information and no need for a priori assignment of weights for the variables involved, the decision for the number of the PCs to be retained should depend on the criticality of a project. If only strategically critical items were studied, it might be decided to retain in the model 5 PCs in order to have a bigger percentage of information for evaluation. The hierarchy of the agency should determine the above-mentioned criticality, by participating actively in decisions for the procurement procedures, following [39] who mentioned management's involvement as one of the leading issues in procurement strategy.

Limitations and Future Research Directions

We are obliged to take into account limitations set in the National Safety Regulations and Statistical confidentiality-Principle No 5 of the European Statistics Code of Practice transposed into Greek Law via [60], thus all our sources/references are available unrestricted. Nevertheless, since this study has not answered all questions concerning this problem, potential future inquiries would be to replicate this study by using a less austere level of significance with respective interpretation/analysis, the development of a procedure for the managers to decide the number of PCs to retain, and the evaluation of the results/conclusions that order performance procedure provides in cases of domestic and international aspects of logistics, if these are examined separately.

Acknowledgements Opinions and results in this paper do not reflect the official position of the Hellenic Armed Forces. Sincere gratitude is hereby extended to the reviewers for their helpful comments.

References

1. Athansopoulos D (1999), *Descriptive Statistics*, Stamoulis publications (in Greek), Piraeus, Greece.
2. Amiri M et al (2008), *Supplier Selection via Principal Components Analysis: An Empirical Examination*, Journal of Applied Sciences 8 20:3715–3720.
3. Apte AU et al (2011), *An optimization approach to strategic sourcing: A case study of the United States Air Force*, Journal of Purchasing & Supply Management 17:222–230
4. Bae H (2012), *The Influencing Factors of Logistics Integration and Customer Service Performance for Value Creation of Port Logistics Firms*, The Asian Journal of Shipping and Logistics 28:345–368.
5. Bidalis M (2009), *Logistics*, 2009 Kleidarithmos Publications (in Greek), Athens, Greece.
6. Bottani E,Rizzi A (2008), *An adapted multi-criteria approach to suppliers and product selection-An application oriented to lead-time reduction*, Int.J. Production Economics 111:763–781.
7. Chad W,Golicic G (2010), *Evaluating buyer-supplier relationship-performance spirals: A longitudinal study*, Journal of Operations Management 28:87–100.
8. Cheng C et al (1999), *Evaluating attack helicopters by AHP based on linguistic variable weights*, The European Journal of Operational Research 116:423–435.
9. Cheraghi H et al (2004), *Critical Success Factors For Supplier Selection: An Update*, The European Journal of Operational Research 116:423–435.
10. Chopra S, Meindl P (2001), *Supply Chain Management 1st Ed.*, Pearson Education Inc, New Jersey, USA.
11. Dagdeviren et al (2009), *Weapon Selection using the AHP and TOPSIS methods under Fuzzy environment*, Expert Systems with Applications 36:8143–8151.
12. Defence Acquisition University (2010), *Defense Acquisition Guidebook*, USA (http://www.dau.mil).
13. European Directive 47 (2012), *Commission Directive 47/2012 Annex List of Defence Related Products*,
14. Chopra S, Meindl P (2004), *Supply Chain Management 2nd Ed.*, Pearson Education Inc, New Jersey, USA.
15. Gattorna J (2010), *Dynamic Supply Chain-Delivering Value Through People 2nd Ed*, Pearson Education Limited, Harlow CM20 2JE, United Kingdom.
16. Georgakakos G (2002), *Statistical Process Control*, Hellenic Open University press (in Greek), Patras, Greece.
17. Goodwin P, Wright G (2004), *Decision Analysis for Management Judgement*, John Wiley & Sons Ltd, The Atrium, West Sussex, England.
18. Guler E (2008), *Incorporating multi-criteria considerations into supplier selection problem using Analytical Hierarchy Process: A case Study*, Journal of Yasar University 3 12:1787–1810.
19. Gupta S et al (2001), *Corruption and military spending*, European Journal of Political Economy 17:749–777.
20. Hellenic MoD (2002), *Greek Land Forces Doctrine (GLFD)*, Hellenic Army General Staff Publications (in Greek),Athens, Greece.
21. Ho W et al (2010), *Multi-criteria decision making approaches for supplier evaluation and selection: A literature review*, European Journal of Operational Research 202:16–24.

22. Hsu C et al (2006), *Supplier Selection Construct: Instrument development and validation*, The International Journal of logistics Management 17 2:213–239.
23. Jahns C, Moser R (2005), *Benchmarking your supply organization: The supply Health Check*,The 90th Annual International Supply Management Conference.
24. Johnson R, Wichern D (2007), *Applied Multivariate Statistical Analysis*, Pearson Prentice Hall, New Jersey, USA.
25. Joint Logistics Commanders-JLC/Department of the US Navy (2005), *ACSIMH-Aviation Critical Safety item Management Handbook*, Patuxent River, USA.
26. Hsu C et al (2006), *Multivariate Statistical Analysis*, Stamoulis publications (in Greek), Athens, Greece.
27. Ko J, Newcomer K (2008), *Transforming Public and Non-profit Organisation–Stewardship for Leading Change*, Management Concepts, Vienna, VA.
28. Koutras V et al (2013), *Polynomial Logistic distribution and its applications in finance*, Journal of Communications in Statistics: Theory and Methods (accepted for publication).
29. Koutrouvelis J (2000), *Probabilities and Statistics II*, Hellenic Open University press (in Greek), Patras, Greece.
30. Laios L (2010), *Supply Management*, Humantec Publications (in Greek), Piraeus, Greece.
31. Lee E-K et al (2001), *Supplier Selection and Management System Considering Relationships in Supply Chain Management*, IEEE transactions on Engineering management 48 No 3.
32. Lee J et al (2010), *A hybrid approach of goal programming for weapon systems selection*, Journal of Computers & Industrial Engineering 58:521–527.
33. Liu F, Hai HL (2005), *The voting analytic hierarchy process method for selecting supplier*, International Journal of Production Economics 97 (3):308–317.
34. Logothetis N (2001, *Procedures and Techniques for the Continuous Improvement of Quality*, Hellenic Open University press (in Greek), Patras, Greece.
35. Lysons K, Farrinton B (2006), *Purchasing and Supply Chain Management 7th Ed.*, Pearson Education Limited, Essex, England.
36. Mc Ginnis M (1995), *Building Better Performance Measures*, NAPM Insights 6:50.
37. Midi H, Shabbak A (2011), *Robust multivariate control charts to detect small shifts in mean*, Journal of Mathematical Problems in Engineering dx.doi.org/10.1155/2011/923463.
38. Monczka R et al (2010), *Purchasing and Supply Chain Management 4th Ed*, Cengage Learning EMEA, Andover, United Kingdom.
39. Monczka R, Petersen K (2009), *Supply Strategy Implementation: Current State and Future Opportunities*, CAPS Research, Arizona State University,USA.
40. Mosxouris S, Giannakopoulos D (2013), *Logistics Management & Strategy*, Rosilli publications (in Greek), Athnes,Greece.
41. NATO Support Agency (2012), *FD 251-01 NAMSA Procurement Regulations Functional Directive 251-01 Amendment 6*, NSPA, Capellen, G.D. of Luxembourg.
42. Nyberg H et al (2006), *Multivariate analysis applied to test procedure for determining gun propelling charge weight Part I. Preliminary analysis of the data set*, The Journal of Chemometrics and Intelligent Laboratory Systems 87:131–138.
43. Papaioannou T, Ferentinos K, (2000), *Mathematical Statistics*, Stamoulis Publications (in Greek),Athens, Greece.
44. Petroni A, Braglia M, (2000), *Vendor Selection Using Principal Component Analysis*, The Journal of Supply Chain Management: A Global Review of Purchasing and Supply 36 2:63–69.
45. Rao CR (1997), *Statistics and Truth-Putting Chance to Work*, World Scientific,Singapore.
46. Shen L et al (2012), *A fuzzy multi criteria approach for evaluating green supplier's performance in green supply chain with linguistic preferences*, Resources Conservation and Recycling 74:170–179.
47. Sen CG et al (2010), *Pre selection of suppliers through an integrated fuzzy analytic hierarchy process and max-min methodology*, International Journal of Production Research 48 (6):1603–1625.
48. Strain J, Preece D (1999), *Project management and the integration of human factors in military system procurement*, International Journal of Project Management 17(5): 283–292.

49. Stueland V (2004), *Supplier Evaluations: Best Practices and Creating or Improving Your Own Evaluation*, 89th Annual International Supply Management Conference.
50. Tadelis S (2012), *Public procurement design: Lessons from the private sector*, The International Journal of Industrial Organization 30:297–302.
51. UK MoD Development Concepts and Doctrine Centre-DCDC (2007), *Logistics for Joint Operations*, Joint Doctrine Publication 4/00, 3rd Edition.
52. UK MoD JSP 886 (2010), *The management of equipment procured under urgent operational requirement arrangements*, Volume 3 Part 10.
53. UK NAO-National Audit Office (2013), *Improving Government Procurement*, London, United Kingdom.
54. USA Government Accountability Office-GAO (1998), *Best Practices-DoD can help suppliers contribute more to weapon system programs*, GAONSIAD-98-87, Washington, USA.
55. USA Government Accountability Office-GAO (2005), *Framework for Assessing the Acquisition Function at federal Agencies*, GAO-05-218G, Washington, USA.
56. USA Government Accountability Office-GAO (2010), *DoD's High Risk Areas*, GAO-10-929T, Washington, USA.
57. Wong L et al (2003), *Military Leadership: A context specific review*, The Leadership Quarterly 14:657–692.
58. Yayha S, Kingsman B (1999), *Vendor Rating for an entrepreneur development programme: a case study using the AHP method*, Journal of the Operations Research Society 50:916–1030.
59. Greek Law 3871 (2010), , Available unrestricted in http://www.et.gr, last assessed in Jun 2013.
60. Greek Law 3871 (2010), , Available unrestricted in http://www.et.gr, last assessed in Jun 2013.
61. Greek Law 3871 (2011), , Available unrestricted in http://www.et.gr, last assessed in Jun 2013.

On the Design of Agent Agreement Protocol using Linear Error-Correcting Codes

Panayotis E. Nastou, Panos Pardalos, Paul Spirakis, and Yannis C. Stamatiou

Abstract In a number of situations, it is necessary for two agents who may have never communicated in the past to, jointly, create a shared information item which can serve as a basis for subsequent protocols that the agents may wish to execute (e.g., negotiation or encryption protocols). One way to create this shared piece of information is to have the two agents start with one random bit string each and then engage in a protocol that enables them to transform, gradually, bit differences (in their strings) into bit agreements. In a previous work, an efficient protocol was proposed which was based on the use of the Extended Golay error-correcting code in order to locate and "correct" bit differences. In this work we generalize this protocol in order to use any generic error-correcting code and derive theoretical performance bounds on the efficiency, based on the characteristics of the employed code. The proposed generalized protocol is fair, in that the final strings (which have the same bits in the majority of positions) depend on the strings possessed by both agents while each agent contributes to the same degree in the formation of these

P.E. Nastou (✉) • P. Pardalos
Center for Applied Optimization, University of Florida, Gainesville, USA
Department of Industrial and Systems Engineering, University of Florida, Gainesville, USA
e-mail: pardalos@ise.ufl.edu

P.E. Nastou
Department of Mathematics, University of Aegean, Samos, Greece
e-mail: pnastou@aegean.gr; pnastou@ufl.edu

Y.C. Stamatiou
Department of Business Administration, University of Patras, Patra, Greece
e-mail: stamatiu@ceid.upatras.gr

P. Spirakis • Y.C. Stamatiou
Computer Technology Institute and Press ("Diophantus"), Patras, 26504, Greece

P. Spirakis
Computer Science Department, University of Liverpool, UK
e-mail: P.Spirakis@liverpool.ac.uk

N.J. Daras (ed.), *Applications of Mathematics and Informatics in Science and Engineering*, Springer Optimization and Its Applications 91, DOI 10.1007/978-3-319-04720-1_24,

strings. Finally, the proposed protocol is lightweight (both computationally and with respect to message exchanges) and, thus, can be implemented in embedded systems and resource limited devices.

Introduction

In a number in real-life situations that involve two negotiating agents it is required, by the agents as a first step, to establish a shared piece of information. Furthermore, the two agents may have never met each other in the past and, thus, they have not had the opportunity to establish this information beforehand. This problem is an instance of a wide class of problems termed *agreement problems*. The general setting of these problems is that there are two (possibly not knowing each other) communicating agents that wish to engage in a protocol in order to reach *agreement* in the sense of establishing a shared information item after the end of the protocol. This information item will be based on their initial, separate information items (which need not be related to each other) and the interactions that take place during the protocol execution.

As an example of this class of problems, the two communicating agents may have as a goal the establishment of communications channel using a shared key encryption scheme. Consequently, they would, first, need to create a shared information item that will serve as the encryption/decryption key. This shared key establishment process, which is part of the *key management problem*, is important in any application that requires the creation of secure communications channels over unprotected networks (see, for instance, [8], e.g., for a concise survey on various key management schemes for ad-hoc network devices). In other application scenarios, the two agents may wish to only reach an agreement on a common piece of knowledge. This knowledge may be, for example, the expectation of a random variable defined over date that follows a specific probability distribution (see, for instance, [1] for a comprehensive discussion of agreement problems and their computational complexity properties).

The problem we consider in this paper is simpler but retains the basic feature of all these problems, i.e., the establishment of a shared piece of information. More specifically, the goal of the agents we consider is the establishment of a common bit string, with no special properties, using simple bit operations. These are arithmetic and logical operations while the agents also perform operations involving look-ups in small bit transformation tables.

In brief, in the set-up we consider, there are two communicating agents (possibly not known to each other) each of whom, initially, possess an arbitrary string. The agents' goal is the establishment of a shared piece of information, out of their initial strings, by executing a protocol that is based on the exchange of short messages and the performance of (locally, at each agent) simple bit operations. The main idea behind the protocol we propose is that at each step the agent whose turn is takes a random sample from its local bit string and encodes it using an error-correcting code, which is known and used by both agents. Then the agent sends the encoding to the other agent as well as the positions from the local bit string from where it took the sample. After receiving this message, the other agent looks at the same positions at its local bit string and uses the error-correcting encoding in order to deduce whether its local substring is close to the sample taken by the other agent. In the case of sufficient proximity, as determined by the error-correcting code, the agent deduces that the examined substring should be the same as the substring encoded and sent by the other agent and, thus, "corrects" this substring locally, sending a message to the other agent to "correct" its corresponding substring too.

In [2] the authors used the Extended Golay error-correcting code and demonstrated, with simulations, that the involved agents reach, fast, a high percentage of bit positions, in their local strings, in which their contents agree, i.e., they have the same bit values. In this paper we generalize this work by deriving theoretical results with respect to efficiency using arbitrary error-correcting codes. Then anyone wishing to implement the protocol can instantiate the error-correcting code, plug in the code's parameters into our theoretical estimates and derive estimates with respect to the quality of the protocol before implementation. In general, the resulting family of protocols (parameterized by the chosen error-correcting code and its features) can work fast in resource limited devices as well, while it is possible that the required error-correcting code is already implemented in the device to support its communication module, which makes the implementation effort of the agreement protocol much easier.

Related Works and Our Approach

An essential component of our protocol is the identification of proximity between two substrings sampled from the strings the agents possess. In the case of sufficient proximity, the two agents can proceed to update the substrings, locally, so as to increase the number of places where their strings agree. In this section we examine other works which are based on the proximity concept and show how our work differs from them.

A recently proposed approach that can be employed to identify bit string proximity is based on the concepts of *secure* and *fuzzy* sketches. As discussed in [3], an $(M, m, \tilde{m}, t)$-*secure sketch* is a randomized mapping $SS : M \rightarrow \{0, 1\}^*$, to which there corresponds a deterministic recovery function Rec which allows the recovery of an input w from its sketch $SS(w)$ and any vector w' sufficiently close to w (i.e., $\forall w, w' \in M$ that satisfies $dis(w, w') \leq t$, it holds $Rec(w', SS(w)) = w$) and for all random variables W over M with min-entropy m, $\tilde{\mathrm{H}}_\infty(W|SS(W)) \geq \tilde{m}$. Dodis, Ostrovsky, Reyzin, and Smith addressed this problem in [3] from a security perspective. They defined a cryptographic primitive, the *secure sketch s*, which produces information about an input w which allows its precise reconstruction from any sufficiently close value w'. Two such constructions were proposed in the authors of [3]. They, first, considered the "fuzzy vault" construction proposed by

Juels and Sudan in [4] in the secure sketch context and then converted it into a fuzzy extractor. This was achieved by bounding the entropy loss from w, given s. Then, they observed that the produced information about w reduces its entropy only minimally (leaking little information about w) and that in the case of a small set of values, so that the set can be encoded by its characteristic vector, they provided the cryptographic construction PinSketch. This forms a secure sketch based on a BCH code, that maintains the exponential improvements in sketch size and running time while at the same time, it handles the variable set size.

Bringer, Chabanne, and Kindarj showed in [6] how to apply secure sketches to *cancelable biometrics* so as to have the advantage of both constructs. They, further, propose an algorithm that provides a good performance for data obtained from a fingerprint database by combining several sketching techniques and a cancelable transformation. In order to archive better performance, they employ the coding/decoding scheme proposed in [5].

There are, also, other works that employ constructions more complex than the secure sketches. An implementation of a biometric authentication technique that uses the fuzzy extractor proposed in [3] is described in [7]. In this work, the adopted error-correcting code is the shortened Reed-Solomon $[9600, 1920, 7681]_{2^{14}}$, which also is a complex code (even more than a Reed-Muller and a BCH code). Works following similar approaches require unnecessarily much computational effort for the goals of our simple agreement problem, in which our focus is not secrecy but a fast convergence to agreement, i.e., a shared piece of information.

In this paper, we generalize the work in [2] in which a simple error-correcting code, the Extending Golay code, was employed to reach agreement between the agents, dispensing with the need for the employment of complex error correction codes and the secrecy requirement. In the present work we generalize our findings, theoretically, by providing theoretical estimates on the efficiency (number of exchanged messages) of a family of agreement protocols following the one proposed in [2], parameterized by the error-correcting code they employ. We, thus, provide a general performance estimate for the agreement problem as a function of the chosen error-correcting code.

Coding Theory

Coding theory provides concepts and mechanisms to communicating parties so as to be able to detect errors appeared during transmission and to correct a certain number of errors. In this section, the fundamental concepts and mechanisms of coding theory are introduced.

Basic Concepts

A finite set $A = \{a_1, \ldots, a_q\}$ is called the alphabet of a code. The set of all sequences of symbols of length n over the code alphabet A is denoted as A^n while any subset $Z \neq \emptyset$ of A^n is called a *q-ary block code* of length n and it is referred as an $[n, |Z|]$-code. For example, a *binary code* is a code that its alphabet is $A = \{0, 1\}$ and a *ternary code* is the code with alphabet $A = \{0, 1, 2\}$. The size q of the code alphabet is called the *radix* of the code Z and each element of Z is called a *codeword* [9]. Since there are $|Z|$ codewords, the q-ary code can be used to encode just $|Z|$ source data words. If $k = \log_q |Z|$ is the number of alphabet symbols used to represent each source data word, the *rate* R of $[n, |Z|]$-code is defined by $R = \frac{k}{n}$ which actually is a ratio of the size of a source data word to the size of a codeword. Alternatively, a q-ary code Z can be denoted by $[n, k]$ which is the notation that will be used in the rest of the paper.

Let D be the set of all possible data words that are to be transmitted by an entity. A function $enc : D \rightarrow Z$ that assigns to every data word $x \in D$ a codeword y is called the *encoding function*. A codeword y is transmitted through a channel and the received word r can be considered as the output of the channel. A function dec that answers if a channel output word is a codeword or not is called the *decision or decoding scheme* of the code. If the channel output word is not a codeword then a *decoding error* has happened.

If $y_1, y_2 \in A^n$, the function $d : A^n \times A^n \rightarrow N$ that assigns to its pair (y_1, y_2) the number of positions that y_1 and y_2 differ is the *Hamming distance* $d(y_1, y_2)$ between y_1 and y_2. The set A^n and the function d form a metric space [9]. The *minimum distance* of a code Z is defined to be $d(Z) = \min_{y_1, y_2 \in Z} d(y_1, y_2)$ and the code Z can be defined by $[n, k, d]$. If whenever at least one and at most t errors occurred on a codeword, the resulting word is not a codeword then the code Z is a *t-error-detecting*. Besides, a code Z is an *exactly t-error-detecting* if it is *t-error-detecting* and not a *(t + 1)-error-detecting*. Based on these definitions, it is easy to prove that a code Z is a *t-error-detecting*, if and only if, $d(Z) = t + 1$.

Let us review now the correcting capabilities of a code Z. If a codeword y is transmitted through a channel and r is the output word of the channel then the number of errors occurred is determined by $d(y, r)$. Since the smaller the distance $d(y, r)$ the higher the probability of the output word r to be a codeword, it is obvious that the decoding scheme should provide the closest codeword to r. This scheme is known as the *minimum distance decoding* scheme. Thus, if the decoding scheme described previously is able to correct t or less errors in any codeword, then the code Z is a *t-error-correcting* code. Besides, a code Z is an *exactly t-error-correcting* if it is *t-error-correcting* and not a *(t + 1)-error-correcting*. In [9], it is proven that $d(Z) =$ p if and only if the code Z is $\lfloor (p - 1)/2 \rfloor$-error-correcting. Moreover, it is easy to examine that a code that is *exactly t-error-correcting* can correct t errors and can simultaneously detect $t + 1$ errors.

For a positive integer m, the set of all codewords y at a distance m from a codeword x, i.e., $d(x,y) \leq m$, is called as the sphere of radius m about x, and it is denoted by $S_q(x,m)$. Two important code parameters are the *packing radius* $pr(Z)$ and the covering radius $cr(Z)$. The packing radius $pr(Z)$ is the largest integer m for which all codewords' spheres are disjoint while the covering radius $cr(Z)$ is the smallest integer m for which all the defined codewords' spheres cover A^n, i.e., every word of A^n is an element of a code sphere [9]. A code is *perfect* if $pr(C) = cr(C)$ holds which means that all defined spheres are disjoint and they cover the whole word space A^n.

Linear Codes

Let the code alphabet be the finite field F_q where $q = p^l$ and p prime number, i.e., $A = F_q$. Then the space of all possible sequences over A of length n denoted by $A^n = F_q^n = V(n,q)$ is a vector space. A code $Z \subset V(n,q)$ is *linear* if it forms a subspace of $V(n,q)$. If the dimension of Z is k over $V(n,q)$, then there are k linear independent vectors or codewords that form a basis of Z. Consequently, if $\{z_1, z_2, \ldots, z_k\}$ is a basis of Z, every codeword of Z can be expressed as a linear combination of the codewords of the basis, i.e., $z = \sum_{i=1}^{k} \lambda_i z_i$ where $\lambda_i \in F_q$. Since every λ_i can take q values, $|Z| = q^k$ and if the minimum distance between any two codewords is d, then Z is considered as an $[n, k, \mathrm{d}]$-*code*.

Moreover, the size of the basis determines also the space of the source words which is $V(k,q)$ with size $|V(k,q)| = q^k$. The codewords of a basis of Z can be arranged as the rows of a $k \times n$ matrix G which is called *generator matrix* of Z, and each codeword in Z is precisely the linear combination of Gs rows, that is $Z = \{x \cdot G | x \in V(k,q)\}$. If G is of the form $G = [I_k | B]$ where I_k is the identity matrix of size k and B is a $k \times (n-k)$ matrix, it is said to be in *standard form.*

The set of all words $y \in V(n,q)$ that satisfy the equality $yG^\tau = 0$ is a linear $[n,(n-k)]$ code and is called the dual code of Z and it is denoted by Z'. Actually, Z' contains the words of the space $V(n,q)$ that are orthogonal with every element of Z. Let $H = [-B^\tau | I_{n-k}]$. It holds that $HG^\tau = 0$ which means that H is the generator matrix of Z'. Moreover, the matrix H is also named as *parity check* matrix of Z since $y \in Z$, if and only if, $yH^\tau = 0$. The parity check matrix H of a linear code Z is used in the design of an effective decoding scheme. For any $y \in V(n,q)$, the word yH^τ is called the *syndrome* of the word y. A word $y \in V(n,q)$ is a codeword of Z, if and only if, its syndrome is 0. The basic idea of a decoding scheme is to find the codeword z that has the smallest distance from the received word. Thus, when a codeword y is received, its syndrome is computed. Then find the smallest word $a \in y + Z = \{y + z | z \in Z\}$ that has the same syndrome with y and the corresponding codeword is $z = y - a$. This process is called *the syndrome decoding*.

The most important subclass of linear codes is the class of *cyclic codes* which provides codes with sophisticated decoding techniques. A linear code $Z = [n, k, d]$ is cyclic, if $z = z_0 z_1 \dots z_{n-1} \in Z$ implies also that $z^1 = z_{n-1} z_0 z_1 \dots z_{n-2} \in Z$ [9]. Every codeword $z = z_0 z_1 \dots z_{n-1}$ is associated with a polynomial $z(x) = z_0 + z_1 x + \dots + z_{n-1} x^{n-1}$ of degree $n - 1$. The codeword z^1 can be obtained by multiplying $z(x)$ by x modulo $x^n - 1$. In cyclic codes, those polynomials $z(x)$ are multiples of a unique polynomial $g(x)$ called the *generator polynomial* which is designed by specifying its roots and divides $x^n - 1$. Let $f_i(x)$, $1 \leq i \leq e$ be the irreducible factors of $x^n - 1$. Then $g(x) = \prod_{i \in L \subset \{1,2,\dots,e\}} f_i(x)$. The dimension of a binary cyclic code is given by $k = n - deg(g(x))$ and the generator matrix is of the form $G = [I_k | B]$ where B is a $k \times (n - k)$ matrix. The elements of each row of B are the coefficients of the polynomials $x^{n-i} \ mod \ g(x)$ where $1 \leq i \leq k$.

The parity check polynomial $h(x)$ associated with the parity check matrix H is related to the generator polynomial as follows:

$$h(x) = \frac{x^n + 1}{g(x)} = h_0 + h_1 x + \dots h_k x^k.$$

Then the parity check matrix H is a $(n - k) \times n$ matrix. Each row of H is the binary vector of length n where its elements correspond to the coefficients of the polynomial $x^i h(x) mod(x^n - 1)$, with $0 \leq i \leq n - k - 1$.

Families of Codes

The family of Hamming Codes, discovered by Hamming in 1950, and the Golay codes, discovered by Golay in 1948, are probably the most well-known of all error-correcting codes. The Hamming Codes are exactly one-error-correcting codes with parameters $r > 0$, $n = \frac{q^r - 1}{q - 1}$, $k = n - r$ and $d = 3$. Moreover, all binary codes of this family are linear, perfect and cyclic.

In 1948, Marcel Golay introduced the error-correcting codes $\mathcal{G}_{11}$, $\mathcal{G}_{12}$, $\mathcal{G}_{23}$ and $\mathcal{G}_{24}$. The code $\mathcal{G}_{24}$ is the *Extended Golay code* which is a binary linear and cyclic $[24, 12, 8]$-code. It is an exactly ternary error-correcting code and can detect any 4 errors. The length of the codeword is $n = 24$-bit, the length of the data word is $k = 12$-bit and the minimum distance $d = 8$. The $\mathcal{G}_{24}$ is obtained by the perfect binary linear code $\mathcal{G}_{23}$ by adding a parity bit. The Extended Golay code generator matrix is $G = [I_{12} | B]$, where B can be defined by:

$$B = \begin{bmatrix} 0&1&1&1&1&1&1&1&1&1&1&1 \\ 1&1&1&0&1&1&1&0&0&0&1&0 \\ 1&1&0&1&1&1&0&0&0&1&0&1 \\ 1&0&1&1&1&0&0&0&1&0&1&1 \\ 1&1&1&1&0&0&0&1&0&1&1&0 \\ 1&1&1&0&0&0&1&0&1&1&0&1 \\ 1&1&0&0&0&1&0&1&1&0&1&1 \\ 1&0&0&0&1&0&1&1&0&1&1&1 \\ 1&0&0&1&0&1&1&0&1&1&1&0 \\ 1&0&1&0&1&1&0&1&1&1&0&0 \\ 1&1&0&1&1&0&1&1&1&0&0&0 \\ 1&0&1&1&0&1&1&1&0&0&0&1 \end{bmatrix}.$$

It is a self-dual code which means that the parity check matrix of the code is $H = G^T$.

Another family of binary linear codes that have a good practical value and decoding properties is the family of Reed-Muller codes. A r-th order binary Reed-Muller code of length 2^m, denoted by $RM(r, m)$ is an $[2^m, k, 2^{m-r}]$ error-correcting code where k is given by

$$k = \sum_{i=0}^{r} \binom{m}{r}.$$

Let m boolean variables and f a boolean function on those variables. The values of the function for every combination of 2^m possible combinations form a binary vector associated with the function. It is known that a function can be expressed in disjunctive normal form (DNF) as follows [10]

$$f = 1 + \sum_{i=1}^{m} a_i x_i + \sum_{\substack{j,i=1 \\ i \neq j}}^{m} a_{ij} x_i x_j + \ldots + a_{1,2,\ldots m} x_1 x_2 \ldots x_m.$$

Actually, the parameter k counts all possible Boolean functions of degree at most r in m variables. The generator matrix G is constructed easily by setting the vectors of length 2^m that corresponds to each of the k boolean functions at the rows of the matrix. It is obvious that the length of a data word that is to be coded should be k. The dual code of $RM(r, m)$ code is the $RM(m-r-1, m)$ [10]. Thus, the generator matrix of $RM(m-r-1, m)$ is the parity check matrix of $RM(r, m)$.

The cyclic q-ary BCH codes, discovered by Bose, Chaudhuri and Hocquenghem in 1960 and 1959 respectively, have a great practical importance for error-correction. A binary BCH code of $d \geq 2t + 1$ is a cyclic code whose generator polynomial $g(x)$ has $2t$ consecutive roots b^e, $b^{e+1}, \ldots b^{e+2t-1}$, $n = 2^m - 1 = lcm(n_e, n_{e+1}, \ldots, n_{e+2t-1})$ where n_{e+i} is the order of the element b^{e+i} of F_{2^m}, $0 \leq i \leq 2t - 1$ and $k = n - deg(g(x))$.

A subclass of BCH codes are the Reed-Solomon codes which are BCH codes of length $q-1$ where q is a prime number. Although it is not a binary code, there are ways to map a q-ary Reed Solomon code to a binary code. A Reed-Solomon code can be used for burst-error-correction, which could be very useful in our case when the similarity of two bit strings is very small, i.e., when there is a large number of positions in which they differ. This large number of differences can be considered as burst errors.

Agent Agreement Protocol Based on Error-Correcting Codes

Two agents A_1 and A_2 contain the bit strings m_1 and m_2 of equal length N and of unknown similarity S respectively. An $[n, k, d]$-code Z is considered to support the protocol, which means that the size of the codeword is n, the size of the data word is k while the number of bit errors of a codeword that can be corrected are $t = \lfloor \frac{d-1}{2} \rfloor$.

The Protocol Process

Initially, each of the two agents sets a parameter named $agreedBits$ to zero. This parameter counts the number of bits that the two agents agree at each step. Moreover, a set of N bit flags, one flag per bit location of the strings, is retained by each agent and initially, all flags are set to zero (i.e., total non-agreement). An agent sets the flag of a bit location to 1 when it agrees with its peer on the value of this bit and increases the $agreedBits$ parameter by 1.

One of them becomes the sender and selects at random from its string, a k-bit subset. These data bits are either marked as agreed bits or are unmarked. The basic idea is that the sender after encoding this k-bit piece of data using the Z code, sends only the control bits and the indices of the selected bits to the receiver. On the other hand, the receiver after receiving the transmitted control bits and the indices of the data bits, collects its data bits determined by the received indices and constructs a codeword consisting of the receiver data bits and the sender control bits. Based on the transmitted control bits, the receiver should be able to determine the number of its data bits that differ from the corresponding data bits of the sender. Conceptually, we consider the constructed word as the sender's codeword where some or none of the data bits have been changed.

If e is the error vector of a codeword r, i.e., $r = c \oplus e$, then the syndrome can be calculated by

$$s = rG^T = eG^T.$$

Since in the above concept only the data bits could be modified and the generator matrix G has the form $G = [I_k|B]$, the error vector e can be written $e = e_1 0$, where e_1 is a k-bit error vector of the data part and 0 is the k-bit zero vector.

All syndromes that correspond to all k-bit values of weight $1, 2$ and up to t that e_1 could take are calculated and stored in a lookup table named $validSyndromes$. These precalculated values are used by the receiver in order to determine if the constructed codeword could be corrected. The $validSyndromes$ lookup table contains $VS_Z = \binom{k}{1} + \binom{k}{2} + \ldots \binom{k}{t}$ integers of k-bit long.

Upon the construction of the codeword described above, the receiver performs the error detection procedure. If there are no errors, the receiver sends a control message to the sender that they agree. Both agents increase the parameter $agreedBits$ by j, where j is the number of bits of the examined k data bits that the two agents have not marked them as agreed bits in a previous communication step $(0 \leq j \leq k)$. The size of the control message is just one byte.

However, if the error detection procedure determines that there are errors, then the receiver calculates the syndrome of the constructed codeword and searches it in the table $validSyndromes$. If the search fails, which means that more than t bit differences occur, the receiver sends a control message to the sender determining the failure of bit agreement and no action is taking place on both sides. In case that the search succeeds, which means that the bit differences are less than or equal to t, the receiver sends a control message to the sender that they agree, the receiver performs the decoding procedure to retrieve the corrected data bits and updates its string. Both the sender and the receiver increase the parameter $agreedBits$ by j as above. It is obvious that in this step the receiver's string converges to the sender's string. In the next protocol step, the two agents change roles and thus the protocol is fair in the sense that there is a negotiation between the two agents. The protocol continues while the $agreedBits$ parameter is less than string length N. At the end of the protocol execution both agents obtain the same string of length N. The above computation and communication steps of the protocol are presented in Fig. 1.

At each step, the agents exchange only two messages. The size of the first message is $k + 1$ integers (k for the bit locations and one for the control bits of the codeword) or $2(k + 1)$ bytes (16-bit integers) while the second message is only one byte. The computation tasks of the sender are the random generation of k bit locations, which depends on the performance of the used random number generator, and the computation of the codeword where n dot products of k-bit vectors are computed.

Moreover, at each step, the receiver performs the error detection procedure which is based on counting the number of 1s of the constructed word and computing the syndrome of the codeword which demands the calculation of k dot products of n-bit vectors. In case the procedure detects errors, the receiver searches for the already computed syndrome in the table $validSyndromes$, and if it exists, the receiver decodes the constructed word and modifies its bit string. It is obvious that the operations that are involved in the protocol are simple and fundamental which means that every processor, either limited or powerful, supports them. As for the storage involved in the above computations $2 \times VS_Z$ bytes are needed for the lookup table $validSyndromes$.

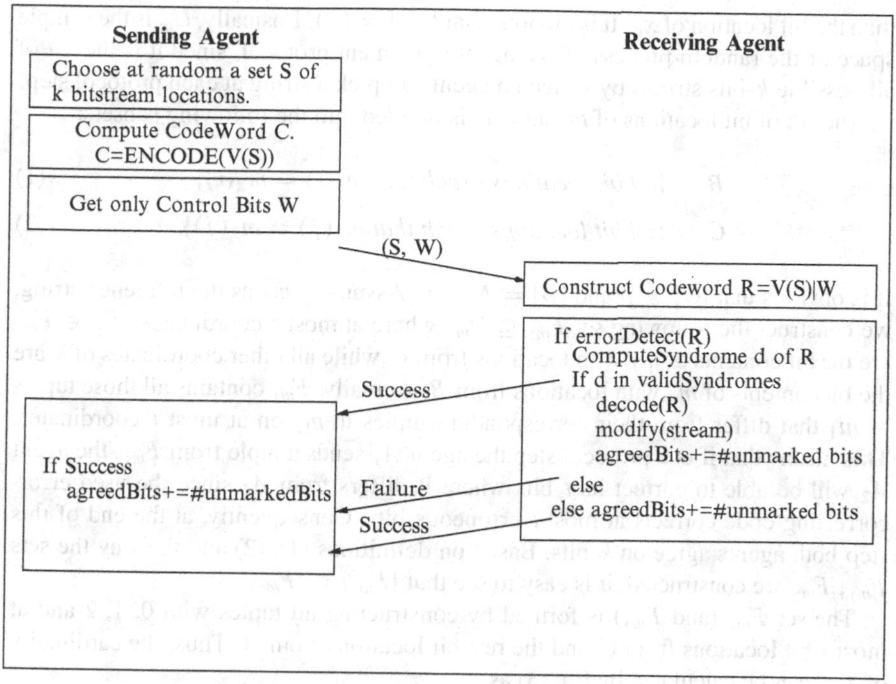

Fig. 1 A step of the protocol using the correcting mechanism

In [2], the use of the Extended Golay error-correcting code in the above generic agreement process along with its simulation results are presented. In the next section, the performance of the protocol is analyzed and a lower bound for the number of protocol steps is given as function of the bit string length N, the similarity S between the two bit strings and the error-correcting code parameters n, k and d.

Performance Analysis of the Protocol

Since $0 \leq S \leq 1$ is the similarity of the strings m_1 and m_2 of equal length N, it is defined as $p = \lceil N * (1 - S) \rceil$ the number of bit locations that they differ. Both agents start the protocol steps described in section "The Protocol Process" using an $[n, k, d]$-code in order to reach an agreement which means that bit strings m_1 and m_2 should be identical at the end of protocol execution.

Since at each protocol step k-bits are picked by bit string m randomly, we define U_m as the set of all possible k-bit strings formed by selecting k bits from m and ordering them according to their locations in m (ascending order), i.e., $U_m = \{(x_1, \ldots, x_i, \ldots, x_k)\}$ where x_i is the content of a bit location of m, and x_i corresponds to a bit location that is greater than the bit location of x_1, and less than the bit location of x_k. It is obvious that $|U_m| = \binom{N}{k}$. Basically, U_m is the sample space of the random process of the agent agreement protocol, since it is the set of all possible k-bits strings by which an agent can pick a string at each protocol step.

The set of bit locations of m_1 and m_2 is divided into the following subsets:

$$B = \{all\ bit\ locations\ i\ such\ that\ m_1(i) = m_2(i)\} \tag{1}$$

$$C = \{all\ bit\ locations\ j\ such\ that\ m_1(j) \neq m_2(j)\} \tag{2}$$

It is obvious that $|C| = p$ and $|B| = N - p$. Assuming m_1 as the reference string, we construct the following set $F_{m_1} \subseteq U_{m_1}$ where at most t coordinates of $x \in F_{m_1}$ are the bit contents of m_1 with locations from C while all other coordinates of x are the bit contents of m_1 with locations from B. Actually, F_{m_1} contains all those tuples of m_1 that differ from their corresponding tuples in m_2 on at most t coordinates. This means that if at a protocol step the agent A_1 sends a tuple from F_{m_1}, the agent A_2 will be able to correct its t bits where it differs from A_1 since the used error-correcting code corrects at most t erroneous bits. Consequently, at the end of this step both agents agree on k bits. Based on definitions (1), (2) and the way the sets F_{m_1}, F_{m_2} are constructed, it is easy to see that $|F_{m_1}| = |F_{m_2}|$.

The set F_{m_1} (and F_{m_2}) is formed by constructing all tuples with 0, 1, 2 and at most t bit locations from C and the rest bit locations from B. Thus, the cardinality of F_{m_1} can be calculated by Eq. (3) as

$$|F_{m_1}| = \sum_{i=0}^{t} \binom{p}{i} \binom{N-p}{k-i}. \tag{3}$$

Since U_{m_1} is the sample space of the protocol process, the set $F'_{m_1} = U_{m_1} - F_{m_1}$ contains all those k tuples where more than t coordinates correspond to bit locations that are from C which means that the used error-correcting code can not correct codewords that were computed by such tuples. A protocol step where an agent has picked a tuple from F'_{m_1} does not improve the agreement process since the recipient agent cannot correct its k tuple that corresponds to the received tuple. The cardinality of F'_{m_1} can be calculated by Eq. (4)

$$|F'_{m_1}| = \sum_{i=t+1}^{k} \binom{p}{i} \binom{N-p}{k-i}. \tag{4}$$

Actually, the sets F_{m_1} and F'_{m_1} form a partition of the protocol sample space U_{m_1} where tuples from F_{m_1} are considered as "good" tuples since they improve the agreement process while tuples from F'_{m_1} are considered as "bad" tuples since they do not improve the agreement process. It is obvious that if t is large ($t \to k$) which means that the error-correcting code is strong since it can correct codewords with large number of errors, the number of "good" tuples is large. This means that the number of protocol steps that are unsuccessful (in the sense that no bit correction happens and thus no improvement of the agreement process) is minimized. Thus, the number of protocol steps that are needed for the agreement process to terminate is minimal. Theoretically, a lower bound for the number of the protocol steps can be obtained if $k = t$, but this is impractical since there is no such error-correcting code. In the case of a large t, the error-correcting code guarantees that the agents will reach an agreement even if their initial bit strings are completely different (Reed-Solomon error-correcting code can handle burst errors, as it was analyzed previously). However, an error-correcting step can be applied only if $N - p \geq k - t \Rightarrow p < N - k + t$ holds, i.e., if there is at least one k tuple where the error-correcting code can have an effect.

On the other hand, if p is increased (i.e., the number of bits that the agents differ is increased) the terms in the summation of Eq. (3) are decreased while the terms of the summation of Eq. (4) are increased. Thus, the number of "bad" tuples is increased and the number of "good" tuples is decreased which means that the number of unsuccessful protocol steps is increased and the protocol performance is decreased. However, if p is decreased, the number of "bad" tuples is decreased while the number of "good" tuples is increased.

If at a protocol step, a tuple from F_{m_1} or F_{m_2} is picked, the number of agreed bits increases at most by $i \leq k$ where i is the number of bits that have not been agreed on in previous steps. Moreover, if j bits were corrected at a protocol step then p is decreased by j and thus at the next protocol step, the probability to pick a "good" tuple increases since according to the previous analysis the cardinality of F_{m_1} and F_{m_2} increases.

Since at each protocol step, the two agents exchange 2 short messages, the minimal number of protocol steps should be found. Actually, the agreement process has to reduce p greedily so that the agents can reach an agreement as fast as possible. The ideal case is at each protocol step each agent can pick a tuple $x \in F_{m_1}$ or $x \in F_{m_2}$ where exactly t coordinates are from C and the others from B. This is because the receiving agent will correct t bits, which is the maximum number that the selected code could correct, and thus both agents will not only agree on k bits, but they will also have resolved their discord on the maximum number of bits that they could solve at each step. Additionally, if x_i and x_j are the tuples picked at the i-th and the j-th protocol steps respectively where $j > i$ and L_i and L_j are the subsets of the corresponding bit locations in m_1 and m_2, the ideal case is L_i and L_j to be disjoint. Consequently, under these conditions each step provides the maximum effect.

Based on the conditions described above, there are $l = \lceil \frac{p}{t} \rceil$ discrete t-combinations of C. Thus, after l steps, the two agents would have resolved their differences. If $N \leq l \times k$, the two agents have checked all their bits and reached an agreement. However, if $N > l \times k$ although the agents have resolved their differences, they did not check all their bits, consequently they need at least $\lceil \frac{N - l \times k}{k} \rceil$ more steps to assure that they agree. Since the protocol parameters are the bit string length N, the similarity S of the initial bit strings and the error-correcting code parameters k and t, a relation between those parameters that satisfies the above inequalities is derived as follows:

$$
\begin{aligned}
N \leq l \times k \leq \left\lceil \frac{p}{t} \right\rceil k &< \left(\frac{p}{t} + 1\right) k \\
\Rightarrow \frac{t}{k}(N - k) &< p = \lceil N(1 - S) \rceil \\
\Rightarrow \frac{t}{k}(N - k) &< N(1 - S) + 1 \\
\Rightarrow \frac{t}{k} - \frac{(t + 1)}{N} &< 1 - S \\
\Rightarrow S &< 1 - \frac{t}{k} + \frac{(t + 1)}{N}.
\end{aligned}
\tag{5}
$$

Having the size of the bit string and the error-correcting parameters, Inequality (5) defines a threshold for the initial bit string similarity that the number of protocol steps is minimized. Based on the above analysis, the minimum number of protocol steps is calculated as follows:

$$
M = \begin{cases} \left\lceil \frac{p}{t} \right\rceil & \text{if } S < 1 - \frac{t}{k} + \frac{(t+1)}{N} \\ \left\lceil \frac{p}{t} \right\rceil + \left\lceil \frac{N - k * \lceil \frac{p}{t} \rceil}{k} \right\rceil & \left(1 - \frac{t}{k} + \frac{(t+1)}{N}\right) \leq S \leq 1. \end{cases}
$$

For a similarity close to 50 %, which is the most frequent case, the number of steps could not exceed $2N$.

Conclusions and Future Work

In this paper we considered the problem of *agreement* between two communicating agents, in the sense of the establishment of a shared piece of information, or common knowledge between them, based on some initial information they may possess, independently of each other.

In this paper we generalized the basic protocol proposed in [2] in which a Golay error-correcting code was employed in order to help the agents "correct," fast, disagreements between their initial information items. In the end, they converge to information items with few differences, which depend to some extent on the initial information, achieving a fair solution that is not biased towards the information of only one of the agents. In the present work the protocol is expressed without any specific error-correcting code in mind. Instead, the protocol is given only the parameters that characterize the error-correcting code (e.g., number of bits, number of detected and corrected errors, etc.) and achieves agreement, using the code as a black-box, after a number of steps that are expressed as a function of the error-correcting code parameters in a uniform manner for all codes. In this way, the agents are free to use any error-correcting code, knowing in advance an estimate of the effort that it is required to reach agreement. This code may be even built-in in the operation of the agents for other purposes (e.g., correcting communications errors).

As a direction of future research, one may extend the proposed protocol to work with any number of agents. This would entail the definition of an appropriate interaction sequence among them (e.g., pairwise, for all possible pairs and then again from the beginning) in order to reach mutual agreement at, roughly, the same number of steps. As another (perhaps more demanding) direction, one may consider deriving a theoretical estimate of the average number of steps required parameterized by the probability distribution function of their initial information items as well as the random decisions (samplings) that take place during the execution of the protocol.

References

1. S. Aaronson, "The Complexity of Agreement", University of California, Berkeley, arXiv:cs.cc/406061 v1, 2004.
2. Panayotis E. Nastou, Paul Spirakis, Yiannis C. Stamatiou and Christina Vichou, "Agent Agreement Protocols based on Golay Error-Correcting Code", *Proceedings of the 4th International Conference on Information, Intelligence, Systems and Applications*, Pireous, Greece, July 2013.
3. Y. Dodis, R. Ostrovsky, L. Reyzin, and A. Smith, "Fuzzy Extractors: how to generate strong keys from biometrics and other noisy data", SIAM J. on Comp., 38(1), pp. 97–139, 2008.
4. A. Juels and M. Sudan, "A fuzzy vault scheme", Designs, Codes and Cryptography, 38(2), pp. 237–257, 2006.
5. J. Bringer, H. Chabanne, G. Cohen, B. Kindarji and G. Zémor, "Optimal iris fuzzy sketches", IEEE 1st International Conf. on Biometrics: Theory, Applications and Systems, 2007.
6. J. Bringer, H. Chabanne and B. Kindarji, "The best of both worlds: Applying secure sketches to cancelable biometrics", Science of Computer Programming, 74(1), pp. 43–51, 2008.
7. S. Cimato, M. Gamassi, V. Piuri, R. Sassi and F. Scotti, "Privacy-aware biometrics: Design and implementation of a multimodal verification system", IEEE Computer Security Applications Conference, ACSAC, pp. 130–139, 2008.
8. R. Dalal, Y. Singh, and M. Khari, "A Review on Key Management Schemes in MANET", International Journal of Distributed and Parallel Systems, Vol. 3, No. 4, 2012.
9. S. Roman, Coding and Information Theory. Springer Verlag, 1992.
10. Robert Morelos-Zaragosa, The Art of Error-Correcting Coding John Wiley and Sons, Inc., 2006.

The Byzantine Generals Problem in Generic and Wireless Networks

Chris Litsas, Aris T. Pagourtzis, and Dimitris Sakavalas

Abstract In this chapter we consider the design of Secure Broadcast protocols in generic networks of known topology. Studying the problem of *Secure Message Transmission* (SMT) proves essential for achieving Broadcast in incomplete networks. We present a polynomial protocol that achieves parallel secure message transmissions between any two sets of nodes of an incomplete network provided that the weakest connectivity conditions which render the Broadcast problem solvable hold. Using the above, we show that the SMT protocol can be used as a subroutine for the simulation of any known protocol for complete networks, which leads us to protocols for generic networks which remain polynomial with respect to the measures of consideration. We extend our result to the case of wireless networks by exploiting the fact that participants are committed to perform local broadcasts, which greatly facilitates achieving an agreement.

Introduction

A fundamental problem in distributed networks is Secure Broadcast, in which the goal is to distribute a message correctly despite the presence of Byzantine faults. In particular, an adversary may control several nodes and is able to make them deviate from the protocol arbitrarily by stopping, rerouting, or even altering a message that they should normally relay intact to a certain node. In general, agreement problems have been primarily studied in complete networks under the threshold adversary model where a bound t is assumed over the number of corrupted players. The importance of achieving agreement in an adversarial environment simultaneously increases with the importance of the applications that demands it.

C. Litsas • A.T. Pagourtzis (✉) • D. Sakavalas
Department of Computer Science, School of Electrical and Computer Engineering, National Technical University of Athens (NTUA), Heroon Politechniou 9, 15780 Zographou, Greece
e-mail: chlitsas@central.ntua.gr; pagour@cs.ntua.gr; sakaval@corelab.ntua.gr

N.J. Daras (ed.), *Applications of Mathematics and Informatics in Science and Engineering*, Springer Optimization and Its Applications 91, DOI 10.1007/978-3-319-04720-1_25,

As an example consider crucial communications during military operations. The above problem was first introduced in [1] under the meaningful name *Byzantine Generals Problem*.

Problem Definition

In a battlefield the commander (dealer) D may wish to send a command—attack or retreat—to his generals. The generals now have to obey his command. The problem is not as easy as is seems because there is always the possibility that some of the generals are corrupted. To make things worse the dealer may be corrupted as well and send different messages to his generals. In the latter case it is, at least, desirable to find a way so that all honest generals take exactly the same decision in order to avoid a disastrous army breakup.

The participants of a distributed protocol (the generals and the commander) are also referred as *players*. The formal definition of the problem follows:

Definition 1 (Secure Broadcast /Byzantine Generals). Consider a set of n players $V = \{v_1, v_2, \ldots, v_n\}$ and let $D \in V$ be the dealer. D sends a message to the other players and also the players are able to communicate in pairs over authenticated channels. In the end each player v_i takes a decision d_i. The problem of secure broadcast is:

- In case that D is honest, i.e., he sends the same message m to every player, every honest player must decide on this message m.
- In case D is corrupted, i.e., he sends different messages to his players, then all honest players must decide on a common message.

For any two members $u, v \in V$ there are two possibilities: they can either communicate with each other over an authenticated channel, or they may have no connection at all. Obviously, the set V together with the set of connections between its elements forms a graph. Finally, without loss of generality (see [2]), we may restrict our study to messages $m \in \{0, 1\}$.

Earlier Work and Our Contribution

The problem was introduced by Lamport, Shostak and Pease in [1], where it was proven that there is a solution if and only if $t < n/3$ (resiliency), where t is the number of corrupted players. In terms of complexity we consider the number of rounds that are needed for the completion of the protocol as well as the total size of the messages exchanged during the protocol. We are also concerned about the complexity of the local computations performed by each participant individually. Since the introduction of the problem, several protocols have been developed to solve it. For complete graphs that have mainly been studied in the literature, there are polynomial protocols with respect to each of the three considered measures which are optimal with respect to the resiliency [3, 4]. On the other hand, not much work has been done in the direction of generic (incomplete) networks. A first protocol was presented by Dolev in [5], where it was proved that the graph must be k-connected, with $t < k/2$ (equivalently $k \geq 2t + 1$). A drawback of that protocol is that it requires exponential computational complexity at each node.

In this chapter we present a protocol that achieves secure transmission from any node to any other node of an incomplete network provided that it is at least $(2t+1)$-connected. This protocol is partially based on the classic Dolev's protocol [5]. In our case though, the computational complexity is polynomial. Furthermore, we show how this protocol can be executed in parallel in order to achieve secure transmission from any set of nodes to any other set of nodes of the graph. Using the above techniques we show that the protocol can be used as a subroutine for the simulation of any known protocol for complete networks. Combining the above with, e.g., the protocol [3] yields a protocol for generic networks which remains polynomial with respect to the three measures that we are interested in. We further modify our protocol in order to develop a protocol especially designed for wireless networks. It is worth noting that most of previous work in the area of wireless networks has considered the problem only over very special network topologies, e.g., grid networks [6]. To the best of our knowledge our protocol is the first efficient one for secure broadcast in generic wireless networks.

Notation

Hereafter the paths of the graph will be represented by strings of the form $\sigma_i \in V^*$. The neighborhood of a node v will be denoted by $\mathcal{N}(v)$. Let $\sigma \in V^*$ and $w \in V$ then, the *order of* w *in string* σ is defined to be

$$ord_w(\sigma) = \begin{cases} |\sigma_1| & \text{if } \sigma = \sigma_1 w \sigma_2, \sigma_1, \sigma_2 \in V^* \\ -1 & \text{if } w \notin \sigma. \end{cases}$$

Solving the Problem in Generic Networks

The protocols (e.g., [1]) that solve the problem in complete networks operate in rounds. Every single round consists of two phases: the *communication phase* where messages are exchanged in parallel between nodes and the *local computations phase* where every node processes the information it has received.

In order to reduce the problem of secure broadcast in general networks to the problem in complete networks, we need a sub-protocol to simulate authenticated message exchange between any two nodes of the network. Next we present an algorithm that implements the above task:

(Input:) Nodes v, u and security parameter t.
(Objective:) Authenticated transmission of message m from v to u.

1. If $u \in \mathcal{N}(v)$, then node v sends the message m to node u, which decides on this value.
2. Else
   a. Every node $w \in V$ calculates the same set $\mathcal{P} = \{p_1, p_2, \ldots, p_{2t+1}\}$ of $2t+1$ disjoint paths from v to u (*valid paths*).
   Every node $w \in V \setminus u$ stores at most one single path $p_w \in \mathcal{P}$, which is the one that contains its name. Node u stores the set $\mathcal{P}$ of *valid paths*.
   b. **Round 0**: v sends the message m_v to every one of his neighbors that happens to be a starting node of one of the disjoint paths of $\mathcal{P}$.
   c. For $i = 1 \ldots \max_{p \in \mathcal{P}} |p|$
   Round i : Every node $w \in V \setminus \{v, u\}$ with $p_w \neq \emptyset$ that received messages in round $(i-1)$ performs (all nodes in parallel):
   If w received in the previous step the message m' from node x, s.t.
   $$\left(ord_x(p_w) = i - 1\right) \wedge \left(ord_w(p_w) = i\right)$$
   then w relays m' to the next node in p_w.
   d. Node u finds the majority of the values he received through valid paths. If there are at least $t+1$ identical values (absolute majority), he decides on this value. If the majority is less than $t+1$ (relative majority) then u decides on a default value $\perp$.

We demand that each node calculates the same set $\mathcal{P}$. We achieve this after forcing each node to run an appropriate variation of the max-flow algorithm [7] with the same input, thus also the same output.

Finally, u receives at most $2t+1$ disjoint paths. In case that v is honest, then given that there are at most t corrupted players, we get that there are at least $t+1$ paths consisting purely of honest players. Consequently, the majority of the received values is the message m. Node u may decide on $\perp$ only if v is corrupted.

Complexity

For the length of the longest computed path it holds that $\max_{p \in \mathcal{P}} |p| \leq n - 2t$ because in the worst case each of the $2t$ paths may consist of one internal node; thus the remaining path is possible to contain all the rest nodes of the graph. Thus,

$$RC = \max_{p \in \mathcal{P}} |p| \leq n - 2t \Rightarrow RC = O(n)$$

Let $\mathcal{M}_w$ be the set of messages received by w during the protocol. Every message sent consists of 1 bit. Each $w \in V \setminus u$ receives a message at most once, since he only has to accept messages in round $ord_w(p_w) - 1$ from the node dictated by p_w. Similarly node u will receive a total of $2t+1$ messages. Thus, the algorithm has bit complexity,

$$BC = \sum_{w \in V} \sum_{m \in \mathcal{M}_w} |m| = \sum_{w \in V \setminus u} \sum_{m \in \mathcal{M}_w} |m| + \sum_{m \in \mathcal{M}_w} |m| \leq$$

$$\leq n + (2t + 1) \Rightarrow BC = O(n)$$

The local computations complexity for each node is bounded by the complexity of the algorithm used for the formation of the set $\mathcal{P}$ (essentially by the complexity of the max-flow algorithm), i.e., $LCC = O(n^3)$.

A Protocol for Generic Networks

Observe that in every network a protocol for multiple message transmissions between every possible pair (u, v) of nodes may be created by executing the 2-node transmission protocol in parallel. Using this observation we can simulate any protocol for complete networks, e.g., [3], by replacing every communication round with a phase of parallel execution of the 2-node protocol. We next give a modification of Protocol that operates in parallel for multiple sender–receiver pairs set $\mathcal{B}$.

(Input:) Set $\mathcal{B}$ of node-pairs, security parameter t.
(Objective:) Authenticated transmission between every pair in $\mathcal{B}$.

1. For every $(v, u) \in \mathcal{B}$
   - If $u \in \mathcal{N}(v)$ then node v sends the message m_v to node u which decides on this value.
     $\mathcal{B} := \mathcal{B} \setminus (v, u)$

Precomputation

2. **Initialize:** $\mathcal{P} = \emptyset$
   For every $(v, u) \in \mathcal{B}$
   a. Every $w \in V$ computes the same set $P_{v,u}$ of $2t + 1$ disjoint paths connecting the pair (v, u)
   b. $\mathcal{P} := \mathcal{P} \cup P_{u,v}$

   Every node $w \in V$ computes and stores the set $\mathcal{P}_w = \{p \in \mathcal{P} | w \in p\}$

Message Transmission

3. **Round 0**: For every $(v, u) \in \mathcal{B}$
   v sends the message (m_v, v, u) to every one of his neighbors that happens to be a starting node of one of the disjoint paths of $P_{v,u}$.

4. For $i = 1 \ldots \max_{p \in \mathcal{P}} |p|$
   Round i : Every node $w \in V$ with $\mathcal{P}_w \neq \emptyset$ that received messages in round $(i-1)$ performs (all nodes in parallel):
   w accepts each message (m, v, u) received from node x, provided that $\exists p \in \mathcal{P}_w \cap P_{v,u}$ s.t.
   $$\big(ord_x(p) = i - 1\big) \wedge \big(ord_w(p) = i\big)$$
   and then w relays each of the accepted messages (m, v, u) to the next node according to p.

Decision

5. For every $(v, u) \in \mathcal{B}$
   u finds the majority of the values he received through valid paths of $P_{v,u} \subset \mathcal{P}_u$. If there are at least $t + 1$ identical values (absolute majority) he decides on this value for m_v. If the majority is less than $t + 1$ (relative majority) then u decides on a default value $\bot$.

Complexity

Due to the parallel transmissions, in protocol 1.2.1, the number of rounds remains at most $n - 2t$, but the bit complexity is now $O(z \cdot n \log n)$, where $z = |\mathcal{B}|$, and the factor $\log n$ is due to the players' names included in the message. Finally, the local computations complexity of each node is $O(z \cdot n^3)$ for the nodes to compute the set of disjoint paths for every given pair.

Given a protocol for complete networks with *round complexity* r, *bit complexity* b and *local computations complexity* c, after the simulation of the communication phase we get a protocol for the generic network model with round complexity,

$$RC = O(r \cdot (n - 2t))$$

due to the r executions of the *multi-node transmissions* protocol. Bit complexity,

$$BC = O(r \cdot n^3 \log n) \text{ and } BC = O(b \cdot n \log n)$$

because of the r executions of the protocol 1.2.1, or b executions of protocol 1.2 for b bits to be transmitted over pairs, including the players' names. Finally the local computations complexity will be,

$$LCC = O(c + n^5)$$

as the paths between every possible pair can be precomputed in the beginning of the protocol and not in every round.

Broadcast in Wireless Networks

A large class of applications involves wireless networks in which nodes possessing radio transmitting/receiving devices are spread out on some physical surface (terrain), and two nodes can communicate if they are within transmission range of each other and signal interference is low. A common abstraction is to consider the network as a graph, and assume (collision assumption) that communication is possible if a node receives a message from only one neighbor in a certain time-slot.

Assumptions

We consider a synchronous wireless network which provides authenticated communication between neighboring nodes and in which the collision assumption holds. We also assume that all nodes are incapable of deviating from the given transmission schedule imposed by the protocol. Finally we assume that there are at least $(2t + 1)$ disjoint paths connecting D with v, $\forall v \notin \mathcal{N}(D)$.

We observe that the dealer D in a radio network is committed to behave honestly during the transmission of his message m. This is due to the fact that every message he transmits is received by all $v \in \mathcal{N}(D)$. Since communication channels are authenticated, every honest neighbor will correctly decide on m.

Obviously, Byzantine Generals problem is simplified in radio networks since the honesty of the dealer yields a 1-round solution in a complete network. Specifically, in this round D sends the message m to every player v and each v accepts the value m that he receives. Therefore, in a generic radio network the problem reduces to every honest player correctly receiving the message of the dealer. The transmission of the message to all the players can be achieved with an appropriate modification of the multi-node transmission protocol.

A Protocol for Wireless Networks

Exploiting the specific properties of wireless networks, we can properly modify the results of the previous section in order to adapt them to the certain context. Namely, a protocol which achieves broadcast in wireless networks can be designed using the multi-node transmission protocol for which the sender–receiver pair set $\mathcal{B}$ consists of all the pairs (D, v), $\forall v \in V \setminus \mathcal{N}(D)$.

In the first round of the protocol, due to the previous observation regarding the dealer's honesty, it suffices that D sends message m to all its neighbors and each $w \in \mathcal{N}(D)$ decides on value m.

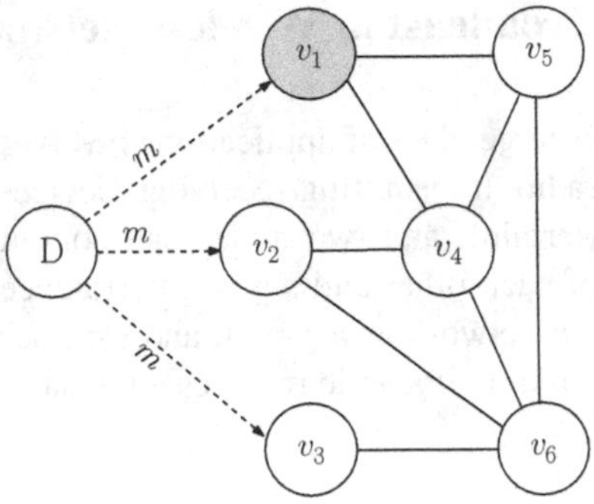

Fig. 1 Dealer D broadcasts message m and player v_1 is corrupted

In order for the message to be transmitted correctly across the network, we follow the message transmission rules of the multi-node transmission protocol with the difference that every player concatenates all the messages (including the names of the corresponding receiver nodes) that need to be relayed to all its neighbors and locally broadcasts them.

Moreover, due to the collision assumption the rounds of the multi-node transmission protocol will be replaced by phases, which consist of numerous rounds in order for every player to transmit separately.

Example

Below (Fig. 1) we give an example to illustrate the *Wireless Broadcast* protocol.

Each $v \in V$ precomputes the sets of disjoint paths $P_4, P_5.P_6$.

$$P_4 = \{Dv_1v_4, Dv_2v_4, Dv_3v_6v_4\},$$
$$P_5 = \{Dv_1v_5, Dv_2v_4v_5, Dv_3v_6v_5\},$$
$$P_6 = \{Dv_1v_4v_6, Dv_2v_6, Dv_3v_6\}$$

Initially D transmits (m, D) to all $v \in \mathcal{N}(v)$. Subsequently the protocol is completed in 3 phases.

1st Phase

Players v_1, v_2, v_3 decide on value m received from the dealer and each transmits in a separate round m', m, m, respectively, along with the corresponding receiver nodes implied by P_4, P_5, P_6.

2nd Phase

According to the computed paths, player v_4 transmits m, m' along with the names of the corresponding receiver nodes v_5, v_6, in order for v_5, v_6 to receive messages m, m' respectively. Similarly v_6 transmits m to relay value m to v_5.

3rd Phase

Finally players v_4, v_5, v_6 compute the $majority(m, m, m') = m$, of the messages received through valid paths (P_4, P_5, P_6) and decide on value m.

Observations

In the wireless network model there is no need for the classic bounds for resiliency ($t < n/3$) and connectivity ($t < k/2$) to hold. Instead, the connectivity bound can be replaced by the weaker assumption that there are at least $(2t + 1)$ disjoint paths connecting D with $v, \forall v \in V \setminus \mathcal{N}(D)$.

The necessity of this assumption, in case we want to avoid further transmission of messages between pairs of players (which would increase the number of rounds significantly), is guaranteed by the results of [8].

Due to the collision assumption each player must transmit in a separate round. In order to minimize the number of rounds, each player w concatenates all messages to be relayed by him (of those he received in the previous phase) and transmits them to $\mathcal{N}(w)$ with a single transmission.

The space requirements for each node w include the storage of the set $\mathcal{P}_w$, for which we observe that,

$$|\mathcal{P}_w| \leq (n - |\mathcal{N}(D)| - 2) + (2t + 1) \leq n - 2 \Rightarrow |\mathcal{P}_w| = O(n)$$

because node w will store at most one path for each $v \in V \setminus (\mathcal{N}(D) \cup D \cup w)$ and $2t + 1$, in which it is the last node.

Complexity

As before, $\max_{p \in \mathcal{P}} |p| \leq n - 2t$ phases are needed for the messages to be relayed over the longest possible paths. Each phase i includes *rounds*(i)—the number of rounds for all the players that need to relay a message to transmit. In conclusion:

$$RC = \sum_{i=1}^{\max_{p \in \mathcal{P}} |p|} rounds(i) \leq \sum_{i=1}^{n-2t} rounds(i) \leq \sum_{i=1}^{n-2t} n =$$

$$= n \cdot (n - 2t) \Rightarrow RC = O(n^2)$$

Let $\mathcal{M}_v$ be the set of messages received by v during the protocol, then

$$BC = \sum_{v \in V} \sum_{m \in \mathcal{M}_v} |m| \overset{*}{\leq} \sum_{v \in V} \sum_{m \in \mathcal{M}_v} c \cdot n \log n \overset{**}{\leq}$$
$$\leq \sum_{v \in V} c \cdot n^2 \log n = c \cdot n^3 \log n \Rightarrow BC = O(n^3 \log n)$$

(∗) In the worst case every concatenated message player w receives will contain one sub-message for every other player containing its name; therefore $\forall m, |m| = O(n \log n)$.

(∗∗) In total, player w will accept $|\mathcal{P}_w|$ messages, thus $|\mathcal{M}_v| \leq n$.

Finally the local computations complexity for each node is bounded by the complexity of the modified max-flow algorithm used for the computation of disjoint paths between the $n - 2t - 2$ pairs $(D, v)_{v \in V \setminus \mathcal{N}(D)}$,

$$LCC = O(n^4)$$

As can be observed, the resulting protocol for wireless networks is more efficient, in all aspects, than the ones proposed for incomplete networks. This is due to the fact that one can take advantage of the local broadcasts performed by the players in order to simplify the problem.

Conclusion

Generic Networks

In this chapter we present a polynomial protocol that achieves secure transmission of messages from any set of nodes to any other set of nodes of an incomplete known-topology network provided that it is at least $(2t + 1)$-connected. Using the above, it is shown that the protocol can be used as a subroutine for the simulation of any known protocol for complete networks, thus we obtain solutions for the broadcast problem in incomplete networks which remain polynomial with respect to the three measures of interest. Essentially we present a reduction from the broadcast problem in generic networks to the problem in the complete network model.

Further research on the Broadcast problem may include the consideration of networks of unknown topology (*ad-hoc*). One could solve the problem of discovering the network topology in the presence of byzantine faults as stated in [10] in order to apply the presented algorithm in the *ad-hoc* model. Another direction is to design protocols specifically designed for the model of unknown network topology by investigating local criteria through which the broadcast will be achieved as presented in [9].

Wireless Networks

We developed a broadcast protocol specially tailored for wireless networks adopting the assumptions that the underlined communication network is of known topology and that all nodes are incapable of deviating from the given transmission schedule imposed by the protocol. The relaxation of these assumptions is a promising and interesting research goal due to the extensive use of *ad-hoc* networks and the emergence of practical applications which involve the presence of powerful adversaries which can deviate from the given schedule and create unexpected collisions. One could attempt to devise special tools, such as selective transmission schedules, taking into account the new requirements imposed by security considerations; for example, a message must be transmitted through several disjoint paths, in order to neutralize the influence of malicious participants.

References

1. L. Lamport, R. E. Shostak and M. C. Pease, *The Byzantine Generals Problem.* ACM Trans. Program. Lang. Syst. pp. 382–401. 1982.
2. B. A. Coan, *Achieving consensus in fault-tolerant distributed computer systems: protocols, lower bounds, and simulations*, Massachusetts Institute of Technology, 1987.
3. J. A. Garay and Y. Moses, *Fully Polynomial Byzantine Agreement for* $n > 3t$ *Processors in* $t + 1$ *Rounds.* SIAM J. Comput. pp. 247–290 1998.
4. A. Bar-Noy, D. Dolev, C. Dwork and H. R. Strong, *Shifting gears: Changing algorithms on the fly to expedite Byzantine agreement.* Information and Computation pp. 205–233 1992.
5. D. Dolev, *The Byzantine Generals Strike Again.* J. Algorithms pp. 14–30 1982.
6. C. Y. Koo, V. Bhandari, J. Katz and N. H. Vaidya, *Reliable broadcast in radio networks: the bounded collision case.* PODC pp. 258–264 2006.
7. R. K. Ahuja, and T. L. Magnanti, and J. B. Orlin, *Network flows: theory, algorithms, and applications.* Prentice-Hall, Inc. 1993.
8. D. Dolev, C. Dwork, O. Waarts and M. Yung, *Perfectly Secure Message Transmission.* J. ACM, pp. 17–47, 1993.
9. A. Pelc and D. Peleg, *Broadcasting with locally bounded Byzantine faults.* Inf. Process. Lett., pp. 109–115, 2005.
10. Mikhail Nesterenko and Sébastien Tixeuil, *Discovering Network Topology in the Presence of Byzantine Faults.* IEEE Trans. Parallel Distrib. Syst. vol. 20, pp. 1777–1789, 2009.

A Multidimensional Hilbert-Type Integral Inequality Related to the Riemann Zeta Function

Michael Th. Rassias and Bicheng Yang

Abstract In this chapter, using methods of weight functions and techniques of real analysis, we provide a multidimensional Hilbert-type integral inequality with a homogeneous kernel of degree 0 as well as a best possible constant factor related to the Riemann zeta function. Some equivalent representations and certain reverses are obtained. Furthermore, we also consider operator expressions with the norm and some particular results.

Keywords Hilbert-type integral inequality • Hilbert-type integral operator • Riemann zeta function • Gamma function • Weight function

2000 Mathematics Subject Classification: 11YXX, 26D15, 47A07, 37A10, 65B10

Introduction

If $p > 1, \frac{1}{p} + \frac{1}{q} = 1, f(x), g(y) \geq 0, f \in L^p(\mathbf{R}_+), g \in L^q(\mathbf{R}_+)$,

$$||f||_p = \left\{ \int_0^\infty f^p(x)dx \right\}^{\frac{1}{p}} > 0,$$

M.Th. Rassias (✉)
Department of Mathematics, ETH-Zentrum, CH-8092 Zurich, Switzerland
e-mail: michail.rassias@math.ethz.ch

B. Yang
Department of Mathematics, Guangdong University of Education, Guangzhou, Guangdong 510303, P. R. China
e-mail: bcyang@gdei.edu.cn

N.J. Daras (ed.), *Applications of Mathematics and Informatics in Science and Engineering*, Springer Optimization and Its Applications 91, DOI 10.1007/978-3-319-04720-1_26,

$||g||_q > 0$, then we have the following Hardy–Hilbert's integral inequality (cf. [1]):

$$\int_0^\infty \int_0^\infty \frac{f(x)g(y)}{x+y} dxdy < \frac{\pi}{\sin(\pi/p)} ||f||_p ||g||_q, \tag{1}$$

where the constant factor $\frac{\pi}{\sin(\pi/p)}$ is the best possible.

If $a_m, b_n \geq 0,\ a = \{a_m\}_{m=1}^\infty \in l^p$, $b = \{b_n\}_{n=1}^\infty \in l^q$, where

$$||a||_p = \{\sum_{m=1}^\infty a_m^p\}^{\frac{1}{p}} > 0,\ ||b||_q > 0,$$

then we still have the following discrete variant of the above inequality with the same best constant $\frac{\pi}{\sin(\pi/p)}$, that is

$$\sum_{m=1}^\infty \sum_{n=1}^\infty \frac{a_m b_n}{m+n} < \frac{\pi}{\sin(\pi/p)} ||a||_p ||b||_q. \tag{2}$$

Inequalities (1) and (2) are important in mathematical analysis and its applications (cf. [1–7]).

In 1998, by introducing an independent parameter $\lambda \in (0, 1]$, Yang [8] presented an extension of (1) for $p = q = 2$. In 2009 and 2011, Yang [4, 5] provided some extensions of (1) and (2) as follows:

If $\lambda_1, \lambda_2, \lambda \in \mathbf{R}, \lambda_1 + \lambda_2 = \lambda, k_\lambda(x, y)$ is a nonnegative homogeneous function of degree $-\lambda$, with

$$k(\lambda_1) = \int_0^\infty k_\lambda(t, 1) t^{\lambda_1 - 1} dt \in \mathbf{R}_+,$$

and

$$\phi(x) = x^{p(1-\lambda_1)-1}, \psi(y) = y^{q(1-\lambda_2)-1}, f(x), g(y) \geq 0,$$

$$f \in L_{p,\phi}(\mathbf{R}_+) = \left\{ f : ||f||_{p,\phi} := \left\{ \int_0^\infty \phi(x)|f(x)|^p dx \right\}^{\frac{1}{p}} < \infty \right\},$$

$$g \in L_{q,\psi}(\mathbf{R}_+), ||f||_{p,\phi}, ||g||_{q,\psi} > 0,$$

then we have

$$\int_0^\infty \int_0^\infty k_\lambda(x, y) f(x) g(y) dxdy < k(\lambda_1) ||f||_{p,\phi} ||g||_{q,\psi}, \tag{3}$$

where the constant factor $k(\lambda_1)$ is the best possible. Moreover, if $k_\lambda(x, y)$ is finite and

$$k_\lambda(x, y)x^{\lambda_1 - 1}(k_\lambda(x, y)y^{\lambda_2 - 1})$$

is decreasing with respect to $x > 0$ ($y > 0$), then for $a_m, b_n \geq 0$,

$$a \in l_{p,\phi} = \left\{ a : ||a||_{p,\phi} := \left\{ \sum_{n=1}^{\infty} \phi(n)|a_n|^p \right\}^{\frac{1}{p}} < \infty \right\},$$

$$b = \{b_n\}_{n=1}^{\infty} \in l_{q,\psi}, \ ||a||_{p,\phi}, ||b||_{q,\psi} > 0,$$

we have

$$\sum_{m=1}^{\infty} \sum_{n=1}^{\infty} k_\lambda(m, n) a_m b_n < k(\lambda_1)||a||_{p,\phi}||b||_{q,\psi}, \tag{4}$$

where the constant factor $k(\lambda_1)$ is still the best possible.

Clearly, for

$$\lambda = 1, k_1(x, y) = \frac{1}{x + y}, \ \lambda_1 = \frac{1}{q}, \lambda_2 = \frac{1}{p},$$

(3) reduces to (1), while (4) reduces to (2). Some further results including a few multidimensional Hilbert-type integral inequalities are provided in [9–19].

In this chapter, using methods of weight functions and techniques of real analysis, we present a new multidimensional Hilbert-type integral inequality with a homogeneous kernel of degree 0 as well as a best possible constant factor related to the Riemann zeta function and the Gamma function, which is an extension of the double case as follows:

$$\int_0^\infty \int_0^\infty \left(\coth\left(\frac{x}{y}\right) - 1 \right) f(x)g(y)dxdy < \frac{\Gamma(\sigma)}{2^{\sigma-1}}\zeta(\sigma)||f||_{p,\varphi}||g||_{q,\psi}, \tag{5}$$

where $\zeta(\cdot)$ is the Riemann zeta function and $\Gamma(\cdot)$ is the Gamma function (cf. [20, 22]). Some equivalent forms and reverses are obtained. Furthermore, we also consider the operator expressions with the norm and certain particular results. For a number of fundamental properties of the Riemann zeta function and the Gamma function, especially in Analytic Number Theory and related subjects, the reader is referred to [21–27, 31].

Some Lemmas

If $m, n \in \mathbf{N}$ ($\mathbf{N}$ is the set of positive integers), $\alpha, \beta > 0$, we define

$$||x||_\alpha := \left(\sum_{k=1}^{m} |x_k|^\alpha\right)^{\frac{1}{\alpha}} \quad (x = (x_1, \cdots, x_m) \in \mathbf{R}^m),$$

$$||y||_\beta := \left(\sum_{k=1}^{n} |y_k|^\beta\right)^{\frac{1}{\beta}} \quad (y = (y_1, \cdots, y_n) \in \mathbf{R}^n).$$

Lemma 1. *If $s \in \mathbf{N}, \gamma, M > 0, \Psi(u)$ is a nonnegative measurable function defined in $(0, 1]$, and*

$$D_M^s := \left\{x \in \mathbf{R}_+^s : 0 < u = \sum_{i=1}^{s} \left(\frac{x_i}{M}\right)^\gamma \leq 1\right\},$$

then we have (cf. [7])

$$\int \cdots \int_{D_M^s} \Psi\left(\sum_{i=1}^{s} \left(\frac{x_i}{M}\right)^\gamma\right) dx_1 \cdots dx_s = \frac{M^s \Gamma^s\left(\frac{1}{\gamma}\right)}{\gamma^s \Gamma\left(\frac{s}{\gamma}\right)} \int_0^1 \Psi(u) u^{\frac{s}{\gamma}-1} du. \tag{6}$$

Lemma 2 (See [18]). *If $s \in \mathbf{N}, \gamma > 0$, and $\varepsilon \geq 0$, then*

$$\int \cdots \int_{\{x \in \mathbf{R}_+^s : ||x||_\gamma \geq 1\}} ||x||_\gamma^{-s-\varepsilon} dx_1 \cdots dx_s = \begin{cases} \frac{\Gamma^s(\frac{1}{\gamma})}{\varepsilon \gamma^{s-1} \Gamma\left(\frac{s}{\gamma}\right)}, & \varepsilon > 0 \\ \infty, & \varepsilon = 0 \end{cases}. \tag{7}$$

Definition 1. For $x = (x_1, \cdots, x_m) \in \mathbf{R}_+^m, y = (y_1, \cdots, y_n) \in \mathbf{R}_+^n, \sigma > 1$, we define two weight functions $\omega(\sigma, y)$ and $\varpi(\sigma, x)$, as follows

$$\omega(\sigma, y) := ||y||_\beta^{-\sigma} \int_{\mathbf{R}_+^m} \left(\coth \frac{||x||_\alpha}{||y||_\beta} - 1\right) \frac{dx}{||x||_\alpha^{m-\sigma}}, \tag{8}$$

$$\varpi(\sigma, x) := ||x||_\alpha^{\sigma} \int_{\mathbf{R}_+^n} \left(\coth \frac{||x||_\alpha}{||y||_\beta} - 1\right) \frac{dy}{||y||_\beta^{n+\sigma}}, \tag{9}$$

where $\coth u = (e^u + e^{-u})/(e^u - e^{-u})$ is the hyperbolic cotangent function (cf. [28]).

By (6), setting

$$v = \frac{M u^{\frac{1}{\alpha}}}{||y||_\beta},$$

we find

$$\begin{aligned}
\omega(\sigma, y) &= ||y||_\beta^{-\sigma} \lim_{M\to\infty} \int_{D_M^m} \left(\coth \frac{||x||_\alpha}{||y||_\beta} - 1\right) \frac{dx}{||x||_\alpha^{m-\sigma}} \\
&= ||y||_\beta^{-\sigma} \lim_{M\to\infty} \int_{D_M^m} \frac{\coth \frac{M}{||y||_\beta} \left[\sum_{i=1}^m \left(\frac{x_i}{M}\right)^\alpha\right]^{\frac{1}{\alpha}} - 1}{M^{m-\sigma} \left[\sum_{i=1}^m \left(\frac{x_i}{M}\right)^\alpha\right]^{\frac{m-\sigma}{\alpha}}} dx \\
&= ||y||_\beta^{-\sigma} \lim_{M\to\infty} \frac{M^m \Gamma^m \left(\frac{1}{\alpha}\right)}{\alpha^m \Gamma\left(\frac{m}{\alpha}\right)} \int_0^1 \frac{\coth\left(\frac{M}{||y||_\beta}\right) u^{\frac{1}{\alpha}} - 1}{M^{m-\sigma} u^{\frac{m-\sigma}{\alpha}}} u^{\frac{m}{\alpha}-1} du \\
&= ||y||_\beta^{-\sigma} \lim_{M\to\infty} \frac{M^\sigma \Gamma^m \left(\frac{1}{\alpha}\right)}{\alpha^m \Gamma\left(\frac{m}{\alpha}\right)} \int_0^1 \left(\coth\left(\frac{M}{||y||_\beta}\right) u^{\frac{1}{\alpha}} - 1\right) u^{\frac{\sigma}{\alpha}-1} du \\
&= \frac{\Gamma^m\left(\frac{1}{\alpha}\right)}{\alpha^{m-1} \Gamma\left(\frac{m}{\alpha}\right)} \int_0^\infty (\coth v - 1) v^{\sigma-1} dv,
\end{aligned}$$

and in view of the Lebesgue term by term theorem (cf. [29]), it follows

$$\begin{aligned}
\int_0^\infty (\coth v - 1) v^{\sigma-1} dv &= \int_0^\infty \left(\frac{e^v + e^{-v}}{e^v - e^{-v}} - 1\right) v^{\sigma-1} dv \\
&= \int_0^\infty \frac{2e^{-2v} v^{\sigma-1}}{1 - e^{-2v}} dv \\
&= 2\int_0^\infty \sum_{k=1}^\infty e^{-2kv} v^{\sigma-1} dv \\
&= 2\sum_{k=1}^\infty \int_0^\infty e^{-2kv} v^{\sigma-1} dv \\
&= 2\sum_{k=1}^\infty \frac{1}{(2k)^\sigma} \Gamma(\sigma) \\
&= \frac{\Gamma(\sigma)}{2^{\sigma-1}} \zeta(\sigma), \qquad (10)
\end{aligned}$$

where

$$\zeta(\sigma) = \sum_{k=1}^\infty \frac{1}{k^\sigma}, \ \sigma > 1.$$

Lemma 3. *For $\sigma, \tilde{\sigma} > 1$, we have*

$$\omega(\sigma, y) = K_2(\sigma) := \frac{\Gamma^m\left(\frac{1}{\alpha}\right)}{\alpha^{m-1}\Gamma\left(\frac{m}{\alpha}\right)} \frac{\Gamma(\sigma)}{2^{\sigma-1}}\zeta(\sigma)(y \in \mathbf{R}_+^n), \tag{11}$$

$$\varpi(\sigma, x) = K_1(\sigma) := \frac{\Gamma^n\left(\frac{1}{\beta}\right)}{\alpha^{n-1}\Gamma\left(\frac{n}{\beta}\right)} \frac{\Gamma(\sigma)}{2^{\sigma-1}}\zeta(\sigma)(x \in \mathbf{R}_+^m), \tag{12}$$

$$\begin{aligned} w(\tilde{\sigma}, y) &:= ||y||_\beta^{-\tilde{\sigma}} \int_{\{x \in \mathbf{R}_+^m : ||x||_\alpha \geq 1\}} \left(\coth \frac{||x||_\alpha}{||y||_\beta} - 1\right) \frac{dx}{||x||_\alpha^{m-\tilde{\sigma}}} \\ &= K_2(\tilde{\sigma})\left[1 - \theta_{\tilde{\sigma}}(||y||_\beta)\right], \end{aligned}$$

and

$$\begin{aligned} \theta_{\tilde{\sigma}}(||y||_\beta) &:= \frac{2^{\tilde{\sigma}-1}}{\Gamma(\tilde{\sigma})\zeta(\tilde{\sigma})} \int_0^{||y||_\beta^{-1}} (\coth v - 1)v^{\tilde{\sigma}-1} dv \\ &= O(||y||_\beta^{-\tilde{\eta}}) \ (\tilde{\eta} > 0; y \in \mathbf{R}_+^n). \end{aligned} \tag{13}$$

Proof. By (10), we obtain (11) and similarly, we get (12).
By (6) for

$$\Psi(u) = 0 \ (u \in (0, 1/M^\gamma)),$$

we find

$$\begin{aligned} w(\tilde{\sigma}, y) &= \frac{\Gamma^m\left(\frac{1}{\alpha}\right)}{\alpha^{m-1}\Gamma\left(\frac{m}{\alpha}\right)} \int_{||y||_\beta^{-1}}^{\infty} (\coth v - 1)v^{\tilde{\sigma}-1} dv \\ &= \frac{\Gamma^m\left(\frac{1}{\alpha}\right)}{\alpha^{m-1}\Gamma\left(\frac{m}{\alpha}\right)} \\ &\quad \times \left[\int_0^\infty (\coth v - 1)v^{\tilde{\sigma}-1} dv - \int_0^{||y||_\beta^{-1}} (\coth v - 1)v^{\tilde{\sigma}-1} dv\right] \\ &= \frac{\Gamma^m\left(\frac{1}{\alpha}\right)}{\alpha^{m-1}\Gamma\left(\frac{m}{\alpha}\right)} \frac{\Gamma(\tilde{\sigma})}{2^{\tilde{\sigma}-1}}\zeta(\tilde{\sigma})\left[1 - \theta_{\tilde{\sigma}}(||y||_\beta)\right]. \end{aligned}$$

Considering a constant $\gamma \in (1, \tilde{\sigma})$, we obtain

$$\lim_{v \to 0+} (\coth v - 1)v^\gamma = \lim_{v \to 0+} \frac{2v^\gamma}{e^{2v} - 1} = \lim_{v \to 0+} \frac{2\gamma v^{\gamma-1}}{2e^{2v}} = 0$$

and

$$\lim_{v\to\infty} (\coth v - 1)v^{\gamma} = 0.$$

There exists a constant $L > 0$, such that

$$\coth v - 1 \le L v^{-\gamma}.$$

Setting $\tilde{\eta} := \tilde{\sigma} - \gamma$ (> 0), it follows

$$0 \le \theta_{\tilde{\sigma}}(||y||_\beta) \le \frac{2^{\tilde{\sigma}-1}L}{\Gamma(\tilde{\sigma})\zeta(\tilde{\sigma})} \int_0^{||y||_\beta^{-1}} v^{\tilde{\eta}-1} dv = \frac{2^{\tilde{\sigma}-1}L}{\Gamma(\tilde{\sigma})\zeta(\tilde{\sigma})\tilde{\eta}} \frac{1}{||y||_\beta^{\tilde{\eta}}},$$

and then

$$\theta_{\tilde{\sigma}}(||y||_\beta) = O(||y||_\beta^{-\tilde{\eta}}) \;\; (y \in \mathbf{R}_+^n).$$

This completes the proof of the lemma. □

Lemma 4. *By the assumptions of Definition 1, if* $p \in \mathbf{R}\backslash\{0, 1\}$, $\frac{1}{p} + \frac{1}{q} = 1$,

$$f(x) = f(x_1, \cdots, x_m) \ge 0, \;\; g(y) = g(y_1, \cdots, y_n) \ge 0,$$

then

(i) for $p > 1$, *we have the following inequality:*

$$J_1 := \left\{ \int_{\mathbf{R}_+^n} \frac{||y||_\beta^{-p\sigma-n}}{[\omega(\sigma, y)]^{p-1}} \left[\int_{\mathbf{R}_+^m} \left(\coth \frac{||x||_\alpha}{||y||_\beta} - 1 \right) f(x) dx \right]^p dy \right\}^{\frac{1}{p}}$$
$$\le \left\{ \int_{\mathbf{R}_+^m} \varpi(\sigma, x) ||x||_\alpha^{p(m-\sigma)-m} f^p(x) dx \right\}^{\frac{1}{p}}, \tag{14}$$

(ii) for $0 < p < 1$ *or* $p < 0$, *we obtain the reverses of (14).*

Proof. (i) For $p > 1$, by Hölder's inequality with weight (cf. [30]), it follows

$$\int_{\mathbf{R}_+^m} \left(\coth \frac{||x||_\alpha}{||y||_\beta} - 1 \right) f(x) dx$$
$$= \int_{\mathbf{R}_+^m} \left(\coth \frac{||x||_\alpha}{||y||_\beta} - 1 \right) \left[\frac{||x||_\alpha^{(m-\sigma)/q}}{||y||_\beta^{(n+\sigma)/p}} f(x) \right] \left[\frac{||y||_\beta^{(n+\sigma)/p}}{||x||_\alpha^{(m-\sigma)/q}} \right] dx$$
$$\le \left\{ \int_{\mathbf{R}_+^m} \left(\coth \frac{||x||_\alpha}{||y||_\beta} - 1 \right) \frac{||x||_\alpha^{(m-\sigma)(p-1)}}{||y||_\beta^{n+\sigma}} f^p(x) dx \right\}^{\frac{1}{p}}$$

$$\times \left\{ \int_{\mathbf{R}_+^m} \left(\coth \frac{||x||_\alpha}{||y||_\beta} - 1 \right) \frac{||y||_\beta^{(n+\sigma)(q-1)}}{||x||_\alpha^{m-\sigma}} dx \right\}^{\frac{1}{q}}$$

$$= [\omega(\sigma, y)]^{\frac{1}{q}} ||y||_\beta^{\frac{n}{p}+\sigma}$$

$$\times \left\{ \int_{\mathbf{R}_+^m} \left(\coth \frac{||x||_\alpha}{||y||_\beta} - 1 \right) \frac{||x||_\alpha^{(m-\sigma)(p-1)}}{||y||_\beta^{n+\sigma}} f^p(x) dx \right\}^{\frac{1}{p}}. \quad (15)$$

Then by Fubini's theorem (cf. [29]), we have

$$J_1 \le \left\{ \int_{\mathbf{R}_+^n} \left[\int_{\mathbf{R}_+^m} \left(\coth \frac{||x||_\alpha}{||y||_\beta} - 1 \right) \frac{||x||_\alpha^{(m-\sigma)(p-1)}}{||y||_\beta^{n+\sigma}} f^p(x) dx \right] dy \right\}^{\frac{1}{p}}$$

$$= \left\{ \int_{\mathbf{R}_+^m} \left[\int_{\mathbf{R}_+^n} \left(\coth \frac{||x||_\alpha}{||y||_\beta} - 1 \right) \frac{||x||_\alpha^{(m-\sigma)(p-1)}}{||y||_\beta^{n+\sigma}} dy \right] f^p(x) dx \right\}^{\frac{1}{p}}$$

$$= \left\{ \int_{\mathbf{R}_+^m} \varpi(\sigma, x) ||x||_\alpha^{p(m-\sigma)-m} f^p(x) dx \right\}^{\frac{1}{p}}. \quad (16)$$

Hence, (14) follows.

(ii) For $0 < p < 1$ or $p < 0$, by the reverse Hölder inequality with weight (cf. [30]), we obtain the reverse of (15). Then by Fubini's theorem, we can still obtain the reverse of (14) and thus the lemma is proved.

□

Lemma 5. *By the assumptions of Lemma 4,*

(i) for $p > 1$, we have the following inequality equivalent to (14):

$$I := \int_{\mathbf{R}_+^n} \int_{\mathbf{R}_+^m} \left(\coth \frac{||x||_\alpha}{||y||_\beta} - 1 \right) f(x) g(y) dx dy$$

$$\le \left\{ \int_{\mathbf{R}_+^m} \varpi(\sigma, x) ||x||_\alpha^{p(m-\sigma)-m} f^p(x) dx \right\}^{\frac{1}{p}}$$

$$\times \left\{ \int_{\mathbf{R}_+^n} \omega(\sigma, y) ||y||_\beta^{q(n+\sigma)-n} g^q(y) dy \right\}^{\frac{1}{q}}, \quad (17)$$

(ii) for $0 < p < 1$ or $p < 0$, we have the reverse of (17) equivalent to the reverses of (14).

Proof. (i) For $p > 1$, by Hölder's inequality (cf. [30]), it follows that

$$I = \int_{\mathbf{R}_+^n} \frac{||y||_\beta^{\frac{n}{q}-(n+\sigma)}}{[\omega(\sigma, y)]^{\frac{1}{q}}} \left[\int_{\mathbf{R}_+^m} \left(\coth \frac{||x||_\alpha}{||y||_\beta} - 1 \right) f(x) dx \right]$$
$$\times \left[[\omega(\sigma, y)]^{\frac{1}{q}} ||y||_\beta^{(n+\sigma)-\frac{n}{q}} g(y) \right] dy$$
$$\leq J_1 \left\{ \int_{\mathbf{R}_+^n} \omega(\sigma, y) ||y||_\beta^{q(n+\sigma)-n} g^q(y) dy \right\}^{\frac{1}{q}}. \tag{18}$$

Then by (14), we obtain (17).

On the other hand, assuming that (17) is valid, we set

$$g(y) := \frac{||y||_\beta^{-p\sigma-n}}{[\omega(\sigma, y)]^{p-1}} \left(\int_{\mathbf{R}_+^m} \left(\coth \frac{||x||_\alpha}{||y||_\beta} - 1 \right) f(x) dx \right)^{p-1}, y \in \mathbf{R}_+^n.$$

Then it follows that

$$J_1^p = \int_{\mathbf{R}_+^n} \omega(\sigma, y) ||y||_\beta^{q(n+\sigma)-n} g^q(y) dy .$$

If $J_1 = 0$, then (14) is trivially valid; if $J_1 = \infty$, then by (16), relation (14) reduces to the form of an equality($= \infty$). Suppose that $0 < J_1 < \infty$. By (17), we have

$$0 < \int_{\mathbf{R}_+^n} \omega(\sigma, y) ||y||_\beta^{q(n+\sigma)-n} g^q(y) dy = J_1^p = I$$
$$\leq \left\{ \int_{\mathbf{R}_+^m} \varpi(\sigma, x) ||x||_\alpha^{p(m-\sigma)-m} f^p(x) dx \right\}^{\frac{1}{p}}$$
$$\times \left\{ \int_{\mathbf{R}_+^n} \omega(\sigma, y) ||y||_\beta^{q(n+\sigma)-n} g^q(y) dy \right\}^{\frac{1}{q}} < \infty.$$

Therefore,

$$J_1 = \left\{ \int_{\mathbf{R}_+^n} \omega(\sigma, y) ||y||_\beta^{q(n+\sigma)-n} g^q(y) dy \right\}^{\frac{1}{p}}$$
$$\leq \left\{ \int_{\mathbf{R}_+^m} \varpi(\sigma, x) ||x||_\alpha^{p(m-\sigma)-m} f^p(x) dx \right\}^{\frac{1}{p}},$$

and then (14) follows. Hence, (14) and (17) are equivalent.

(ii) For $0 < p < 1$ or $p < 0$, similarly, we obtain the reverse of (17) which is equivalent to the reverse of (14) and thus the lemma is proved. □

Main Results and Operator Expressions

Let

$$\Phi(x) := ||x||_{\alpha}^{p(m-\sigma)-m},\ \Psi(y) := ||y||_{\beta}^{q(n+\sigma)-n} (x \in \mathbf{R}_{+}^{m}, y \in \mathbf{R}_{+}^{n}),$$

by Lemmas 3, 4, and 5, we obtain

Theorem 1. *Suppose that* $\alpha, \beta > 0$, $\sigma > 1$, $p \in \mathbf{R}\backslash\{0,1\}$, $\frac{1}{p} + \frac{1}{q} = 1$,

$$f(x) = f(x_1, \cdots, x_m) \geq 0,\ g(y) = g(y_1, \cdots, y_n) \geq 0,$$

$$0 < ||f||_{p,\Phi} = \left\{ \int_{\mathbf{R}_{+}^{m}} \Phi(x) f^{p}(x) dx \right\}^{\frac{1}{p}} < \infty,$$

and

$$0 < ||g||_{q,\Psi} = \left\{ \int_{\mathbf{R}_{+}^{n}} \Psi(y) g^{q}(y) dy \right\}^{\frac{1}{q}} < \infty.$$

(i) If $p > 1$, *then we have the following equivalent inequalities with the best possible constant factor* $K(\sigma)$, *that is*

$$I = \int_{\mathbf{R}_{+}^{n}} \int_{\mathbf{R}_{+}^{m}} \left(\coth \frac{||x||_{\alpha}}{||y||_{\beta}} - 1 \right) f(x) g(y) dx dy < K(\sigma) ||f||_{p,\Phi} ||g||_{q,\Psi}, \tag{19}$$

and

$$J := \left\{ \int_{\mathbf{R}_{+}^{n}} ||y||_{\beta}^{-p\sigma-n} \left(\int_{\mathbf{R}_{+}^{m}} \left(\coth \frac{||x||_{\alpha}}{||y||_{\beta}} - 1 \right) f(x) dx \right)^{p} dy \right\}^{\frac{1}{p}} < K(\sigma) ||f||_{p,\Phi}, \tag{20}$$

where

$$K(\sigma) = \left[\frac{\Gamma^{n}\left(\frac{1}{\beta}\right)}{\beta^{n-1} \Gamma\left(\frac{n}{\beta}\right)} \right]^{\frac{1}{p}} \left[\frac{\Gamma^{m}\left(\frac{1}{\alpha}\right)}{\alpha^{m-1} \Gamma\left(\frac{m}{\alpha}\right)} \right]^{\frac{1}{q}} \frac{\Gamma(\sigma)}{2^{\sigma-1}} \zeta(\sigma). \tag{21}$$

(ii) *If* $0 < p < 1$ *or* $p < 0$, *then we still have the equivalent reverses of (19) and (20) with the same best constant factor* $K(\sigma)$.

Proof. (i) For $p > 1$, by the given conditions, we can prove that (15) becomes a strict inequality. Otherwise if (15) takes the form of equality, then there exist constants A and B, which are not all zero, such that for a.e. $y \in \mathbf{R}_+^n$, it holds:

$$A \frac{||x||_\alpha^{(m-\sigma)(p-1)}}{||y||_\beta^{n+\sigma}} f^p(x) = B \frac{||y||_\beta^{(n+\sigma)(q-1)}}{||x||_\alpha^{m-\sigma}} \text{ a.e. in } x \in \mathbf{R}_+^m. \tag{22}$$

If $A = 0$, then it follows that $B = 0$, which is impossible.
If $A \neq 0$, then (22) reduces to

$$||x||_\alpha^{p\ (m-\sigma)-m} f^p(x) = \frac{B||y||_\beta^{q(n+\sigma)}}{A||x||_\alpha^m} \text{ a.e. in } x \in \mathbf{R}_+^m,$$

which contradicts $0 < ||f||_{p,\Phi} < \infty$.
In fact by (7), it follows

$$\int_{\mathbf{R}_+^m} ||x||_\alpha^{-m} dx = \infty.$$

Hence (14) still assumes the form of strict inequality. By Lemma 3 and Lemma 4, we deduce (20).
Similarly to (18), we still have

$$I \leq J \left\{ \int_{\mathbf{R}_+^n} ||y||_\beta^{q(n+\sigma)-n} g^q(y) dy \right\}^{\frac{1}{q}}. \tag{23}$$

Then by (23) and (20), we obtain (19). It is evident by Lemma 5 and the assumptions that the relations (19) and (18) are also equivalent.
For $0 < \varepsilon < p(\sigma - 1)$, we define $\tilde{f}(x), \tilde{g}(y)$ as follows

$$\tilde{f}(x) := \begin{cases} 0, & 0 < ||x||_\alpha < 1, \\ ||x||_\alpha^{\sigma - \frac{\varepsilon}{p} - m}, & ||x||_\alpha \geq 1, \end{cases}$$

$$\tilde{g}(y) := \begin{cases} 0, & 0 < ||y||_\beta < 1, \\ ||y||_\beta^{-\sigma - \frac{\varepsilon}{q} - n}, & ||y||_\beta \geq 1. \end{cases}$$

Then for $\tilde{\sigma} = \sigma - \frac{\varepsilon}{p}$, by (7), we derive

$$0 \leq \int_{\{y \in \mathbf{R}_+^n : ||y||_\beta \geq 1\}} ||y||_\beta^{-n-\varepsilon} O(||y||_\beta^{-\tilde{\eta}}) dy$$

$$\leq L^* \int_{\{y \in \mathbf{R}_+^n : ||y||_\beta \geq 1\}} ||y||_\beta^{-n-(\varepsilon+\tilde{\eta})} dy$$

$$= L^* \frac{\Gamma^n\left(\frac{1}{\beta}\right)}{(\varepsilon + \tilde{\eta})\beta^{n-1}\Gamma\left(\frac{n}{\beta}\right)} < \infty,$$

and in view of (7) and (13), it follows that

$$||\tilde{f}||_{p,\Phi}||\tilde{g}||_{q,\Psi}$$

$$= \left\{\int_{\{x \in \mathbf{R}_+^m : ||x||_\alpha \geq 1\}} ||x||_\alpha^{-m-\varepsilon} dx\right\}^{\frac{1}{p}} \left\{\int_{\{y \in \mathbf{R}_+^n : ||y||_\beta \geq 1\}} ||y||_\beta^{-n-\varepsilon} dy\right\}^{\frac{1}{q}}$$

$$= \frac{1}{\varepsilon} \left\{\frac{\Gamma^m\left(\frac{1}{\alpha}\right)}{\alpha^{m-1}\Gamma\left(\frac{m}{\alpha}\right)}\right\}^{\frac{1}{p}} \left\{\frac{\Gamma^n\left(\frac{1}{\beta}\right)}{\beta^{n-1}\Gamma\left(\frac{n}{\beta}\right)}\right\}^{\frac{1}{q}},$$

and

$$\tilde{I} := \int_{\mathbf{R}_+^n} \int_{\mathbf{R}_+^m} \left(\coth \frac{||x||_\alpha}{||y||_\beta} - 1\right) \tilde{f}(x)\tilde{g}(y) dx dy$$

$$= \int_{\{y \in \mathbf{R}_+^n : ||y||_\beta \geq 1\}} ||y||_\beta^{-n-\varepsilon} w(\tilde{\sigma}, y) dy$$

$$= K_2(\tilde{\sigma}) \int_{\{y \in \mathbf{R}_+^n : ||y||_\beta \geq 1\}} ||y||_\beta^{-n-\varepsilon} \left(1 - O(||y||_\beta^{-\tilde{\eta}})\right) dy$$

$$= \frac{1}{\varepsilon} K_2(\tilde{\sigma}) \left[\frac{\Gamma^n\left(\frac{1}{\beta}\right)}{\beta^{n-1}\Gamma\left(\frac{n}{\beta}\right)} - \varepsilon O_{\tilde{\sigma}}(1)\right].$$

If there exists a constant $K \leq K(\sigma)$, such that (19) is valid when replacing $K(\sigma)$ by K, then we obtain

$$\frac{\Gamma^m\left(\frac{1}{\alpha}\right)}{\alpha^{m-1}\Gamma\left(\frac{m}{\alpha}\right)}\frac{\Gamma(\tilde{\sigma})}{2^{\tilde{\sigma}-1}}\zeta(\tilde{\sigma})\left[\frac{\Gamma^n\left(\frac{1}{\beta}\right)}{\beta^{n-1}\Gamma\left(\frac{n}{\beta}\right)} - \varepsilon O_{\tilde{\sigma}}(1)\right]$$

$$\leq \varepsilon \tilde{I} < \varepsilon K\|\tilde{f}\|_{p,\Phi}\|\tilde{a}\|_{q,\Psi}$$

$$= K\left\{\frac{\Gamma^m\left(\frac{1}{\alpha}\right)}{\alpha^{m-1}\Gamma\left(\frac{m}{\alpha}\right)}\right\}^{\frac{1}{p}}\left\{\frac{\Gamma^n\left(\frac{1}{\beta}\right)}{\beta^{n-1}\Gamma\left(\frac{n}{\beta}\right)}\right\}^{\frac{1}{q}},$$

and thus $K(\sigma) \leq K(\varepsilon \to 0^+)$.

Hence $K = K(\sigma)$ is the best possible constant factor of (19).

By the equivalency, we can prove that the constant factor $K(\sigma)$ in (20) is the best possible. Otherwise, by (23) we would reach a contradiction to the fact that the constant factor $K(\sigma)$ in (19) is the best possible.

(ii) For $0 < p < 1$ or $p < 0$, similarly, we can still obtain the equivalent reverses of (19) and (20) with the best constant factor. This completes the proof of the theorem.

□

Corollary 1. *Let the assumptions of Theorem 1 be fulfilled, and additionally,*

$$0 < \|f\|_1 := \int_{\mathbf{R}_+^m} f(x)dx < \infty \quad \text{and} \quad 0 < \|g\|_1 := \int_{\mathbf{R}_+^n} g(y)dy < \infty.$$

Then,

(i) if $p > 1$, then we have the following equivalent inequalities with the best possible constant factor $K(\sigma)$, that is

$$\int_{\mathbf{R}_+^n}\int_{\mathbf{R}_+^m} \coth\frac{\|x\|_\alpha}{\|y\|_\beta} f(x)g(y)dxdy < \|f\|_1\|g\|_1 + K(\sigma)\|f\|_{p,\Phi}\|g\|_{q,\Psi}, \tag{24}$$

$$\left\{\int_{\mathbf{R}_+^n}\|y\|_\beta^{-p\sigma-n}\left(\int_{\mathbf{R}_+^m}\coth\frac{\|x\|_\alpha}{\|y\|_\beta}f(x)dx - \|f\|_1\right)^p dy\right\}^{\frac{1}{p}} < K(\sigma)\|f\|_{p,\Phi}; \tag{25}$$

(ii) if $0 < p < 1$ or $p < 0$, then we still have the equivalent reverses of (24) and (25) with the same best constant factor $K(\sigma)$.

For $m = n = \alpha = \beta = 1$ in Theorem 1 and Corollary 1, we obtain

Corollary 2. *Suppose that $\sigma > 1$, $p \in \mathbf{R}\backslash\{0, 1\}$, $\frac{1}{p} + \frac{1}{q} = 1$,*

$$\varphi(x) := x^{p(1-\sigma)-1},\ \psi(y) := y^{q(1+\sigma)-1}\ (x, y > 0),\ f(x) \geq 0,\ g(y) \geq 0,$$

as well as

$$0 < ||f||_{p,\varphi} < \infty \text{ and } 0 < ||g||_{q,\psi} < \infty.$$

Then,

(i) for $p > 1$, we have (5) and the following equivalent inequality with the best possible constant factor

$$\frac{\Gamma(\sigma)}{2^{\sigma-1}}\zeta(\sigma),$$

that is

$$\left\{\int_0^\infty y^{-p\sigma-1}\left[\int_0^\infty \left(\coth\frac{x}{y}-1\right) f(x)dx\right]^p dy\right\}^{\frac{1}{p}} < \frac{\Gamma(\sigma)}{2^{\sigma-1}}\zeta(\sigma)||f||_{p,\varphi}; \tag{26}$$

(ii) for $0 < p < 1$ or $p < 0$, we obtain the equivalent reverses of (5) and (26) with the same best constant factor.

Moreover, if

$$0 < ||f|| := \int_0^\infty f(x)dx < \infty \text{ and } 0 < ||g|| := \int_0^\infty g(y)dy < \infty,$$

then

(i) for $p > 1$, we have the following equivalent inequalities with the best possible constant factor

$$\frac{\Gamma(\sigma)}{2^{\sigma-1}}\zeta(\sigma),$$

that is

$$\int_0^\infty \int_0^\infty \coth\frac{x}{y} f(x)g(y)dxdy < ||f||\,||g|| + \frac{\Gamma(\sigma)}{2^{\sigma-1}}\zeta(\sigma)||f||_{p,\varphi}||g||_{q,\psi}, \tag{27}$$

$$\left\{\int_0^\infty y^{-p\sigma-1}\left[\int_0^\infty \coth\frac{x}{y} f(x)dx - ||f||\right]^p dy\right\}^{\frac{1}{p}} < \frac{\Gamma(\sigma)}{2^{\sigma-1}}\zeta(\sigma)||f||_{p,\varphi}, \tag{28}$$

(ii) for $0 < p < 1$ or $p < 0$, we obtain the equivalent reverses of (27) and (28) with the same best constant factor.

By the assumptions of Theorem 1 for $p > 1$, in view of $J < K(\sigma)||f||_{p,\Phi}$, we define:

Definition 2. A multidimensional Hilbert-type integral operator

$$T : L_{p,\Phi}(\mathbf{R}_+^m) \to L_{p,\Psi^{1-p}}(\mathbf{R}_+^n) \tag{29}$$

is defined as follows:

For $f \in L_{p,\Phi}(\mathbf{R}_+^m)$, there exists a unique representation

$$Tf \in L_{p,\Psi^{1-p}}(\mathbf{R}_+^n),$$

satisfying

$$(Tf)(y) := \int_{\mathbf{R}_+^m} \left(\coth \frac{||x||_\alpha}{||y||_\beta} - 1\right) f(x)dx \ (y \in \mathbf{R}_+^n). \tag{30}$$

For $g \in L_{q,\Psi}(\mathbf{R}_+^n)$, we define the following formal inner product of Tf and g as follows:

$$(Tf, g) := \int_{\mathbf{R}_+^n} \int_{\mathbf{R}_+^m} \left(\coth \frac{||x||_\alpha}{||y||_\beta} - 1\right) f(x)g(y)dxdy. \tag{31}$$

Then by Theorem 1 for

$$p > 1, \ 0 < ||f||_{p,\Phi}, \ ||g||_{q,\Psi} < \infty,$$

we have the following equivalent inequalities:

$$(Tf, g) < K(\sigma)||f||_{p,\Phi}||g||_{q,\Psi}, \tag{32}$$

and

$$||Tf||_{p,\Psi^{1-p}} < K(\sigma)||f||_{p,\Phi}. \tag{33}$$

It follows that the operator T is bounded with

$$||T|| := \sup_{f(\neq\theta)\in L_{p,\Phi}(\mathbf{R}_+^m)} \frac{||Tf||_{p,\Psi^{1-p}}}{||f||_{p,\Phi}} \leq K(\sigma).$$

Since the constant factor $K(\sigma)$ in (33) is the best possible, we obtain

$$||T|| = K(\sigma) = \left[\frac{\Gamma^n\left(\frac{1}{\beta}\right)}{\beta^{n-1}\Gamma\left(\frac{n}{\beta}\right)}\right]^{\frac{1}{p}} \left[\frac{\Gamma^m\left(\frac{1}{\alpha}\right)}{\alpha^{m-1}\Gamma\left(\frac{m}{\alpha}\right)}\right]^{\frac{1}{q}} \frac{\Gamma(\sigma)}{2^{\sigma-1}}\zeta(\sigma). \tag{34}$$

Acknowledgements The authors wish to express their thanks to Professors Tserendorj Batbold, Mario Krnic, and Jichang Kuang for their careful reading of the manuscript and for their valuable suggestions.

M. Th. Rassias: This work is supported by the Greek State Scholarship Foundation (IKY).

B. Yang: This work is supported by 2012 Knowledge Construction Special Foundation Item of Guangdong Institution of Higher Learning College and University (No. 2012KJCX0079).

References

1. Hardy G. H., Littlewood J. E., Pólya G., Inequalities. Cambridge University Press, Cambridge, 1934
2. Mitrinović D. S., Pečarić J. E., Fink A. M., Inequalities Involving Functions and their Integrals and Derivatives. Kluwer Acaremic Publishers, Boston, 1991
3. Batbold Ts., Adiyasuren V., Castillo R. E., Extension of reverse Hilbert-type inequality with a generalized homogeneous kernel. Rev. Colombiana Mat. **45**(2), 187-195 (2011)
4. Yang B. C., Hilbert-Type Integral Inequalities. Bentham Science Publishers Ltd., Sharjah, 2009
5. Yang B. C., Discrete Hilbert-Type Inequalities. Bentham Science Publishers Ltd., Sharjah, 2011
6. Yang B. C., The Norm of Operator and Hilbert-Type Inequalities. Science Press, Beijing, 2009, China
7. Yang B. C., Hilbert-type Integral operators: norms and inequalities, In: Nonlinear Analysis, Stability, Approximation, and Inequalities (P. M. Pardalos, et al.). Springer, New York, 2012, pp. 771–859
8. Yang B. C., On Hilbert's integral inequality. Journal of Mathematical Analysis and Applications, **220**, 778–785(1998)
9. Yang B. C., Brnetić I, Krnić M., Pečarić J. E., Generalization of Hilbert and Hardy-Hilbert integral inequalities. Math. Ineq. and Appl., **8**(2), 259–272 (2005)
10. Krnić M., Pečarić J. E., Hilbert's inequalities and their reverses, Publ. Math. Debrecen, **67**(3–4), 315–331 (2005)
11. Rassias M. Th., Yang B. C., On half - discrete Hilbert's inequality, Applied Mathematics and Computation, to appear
12. Azar L., On some extensions of Hardy-Hilbert's inequality and Applications. Journal of Inequalities and Applications, 2009, No. 546829
13. Arpad B., Choonghong O., Best constant for certain multilinear integral operator. Journal of Inequalities and Applications, 2006, No. 28582
14. Kuang J. C., Debnath L., On Hilbert's type inequalities on the weighted Orlicz spaces. Pacific J. Appl. Math., **1**(1), 95–103 (2007)
15. Zhong W. Y., The Hilbert-type integral inequality with a homogeneous kernel of $-\lambda$-degree. Journal of Inequalities and Applications, 2008, No. 917392
16. Hong Y., On Hardy-Hilbert integral inequalities with some parameters. J. Ineq. in Pure & Applied Math.,**6**(4), Art. 92, 1–10 (2005)
17. Zhong W. Y., Yang B. C., On multiple Hardy-Hilbert's integral inequality with kernel. Journal of Inequalities and Applications, Vol. 2007, Art.ID 27962, 17 pages, doi: 10.1155/ 2007/27
18. Yang B. C., Krnić M., On the norm of a multidimensional Hilbert-type operator, Sarajevo Journal of Mathematics, **7**(20), 223–243(2011)
19. Li Y. J., He B., On inequalities of Hilbert's type. Bulletin of the Australian Mathematical Society, **76**(1), 1–13 (2007)
20. Edwards H. M., Riemann's Zeta Function. Dover Publications, New York, 1974
21. Milovanović G. V., Rassias M. Th. (eds.), Analytic Number Theory, Approximation Theory and Special Functions. Springer, New York, to appear
22. Apostol T. M., Introduction to Analytic Number Theory. Springer – Verlag, New York, 1984

23. Erdős P., Suranyi J., Topics in the Theory of Numbers. Springer – Verlag, New York, 2003
24. Hardy G. H., Wright E. W., An Introduction to the Theory of Numbers. 5th edition, Clarendon Press, Oxford, 1979
25. Iwaniec H., Kowalski E., Analytic Number Theory. American Mathematical Society, Colloquium Publications, Volume 53, Rhode Island, 2004
26. Landau E., Elementary Number Theory. 2nd edition, Chelsea, New York, 1966
27. Miller S. J., Takloo – Bighash R., An Invitation to Modern Number Theory. Princeton University Press, Princeton and Oxford, 2006
28. Zhong Y. Q., On Complex Functions. Higher Education Press, Beijing, China, 2004
29. Kuang J. C., Introduction to Real Analysis. Hunan Education Press, Chansha, China, 1996
30. Kuang J. C., Applied Inequalities. Shangdong Science Technic Press, Jinan, China, 2004
31. Rassias M. Th., Problem–Solving and Selected Topics in Number Theory: In the Spirit of the Mathematical Olympiads (Foreword by Preda Mihăilescu). Springer, New York, 2011

Robustness of Fictitious Play in a Resource Allocation Game

Michalis Smyrnakis

Abstract Nowadays it is well known that decentralised optimisation tasks can be represented as so-called "potential games". An example of a resource allocation problem that can be cast as a game is the "vehicle-target assignment problem" originally proposed by Marden et al.

In this article we use fictitious play as "negotiation" mechanism between the agents, and we examine its robustness in the case where a fraction of non-cooperative players, s, choose a random action. This addresses situations in which there is, e.g., a malfunction of some units. In our simulations we consider cases where the non-cooperative agents communicate their proposed action to the other agents and cases in which they do not announce their actions (e.g. in the case of a breakdown of communication). We observe that the performance of fictitious play is the same as if all players were able to fully coordinate, when the fraction of the non-coordinating agents, s, is smaller than a critical value $\tilde{s}$. Moreover in both cases, where non-cooperative agents shared and did not share their action with others, the critical value was the same. Above this critical value the performance of fictious play is always affected. Also even in the case where only the 40% of the agents manage to cooperate and share their information the final reward is the 85% of the reference case's reward, where every agent cooperates.

Introduction

Multi-agent systems find an increasing number of applications. Decentralised optimisation is an important component of these tasks, where agents should coordinate in order to achieve a common goal. Ad-hoc sensor networks [1], transportation [2],

---

M. Smyrnakis (✉)
Department of Automatic Control and Systems Engineering Mappin Street, S1 3JD, Sheffield, UK
e-mail: m.smyrnakis@sheffield.ac.uk

N.J. Daras (ed.), *Applications of Mathematics and Informatics in Science and Engineering*, Springer Optimization and Its Applications 91, DOI 10.1007/978-3-319-04720-1_27,

security defence tasks [3], smart grids [4, 5] and resource allocation [6, 7] are examples of applications where multi-agent systems are used. These applications share common characteristics such as high computational demand and communication restrictions between the agents. Therefore the corresponding optimisation tasks should be performed decentrally.

Game theory and in particular potential games provide the mathematical framework for decentralised optimisation problems. In a game-theoretic formalisation the players of the game are the agents of the decentralised optimisation task, the optimisation function becomes the utility function of the game and players actions are the actions of the agents in the decentralised optimisation task. Thus the task of finding a local or a global optimum in the optimisation problem is equivalent to search for the Nash equilibrium of a game and therefore algorithms from game-theoretic learning literature can be used to solve decentralised optimisation tasks.

A widely used game-theoretic learning algorithm that can be used as a coordination mechanism in decentralised optimisation tasks is fictitious play. It is an iterative learning algorithm, where players maintain some beliefs about their opponent strategies and based on these beliefs they choose the action that maximises their expected reward. Fictitious play, in contrast to other learning algorithms such as WOLF [8], genetic algorithms [9] and ant colonies algorithms [10] which are heuristic algorithms, has been proved that it converges to the Nash equilibrium of various classes of games and among the others in potential games, hence it converges to the optimum of the decentralised optimisation tasks.

In this paper we are going to consider the case where due to some hardware malfunction or communication breakdown, a fraction of agents s will not be able to perform their tasks and hence they will not be able to coordinate with other agents. In order to study the effects of s in the fictitious play's performance we compare the global reward that agents gain if a fraction s of them cannot coordinate with the others and choose their actions randomly, with a reference reward. We will use as a reference reward, the reward the players will gain in the case where all the agents can fully coordinate.

The rest of the paper is organised as follows: In section "Background" we briefly describe some basic definitions of game theory and fictitious play. In section "Simulation Scenario" we present the simulation scenario that we are going to use, and section "Simulation Results" includes the results that we obtained. In the final part of our paper we summarise our conclusions.

Background

Decentralised optimisation problems can be naturally expressed as strategic form games. In strategic form games players choose their actions instantly and their reward can be represented in a matrix form. A strategic form game Γ consists of a set of players $1, 2, \ldots, \mathcal{I}$. Each player i has a set of available actions $x^i \in X^i$, and the combined action of all players defines their joint action,

$x = (x^1, \dots x^{\mathcal{I}}) \in X = X^1 \times \dots \times X^{\mathcal{I}}$. Finally each player i has a utility function that is a mapping from the joint action space to the real numbers $r^i : X \rightarrow \mathbf{R}$.

The mixed strategy of a player i, y^i, is defined as an element of the set of all the probability distributions over the action space of player i, Δ^i. Then similarly to the joint actions the joint mixed strategy y is defined as an element of the set product $\Delta = \Delta^1 \times \dots \times \Delta^{\mathcal{I}}$. The special case, where a Player i, puts all the mass of his probability distribution in a single action, and thus he selects this action with probability one is called pure strategy of Player i.

For convenience of notation we will often write the joint action as $x = (x^i, x^{-i})$, where $-i$ denotes all the players but i, and analogously for the joint strategy $y = (y^i, y^{-i})$. The expected reward of a player i when he chooses a strategy y^i and his opponents choose the joint strategy y^{-i} is $r^i(y^i, y^{-i})$.

One of the most common decision rules that a player can use in order to choose his actions is best response. A player who selects his actions based on the best response decision rule chooses the action that maximises his expected payoff given the joint mixed strategy of his opponents. More formally, Player i, when his opponents' joint mixed strategy is y^{-i} then his best response is defined as:

$$BR^i(y^{-i}) = \underset{y^i \in \Delta^i}{\operatorname{argmax}} \quad r^i(y^i, y^{-i}). \tag{1}$$

Nash in [11], based on Kakutani's fixed point theorem, proved that every game has at least one equilibrium point which corresponds to a joint mixed strategy $\hat{y}$ that is a fixed point of the best response correspondence, $\hat{y}^i \in BR^i(\hat{y}^{-i}) \forall i$. A joint mixed strategy $\hat{y}$ is a Nash equilibrium when

$$r^i(\hat{y}^i, \hat{y}^{-i}) \leq r^i(x^i, \hat{y}^{-i}) \qquad \text{for all } i, \text{ for all } x^i \in X^i. \tag{2}$$

Equation (2) implies that if a strategy $\hat{y}$ is a Nash equilibrium, then it is not possible for a player to increase his reward by unilaterally changing his strategy. When all the players in a game select equilibrium actions using pure strategies then the equilibrium is referred as pure strategy Nash equilibrium.

In decentralised optimisation tasks the global utility must be an aggregation of each agent's utility [12, 13]. One of the desired properties of the individual utilities is the monotonic relation to the global utility. This suggests that an action which improves or reduces the utility of an individual should accordingly increase or reduce the global utility. This relation is satisfied by the utility function of potential games. In particular a game Γ is a potential game if its reward function has the following property [14]:

$$r^i(x^i, x^{-i}) - r^i(\tilde{x^i}, x^{-i}) = \phi(x^i, x^{-i}) - \phi(\tilde{x^i}, x^{-i})$$

where ϕ is a potential function and the above equality stands for every player i, for every action $x^{-i} \in X^{-i}$, and for every pair of actions x^i, $\tilde{x^i} \in X^i$. The potential function depicts the changes in the players' payoffs when they unilaterally change their actions. Every potential game has at least one pure strategy Nash equilibrium [14], and thus there is at least one equilibrium that corresponds to a joint action x.

Wonderful life utility [15, 16] is a method to design the individuals' utility functions of a potential game so that the global utility function of a decentralised optimisation problem will act as a potential function. Player is utility when a joint action $x = (x^i, x^{-i})$ is played is the difference in global utility obtained by the player selecting action s^i in comparison with the global utility that would have been obtained if i had selected an arbitrarily chosen reference action x_0^i:

$$r^i(x^i, x^{-i}) = r_g(x^i, x^{-i}) - r_g(x_0^i, x^{-i})$$

where r_g is the global utility function.

Fictitious Play

Fictitious play [17] is the canonical example of iterative game-theoretic learning algorithms. Players repeatedly play the game Γ and in each iteration of the game they update their beliefs about their opponents' mixed strategies using the knowledge they have acquired from the previous iterations of the game and they use best response, Eq. (1), to choose their actions. In particular at the initial iteration of the game every player maintains some arbitrary, non-negative weights, κ_0 for each of his opponents. After playing the first iteration of the game players observe their opponents' actions and update their weight function. In particular a Player i updates the weight function that he maintains about his opponent, Player j, as follows [18]:

$$\kappa_t^i(x^j) = \kappa_{t-1}^i(x^j) + \begin{cases} 1 \text{ if } & x_{t-1}^j = x^j \\ 0 \text{ otherwise} \end{cases} \tag{3}$$

Based on these weights Player i then estimates Player js strategy using the following equation:

$$y_t^j(x^j) = \frac{\kappa_t^i(x^j)}{\sum_{x^j \in X^j} \kappa_t^i(x^j)} \tag{4}$$

This can be also written as:

$$y_t^j(x^j) = \left(1 - \frac{1}{t^j}\right) y_{t-1}^j(x^j) + \frac{1}{t^j} I_{x_t^j = x^j} \tag{5}$$

where $t^j = t + \sum_{x^j \in X^j} \kappa_0^j(x^j)$. Players, based on their estimations about their opponents' joint mixed strategy, use the best response decision rule to choose the action that maximises their expected reward.

In fictitious play, players assume that their opponents use the same mixed strategy in every iteration of the game. We can also see this from Eq. (5), where all the actions have the same weight in the estimation of the opponent's mixed strategy, even if they have been observed at the initial iterations of the game. Under the assumption that the distribution of the opponents' mixed strategy follows a multinomial distribution the maximum likelihood estimation of its parameters can be obtained by using Eq. (4). Bayesian methods and in particular maximun a-posteriory probability [18] can be used in order to obtain an alternative estimation to maximum likelihood. In particular if players use a Dirichlet prior distribution for the strategy of their opponents and evaluate the parameters of the posterior distribution using the maximum a-posteriori probability estimation, the same estimation of opponents' mixed strategy is the same with the maximum likelihood estimates. In [19] it was proved that if the "moderation" process [20] is used instead of the maximum a-posteriori probability estimator, the estimation of the opponent's strategy is also similar to Eq. (4).

It has been proved that fictitious play converges to the Nash equilibrium of $2 \times n$ games [21], 2×2 games with generic payoff [22], zero sum games [23], games that can be solved using iterative dominance [24] and potential games [14]

Simulation Scenario

In our simulations we consider the following resource allocation task which can be expressed as a potential game. Following [16] we assume that in an area A there are I vehicles that should destroy the J "hostile" targets. Each target has different attributes, hence the reward that agents receive if they destroy target j or target $\tilde{j}$ will be different. Similarly vehicles have different characteristics and thus the probability a vehicle i has to destroy target j is different from the probability the vehicle $\tilde{i}$ has to destroy the same target. Moreover each vehicle can choose only one target to destroy but a target might be engaged by more than one vehicle. This resource allocation task is cast as a potential game Γ where each vehicle represents a player of the game with available actions the targets that the player can engage. We will write p_{ij} for the probability of a player i to destroy a target j and v_j for the value of target j. The utility that a target j produces is the product of its value with the probability that it will be destroyed by all the vehicles who engage it. More formally we can write:

$$r_j = v_j * \left(1 - \prod_{i:x^i=j} (1 - p_{ij})\right) \tag{6}$$

Because the probability of a vehicle i to destroy a target j is independent of other agents' actions we can express the global utility that players receive at the end of each iteration of fictitious play as the expected value of the destroyed targets.

$$r_{\text{global}} = \sum_j r_j. \tag{7}$$

Player is individual expected reward if he chooses to engage target j is calculated using wonderful life utility as:

$$r^i(x^i = j, x^{-i}) = \sum_j r_j(x^i = j, x^{-i}) - r_j(x_0^i, x^{-i})$$

where x_0^i was set to be the greedy action of player i: $x_0^i = \underset{j}{\text{argmax}}\ v_j\, p_{ij}$.

Simulation Results

In the simulation scenario that we described in the previous section it is possible that some vehicles will not be able to coordinate with the other vehicles, either because of a malfunction or a communication breakdown. Therefore there will be a fraction of non-cooperative agents, s, who will choose their actions randomly instead of using fictitious play in order to choose their actions. We study the robustness of fictitious play in this scenario when the non-cooperative agents inform the other agents about the action they choose (communicating agents) and the case they are not exchanging any information with other agents (non-communicating agents).

We use as a reference utility the global utility that corresponds to the final iteration's joint action of the instance whereas all the players use fictitious play as coordination mechanism. We then evaluate the change in the utility of the instances where a fraction s of non-coordination agents choose their action at random with respect to the reference utility.

In our simulations, in an area A we place 300 vehicles that could engage one of the 150 targets that appear in A. The probability p_{ij} that a vehicle i destroys a target j and the value of each vehicle v_j is uniformly chosen, $0 \leq p_{ij} \leq 1$ and $0 \leq v_j \leq 1$, respectively. We examine the robustness of fictitious play when $0 < s < 1$ and we measure the change from the reference utility using the following equation:

$$r_{\text{ch}} = \frac{r_{\text{global}}}{r_{\text{global}}(\text{reference})}. \tag{8}$$

The results we present are the average over 50 learning episodes, and in each learning episode agents "negotiated" 50 times. Figures 1 and 2 depict the results for both cases of communicating and non-communicating agents, respectively.

In both cases we observe that there is a critical point $\tilde{s}$ where the outcome of fictitious play is not influenced by the agents who do not coordinate. As expected in the case of non-communicating agents, the final reward of fictitious play starts to decrease for smaller values of s than the case of communicating agents. This is because they use less information than the communicating agents. Nevertheless the final reward of fictitious play, as it is depicted in Figs. 1 and 2, is independent of s, when $s \leq 0.3$.

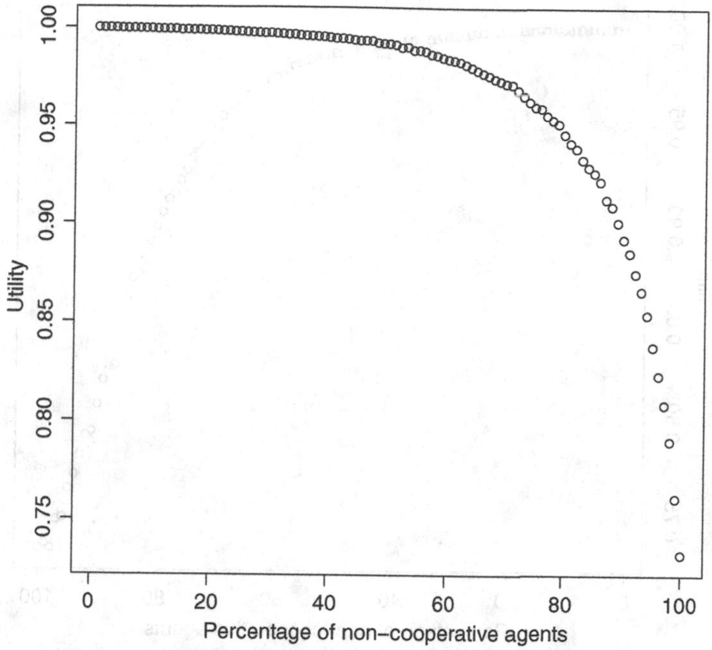

Fig. 1 Changes in r_{ch} when players non-coordinating agents communicate their actions with the other agents. The x axis represents the percentage of non-coordinating agents and the y axis the change in the utility as it is evaluated using Eq. (8)

We also examine the relation between the changes in fictitious play rewards and the number of targets that vehicles can choose to engage. We denote β the ratio of the number of targets over the number of vehicles, $\beta = \frac{\text{\# of targets}}{\text{\# of sensors}}$. We examine how changes in the values of β and s affect the value of r_{ch}. In our simulations we allow $0 \leq s \leq 1$ and $0 \leq \beta \leq 1$. The results we present are the average over 50 learning episodes, where we assume that 300 vehicles should coordinate and each learning episode run for 50 iterations. The probability p_{ij} that a vehicle i destroys a target j and the value of each vehicle v_j are uniformly chosen, $0 \leq p_{ij} \leq 1$ and $0 \leq v_j \leq 1$, respectively. Figures 3 and 4 depict the results that we obtained for the case where agents shared information with other agents and for the case that they do not exchange information, respectively.

We observe that when $b \leq 0.15$ the outcome of Eq. (8) does not depend on s in both cases. Also when $s \leq 0.3$, the performance of fictitious play is the same as if everyone were able to fully cooperate and communicate independently of β. Nevertheless above this critical value of s, $s > 0.3$, r_{ch} is reducing as the value of β is increasing. Moreover, independently of β, even if the 60% of agents do not cooperate and share their information, then the final reward will be the 85% of the reward of the case where everyone cooperates. In the case of players that do not share their information the final reward is the 75% of the reference reward.

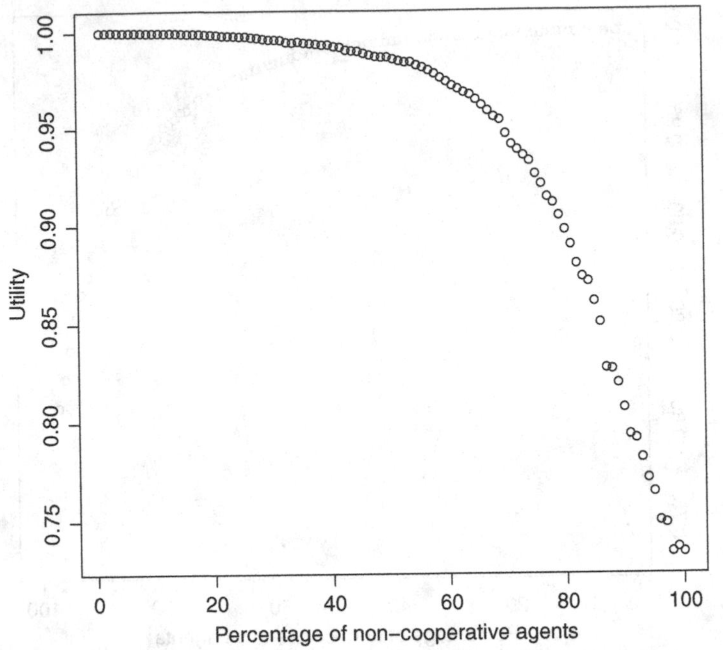

Fig. 2 Changes in r_{ch} when players non-coordinating agents do not communicate their actions with the other agents. The x axis represents the percentage of non-coordinating agents and the y axis the change in the utility as it is evaluated using Eq. (8)

The areas where r_{ch} is maximised, the light shaded areas of Figs. 3 and 4, are different. This indicates that r_{ch} value decreases faster in the case of non-communicating agents than the case of communicating agents. In order to verify this we used Wilcoxon two-sample test. We use as null hypothesis that the sample of the two cases comes from the same distribution, and therefore the differences we observe in Figs. 3 and 4 are not statistically significant, against the alternative hypothesis that the samples are from different distributions. We use Wilcoxon two-sample test instead of its parametric alternative because based on the Shaphiro–Wilks normality test we observe that the two samples do not follow a normal distribution. The p-value of the Wilcoxon two-sample test is smaller than 0.001 and therefore we reject the null hypothesis. Nevertheless if we narrow our sample and take into account only the cases where $0 \leq s \leq 0.4$ the two samples are not statistically different, independently of β, since the p-value of the Wilcoxon two pair test is 0.2427 and therefore we cannot reject the null hypothesis that the two samples are from the same distribution. Thus, independently of β, when $s \leq 0.4$, the value of $r_c h$ in both cases where agents share and cannot share their actions with the other agents will not be statistically different.

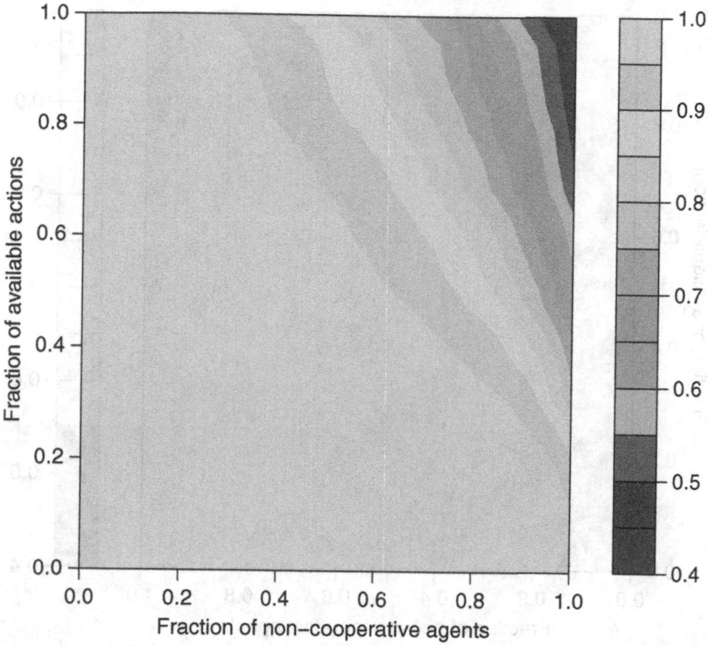

Fig. 3 Contour plot for the case of communicating agents. The x axis represents s as a fraction of the total number of people and the y axis β. Dark colours in the contour plot denote lower performance of fictitious play than the reference utility

Conclusions

Decentralised optimisation tasks can be seen as the task of finding the Nash equilibria of a potential game. Fictitious play is the canonical example in game-theoretic learning literature. We studied the robustness of fictitious play algorithm in cases where some non-cooperative players decide to use a random action instead of using fictitious play to "negotiate" and choose an action that maximises their expected reward. We used the resource allocation problem that proposed by [16] as simulation scenario and observed how the fraction of non-cooperative agents s, and the ratio of the number of actions over the number of players β, affect the outcome of the fictitious play algorithm. We observe that there is always a critical value of s, $\tilde{s}$, where up to this fraction of non-cooperating agents the performance of fictitious play is the same as if everyone were able to fully cooperate and communicate. Above the critical value, $\tilde{s}$, the final reward of fictitious play is affected. The critical value $\tilde{s}$ we observed in our simulations for both cases, where non cooperative agents shared and did not share their action with others, was $\tilde{s} = 0.3$. Moreover in the case where $s = 0.6$ the final reward will be the 85% of the reward of the case where everyone cooperates in the case of agents who share their actions and 75% for agents who choose not to share their actions.

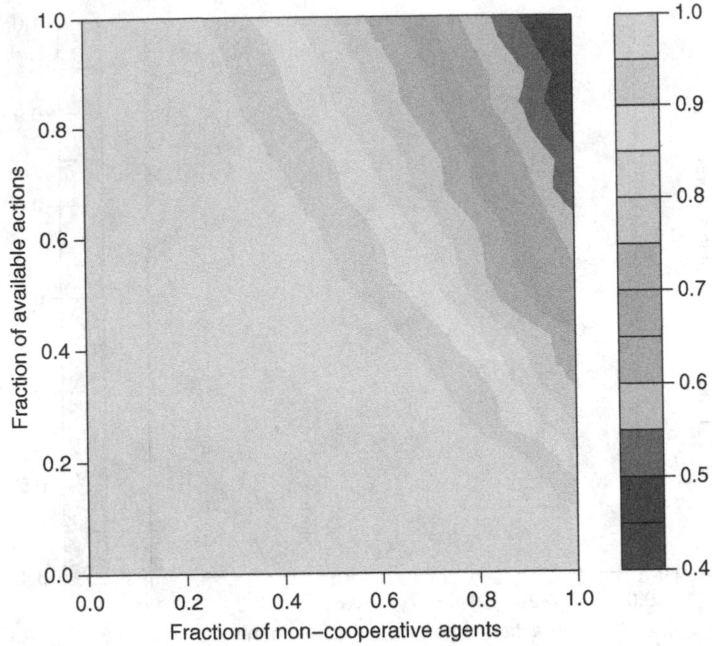

Fig. 4 Contour plot for the case of non-communicating agents. The x axis represents s as a fraction of the total number of people and the y axis β. Dark colours in the contour plot denote lower performance of fictitious play than the reference utility

Summarising, fictitious play is robust in the game we examined and its outcome is not affected from the number of targets and the number of non-coordinating agents if $s \leq 0.3$ even if they cannot share their actions. We believe that, even though our results are based only on simulations, our findings are applicable to a number of optimisation and resource allocation problems with decentralised command structure and in which a fraction of units may fail to cooperate to act towards a common goal and/or in which communication structures may break down.

Acknowledgements This work is supported by The Engineering and Physical Sciences Research Council EPSRC (grant number EP/I005765/1).

References

1. Kho, J., Rogers, A. and Jennings, N. R. (2009) Decentralise control of adaptive sampling in wireless sensor networks. *ACM Transactions on Sensor Networks*, **5** (3).
2. van Leeuwen, P. (2002) Scheduling aircraft using constraint satisfaction. *In Electronic Notes in Theoretical Computer Science.* Elsevier, 252–268.
3. Zhang, Y. and Meng, Y. (2010) A decentralized multi-robot system for intruder detection in security defense. *In Intelligent Robots and Systems (IROS).*

4. Kallitsis, M.G., Michailidis, G. and Devetsikiotis, M. (2011) A decentralized algorithm for optimal resource allocation in smartgrids with communication network externalities. *Smart Grid Communications (SmartGridComm), 2011.*
5. Rogers, A., Ramchurn, S. and Jennings, N.R. (2012) Delivering the smart grid: Challenges for autonomous agents and multi-agent systems research. *Twenty-Sixth AAAI Conference on Artificial Intelligence (AAAI-12).*
6. Kitano, H., Todokoro, S., Noda, I., Matsubara, H., Takahashi, T., Shinjou, A. and Shimada, S. (1999) Robocup rescue: Search and rescue in large-scale disaster as a domain for autonomous agents research. *In IEEE International Conference on Systems, Man, and Cybernetics (SMC '99)*, **6**, 739–743.
7. Stranjak, A., Dutta, P.S., Rogers, A. and Vytelingum, P.V. (2008) A multi-agent simulation system for prediction and sceduling of aero engine overhaul. *In Proceedings of the 7th International Conference on Autonomous Agents and Multiagent Systems (AAMAS '08)*
8. Bowling, M., anf Veloso, M. (2002). Multiagent learning using a variable learning rate. *Artificial Intelligence*, **136** (2), 215–250.
9. Herrmann, J. W. (1999). A genetic algorithm for minimax optimization problems. *In Evolutionary Computation on Proceedings of the 1999 Congress.*
10. Dorigo, M., Birattari, M., and Stutzle, T. (2006) Ant colony optimization. *Computational Intelligence Magazine, IEEE*, **1**(4), 28–39.
11. Nash, J. (1950) Equilibrium Points in n-Person Games *Proceedings of the National Academy of Science, USA*, **36**, 48–49.
12. Bertsekas, D. (1982) Distributed dynamic programming. *IEEE Transactions on Automatic Control*, **27**(3), 610–616.
13. Silva, C.A., Sousa, J.M.C., Runkler, T.A. and Sa da Costa, J.M.G. (2009) Distributed supply chain management using ant colony optimization. *European Journal of Operational Research*, **199**(2), 349–358.
14. Monderer, D. and Shapley, L. (1996) Potential Games. *Games and Economic Behavior*, **14**, 124–143.
15. Wolpert, D. and Tumer, K. (1999) An overview of collective intelligence. *In J. M. Bradshaw, editor, Handbook of Agent Technology*. AAAI Press/MIT Press.
16. Arslan, G., Marden, J. and Shamma, J. (2007) Autonomous Vehicle-Target Assignment: A Game Theoretical Formulation. *Journal of Dynamic Systems, Measurement, and Control*, **129**, 584–596.
17. Brown, G.W. (1951) Iterative Solutions of Games by Fictitious Play. *In Activity Analysis of Production and Allocation, T.C. Koopmans (Ed.)*. New York: Wiley.
18. Fudenberg, D. and Levine, D. (1998) *The theory of Learning in Games*. The MIT Press
19. Rezek, I., Leslie, D. S., Reece, S. and Roberts, S. J., Rogers, A., Dash, R. K. and Jennings, N.R. (2008) On Similarities between Inference in Game Theory and Machine Learning. *Journal of Artificial Intelligence Research*, **33**, 259–283.
20. MacKay, D. J. (1992) The evidence framework applied to classification networks. *Neural computation*, **4**(5), 720–736.
21. Berger U. (2005) Fictitious play in 2xn games. *Journal of Economic Theory*, **120** (2), 139–154.
22. Miyasawa, K. (1961) On the convergence of learning process in a 2x2 non-zero-sum two person game. *Research Memorandum No 33, Princeton Unniversity.*
23. Robinson, J. (1951) An iterative Method of solving a game. *Annals of Mathematics*, **54**, 269–301.
24. Nachbar, J. (1990) "Evolutionary" selection dynamics in games: Convergence and limit properties. *International Journal of Game Theory*, **19**, 58–89.

Printed in the United States
By Bookmasters